Qualitative and Nutritional Improvement of Cereal-Based Foods and Beverages

Qualitative and Nutritional Improvement of Cereal-Based Foods and Beverages

Editors

Antonella Pasqualone
Carmine Summo

MDPI • Basel • Beijing • Wuhan • Barcelona • Belgrade • Manchester • Tokyo • Cluj • Tianjin

Editors
Antonella Pasqualone
Department of Soil, Plant and Food Science (DISSPA), University of Bari Aldo Moro
Italy

Carmine Summo
Department of Soil, Plant and Food Science (DISSPA), University of Bari Aldo Moro
Italy

Editorial Office
MDPI
St. Alban-Anlage 66
4052 Basel, Switzerland

This is a reprint of articles from the Special Issue published online in the open access journal *Foods* (ISSN 2304-8158) (available at: https://www.mdpi.com/journal/foods/special_issues/cereal_foods).

For citation purposes, cite each article independently as indicated on the article page online and as indicated below:

LastName, A.A.; LastName, B.B.; LastName, C.C. Article Title. *Journal Name* **Year**, *Volume Number*, Page Range.

ISBN 978-3-0365-0706-4 (Hbk)
ISBN 978-3-0365-0707-1 (PDF)

Cover image courtesy of Antonella Pasqualone.

Contents

About the Editors

Antonella Pasqualone is an Associate Professor of Food Science and Technology at the University of Bari, Italy. Scientifically responsible for many national and international projects, she published 186 articles in refereed international scientific journals, with an h-index of 33 (February 2021). Her main research interests are in the field of cereal science and technology: (i) improvement of the nutritional quality of cereal-based foods by integration with sustainable crops such as pulses; (ii) formulation and characterization of functional pasta and bakery products, also gluten-free; (iii) studies on the baking systems of flatbreads; (iv) improvement of bread shelf-life; (v) activity of polyphenol oxidase in durum wheat and correlations with pasta browning; (vi) effect of drying or baking on the volatile profile of pasta and bread; (vii) food upcycling by reusing food industry waste and by-products in cereal-based foods; (viii) rediscovery and qualitative standardization of non-alcoholic fermented cereal beverages and traditional cereal-based foods.

Carmine Summo is an Associate Professor of Food Science and Technology at the University of Bari, Italy. Scientific coordinator and Research Unit member for several research projects, he published 155 articles in refereed international scientific journals, with h-index = 26 (February 2021). His main research interests fall in the field of cereals and legume-based foods: (i) formulation and characterization of functional pasta and bakery products containing bioactive extracts from food industry by-products and waste; (ii) quality characterization of pasta and bakery products, also gluten-free; (iii) improvement of the nutritional quality of cereal-based foods by adding legume flours; (iv) studies on the effect of drying on the volatile compounds of pasta; (v) formulation and characterization of innovative ready-to-eat legume-based foods; (vi) chemical, nutritional and functional characterization of legume varieties to increase the knowledge on the legume properties and avoid the risk of genetic erosion; (vii) studies on the application of air classification to legume flours and on food application of the derived fractions; (viii) production of meat analogues from cereals and pulses flours.

Editorial

Qualitative and Nutritional Improvement of Cereal-Based Foods and Beverages

Antonella Pasqualone * and Carmine Summo

Department of Soil, Plant and Food Science (DISSPA), University of Bari Aldo Moro, Via Amendola, 165/a, I-70126 Bari, Italy; carmine.summo@uniba.it
* Correspondence: antonella.pasqualone@uniba.it

Received: 28 January 2021; Accepted: 3 February 2021; Published: 5 February 2021

Abstract: The main directions of research aimed at nutritional improvement have to face either excesses or deficiencies in the diet. To this end, different strategies may be adopted, such as the reformulation of products, the introduction of functional ingredients, and the application of biotechnology to increase the bioavailability of bioactive compounds. These interventions, however, can alter the physico-chemical and sensory properties of the final products, making it necessary to achieve a balance between nutritional and quality modification. This Special Issue offers readers information on innovative ways to improve the cereal-based foods and beverages, useful for researchers and for industry operators.

Keywords: functional foods; upcycling; byproducts; bioactive compounds; dietary fiber; new quality; pulses; insects; bread; pasta

---

Increased consumer awareness of the effects of food in preventing nutrient-related diseases and maintaining physical and mental well-being, has made nutritional improvement an important goal of the food and beverage industry, including the cereal sector. To this end, different strategies may be adopted, such as the reformulation of products, the introduction of functional ingredients, and the application of biotechnologies to increase the bioavailability of bioactive compounds. These interventions, however, can alter the physico-chemical and sensory properties of the final products, making it necessary to achieve a balance between nutritional and quality modification.

The Special Issue "Qualitative and Nutritional Improvement of Cereal-Based Foods and Beverages" collects 17 original research articles and one review aimed at exploring innovative ways to improve cereal-based foods and beverages, an old—if not ancient—group of products which are still on our table every day.

In these articles, a wide range of very different food products is considered, such as bakery products (including white bread, brown bread, durum wheat bread, tortilla, pizza base, muffins and biscuits), fresh and dry pasta, extruded sticks and instant flours, fortified blended foods and Sunsik, the latter being a traditional Korean beverage [1]. Cereal-based beverages, indeed, hold a long tradition and have become known for their sensory and health-promoting attributes [2].

The main directions of research aimed at nutritional improvement have to face either excess or deficiency in the diet. In the latter case, nutrient-rich foods with long shelf-life are needed to prevent malnutrition, whereas in developed countries it is mostly required to decrease the energy value, sucrose, salt, and increase dietary fiber content of foods to prevent obesity and nutrient-related chronic diseases such as cardiovascular disease, hypertension, and diabetes mellitus. The 2030 Agenda for Sustainable Development, adopted in 2015 by the United Nations, set the goals *"to end hunger, achieve food security and improved nutrition, and promote sustainable agriculture"* (Goal 2) and *"to ensure healthy lives and promote well-being for all at all ages*" (Goal 3), recognizing non-communicable diseases (NCDs) as a major challenge for sustainable development [3]. In addition, patients with obesity and other chronic underlying conditions are at particularly high risk of developing severe COVID-19 complications [4].

The most natural way to improve the nutritional profile of cereal-based foods is to use wholemeal flour, thus retaining all fiber and micronutrients of wheat caryopsis. Wholewheat flour is a valuable raw material, irrespective of the milling system (stone milling or roller milling) [5]. Higher contents of bioactives can be reached if purple wheat, whose debranning fractions are particularly rich in anthocyanins, is used [6]. Fermentation further improves the nutritional features of wholemeal flours. Fresh pasta prepared by mixing semolina with wholewheat sourdough shows higher content of free essential amino acids and phenolic compounds, lower phytic acid content, and higher antioxidant activity, than control pasta where non-fermented wholewheat semolina is used [7].

Another strategy consists in adding locally available ingredients to reformulate existing cereal-based foods, in order to increase the nutritional value while diminishing the risk of genetic erosion of the local crops and reduce imports. In this context, the leaf powder of *Moringa oleifera*, a plant originating in India and Africa, which is rich in proteins, minerals, and phytochemicals, has been proposed as an additional ingredient to improve the nutritional profile of white and brown bread [8] or fresh pasta [9]. The fortification with moringa leaf increases protein and iron content, but makes bread darker lowering the consumer acceptability, although with a minor impact on brown bread. The addition of moringa leaf powder to fresh pasta in the range 5–15% significantly increases the content of polyphenols even at the lowest percentage. Similarly, legumes can be used which are rich in proteins and complement the amino acid profile of cereal-based foods. Flour of Apulian black chickpeas, an autochthonous black-coated chickpea cultivated in Southern Italy, rich in anthocyanins, has been added to various bakery products [10], namely bread, "focaccia"—an Italian traditional bakery product similar to pizza [11], and pizza crust by substituting flour in the 10–40% range. The rheological properties of dough worsen, resulting in harder and darker final products. However, the nutritional features improve in terms of higher contents of fiber, proteins, and bioactive compounds, as well as higher antioxidant activity [10]. In pasta, the addition of chickpea and hemp flour, previously fermented and enzymatically treated, improves the nutritional profile and protein digestibility, and reduces the sensory drawbacks and the antinutrients (tannins, phytates and raffinose) [12].

Red quinoa or Taiwan djulis (*Chenopodium formosanum* Koidz.) can be used to develop sourdough bread (20% djulis sourdough and 80% wheat flour) [13], whereas germinated wheat, in combination with an extract of *Achyranthes aspera* and *Acanthopanax*, two plants used in Asian traditional medicine, has been proposed to fortify Sunsik, a traditional ready-to-drink Korean cereal-based beverage made of roasted brown rice, barley, adlay, oat, and black beans [1]. Amaranth (*Amaranthus hypochondriacus* L.) and flaxseed (*Linum usitatissimum* L.), instead, have been added to corn-based instant-extruded products, to meet the needs of nutritionally balanced ready-to-eat foods. The addition effectively increases lysine, polyunsaturated fatty acids, minerals, and fiber of the end-products [14]. Seed flour of *Brosimum alicastrum* Sw., a Mexican tree locally named "ramón," characterized by high protein, dietary fiber, and micronutrient content, has been added to wheat tortillas improving the healthy features while keeping pliability unaltered, but with a browner color [15].

Insects are another valuable source of proteins which could be used to overcome the challenge of a more sustainable food chain while improving the nutritional profile of end-products. Meal of winged termites (*Macrotermes bellicosus*) has been added to biscuits prepared with sorghum and wheat flour [16]. Sorghum, which can be grown in tropical areas, makes this biscuit formulation viable in sub-Saharan areas, where protein-energy malnutrition is a major health concern. A significant increase of proteins, minerals, and amino acids is achieved, but biscuits become darker and less hard [16]. In the same geographic area, fortified blended foods (corn-soy blends or wheat-soy blends) are used to prepare viscous porridges in supplementary feeding programs [17]. These food aids often result in products too viscous for being fed to infants and young children, therefore it is needed to dilute them, lowering the nutritional value and energy density. The addition of cowpea has been proposed, also in combination with extrusion cooking, to obtain a porridge able to deliver the correct amount of nutrients at lower viscosities [17]. Extruded cooked products are rehydrated easily without cooking [18], which is important to save energy sources.

The use of waste and byproducts from the food industry, where upcycling is becoming increasingly important to improve sustainability, represents another mean for enhancing the nutritional and healthy features of cereal-based foods [19]. Coffee silverskin, a byproduct from the coffee industry rich in dietary fiber, proteins and bioactive compounds, has been used to produce extruded sticks based on cornmeal and amaranth flour [20], whereas almond skins, a byproduct of almond-based confectionery industry, rich in fiber and phenolic compounds, have been effectively considered as a functional ingredient in biscuits, which become more friable but darker [19].

Finally, to meet the needs related to western lifestyle, the production of muffins containing agave syrup instead of sucrose [21], and the reduction of the sodium content of bread by using a natural low-sodium sea salt [22], have been proposed.

To sum up, this Special Issue gives an interesting contribution to the field and offers readers information on several ways to innovate and improve the cereal-based foods and beverages, which can be useful both for researchers and for industry operators. In most cases, the reformulation with additional ingredients alters the sensory properties, therefore raising the need of communicating a "new quality" to consumers to explain that the differences with conventional counterparts are largely compensated by improved nutritional features.

Author Contributions: A.P. writing—original draft preparation; A.P., C.S. writing—review and editing; A.P., C.S. funding acquisition. All authors have read and agreed to the published version of the manuscript.

Funding: This work was funded by: (1) Agropolis Fondation, Fondazione Cariplo, and Daniel and the Nina Carasso Foundation through the "Investissements d'avenir" program with reference number ANR-10-LABX-0001-01, under the "Thought for Food" Initiative (project "LEGERETE"); (2) the Italian Ministry of Education, University and Research (MIUR), program PRIN 2017 (grant number 2017SFTX3Y) "The Neapolitan pizza: processing, distribution, innovation and environmental aspects."

Conflicts of Interest: The authors declare no conflict of interest.

References

1. Kim, B.R.; Park, S.S.; Youn, G.J.; Kwak, Y.J.; Kim, M.J. Characteristics of Sunsik, a Cereal-Based Ready-to-Drink Korean Beverage, with Added Germinated Wheat and Herbal Plant Extract. *Foods* **2020**, *9*, 1654. [CrossRef] [PubMed]
2. Ignat, M.V.; Salanță, L.C.; Pop, O.L.; Pop, C.R.; Tofană, M.; Mudura, E.; Coldea, T.E.; Borșa, A.; Pasqualone, A. Current functionality and potential improvements of non-alcoholic fermented cereal beverages. *Foods* **2020**, *9*, 1031. [CrossRef] [PubMed]
3. United Nations. Resolution Adopted by the General Assembly on 25 September 2015. Transforming Our World: The 2030 Agenda for Sustainable Development. A/Res/70/1. 2015. Available online: https://www.un.org/ga/search/view_doc.asp?symbol=A/RES/70/1&Lang=E (accessed on 21 January 2021).
4. Katzmarzyk, P.T.; Salbaum, J.M.; Heymsfield, S.B. Obesity, noncommunicable diseases, and COVID-19: A perfect storm. *Am. J. Hum. Biol.* **2020**, e23484. [CrossRef]
5. Pagani, M.A.; Giordano, D.; Cardone, G.; Pasqualone, A.; Casiraghi, M.C.; Erba, D.; Blandino, M.; Marti, A. Nutritional Features and Bread-Making Performance of Wholewheat: Does the Milling System Matter? *Foods* **2020**, *9*, 1035. [CrossRef] [PubMed]
6. Parizad, P.A.; Marengo, M.; Bonomi, F.; Scarafoni, A.; Cecchini, C.; Pagani, M.A.; Marti, A.; Iametti, S. Bio-Functional and Structural Properties of Pasta Enriched with a Debranning Fraction from Purple Wheat. *Foods* **2020**, *9*, 163. [CrossRef] [PubMed]

7. Fois, S.; Campus, M.; Piu, P.P.; Siliani, S.; Sanna, M.; Roggio, T.; Catzeddu, P. Fresh Pasta Manufactured with Fermented Whole Wheat Semolina: Physicochemical, Sensorial, and Nutritional Properties. *Foods* **2019**, *8*, 422. [CrossRef] [PubMed]
8. Govender, L.; Siwela, M. The Effect of *Moringa oleifera* Leaf Powder on the Physical Quality, Nutritional Composition and Consumer Acceptability of White and Brown Breads. *Foods* **2020**, *9*, 910. [CrossRef] [PubMed]
9. Rocchetti, G.; Rizzi, C.; Pasini, G.; Lucini, L.; Giuberti, G.; Simonato, B. Effect of *Moringa oleifera* L. Leaf Powder Addition on the Phenolic Bioaccessibility and on In Vitro Starch Digestibility of Durum Wheat Fresh Pasta. *Foods* **2020**, *9*, 628. [CrossRef] [PubMed]
10. Pasqualone, A.; De Angelis, D.; Squeo, G.; Difonzo, G.; Caponio, F.; Summo, C. The Effect of the Addition of Apulian black Chickpea Flour on the Nutritional and Qualitative Properties of Durum Wheat-Based Bakery Products. *Foods* **2019**, *8*, 504. [CrossRef] [PubMed]
11. Pasqualone, A.; Delcuratolo, D.; Gomes, T. Focaccia Italian flat fatty bread. In *Flour and Breads and Their Fortification in Health and Disease Prevention*, 1st ed.; Preedy, V.R., Watson, R.R., Patel, V.B., Eds.; Academic Press: Cambridge, MA, USA, 2011; pp. 47–58. [CrossRef]
12. Schettino, R.; Pontonio, E.; Rizzello, C.G. Use of Fermented Hemp, Chickpea and Milling By-Products to Improve the Nutritional Value of Semolina Pasta. *Foods* **2019**, *8*, 604. [CrossRef] [PubMed]
13. Chung, P.L.; Liaw, E.T.; Gavahian, M.; Chen, H.H. Development and Optimization of Djulis Sourdough Bread Using Taguchi Grey Relational Analysis. *Foods* **2020**, *9*, 1149. [CrossRef] [PubMed]
14. Tobias-Espinoza, J.L.; Amaya-Guerra, C.A.; Quintero-Ramos, A.; Pérez-Carrillo, E.; Núñez-González, M.A.; Martínez-Bustos, F.; Meléndez-Pizarro, C.O.; Báez-González, J.G.; Ortega-Gutiérrez, J.A. Effects of the addition of flaxseed and amaranth on the physicochemical and functional properties of instant-extruded products. *Foods* **2019**, *8*, 183. [CrossRef] [PubMed]
15. Subiria-Cueto, R.; Larqué-Saavedra, A.; Reyes-Vega, M.L.; de la Rosa, L.A.; Santana-Contreras, L.E.; Gaytán-Martínez, M.; Vázquez-Flores, A.A.; Rodrigo-García, J.; Corral-Avitia, A.Y.; Núñez-Gastélum, J.A.; et al. *Brosimum alicastrum* Sw. (Ramón): An Alternative to Improve the Nutritional Properties and Functional Potential of the Wheat Flour Tortilla. *Foods* **2019**, *8*, 613. [CrossRef] [PubMed]
16. Awobusuyi, T.D.; Siwela, M.; Pillay, K. Sorghum–Insect Composites for Healthier Cookies: Nutritional, Functional, and Technological Evaluation. *Foods* **2020**, *9*, 1427. [CrossRef] [PubMed]
17. Chanadang, S.; Chambers IV, E. Determination of the sensory characteristics of traditional and novel fortified blended foods used in supplementary feeding programs. *Foods* **2019**, *8*, 261. [CrossRef] [PubMed]
18. Pasqualone, A.; Costantini, M.; Coldea, T.E.; Summo, C. Use of legumes in extrusion cooking: A review. *Foods* **2020**, *9*, 958. [CrossRef] [PubMed]
19. Pasqualone, A.; Laddomada, B.; Boukid, F.; De Angelis, D.; Summo, C. Use of Almond Skins to Improve Nutritional and Functional Properties of Biscuits: An Example of Upcycling. *Foods* **2020**, *9*, 1705. [CrossRef] [PubMed]
20. Beltrán-Medina, E.A.; Guatemala-Morales, G.M.; Padilla-Camberos, E.; Corona-González, R.I.; Mondragón-Cortez, P.M.; Arriola-Guevara, E. Evaluation of the Use of a Coffee Industry By-Product in a Cereal-Based Extruded Food Product. *Foods* **2020**, *9*, 1008. [CrossRef] [PubMed]
21. Ozuna, C.; Trueba-Vázquez, E.; Moraga, G.; Llorca, E.; Hernando, I. Agave Syrup as an Alternative to Sucrose in Muffins: Impacts on Rheological, Microstructural, Physical, and Sensorial Properties. *Foods* **2020**, *9*, 895. [CrossRef] [PubMed]
22. Arena, E.; Muccilli, S.; Mazzaglia, A.; Giannone, V.; Brighina, S.; Rapisarda, P.; Fallico, B.; Allegra, M.; Spina, A. Development of Durum Wheat Breads Low in Sodium Using a Natural Low-Sodium Sea Salt. *Foods* **2020**, *9*, 752. [CrossRef] [PubMed]

Publisher's Note: MDPI stays neutral with regard to jurisdictional claims in published maps and institutional affiliations.

Article

Characteristics of *Sunsik*, a Cereal-Based Ready-to-Drink Korean Beverage, with Added Germinated Wheat and Herbal Plant Extract

Bo Ram Kim [1], Seung Soo Park [1], Geum-Joung Youn [2], Yeon Ju Kwak [2] and Mi Jeong Kim [1,3,*]

[1] Interdisciplinary Program in Senior Human Ecology, Changwon National University, Changwon 51140, Korea; kinj56@daum.net (B.R.K.); pssll03@naver.com (S.S.P.)

[2] Research Institute of GH Biofarm, Agricultural Corporation Gagopa Healing Food, 177, Samgye-ro, Naeseo-eup, Masanhoewon-gu, Changwon 51219, Korea; ygj@kfoofs.kr (G.-J.Y.); kyjred@kfoods.kr (Y.J.K.)

[3] Department of Food and Nutrition, Changwon National University, Changwon 51140, Korea

* Correspondence: mjkim@changwon.ac.kr; Tel.: +82-55-213-3512

Received: 16 October 2020; Accepted: 6 November 2020; Published: 12 November 2020

Abstract: The purpose of this study was to develop a formulation of *Sunsik* with improved health benefits by adding germinated wheat (GW) and herbal plant extract (HPE) using a response surface methodology (RSM). The central composite experimental design (CCD) was used to evaluate the effects of *Sunsik* with added HPE (2–4%) and GW (10–20%) on total phenolic content (TPC), total flavonoid content (TFC), Trolox equivalent antioxidant capacity (TEAC), 2,2-diphenyl-1-picrylhydrazyl (DPPH) radical scavenging capacity, gamma butyric acid (GABA) content, total color changes (ΔE), browning index (BI), water absorption index (WAI), and water solubility index (WSI). As a result of the CCD, the independent and dependent variables were fitted by the second-order polynomial equation, and the lack of fit for response surface models was not significant except in relation to WSI. The GABA content, TPC, and TEAC were more adequate for a linear model than for a quadratic model, and they might be affected by GW rather than HPE. Alternatively, the TFC, DPPH radical scavenging capacity, WAI, WSI, ΔE, and BI were fitted with quadratic models. The optimum formulation that could improve antioxidant and physicochemical properties was *Sunsik* with 3.5% and 20% added HPE and GW, respectively.

Keywords: cereal-based ready-to-drink beverage; convenient meal replacement (CMR); germinated wheat; response surface methodology (RSM); gamma-amino butyric acid (GABA); antioxidant properties

1. Introduction

Recently, the increase in single-person and double-income households has shifted consumers' eating behaviors toward the increased consumption of home meal replacements (HMRs) or convenient meal replacement (CMRs) [1]. As ready-to-eat foods, CMRs are a more convenient and simpler meal replacement than HMRs, and they could reduce meal preparation and eating time. The CMR market quadrupled from $600 million in 2009 to $2.3 billion 2019. In Korea, the proportion of single-person households is expected to reach 35% of the total population in 2030, and the CMR market is expected to continue to grow.

The types of CMR products are diversifying, such as to include liquid and powder grains, porridges, and cereal bars. Among them, cereal-based beverages are a representative CMR product consumed worldwide because they provide an efficient means to increase the intake of essential nutrients among busy modern people. A few studies investigated the physicochemical and health-conscious properties of various cereal beverages [2,3]. Bembem and Agrahar-Murugkar [2] reported that millet-based ready-to-drink beverages improved radical scavenging activity, total phenolic content (TPC), and viscosity in the geriatric population. In another study, multigrain beverages prepared with barley, oats, buckwheat, and red rice were identified as providing additional health benefits, such as phenolic content and soluble fiber, to consumers [3].

Sunsik has been consumed for a long time as a cereal-based ready-to-drink beverage in Korea. It is made of partially raw or thermal-processed and dried agricultural and marine products [4]. The most common ingredients of *Sunsik* are roasted brown rice, barley, adlay, oat, and black beans [5]. With the recent increase in the demand for healthy foods, much research has reported that additional ingredients, such as various dried vegetables, nuts, and fruits, could be added to *Sunsik* to offer more health-conscious nutrients [6–8]. For example, Park [8] reported that *Sunsik* with added mealworm was higher in antioxidant capacities and in consumer preference than a control *Sunsik*. Regarding the quality of ready-to-drink of *Sunsik*, it should disperse and dissolve well in water or milk within a few minutes. Koh, Jang, and Surh [6] reported that fermented *Sunsik* had a higher soluble solid content, oxidative stability, and amino acids than unfermented *Sunsik*, resulting in an improved solubility and nutrient content. Although several studies reported enhancements in the quality and nutrient content of *Sunsik*, there is limited information on the health benefits of *Sunsik* with added germinated wheat (GW) and herbal plant extract (HPE).

Germination has been identified as an effective processing method to improve the nutritional quality and health-related compounds of cereal [9]. In numerous studies, gamma amino butyric acids (GABA) and phenolic acid compositions were increased as the germination time of wheat increased, suggesting the possibility of GW as a health-conscious ingredient [10–12]. In addition, Dhillon et al. [13] found that the antioxidant activity of and consumer preference for breads were improved when GW flour at 30 °C for 72 h was partially used to make bread. The changes in the physiological and biochemical properties of GW might be due to the activation of endogenous enzymes that break down starch and protein into small molecules [14,15]. The activation of endogenous enzymes may also play a role in increasing the solubility of *Sunsik* with added GW when it mixes with water or milk. In addition, plant herbal medicines, such as *Achyranthes aspera*, safflower seed, and Acanthopanax, have been used for the prevention of various diseases in traditional treatments in Asian countries [16,17]. It is known that safflower seeds are rich in lignin, flavonoid, and serotonin and have excellent effects on bone diseases, such as osteoporosis [18]. As previously published in many studies, the extracts of *A. aspera* and *Acanthopanax* showed a reduced inflammatory effect and antioxidant capacities [19–22]. The above-mentioned herbal plant medicines are used not only for therapeutic purposes, but also by adding them to various foods in the form of extracts to increase the health-related functions in the food matrix, such as noodles, drinks, and cookies [23–26]. The HPEs, including *A. aspera*, safflower seed, and *Acanthopanax*, used in this study confirmed previously the pharmacological effects on osteogenic differentiation in human mesenchymal stem cells [27]. The mixture extracts of herbal plants were freeze-dried and then were used in various food products of Gagopa Healing Food Co., Ltd. (Changwon, Korea).

Currently, *Sunsik* with added GW flour and HPE is not available in the marketplace yet. Thus, if GW and HPE are added to commercial *Sunsik*, which is conveniently used as ready to drink beverage, the new *Sunsik* product might be more beneficial to health. The purpose of this study was to determine the optimum formula amounts of GW flour and HPE powder for new *Sunsik* products as cereal-based ready-to-drink beverages. To determine the optimum formulation of *Sunsik*, the response surface methodology (RSM) was adopted using a central composite experimental design (CCD). The antioxidant capacities, GABA, water absorption index (WAI), water solubility index (WSI), total color changes (ΔE), and browning index (BI) were analyzed to optimize the health-conscious nutrients and quality of *Sunsik*; then, the newly optimized *Sunsik* was compared with control *Sunsik* in terms of various health-conscious and physicochemical properties.

2. Materials and Methods

2.1. Materials

The *Sunsik* and HPE were provided from Gagopa Healing Food Co., Ltd. (Changwon, Korea). The main ingredients of *Sunsik* consisted of 30% barley, 30% brown rice, 20% adlay, 10% black bean, and 10% oat. In general, each cereal was steamed and then dry-roasted. The four roasted cereals were pulverized in a batch for a production of the Sunsik. The Sunsik used in this study is being sold on the market. Gagopa Healing Food Co., Ltd. (Changwon, Korea) found effects of HPE on osteogenic differentiation through preliminary studies, and the results already published [27]. The HPE used in this study is composed of safflower seed (85%), *A. aspera* (5%), manyprickle acanthopanax (5%), and *Kalopanax septemlobus* (5%) [27]. In addition, the GW used in this study was prepared according to preliminary experiments. Anzunbaengi wheat, which was cultivated in Jinju, Korea, was germinated at 17.6 °C for 46.18 h to enhance GABA. After germination, the GW was freeze-dried and then grounded to powder. To develop a cereal-based ready-to-eat beverage to enhance health-related properties, *Sunsik* was formulated with HPE and GW to maximize GABA and antioxidant capacities. The ranges of HPE and GW used in this study were 2–4% and 10–20%, respectively, and the ranges were determined based on samples of five points or more as a result of consumer acceptability (nine-point hedonic scale) of *Sunsik* with added HPE or GW, respectively.

2.2. Experimental Design and Optimization of the Formulation

The amounts of HPE and GW were optimized using a CCD of an RSM [28]. The independent values were studied at five different levels (− α, −1, 0, + 1, and + α), and the actual levels are presented in Table 1.

Table 1. The coded levels and actual values of 13 experiments formulated with a central composite design (CCD).

Experiment No.	Coded Levels		Actual Values		
	X_1 (HPE, g)	X_2 (GW, g)	X_1 (HPE, g)	X_2 (GW, g)	*Sunsik* (g)
1	−1	−1	1	5	44
2	1	−1	2	5	43
3	−1	1	1	10	39
4	1	1	2	10	38
5	α(−)	0	0.79	7.5	41.71
6	α(+)	0	2.21	7.5	40.29
7	0	α(−)	1.5	3.96	44.54
8	0	α(+)	1.5	11.04	37.46
9	0	0	1.5	7.5	41
10	0	0	1.5	7.5	41
11	0	0	1.5	7.5	41
12	0	0	1.5	7.5	41
13	0	0	1.5	7.5	41

Table 1 and they were evaluated to maximize the GABA, total flavonoid content (TFC), TPC, 2,2-diphenyl-1-picrylhydrazyl (DPPH) radical scavenging capacity, Trolox equivalent antioxidant capacity (TEAC), and WSI and to minimize the WAI, ΔE, and BI. The effects of the two independent variables on the responses (Y) were modeled using the response surface regression, and they were predicted by the following Equation (1) [28]:

$$Y_k = \beta_0 + \beta_1 X_1 + \beta_2 X_2 + \beta_{12} X_1 X_2 + \beta_{11} X_1^2 + \beta_{22} X_2^2 \quad (1)$$

where β_0 is a constant, β_1 and β_2 are the linear coefficients, β_{12} is the interaction coefficient, and β_{11} and β_{22} are the quadratic coefficients. X_1 and X_2 are the levels of HPE and GW, respectively. Y_k is the response variable, and each response variable is as follows; Y_1 = GABA (µg/g), Y_2 = TFC (µg CE/g), Y_3 = TPC (µg GE/100g), Y_4 = DPPH (µM TE/100g), Y_5 = TEAC (mM TE/100g), Y_6 = WAI, Y_7 = WSI, and Y_8 = ΔE, Y_9 = BI. To validate the linear or quadratic model, each experimental data of independent variables was compared with the predicted values using the model developed in this study.

2.3. Extraction Procedure of *Sunsik* Samples

In total, 5 g of each *Sunsik* sample was extracted with 80% ethanol at 65 °C for 2 h, and the supernatants obtained by centrifugation (5000 rpm for 30 min) were evaporated to dryness at 45 °C using a nitrogen evaporator (Eyela MG-2200, Tokyo Rikakikai Co. Ltd., Tokyo, Japan). The dried extract was then re-dissolved with 80% ethanol into a final volume of 5 mL. The extract was used to determine the GABA, TEAC, DPPH, TFC, and TPC.

2.4. Gamma-Amino Butyric Acid (GABA)

The GABA contents of the *Sunsik* samples were determined according to the method described in Sharma et al. [29]. In brief, 0.1 mL of each extract was mixed with 0.2 mL of 0.2 M borate buffer and 1 mL of 6% phenol reagent. Then, 0.4 mL of 7.5% sodium hypochlorite was added, and the mixture was boiled for 10 min in a water bath. The samples were immediately cooled for 5 min, and the absorbance was measured using a spectrophotometer (EMC-11D-V Spectrophotometer, EMCLAB Instruments, Duisburg, Germany) at 630 nm. The GABA was used as a standard curve and prepared with a range of concentrations from 0 to 50 mg. Results were expressed as mg/g.

2.5. Total Flavonoid Content (TFC)

TFC was determined using the methods previously described by Dahl [30]. The extract of samples (250 µL) was added to 1.25 mL distilled water, and 70 µL of 5% sodium nitrite was added to the mixture. After 6 min, 150 µL of 10% aluminum chloride was added to the mixture. After 5 min, 0.5 mL of 1 N sodium hydroxide was added to the mixture. The absorbance was measured immediately at 510 nm. Distilled water was used as a blank. Catechin was used as a standard curve and prepared with a range of concentrations from 0 to 2.5 mg. The results were reported as catechin equivalents (CE) µg/g.

2.6. Total Phenolic Content (TPC)

TPC was determined by the method described by de la Rosa et al. [31] with modifications. TPC was measured using the Folin-Ciocalteu method. In total, 100 µL of each extract was added to 2.5 mL of 10% Folin-Ciocalteu reagent, and the mixture was allowed to stand for 2 min. Then, 2 mL of 6% sodium carbonate was added to the mixture, and it was incubated at 50 °C for 15 min in a water bath. The absorbance was measured at 760 nm, and distilled water was used as a blank. Gallic acid was used as a standard curve and prepared with a range of concentrations from 0 to 50 mg. Results were expressed as gallic acid equivalents (GAE) mg/g.

2.7. DPPH Radical Scavenging Capacity

The determination of the effect scavenging of the DPPH radical was based on a procedure previously described by Wong et al. [32]. A 0.1 mM DPPH solution diluted with 100% methanol was prepared. In addition, 0.1 mL of the sample and 1.9 mL of 0.1 mM DPPH were mixed well. The DPPH solution was allowed to stand for 30 min at room temperature in the dark. Then, the absorbance was measured at 515 nm, and 100% methanol was used as a blank. Furthermore, 10 mM Trolox was used as a standard curve and prepared with a range of concentrations from 0 to 500 µM. Results were expressed as µmol of Trolox equivalents (TE) µmol/100 g.

2.8. Trolox Equivalent Antioxidant Capacity (TEAC)

TEAC was performed as described by Simsek and El [33], with modifications. Briefly, an $ABTS^{+}$ stock solution was prepared with 7.4 mM ABTS and 2.6 mM potassium persulfate and mixed. After, the mixture was allowed to stand for 16 h at room temperature in the dark. The $ABTS^{+}$ stock solution was diluted with 100% methanol to an absorbance wavelength of 0.7 at 734 nm. Then, 2960 μL of the $ABTS^{+}$ stock solution was added to 20 μL of the sample, and absorbance was measured after 7 min. Trolox was used as a standard curve and prepared with a range of concentrations from 0 to 1000 μg. Results were expressed as mmol of TE mmol/100 g.

2.9. Water Absorption Index (WAI) and Water Solubility Index (WSI)

The WAI and WSI of the optimized *Sunsik* and control samples were determined using methods previously described by Du et al. [34] with slight modifications. In total, 2.5 g of the sample was added to 30 mL of distilled water and mixed in a shaking water bath at 30 °C for 30 min. Then, the mixture was centrifuged at 3000 rpm for 15 min. The supernatant and remaining sediment from the mixture were weighted. The supernatant was decanted into an aluminum dish and dried at 105°C overnight using a dry oven. The WAI and WSI were calculated as in the following equations, respectively.

$$\mathrm{WAI} = \frac{\mathrm{weightofthesediment\ (g)}}{\mathrm{weightofthesample\ (g)}} \quad (2)$$

$$\mathrm{WSI}(\%) = \frac{weight\ of\ dry\ solids\ from\ the\ supernatant\ (g)}{weight\ of\ the\ sample\ (g)} \times 100 \quad (3)$$

2.10. Color Properties

The color values of the optimized Sunsik and control samples were determined with a CIE Lab system using a color meter (CR-400, Konica minolta sensing Inc., Osaka, Japan). It was calibrated with a white ceramic plate before measuring the sample. The total color changes (ΔE) and browning index (BI) were calculated as follows [35,36]:

$$\Delta E = \sqrt{(L_0^* - L^*)^2 + (a_0^* - a^*)^2 + (b_0^* - b^*)^2} \quad (4)$$

$$BI = [100(X - 0.31)]/0.172 \quad (5)$$

$$X = (a^* + 1.75L^*)/(5.645L^* + a^* - 3.012b^*) \quad (6)$$

where L_0^*, a_0^*, and b_0^* are color parameters for the control and L^*, a^*, and b^* are color parameters for each Sunsik sample.

2.11. Apparent viscosity of Sunsik Samples

The apparent viscosity of the optimized *Sunsik* and control samples was measured using a digital rotary viscometer (WVS-0.1M, DAIHAN Scientific, Gang-Won-Do, Korea). First, 45 g of the sample was placed in a 500-mL beaker, and 300 mL of water or milk was poured in, followed by thorough mixing with a magnetic stirrer (MS-20D, DAIHAN Scientific, Gang-Won-Do, Korea). Finally, the thoroughly mixed sample was poured into a 250-mL beaker (SDS 2400, DONG SUNG science, Gang-Won-Do, Korea) and the viscosity of the sample was measured. When measuring the viscosity, the standard was measured when the torque value was close to 50%.

2.12. Cell Proliferative Effects of Sunsik Samples on Caco-2 and HepG2 Cells

In total, 15 g of the *Sunsik* samples was extracted with 80% ethanol, evaporated to dryness at 45 °C, and re-dissolved in dimethyl sulfoxide (DMSO) according to a previously described method [37]. The Caco-2 (ATCC®HTB-37™, Manassas, USA) cell was cultured in MEM (Hyclone Laboratories Inc.,

South Logan, UT, USA) with 10% or 20% fetal bovine serum (FBS, Welgene, Daegu, Korea) at 37 °C in a humidified incubator with 5% CO_2. The cell proliferation of *Sunsik* extracts was determined by MTT (3-(4,5-dimethylthiazol-2-yl)-2,5-diphenyl tetrazolium bromide) assay. The cells (1 × 10_4/well) were seeded in 96-well plates and then allowed to attach overnight. After overnight, the media included with *Sunsik* extracts were exchanged and incubated for 72 h. After 72 h of incubation, cell proliferation was determined using the MTT Cell Proliferation Assay kit (Roche Ltd., Mannheim, Germany) at 570–655 nm with a SpectraMax®i3 plate reader (Molecular Devices, Sunnyvale, CA, USA).

2.13. Data Analysis

The Design Expert software (version 11, State-Ease Inc., Minneapolis, USA) was used to analyze the experimental data for best fit model equations and to obtain response plots for each response variable. The combination of independent variables generating the highest overall desirability was selected as the optimum formulation. To validate the optimization process, the Sunsik was prepared using the optimum levels of independent variables and analyzed for the selected responses. The absolute residual error (%) was calculated using the experimental and predicted data through the following Equation (7):

$$Absolute\ residual\ error(\%) = \frac{Actual\ value - Predicted\ value}{Actual\ value} \times 100 \quad (7)$$

All experiments were carried out in triplicate, and ANOVA was performed to determine differences among the samples using the XLSTAT software (Addinsoft, Paris, France). When a difference among the samples was identified, the Student Newan–Keul's (SNK) multiple comparison was performed to separate the means.

3. Results and Discussion

3.1. Fitting the Model and Statistical Analysis

The RSM is often used to determine the formulation ratio of a new product in the food industry. In this study, a CCD was applied to determine the optimum formulation of HPE and GW to prepare healthy Sunsik, a cereal-based ready-to-drink Korean beverage. The independent and dependent variables were fitted by linear or quadratic equations, and Table 2 shows the statistical results of the regression coefficients, R^2, adjusted R^2, lack of fit, and p values of the fitted models on analyzed responses by CCD. As shown in Table 2, the lack of fit for response surface models was not significant without the WSI, implying that the response surface models were adequately explained for predicting the relevant responses [28].

Table 2. The regression coefficients, R square, adjusted R square, lack of fit, and *p* values of the fitted models on dependent variables.

		Health Conscious Properties					Physicochemical Properties			
		GABA	TFC	TPC	DPPH	TEAC	WAI	WSI	ΔE	BI
Constant	β_0	2.09	30.99	70.57	106.59	120.16	1.85	48.44	0.2224	20.02
Linear	β_1	0.0170	1.11	−0.32	3.71 **	1.34	0.0068	−0.3332 *	−0.2736 **	0.2590 **
	β_2	0.1031 **	3.03 ***	2.21 *	3.32 **	3.39 **	−0.0196 *	4.52	0.1071 *	0.0150
Quadratic	β_{11}		1.18 *		2.39 **		−0.0394	0.0882 *	−0.0362 ***	−0.1831
	β_{22}		−1.45 *		−3.46 **		0.0140 *	−4.61	0.2708 **	−0.0531
Interaction	β_{12}		−1.47		−2.80		0.0263 **	−2.03	0.1384	0.0061
R^2		0.546	0.889	0.487	0.886	0.563	0.828	0.583	0.952	0.702
Adjusted R^2		0.455	0.809	0.384	0.804	0.476	0.805	0.285	0.917	0.489
Lack of Fit (*p* value)		0.196	0.745	0.052	0.094	0.228	0.533	0.041	0.452	0.193
p value		0.019	0.003	0.004	0.003	0.015	0.013	0.203	0.0002	0.075

*, **, *** significantly differ at $p > 0.05$, $p < 0.01$, and $p < 0.001$, respectively. β_1: herbal plant extract; β_2: germinated wheat.

Among the responses, GABA, TPC, and TEAC were more adequate for a linear model than for a quadratic model. Because the β_2 values of GABA ($p < 0.01$), TPC ($p < 0.05$), and TEAC ($p < 0.01$) differed significantly, the GABA, TPC, and TEAC contents of newly developed Sunsik might be affected by GW rather than HPE. The final equations of GABA, TPC, and TEAC as follows:

$$\text{GABA} = 2.09 + 0.017 \times \text{HPE} + 0.1031 \times \text{GW} \quad (8)$$

$$\text{TPC} = 70.57 - 0.3237 \times \text{HPE} + 2.21 \times \text{GW} \quad (9)$$

$$\text{TEAC} = 120.16 + 1.34 \times \text{HPE} + 3.39 \times \text{GW} \quad (10)$$

As described in Table 2, the TFC, DPPH, WAI, WSI, ΔE, and BI were fitted with quadratic models. The final equations of TFC, DPPH, WAI, WSI, ΔE, and BI were coded as follows:

$$\text{TFC} = 30.99 + 1.11 \times \text{HPE} + 3.03 \times \text{GW} + 1.18 \times \text{HPE} \times \text{GW} - 1.45 \times \text{HPE}^2 - 1.47 \times \text{GW}^2 \quad (11)$$

$$\text{DPPH} = 106.59 + 3.71 \times \text{HPE} + 3.32 \times \text{GW} + 2.39 \times \text{HPE} \times \text{GW} - 3.49 \times \text{HPE}^2 - 2.80 \times \text{GW}^2 \quad (12)$$

$$\text{WAI} = 1.85 + 0.0068 \times \text{HPE} - 0.0196 \times \text{GW} - 0.0394 \times \text{HPE} \times \text{GW} + 0.014 \times \text{HPE}^2 + 0.0263 \times \text{GW}^2 \quad (13)$$

$$\text{WSI} = 48.44 - 0.3332 \times \text{HPE} + 4.52 \times \text{GW} + 0.0882 \times \text{HPE} \times \text{GW} - 4.61 \times \text{HPE}^2 - 2.03 \times \text{GW}^2 \quad (14)$$

$$\Delta\text{E} = 0.2224 - 0.2736 \times \text{HPE} + 0.1071 \times \text{GW} - 0.0362 \times \text{HPE} \times \text{GW} + 0.2708 \times \text{HPE}^2 + 0.1384 \times \text{GW}^2 \quad (15)$$

$$\text{BI} = 20.02 + 0.259 \times \text{HPE} + 0.015 \times \text{GW} - 0.1831 \times \text{HPE} \times \text{GW} - 0.0531 \times \text{HPE}^2 + 0.0061 \times \text{GW}^2 \quad (16)$$

The higher values of R^2 and adjusted R^2 mean desirability of the model to explain the relationships between variables [28]. In this study, the responses with R^2 values of 0.8 or higher were TFC, DPPH, WAI, and ΔE, indicating that the fitted equations adequately describe the effects of adding GW and HPE to Sunsik on each dependent variable.

3.2. Effects of Independent Values on Health-Conscious Properties

The GABA, TFC, and TPC contents and antioxidant capacities (DPPH radical scavenging capacity and TEAC) of differently formulated Sunsik samples by CCD are shown in Table 3. Significant differences among the 13 samples were found in the GABA ($p < 0.01$), TFC ($p < 0.001$), TPC ($p < 0.001$), DPPH ($p < 0.05$), and TEAC ($p < 0.05$) contents. The GABA content, TFC, and TPC are some of the major compounds that contribute to the antioxidant capacities, such as DPPH and TEAC [11,30,38]. The GABA content and TPC were in the ranges of 1.81–2.25 μg/g and 67–76 μg GE/100g, respectively. As shown in Table 2, the GABA content and TPC were significant in the β_2 value ($p < 0.01$ for GABA and $p < 0.05$ for TPC) but not significant in the β_1 value, indicating that the GABA content and TPC of Sunsik with added HPE and GW were influenced by increased GW. These results were also confirmed in the three-dimensional response surface plots of Figure 1a,c.

Table 3. The experimental values of the health-conscious variables for each independent variable.

	Experiment No.			GABA ** (Y_1, μg/g)	TFC *** (Y_2, μg CE/g)	TPC *** (Y_3, μg GE/100 g)	DPPH * (Y_4, μM TE/100 g)	TEAC * (Y_5, mM TE/100 g)
	HPE (g)	GW (g)	*Sunsik* (g)					
1	1	5	44	2.01 ± 0.09 ^{ab}	26 ± 2.30 ^{bc}	71 ± 2.23 ^{cd}	96 ± 3.7 ^{b}	113 ± 3.93 ^{b}
2	2	5	43	2.00 ± 0.08 ^{ab}	24 ± 1.21 ^{c}	67 ± 0.44 ^{f}	102 ± 2.9 ^{ab}	114 ± 3.70 ^{b}
3	1	10	39	2.19 ± 0.13 ^{a}	29 ± 3.68 ^{abc}	76 ± 0.88 ^{a}	95 ± 3.1 ^{ab}	122 ± 5.48 ^{ab}
4	2	10	38	2.14 ± 0.16 ^{a}	32 ± 3.34 ^{ab}	74 ± 0.97 ^{b}	110 ± 9.2 ^{ab}	122 ± 3.30 ^{ab}
5	0.79	7.5	41.71	1.98 ± 0.06 ^{ab}	26 ± 0.92 ^{abc}	69 ± 0.70 ^{de}	96 ± 6.9 ^{b}	116 ± 4.27 ^{b}
6	2.21	7.5	40.29	2.12 ± 0.04 ^{a}	31 ± 2.80 ^{abc}	72 ± 0.09 ^{c}	103 ± 9.4 ^{b}	123 ± 2.86 ^{ab}
7	1.5	3.96	44.54	1.81 ± 0.15 ^{b}	22 ± 2.00 ^{c}	68 ± 0.40 ^{ef}	94 ± 6.9 ^{ab}	119 ± 3.60 ^{ab}

Table 3. *Cont.*

Experiment No.				GABA ** (Y_1, μg/g)	TFC *** (Y_2, μg CE/g)	TPC *** (Y_3, μg GE/100 g)	DPPH * (Y_4, μM TE/100 g)	TEAC * (Y_5, mM TE/100 g)
	HPE (g)	GW (g)	*Sunsik* (g)					
8	1.5	11.04	37.46	2.17 ± 0.08 [a]	33 ± 3.48 [abc]	72 ± 0.88 [c]	107 ± 8.3 [a]	127 ± 5.22 [a]
9	1.5	7.5	41	2.09 ± 0.04 [a]	30 ± 2.54 [ab]	69 ± 0.99 [de]	108 ± 8.8 [ab]	120 ± 2.36 [ab]
10	1.5	7.5	41	2.11 ± 0.11 [a]	30 ± 1.43 [ab]	68 ± 0.02 [ef]	106 ± 1.8 [ab]	123 ± 4.42 [ab]
11	1.5	7.5	41	2.17 ± 0.11 [a]	30 ± 1.90 [abc]	69 ± 0.71 [de]	105 ± 2.5 [ab]	124 ± 4.95 [ab]
12	1.5	7.5	41	2.16 ± 0.20 [a]	33 ± 3.64 [abc]	71 ± 0.07 [cd]	105 ± 3.7 [ab]	122 ± 4.76 [ab]
13	1.5	7.5	41	2.25 ± 0.12 [a]	33 ± 2.32 [abc]	71 ± 0.94 [cd]	109 ± 8.5 [ab]	119 ± 7.35 [ab]

All values are means of three replications ± standard deviation. Values with the same letter(s) within a column are not significantly different. *, **, *** significantly differ at $p > 0.05$, $p < 0.01$, and $p < 0.001$, respectively.

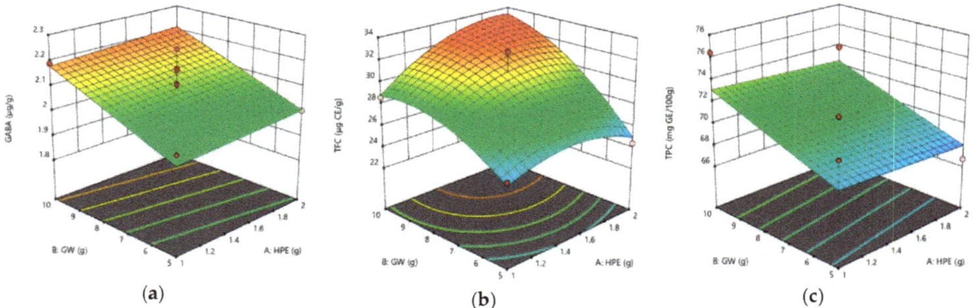

Figure 1. Three-dimensional response surface plots of the GABA content (**a**), TFC (**b**), and TPC (**c**). (GW: germinated wheat; HPE: herbal plant extract; GABA: gamma aminobutyric acid; TFC: total flavonoid content; TPC: total phenolic acid).

Conversely, the addition of HPE and GW had significant quadratic effects ($p < 0.05$ for β_{11} and $p < 0.05$ for β_{22}) on TFC (Table 2). Figure 1b shows the three-dimensional response surface plots of TFC, implying the TFC of *Sunsik* is increased by both HPE and GW.

The antioxidant properties of 13 Sunsik samples corresponding to the experiments generated by the CCD were determined by DPPH and TEAC (Table 3). The DPPH and TEAC values of the samples differed significantly (both $p < 0.05$) and were in the ranges of 96–110 μM TE/100g and 113–127 mM TE/100 g, respectively. As presented in Table 2, the DPPH value was fitted with a quadratic model while TEAC value was fitted with a linear model. The comprehensive effects of the dependent variables (HPE and GW) on the antioxidant properties of Sunsik are represented by the response surface plots in Figure 2.

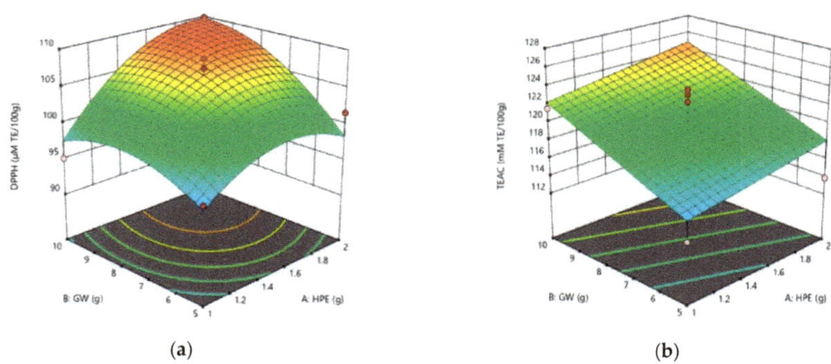

Figure 2. Three-dimensional response surface plots of DPPH (**a**) and TEAC (**b**).

The Sunsik samples with higher antioxidant activities contained relatively high GABA content, TPC, and TFC. These results are in agreement with previous studies [11], which reported a higher antioxidant capacity of the samples containing higher GABA content, TPC, and TFC. The increments of TPC and GABA content in Sunsik samples could be explained by the addition of GW. Chen et al. [39] reported that phenolic contents in GW increased by lignin synthesis during germination. In addition, another study explained that the GABA content in GW increased via the decarboxylation of L-glutamate [11]. Safflower seed, a major material of HPE, has protective effects against osteoporosis and a beneficial effect on atherogenic risk through various phenolic compounds, such as lignin and flavonoids [25]. Recently, the antioxidant, anti-cancer, anti-inflammatory effects of safflower seeds have been identified by a few studies [25,40,41].

3.3. *Effects of Independent Values on Physicochemical Properties*

The WAI and WSI are important parameters in powdered cereal-based beverages, such as Sunsik, which is eaten by dissolving in milk or water. The WAI and WSI values of the Sunsik samples tested in this study are presented in Table 4. The WAI values of the Sunsik samples were in the range of 1.82–1.95 and did not differ significantly (Table 4). Although there was no statistically significant difference in the WAI values of Sunsik samples, they tended to increase as the amount of HPE increased (Figure 3a). The WAI value of reconstituted powder, such as Sunsik examined in this study, might play a role in preventing its dissolution in milk or water [42]. As shown in the WAI results of Table 2, the linear coefficients of HPE (β_1) and GW (β_2) were 0.0018 and −0.0195, respectively, implying that GW in newly formulated Sunsik had a negative effect. The WSI is the amount of soluble components released from the Sunsik samples, and the values ranged from 32% to 59% (Table 4). The WSI values of Sunsik with 1.5 g of added HPE and 11.04 g of added GW were the highest among the samples, suggesting the contribution of GW to the solubility of the newly formulated Sunsik samples (Figure 3b).

Table 4. The experimental values of the physicochemical variables for each independent variable.

Experiment No.				*WAI*	*WSI* (%) ***	ΔE ***	*BI* **
	HPE (g)	**GW (g)**	***Sunsik* (g)**				
1	1	5	44	1.88 ± 0.06	42 ± 1.92 [d]	0.72 ± 0.18 [ab]	19.4 ± 0.35 [c]
2	2	5	43	1.95 ± 0.04	41 ± 1.01 [d]	0.22 ± 0.06 [bc]	20.2 ± 0.38 [ab]
3	1	10	39	1.91 ± 0.02	41 ± 0.49 [d]	1.12 ± 0.24 [abc]	20.0 ± 0.39 [abc]
4	2	10	38	1.82 ± 0.12	40 ± 0.33 [d]	0.48 ± 0.08 [b]	20.1 ± 0.20 [ab]
5	0.79	7.5	41.71	1.86 ± 0.04	41 ± 1.17 [d]	1.13 ± 0.27 [ab]	19.5 ± 0.34 [bc]
6	2.21	7.5	40.29	1.91 ± 0.02	40 ± 1.26 [d]	0.39 ± 0.12 [b]	20.4 ± 0.32 [a]
7	1.5	3.96	44.54	1.92 ± 0.08	32 ± 0.23 [e]	0.43 ± 0.04 [b]	20.3 ± 0.31 [ab]
8	1.5	11.04	37.46	1.89 ± 0.02	59 ± 0.63 [a]	0.56 ± 0.16 [b]	19.9 ± 0.23 [abc]
9	1.5	7.5	41	1.86 ± 0.06	49 ± 1.12 [c]	0.23 ± 0.02 [bc]	20.2 ± 0.23 [ab]
10	1.5	7.5	41	1.89 ± 0.05	53 ± 2.11 [b]	0.06 ± 0.03 [c]	20.2 ± 0.05 [ab]
11	1.5	7.5	41	1.83 ± 0.05	46 ± 1.89 [c]	0.24 ± 0.04 [bc]	20.0 ± 0.11 [abc]
12	1.5	7.5	41	1.83 ± 0.01	48 ± 1.95 [c]	0.29 ± 0.07 [bc]	19.8 ± 0.09 [abc]
13	1.5	7.5	41	1.84 ± 0.08	46 ± 1.93 [c]	0.30 ± 0.06 [bc]	19.9 ± 0.01 [abc]

All values are means of three replications ± standard deviation. Values with the same letter(s) within a column are not significantly different. **, *** significantly differ at $p < 0.01$ and $p < 0.001$, respectively.

Significant differences were observed in the ΔE ($p < 0.001$) and BI ($p < 0.01$) values among the newly formulated Sunsik samples (Table 4), which were in the ranges of 0.22–1.13 and 19.2–20.3, respectively. In the results of the regression coefficients, the HPE addition negatively affected and the GW addition positively affected the ΔE of the newly formulated Sunsik. The three-dimensional response surface plots also showed a similar trend (Figure 3c), indicating that the color of the newly formulated Sunsik was mostly affected by a higher GW amount than HPE amount. Such a result was expected, as more GW (10–20%) was added to Sunsik than HPE (2–4%). The color affects consumer perceptions of various foods or beverages, and color changes or a brown color during processing or cooking might negatively affect consumer preferences [43]. As shown in Figure 3d, the brown color changes of Sunsik were the result of adding HPE. In a preliminary experiment to determine the range of the HPE amount, consumers tended not to prefer Sunsik with more than 4% HPE added due to its darkened color.

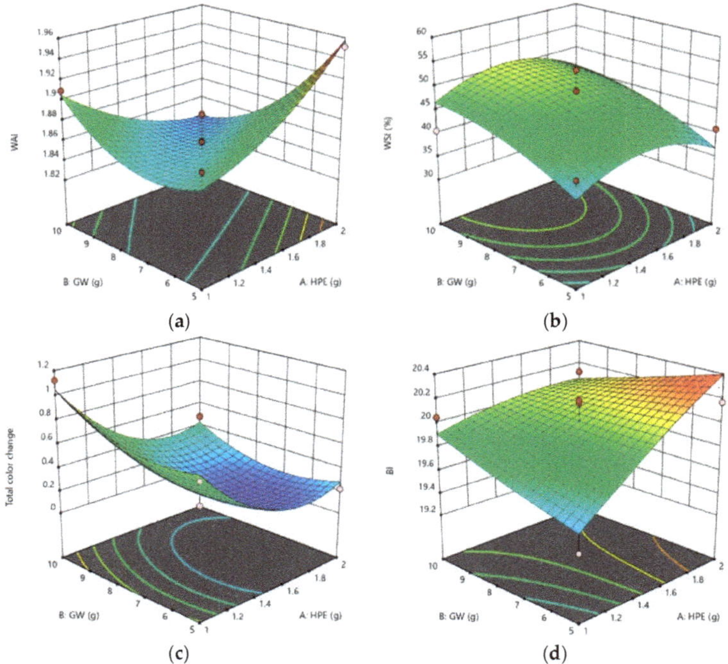

Figure 3. Three-dimensional response surface plots of the WAI (**a**), WSI (**b**), ΔE (**c**), and BI (**d**).

3.4. Optimization and Validation

Cereal-based products like Sunsik are often developed with the addition of two or more ingredients to provide additional health benefits to consumers. In this study, both GW and HPE had a significant effect on the health-related properties and physicochemical characteristics of Sunsik. The additions of GW and HPE in newly formulated Sunsik were response specific. Thus, optimization is needed to attain a formulation with the desired characteristics concerning all the responses.

Sunsik, a cereal-based ready-to-drink beverage, was optimized considering maximized properties, such as GABA, TFC, TPC, DPPH, TEAC, and WSI. By contrast, WAI, ΔE, and BI were minimized in Sunsik products. The optimized formula of Sunsik developed in this study was 10 g of GW, 1.79 g of HPE, and 38.21 g of Sunsik corresponding to the highest desirability of 0.719. In addition, the predicted and actual values for optimized formulations of Sunsik are presented in Table 5. Both the predicted and actual values were compared and were verified using absolute residual error values (Table 5). The errors for the responses were found to be less than 5% without ΔE. This indicated the precision of the developed and optimized regression models for the newly formulated Sunsik products.

Table 5. Predicted and actual values of the optimized *Sunsik* formulation.

Responses	Optimized Formulation			
	Goal	Predicted Values	Actual Values	Error (%)
GABA (Y_1, µg/g)	Maximize	2.21	2.23 ± 0.04	0.9
TFC (Y_2, µg CE g)	Maximize	33.39	33.75 ± 0.25	1.07
TPC (Y_3, µg GE/100g)	Maximize	72.56	73.26 ± 0.46	0.97
DPPH (Y_4, µM TE/100g)	Maximize	110	112 ± 0.58	2.12
TEAC (Y_5, mM TE/100g)	Maximize	124	125 ± 0.58	0.54
WAI (Y_6)	Minimize	1.84	1.80 ± 0.03	2.44
WSI (Y_7)	Maximize	49.2	48.52 ± 1.28	0.25
ΔE (Y_8)	Minimize	0.38	0.25 ± 0.03	11.71
BI (Y_9)	Minimize	20.06	20.42 ± 0.12	1.61

3.5. Health-Conscious and Physicochemical Properties of Optimized Sunsik

Because the purpose of this study was to develop a newly formulated Sunsik containing GW and HPE to provide health benefits over the commercially available Sunsik, various properties of commercial and optimized Sunsik were compared. The health-conscious and physicochemical properties of both Sunsik samples are presented in Table 6. The GABA content, TPC, and TFC might be major constituents contributing to the antioxidant capacities and antiproliferative cancer cells [38]. Significant differences between the commercial and optimized Sunsik samples with respect to the GABA content ($p < 0.001$), TFC ($p < 0.001$), and TPC ($p < 0.001$) were observed (Table 6). The optimized Sunsik contained more GABA (2.23 µg/g) content, TFC (33.75 µg CE/ 100g), and TPC (73.75 µg GE/100g) than commercial Sunsik (GABA: 1.7 µg/g; TFC 19.8 µg CE/100 g; TPC: 54.4 µg GE/100g), confirming health benefits of optimized Sunsik compared to commercial Sunsik.

In addition, the DPPH ($p < 0.001$) and TEAC ($p < 0.001$) of optimized Sunsik, to which 10 g of GW and 1.79 g of HPE were added, increased significantly compared to commercial Sunsik. Numerous studies have been developed new product with more antioxidant or antiproliferative activities to contribute health benefits of consumed products [7,8,38]. According to Kim and Kim [38], cereal products containing higher phenolic or flavonoid contents had higher antioxidant capacities. In this study, optimized Sunsik contained higher TPC, TFC, DPPH, and TEAC values than the commercial Sunsik. Similar trends were observed in terms of the proliferative activities of cancer cells. The relative proliferative effects on Caco-2 and HepG2 cells after treatment with an extract of the samples are shown as the median effective dose (EC_{50}) in Table 6. The EC_{50} values of optimized Sunsik for Caco-2 and HepG2 cells were 45.7 and 35.2 mg/mL, respectively. Commercial Sunsik was relatively high in EC_{50} values of Caco-2 (97.9 mg/mL) and HepG2 (76.2 mg/mL) cells compared to those of optimized Sunsik (Caco-2: 45.7 mg/mL; HepG2: 35.2 mg/mL), indicating relatively low antiproliferative activities. Many studies have reported that foods or beverages with antioxidant activities have cancer-protective effects [37], suggesting that cereal-based beverages could inhibit cancer cell growth. In this study, optimized Sunsik added with GW and HPE showed higher antioxidant capacity and antiproliferative activity than commercial Sunsik.

The WAI, WSI and viscosity of optimized Sunsik with added GW and HPE were compared to commercial Sunsik, and the results are shown in Table 6. The WAI and viscosity of cereal-based beverages are important quality factors [3,4]. According to the finding of Fernandes, Sonawane, and Arya [3], the high absorbing properties in cereal-based beverages resulted in increased viscosity, and high viscosity negatively affected mouthfeel and overall acceptability in sensory tests. According to the results of the current study, the WAI and viscosity of optimized Sunsik with added GW and HPE were less than that of the commercial Sunsik sample. The low WAI and viscosity might contribute to the solubility of Sunsik, which is eaten by dissolving in milk or water, showing higher WSI values in optimized Sunsik than in commercial Sunsik.

Table 6. Health-conscious and physicochemical properties of the optimized Sunsik formulation.

		Commercial *Sunsik*	Optimized *Sunsik*
Health conscious properties	GABA (*μg/g*) ***	1.7 ± 0.09 [b]	2.23 ± 0.04 [a]
	TFC (*μg CE/g*) ***	19.8 ± 1.72 [b]	33.75 ± 0.25 [a]
	TPC (*μg GE/100g*) ***	54.4 ± 3.57 [b]	73.26 ± 0.46 [a]
	DPPH (*μM TE/100g*) ***	77.3 ± 2.06 [b]	112 ± 0.58 [a]
	TEAC (*mM TE/100g*) ***	96.9 ± 3.27 [b]	125 ± 0.58 [a]
	EC_{50} for Caco-2 cell (mg/mL) ***	97.4 ± 4.2 [a]	45.7 ± 1.6 [b]
	EC_{50} for HepG2 cell (mg/mL) ***	76.2 ± 3.8 [a]	35.2 ± 2.5 [b]
Physicochemical properties	WAI ***	3.6 ± 0.03 [a]	1.80 ± 0.03 [b]
	WSI (%) ***	7.4 ± 0.1 [b]	48.52 ± 1.28 [a]
	Apparent viscosity (cP) ***	294 ± 2.87 [a]	47 ± 4.42 [b]

All values are means of three replications ± standard deviation. Values with same letter(s) within a row are not significantly different. *** significantly differ at $p < 0.001$.

4. Conclusions

This study showed that the CCD and RSM could be used to optimize the formulation of *Sunsik*, a cereal-based ready-to-eat beverage. RSM predicted that a *Sunsik* formula of 10 g GW, 1.79 g HPE, and 38.21 g *Sunsik* would provide a better quality with more health-conscious and physicochemical characteristics. The optimized *Sunsik* is characterized by higher GABA, TPC, TFC, DPPH, TEAC, and WAI values than commercial *Sunsik*. The EC_{50} of cancer cells, WAI, and viscosity were low in optimized *Sunsik* compared to commercial *Sunsik*. Overall, *Sunsik* with 10 g of added GW and 1.79 g of added HPE might increase various health-related components and biological activities while maintaining the quality of the cereal-based beverage.

Author Contributions: Conceptualization, M.J.K.; methodology, B.R.K., S.S.P., Y.J.K., G.-J.Y.; investigation, B.R.K., S.S.P.; data curation, B.R.K., S.S.P.; writing—original draft preparation, B.R.K.; writing—review and editing, M.J.K.; supervision, M.J.K. All authors have read and agreed to the published version of the manuscript.

Funding: This research was supported by Changwon National University in 2019~2020.

Conflicts of Interest: The authors declare no conflict of interest.

References

1. Choi, E.; Kim, B.H. A comparison of the fat, sugar, and sodium contents in ready-to-heat type home meal replacements and restaurant foods in Korea. *J. Food Compos. Anal.* **2020**, *92*, 103524. [CrossRef]
2. Bembem, K.; Agrahar-Murugkar, D. Development of millet based ready-to-drink beverage for geriatric population. *J. Food Sci. Technol.* **2020**, *57*, 3278–3283. [CrossRef]
3. Fernandes, C.G.; Sonawane, S.K.; Arya, S.S. Optimization and modeling of novel multigrain beverage: Effect of food additives on physicochemical and functional properties. *J. Food Process. Preserv.* **2019**, *43*, 14151. [CrossRef]
4. Jung, J.-H.; Lee, S. Microbial Growth in Dry Grain Food (Sunsik) Beverages Prepared with Water, Milk, Soymilk, or Honey-Water. *J. Food Sci.* **2010**, *75*, M239–M242. [CrossRef] [PubMed]
5. Lee, E.-J.; Kim, S.-G.; Yoo, S.-R.; Oh, S.-S.; Hwang, I.-G.; Kwon, G.-S.; Park, J.-H. Microbial contamination by Bacillus cereus, Clostridium perfringens, and Enterobacter sakazakii in sunsik. *Food Sci. Biotechnol.* **2007**, *16*, 948–953.
6. Koh, E.; Jang, K.-H.; Surh, J. Improvement of physicochemical properties of cereal based ready-to-eat Sunsik using fermentation with Bionuruk and Bifidobacterium longum. *Food Sci. Biotechnol.* **2014**, *23*, 1977–1985. [CrossRef]
7. Bang, Y.-S.; Jang, E.H.; Chung, H.-J. Quality and physicochemical characteristics of newly developed Sunsik products with germinated brown rice. *Korean J. Food Sci. Technol.* **2017**, *49*, 513–518.

8. Park, K.-H. Quality and characteristics of manufacturing Sunsik with edible insect (mealworm). *Culin. Sci. Hosp. Res.* **2018**, *24*, 13–23.
9. Xu, L.; Wang, P.; Ali, B.; Yang, N.; Chen, Y.; Wu, F.; Xu, X. Changes of the phenolic compounds and antioxidant activities in germinated adlay seeds. *J. Sci. Food Agric.* **2017**, *97*, 4227–4234. [CrossRef] [PubMed]
10. Gawlik-Dziki, U.; Dziki, D.; Nowak, R.; Świeca, M.; Olech, M.; Pietrzak, W. Influence of sprouting and elicitation on phenolic acids profile and antioxidant activity of wheat seedlings. *J. Cereal Sci.* **2016**, *70*, 221–228. [CrossRef]
11. Kim, M.J.; Kwak, H.S.; Kim, S.S. Effects of Germination on Protein, γ-Aminobutyric Acid, Phenolic Acids, and Antioxidant Capacity in Wheat. *Molecules* **2018**, *23*, 2244. [CrossRef] [PubMed]
12. Dziki, D.; Gawlik-Dziki, U.; Różyło, R.; Miś, A. Drying and Grinding Characteristics of Four-Day-Germinated and Crushed Wheat: A Novel Approach for Producing Sprouted Flour. *Cereal Chem. J.* **2015**, *92*, 312–319. [CrossRef]
13. Dhillon, B.; Choudhary, G.; Sodhi, N.S. A study on physicochemical, antioxidant and microbial properties of germinated wheat flour and its utilization in breads. *J. Food Sci. Technol.* **2020**, *57*, 2800–2808. [CrossRef] [PubMed]
14. Chen, Z.; Yu, L.; Wang, X.; Gu, Z.; Beta, T. Changes of phenolic profiles and antioxidant activity in canaryseed (*Phalaris canariensis* L.) during germination. *Food Chem.* **2016**, *194*, 608–618. [CrossRef]
15. Koehler, P.; Hartmann, G.; Wieser, H.; Rychlik, M. Changes of Folates, Dietary Fiber, and Proteins in Wheat as Affected by Germination. *J. Agric. Food Chem.* **2007**, *55*, 4678–4683. [CrossRef]
16. Li, T.S. *Taiwanese Native Medicinal Plants: Phytopharmacology and Therapeutic Values*; CRC Press: Boca Raton, FL, USA, 2006.
17. Park, C.H.; Lee, A.Y.; Kim, J.H.; Seong, S.H.; Jang, G.Y.; Cho, E.J.; Choi, J.S.; Kwon, J.; Kim, Y.O.; Lee, S.W.; et al. Protective Effect of Safflower Seed on Cisplatin-Induced Renal Damage in Mice via Oxidative Stress and Apoptosis-Mediated Pathways. *Am. J. Chin. Med.* **2018**, *46*, 157–174. [CrossRef]
18. Kim, K.-W.; Suh, S.-J.; Lee, T.-K.; Ha, K.T.; Kim, J.-K.; Kim, K.-H.; Kim, D.-I.; Jeon, J.H.; Moon, T.-C.; Kim, C.-H. Effect of safflower seeds supplementation on stimulation of the proliferation, differentiation and mineralization of osteoblastic MC3T3-E1 cells. *J. Ethnopharmacol.* **2008**, *115*, 42–49. [CrossRef]
19. Ambreen, M.; Mirza, S.A. Evaluation of anti-inflammatory and wound healing potential of tannins isolated from leaf callus cultures of *Achyranthes aspera* and *Ocimum basilicum*. *Pak. J. Pharm. Sci.* **2020**, *33*, 361–369.
20. Li, T.; Ferns, K.; Yan, Z.-Q.; Yin, S.-Y.; Kou, J.-J.; Li, D.; Zeng, Z.; Yin, L.; Wang, X.; Bao, H.-X.; et al. Acanthopanax senticosus: Photochemistry and Anticancer Potential. *Am. J. Chin. Med.* **2016**, *44*, 1543–1558. [CrossRef]
21. Wang, H.; Li, D.; Du, Z.; Huang, M.-T.; Cui, X.; Lu, Y.; Li, C.; Woo, S.L.; Conney, A.H.; Zheng, X.; et al. Antioxidant and anti-inflammatory properties of Chinese ilicifolius vegetable (*Acanthopanax trifoliatus* (L.) Merr) and its reference compounds. *Food Sci. Biotechnol.* **2015**, *24*, 1131–1138. [CrossRef]
22. Yadav, E.; Singh, D.; Yadav, P.; Verma, A. Attenuation of dermal wounds via downregulating oxidative stress and inflammatory markers by protocatechuic acid rich n-butanol fraction of *Trianthema portulacastrum* Linn. in wistar albino rats. *Biomed. Pharmacother.* **2017**, *96*, 86–97. [CrossRef] [PubMed]
23. Kim, J.-H.; Park, J.-H.; Park, S.-D.; Choi, S.-Y.; Seong, J.-H.; Moon, K.-D. Preparation and antioxidant activity of health drink with extract powders from safflower (*Carthamus tinctorius* L.) seed. *Korean J. Food Sci. Technol.* **2002**, *34*, 617–624.
24. Kwak, D.-Y.; Kim, J.-H.; Choi, M.-S.; Shin, S.-R.; Moon, K.-D. Effect of hot water extract powder from safflower (*Carthamus tinctorius* L.) seed on quality of noodle. *J. Korean Soc. Food Sci. Nutr.* **2002**, *31*, 460–464.
25. Yu, S.-Y.; Lee, Y.-J.; Kim, J.-D.; Kang, S.-N.; Lee, S.-K.; Jang, J.-Y.; Lee, H.-K.; Lim, J.-H.; Lee, O.-H. Phenolic Composition, Antioxidant Activity and Anti-Adipogenic Effect of Hot Water Extract from Safflower (*Carthamus tinctorius* L.) Seed. *Nutrients* **2013**, *5*, 4894–4907. [CrossRef] [PubMed]
26. Boo, K.-H.; Lee, D.; Jeon, G.L.; Ko, S.H.; Cho, S.K.; Kim, J.H.; Park, S.P.; Hong, Q.; Lee, S.-H.; Lee, D.-S.; et al. Distribution and Biosynthesis of 20-Hydroxyecdysone in Plants of *Achyranthes japonica* Nakai. *Biosci. Biotechnol. Biochem.* **2010**, *74*, 2226–2231. [CrossRef] [PubMed]
27. Da-Sol, K.; Ki-Soo, S.; Min-Sung, P.; Mi-Kyoung, K.; Kyoung-Eun, P.; Yeon-Ju, K.; Geum-Joung, Y.; Hyung-Joon, K.; Moon-Kyoung, b.; Soo-Kyung, B.; et al. The Effect of Natural-Herb Mixture Extract on Osteogenic Differentiation in Human Mesenchymal Stem cells. *Korean J. Oral Maxillofac. Pathol.* **2019**, *43*, 171–178. [CrossRef]

28. Sabokbar, N.; Khodaiyan, F.; Moosavi-Nasab, M. Optimization of processing conditions to improve antioxidant activities of apple juice and whey based novel beverage fermented by kefir grains. *J. Food Sci. Technol.* **2014**, *52*, 1–11. [CrossRef]
29. Sharma, S.; Saxena, D.C.; Riar, C.S. Changes in the GABA and polyphenols contents of foxtail millet on germination and their relationship with in vitro antioxidant activity. *Food Chem.* **2018**, *245*, 863–870. [CrossRef]
30. Dahl, L.K. Salt and hypertension. *Am. J. Clin. Nutr.* **1972**, *25*, 231–244. [CrossRef]
31. De La Rosa, L.A.; Álvarez-Parrilla, E.; Shahidi, F. Phenolic Compounds and Antioxidant Activity of Kernels and Shells of Mexican Pecan (*Carya illinoinensis*). *J. Agric. Food Chem.* **2011**, *59*, 152–162. [CrossRef]
32. Wong, S.P.; Leong, L.P.; Koh, J.H.W. Antioxidant activities of aqueous extracts of selected plants. *Food Chem.* **2006**, *99*, 775–783. [CrossRef]
33. Simsek, S.; El, S.N. In Vitro starch digestibility, estimated glycemic index and antioxidant potential of taro (*Colocasia esculenta* L. Schott) corm. *Food Chem.* **2015**, *168*, 257–261. [CrossRef] [PubMed]
34. Du, S.-K.; Jiang, H.; Yu, X.; Jane, J.-l. Physicochemical and functional properties of whole legume flour. *LWT-Food Sci. Technol.* **2014**, *55*, 308–313. [CrossRef]
35. Marand, M.A.; Amjadi, S.; Marand, M.A.; Roufegarinejad, L.; Jafari, S.M. Fortification of yogurt with flaxseed powder and evaluation of its fatty acid profile, physicochemical, antioxidant, and sensory properties. *Powder Technol.* **2020**, *359*, 76–84. [CrossRef]
36. Zambrano-Zaragoza, M.D.L.L.; Mercado-Silva, E.; Del Real Lopez, A.; Gutiérrez-Cortez, E.; Cornejo-Villegas, M.; Quintanar-Guerrero, D. The effect of nano-coatings with α-tocopherol and xanthan gum on shelf-life and browning index of fresh-cut "Red Delicious" apples. *Innov. Food Sci. Emerg. Technol.* **2014**, *22*, 188–196. [CrossRef]
37. Kim, M.J.; Kim, S.S. Antioxidant and antiproliferative activities in immature and mature wheat kernels. *Food Chem.* **2016**, *196*, 638–645. [CrossRef]
38. Kim, M.J.; Kim, S.S. Utilisation of immature wheat flour as an alternative flour with antioxidant activity and consumer perception on its baked product. *Food Chem.* **2017**, *232*, 237–244. [CrossRef]
39. Chen, Z.; Wang, P.; Weng, Y.; Ma, Y.; Gu, Z.; Yang, R. Comparison of phenolic profiles, antioxidant capacity and relevant enzyme activity of different Chinese wheat varieties during germination. *Food Biosci.* **2017**, *20*, 159–167. [CrossRef]
40. Güner, A.; Kızılşahin, S.; Nalbantsoy, A.; Yavaşoğlu, N.; Ülkü, K. Apoptosis-inducing activity of safflower (*Carthamus tinctorius* L.) seed oil in lung, colorectal and cervix cancer cells. *Biologia* **2020**, *75*, 1465–1471. [CrossRef]
41. Lin, T.-K.; Zhong, L.; Santiago, J.L. Anti-Inflammatory and Skin Barrier Repair Effects of Topical Application of Some Plant Oils. *Int. J. Mol. Sci.* **2017**, *19*, 70. [CrossRef]
42. Affandi, N.; Zzaman, W.; Yang, T.A.; Easa, A.M. Production of Nigella sativa Beverage Powder under Foam Mat Drying Using Egg Albumen as a Foaming Agent. *Beverages* **2017**, *3*, 9. [CrossRef]
43. Spence, C. Background colour & its impact on food perception & behaviour. *Food Qual. Prefer.* **2018**, *68*, 156–166. [CrossRef]

Review

Current Functionality and Potential Improvements of Non-Alcoholic Fermented Cereal Beverages

Maria Valentina Ignat [1], Liana Claudia Salanță [2,*], Oana Lelia Pop [2], Carmen Rodica Pop [2], Maria Tofană [2], Elena Mudura [1], Teodora Emilia Coldea [1], Andrei Borșa [1] and Antonella Pasqualone [3]

[1] Department of Food Engineering, Faculty of Food Science and Technology, University of Agricultural Sciences and Veterinary Medicine Cluj-Napoca, 400372 Cluj-Napoca, Romania; maria.socaci@usamvcluj.ro (M.V.I.); elena.mudura@usamvcluj.ro (E.M.); teodora.coldea@usamvcluj.ro (T.E.C.); andrei.borsa@usamvcluj.ro (A.B.)

[2] Department of Food Science, Faculty of Food Science and Technology, University of Agricultural Sciences and Veterinary Medicine Cluj-Napoca, 400372 Cluj-Napoca, Romania; oana.pop@usamvcluj.ro (O.L.P.); carmen-rodica.pop@usamvcluj.ro (C.R.P.); maria.tofana@usamvcluj.ro (M.T.)

[3] Department of Soil, Plant and Food Sciences, University of Bari 'Aldo Moro', Via Amendola, 165/A, 70126 Bari, Italy; antonella.pasqualone@uniba.it

* Correspondence: liana.salanta@usamvcluj.ro; Tel.: +40-264-596-384

Received: 30 June 2020; Accepted: 30 July 2020; Published: 1 August 2020

Abstract: Fermentation continues to be the most common biotechnological tool to be used in cereal-based beverages, as it is relatively simple and economical. Fermented beverages hold a long tradition and have become known for their sensory and health-promoting attributes. Considering the attractive sensory traits and due to increased consumer awareness of the importance of healthy nutrition, the market for functional, natural, and non-alcoholic beverages is steadily increasing all over the world. This paper outlines the current achievements and technological development employed to enhance the qualitative and nutritional status of non-alcoholic fermented cereal beverages (NFCBs). Following an in-depth review of various scientific publications, current production methods are discussed as having the potential to enhance the functional properties of NFCBs and their safety, as a promising approach to help consumers in their efforts to improve their nutrition and health status. Moreover, key aspects concerning production techniques, fermentation methods, and the nutritional value of NFCBs are highlighted, together with their potential health benefits and current consumption trends. Further research efforts are required in the segment of traditional fermented cereal beverages to identify new potentially probiotic microorganisms and starter cultures, novel ingredients as fermentation substrates, and to finally elucidate the contributions of microorganisms and enzymes in the fermentation process.

Keywords: cereal beverage; fermentation; functional; non-alcoholic; health benefits

1. Introduction

Consumers' lifestyles have changed in recent years and will continue to change by being influenced by globalization, economic growth, rapid advances in food science and technology, and/or lifestyle choices and religious restrictions [1–3]. With regard to consumption behaviour, sensory acceptance is still the main choice criteria for consumers [4–6] and it is strongly dependent on cultural backgrounds, as well as previous sensory exposure to a specific food product [7,8].

Beverages are an optimum vehicle to transport nutrients and bioactive compounds into the body as well as to facilitate their bioavailability. Bioactive compounds, such as phytochemicals (e.g., phytoestrogens, phenolic compounds, flavonoids, carotenoids, etc.), dietary fibre, vitamins, fatty acids, probiotics, and minerals, can be incorporated into beverages. The presence of these compounds offers the prospect of using food as a valuable element in disease prevention strategies, particularly in the early stages of the diseases [9–11]. Worldwide, statistics clearly show the growing trend of functional beverage consumption [12], due to their nutrient contents, convenient packaging, design, ease of transportation and storage, and for their shelf-stable nature [13].

Functional beverages could be classified as dairy based, fruit and vegetable based, legume based, cereal based, coffee, or tea. The functionality traits of these beverages address different needs and lifestyles: to boost energy, to fight the ageing process, fatigue, and stress, or to target diseases [14].

Fermentation is widely used to improve the nutritional value, the digestibility level, shelf life, functional properties, texture, taste, and flavour of the beverages [15–18]. A popular class of fermented beverages are those made from cereals: barley (*Hordeum vulgare* L.), maize (*Zea Mays* L.), millet (*Panicum miliaceum* L.), oats (*Avena sativa* L.), rice (*Oryza glaberrima*/*Oryza sativa*), rye (*Secale cereale*), sorghum (*Sorghum bicolor*), and wheat (*Triticum aestivum* L.) [19]. These cereals are a good fermentation substrate and also pose a potentially functional trait, as they contain nutrients that can be easily assimilated by probiotics [20]. The processing steps could include soaking, sprouting, malting, cooking, grinding, and filtering. The fermentation of cereals is influenced by several factors (e.g., length of fermentation, temperature and pH, moisture content of grain, growth factor requirements, cereal nutrients, etc.) which require control by using technological methods to standardize quality.

Fermented beverages are affordable, and their production involves traditional methods to maintain hygiene conditions, product quality, and security [19,21]. Lactic acid bacteria (LAB) dominate the fermentation process and lead to a low pH, which is incompatible with the development of pathogenic bacteria, thereby increasing the shelf life and product safety [22]. Traditional methods exploit mixed cultures of various potentially beneficial microorganisms, referred to as probiotics [23]. Fermented beverages are considered essential because they serve as vehicles for beneficial microorganisms that play an important role in human health, and also for their nutritional, nutraceutical, and pharmaceutical properties [21–24]. Moreover, it was shown that fermentation improves protein digestibility and the bioavailability of minerals and other micronutrients [25].

The consumption of these beverages has the potential to reduce the adverse health and economic impacts of poor diets. Additionally, among the main benefits of cereal-based beverages, there is the possibility of consumption by vegetarians, vegans, and lactose-intolerant consumers [26]. This paper outlines the current achievements and further development for enhancing the functionality of the non-alcoholic fermented cereal beverage (NFCB) segment.

2. Non-Alcoholic Fermented Cereal Beverage Segment

2.1. An Overview of NFCBs

Reviewing the existent literature regarding some of the most studied NFCBs, it is shown that they are usually based on barley, maize, millet, oats, rice, rye, sorghum, and wheat [27,28]. Cereals are known as important sources of dietary proteins [29], energy, carbohydrates, vitamins, minerals, and fibre (arabinoxylan and β–glucan), but, at the same time, they are deficient in some basic components (essential amino acids, e.g., lysine) [19,30]. Whole grains are known to have important bioactive compounds and high nutritional values and their regular consumption has a positive effect on health [31–33]. The presence of soluble fibre lowers the glycaemic index of the beverage by slowing down its digestion and absorption, whereas phenolic compounds have antioxidant potential and scavenge harmful free radicals in the body [9].

Most of the traditional and currently produced NFCBs are considered functional foods and wholesome nutritional products [22,34,35]. Functional foods have been intensively studied in recent years and are continuously looked at as research efforts turn to processing food matrices, such as cereals, vegetables, and fruits, into medicine-like products.

A more detailed definition describes functional foods as industrially processed or natural foods that, when regularly consumed within a diverse diet at efficacious levels, have potentially positive effects on health outside basic nutrition [36]. This means that such products should provide a therapeutic benefit if consumed regularly within a diverse diet and with the condition of having the main nutrients extracted in a standardized manner and dosage [37].

Originally, cereal fermented beverages were mainly produced due to the need for conserving and utilizing various cereals and crops with affordable financial implications. Some of them, which are now commercially available as soft drinks and non-alcoholic beverages, were traditionally prepared as alcoholic beverages, having higher contents of alcohol [38].

Currently, "non-alcoholic" is a regulatory term and the laws regarding it vary across the globe. For example, EU regulation no. 1169/2011 simply states that a beverage must be labelled as an alcoholic drink if it has an alcoholic strength by volume (ABV) of over 1.2%. Moreover, in Great Britain, an alcohol-free drink has a maximum alcohol content of 0.05 % ABV; in Germany, a beverage is considered alcohol-free if the maximum alcohol limit is below 0.5 % ABV; in Spain, a non-alcoholic beer contains a maximum of 1 % ABV; in France, alcohol-free beers contain a maximum of 1.2 %; in US, non-alcoholic beers are of a maximum 0.5 % ABV; in China, non-alcoholic drinks can be of up to 0.5 % ABV [39]; in Japan, non-alcoholic beverages contain up to 1% ABV [40]. Depending on the country's traditional recipe, *boza*, a lactic acid fermented drink, has an alcohol content of less than 1% ABV in Turkey and up to 7% ABV in Egypt [41], probably due to microorganisms involved in the fermentation process [42].

Fermented drinks were and are obtained using a combination of fermentable substrates, like cereal mixes, fruits, plants, spices, legumes, and vegetables. The possibility of combining various fermentable substrates and of adding supplementary bioactive compounds to the final product encourages the development of different versions of the same original recipes, according to specific resources, health concerns, and nutritional needs. Such enhancements can be seen, for example, in vitamin A-fortified *mahewu* [43] or in establishing the suitable temperature for saccharification and oligosaccharide production efficiency in *amazake*, containing up to 0.5 % ABV [44].

Globally, there are numerous similar non-alcoholic cereal fermented beverages with similar names and profiles targeted for thirst quenching properties, on nutrition added value, cultural significance, and on providing alternatives to alcoholic beverages. Fermented beverages based on cereals are somewhat common around the world as staple foods, particularly in developing countries and are all made in a similar manner, generally using spontaneous microbial cultures [38]. Although there are various non-alcoholic cereal fermented beverages with different chemical profiles and sensory traits, all of them present certain bioactive compounds and therapeutic agents, which make them beneficial for human health. The functional compounds available and detected in NFCBs have variable values, therefore, it is difficult to assess their functional impact when consumed regularly [17]. If cereals lack certain nutrients, then additional food matrices can be added to enhance the final NFCB, as seen in a case of *mahewu* enhanced with *Moringa oleifera* leaf powder for elevating Ca and Fe contents [45]. Another example of a functional NFCB is a non-alcoholic beverage made of green tea and barley malt wort for delivering superior amino acid content [46].

In Table 1, there are several NFCBs listed with their place of origin, constituent cereals, microorganisms, nutritional compounds, and potential health benefits. As previously mentioned by other authors, the functional outcome of such products is strongly connected to the selection of the cereals, microorganisms, fermentation temperature and time, and other additional food matrices [47–49].

Table 1. Non-alcoholic fermented cereal beverages and their microorganisms, nutritional composition, and health benefits.

Beverage/Place of Origin	Cereals	Microorganisms	Functional Compounds	Health Benefits	Ref.
Amazake–Japan	rice *koji*	*Lactobacillus sakei*, *Aspergillus oryzae*;	amino acids; vitamins B1, B2, B6; pantothenic acid, vitamin E, flavonoids, dietary fibre, polysaccharides, sterols;	improves digestion; mitigates hypertension, skin-enhancing action, alleviates liver cirrhosis (200 kcal/150 mL/day/ 12 weeks);	[50–53]
Borș/Borsht–Central and Eastern Europe, Romania	wheat bran, corn flour	*Lactobacillus delbrueckii* ssp. *Delbrueckii*;	lipophilic and hydrophilic antioxidants (tocopherols, tocotrienols), phenolic compounds, *group B vitamins*, vitamin E, alkylresorcinols; lignans;	alleviates respiratory and digestive diseases (indigestion, vomiting), effective management of hepatic and bile diseases, potentially beneficial in cancer treatment;	[54–56]
Boza–Turkey, Greece, Bulgaria, Albania, Romania, Bosnia Herzegovina; South Africa	barley, oats, rye, millet, maize, wheat, rice	*Lactobacillus plantarum*, *Lactobacillus rhamnosus*, *Lactobacillus pentosus*, *Lactobacillus paracasei*, *Lactobacillus fermentum*, *Lactobacillus brevis*, *C. inconspicua*, *C. pararugosa*;	vitamin A, vitamins B1, B2, B6, nicotinamide; Ca, Fe, P, Zn, Na, β-glucan, dietary fibres;	improves gastrointestinal health, stimulates the immune system, decreases cholesterol level;	[38,41,57–61]
Busa–Syria, Egypt, Kenya, Turkistan;	rice or millet	*Lactobacillus* sp. *Saccharomyces* spp.	dietary fibre, amino acids, fatty acids, vitamins B1, B2;	lowers cholesterol and reduces risk of cancer and obesity, lowers blood pressure, beneficial in diabetes;	[42,61]
Gowé–West Africa, Benin	malted and non-malted sorghum, maize	*Lb. fermentum*, *Weissella confusa*, *Weissella kimchii*, *Lactobacillus mucosae*, *Pediococcus acidilactici*, *Pediococcus pentosaceus*;	amino acids (glutamic acid and leucine), minerals (Fe, Ca, Zn, P);	certain lactic acid bacteria strains can help in preventing infections by urogenital pathogens; antimicrobial efficacy;	[62–66]
Kunun-zaki–Nigeria	wheat and sorghum /millet, wheat, malted rice	*Lb. plantarum*, *Lb. fermentum*, *Lactococcus lactis*; *Saccharomysces cerevisiae*;	minerals (Fe, Ca, Mg, K);	provides micro- and macronutrients, improves nutritional status;	[67–70]
Kvass–Lithuania, Russia, Eastern Poland	extruded rye, malted barley	*Lactobacillus casei*, *Lb. sakei*, *P. pentosaceus*, *S. cerevisiae*;	vitamins B1, B3, B2, B6; dietary fibres, Zn, Cu, maltose, maltotriose, glucose, fructose;	modulates metabolism, reduces flatulence, alleviates hyperacidity;	[17,71,72]

Table 1. *Cont.*

Beverage/Place of Origin	Cereals	Microorganisms	Functional Compounds	Health Benefits	Ref.
Mahewu/*Amahewu*–Africa (Botswana, South Africa, Zimbabwe)	maize, sorghum, millet malt or wheat flour	*Lb. brevis, L. casei, L. lactis, Lb. plantarum, S. cerevisiae, S. pombe;*	Na, K, Ca, Fe, Zn, Mn; dietary fibre, carbohydrates, *group B vitamins;*	bacteriostatic and bactericidal properties against enteric pathogens;	[73–76]
Munkoyo–Zambia, Democratic Republic of Congo	maize	*Lb. plantarum,Weissella confusa, L. lactis Enterococcus italicus;*	fibre, vitamins B1, B2, B3, B6, B12, Ca, Fe, Zn, proteins, crude fat;	suppresses diarrhoea; anti-allergen, antimicrobial properties;	[22,77–79]
Obushera–Uganda	sorghum flour or millet, maize	*L. Lactis, Lb. plantarum, Lb. fermentum, Lb. delbrueckii, Weissella confusa;*	proteins, minerals, fibre;	NA	[80]
Oshikundu/*Ontaku*–Namibia	pearl millet meal, sorghum, or pearl millet malt	*Lb. plantarum, L. lactis, Lb. delbrueckii ssp. delbrueckii, Lb. fermentum, Lb. pentosus, Lactobacillus curvatus;*	shikimic acid, maleic acid, phytic acid, succinic acid; vitamins B1, B2; Ca, Cu, Fe, K, Mg, Mn, Na, S, Zn, P;	NA	[81–83]
Pozol–South Eastern Mexico, Central America	maize	*Lb. plantarum, Lb. fermentum, Lb. casei, Lb. delbrueckii, Leuconostoc sp., Bifidobacterium sp., Streptococcus sp., Saccharomyces sp.;*	group B vitamins; dietary fibre;	reduces cholesterol levels; improves gastrointestinal health; bactericidal, bacteriolytic, bacteriostatic activities;	[38,84,85]
Shalgam–Turkey	bulgur flour (wheat)	*Lb. plantarum, Lb. paracasei, Lb. brevis Lb. fermentum, S. cerevisiae;*	β-carotene, group B vitamins, Ca, Na, Fe;	antiseptic agent; probiotic food; regulates the digestive system's pH; diuretic action;	[59,86]
Tobwa (without malt, only LAB)/*Togwa*–East Africa, Tanzania, Zimbabwe	maize, finger millet (togwa)	*Lb. plantarum, Lb. brevis, Lb. fermentum, Lactobacillus cellobiosus, P. pentosaceus, W. confusa;*	amino acids; dietary fibre;; vitamin B2, B9, B12;	antimicrobial activity; enteropathogenic inhibition of *Campylobacter jejuni* and *Escherichia coli*; eases diarrhoea and prevents malnutrition;	[19,87–89]

Some of the NFCBs presented in Table 1 are commercially available, with various scales of production, while others are mainly homemade beverages, produced for individual consumption using traditional methods. *Busa*, *kunun-zaki*, *mahewu*, *munkoyo*, *obushera*, *oshikundu*, *pozol*, and *tobwa* are produced in rural and urban areas and their production is essentially a home-based industry as for now, there is no large-scale factory production. *Amazake* is commercially available in Japan and it is classified as a soft drink [90,91]. *Borş* is also industrially produced and used as a flavour enhancer in Romanian gastronomy or it is consumed as a refreshing drink [55]. *Boza* is one of the most popular Bulgarian beverages and is industrially produced at a large scale in all countries of the Balkan Peninsula [92]. *Kvass* is a traditional fermented Slavic and Baltic beverage and is one of the most popular beverages in Russia, numerous varieties emerged due to its popularity and market demand. Currently, *kvass* production is designed according to traditional processes, with the implementation of modern biotechnological methods [72]. *Shalgam* is a traditional beverage of southern Turkish cities and it is commercialized in various markets of European cities [86]. In Table 2, there are more details presented regarding NFCBs and their sensory properties, pH values, nature of use, fermentation status, and production scale.

Table 2. Non-alcoholic fermented cereal beverages and their sensory properties, nature of use, and fermentation status.

Beverage	Sensory Properties/pH	Nature of Use	The Status of Fermentation/Production Scale	Ref.
Amazake	cloudy appearance, sour-sweet taste; pH ~3.9;	dessert, snack, natural sweetening agent, baby food, salad dressing;	homemade and industrialised;	[90,91,93]
Borş	sour-bitter taste; odour notes: "bran", "yogurt", "goat milk-cheese", "pungent/sour", "ripe/fermented fruit"; pH 3.3–4.2;	used as a souring ingredient in soups, nutritious drink;	homemade and industrialised;	[54–56]
Boza	thick liquid, pale yellow colouring; sweet-sour taste; pH 2.93–3.72;	nutritious food, snack;	homemade and industrialised:	[58–61,94]
Busa	thick homogeneous suspension, light to dark beige; sweet-sour taste; pH 3.4–5.3;	traditionally made and served as an alcoholic drink;	homemade;	[42,95]
Gowé	brown/white colour; sweet, acidic, cereal taste; soft texture; pH ~3.5–4.7;	thirst-quenching and energy drink; children's food;	traditional, small-scale processors;	[65,66]
Kunun-zaki	low viscosity, creamy appearance; sweet-sour taste; pH ~3.8;	refreshing drink; nutritious beverage;	homemade, local producers;	[67,68]
Kvass	slightly cloudy appearance, light-dark brown colour; sweet-sour taste; pH 3.2–4.3;	soft drink;	traditionally homemade; industrialised differently than the traditional approach;	[71,72]
Mahewu/*Amahewu*	creamy colour, sour taste; pH ~3.5;	weaning food for infants, consumed in schools, farms, mines, etc.	homemade, commercially produced in African countries;	[45,74–76]
Munkoyo	slight yellow colour; sweet, mildly sour taste; pH 3.3–4.2;	consumed at household level; energy drink;	homemade;	[22,77–79]
Obushera	moderately thick composition, pale brown colour; sweet and sour taste; pH < 4.5;	thirst quencher, social drink, energy drink, and weaning food;	homemade; commercially relevant types: Obutoko, Obuteire, Ekitiribita;	[80,96]

Table 2. *Cont.*

Beverage	Sensory Properties/pH	Nature of Use	The Status of Fermentation/Production Scale	Ref.
Oshikundu/Ontaku	white colour, milky appearance, sweet taste; pH 3.3–3.7;	a token of welcome and hospitality; consumed at special events and daily social interactions;	homemade; local producers;	[81–83]
Pozol	yellow-brown colour; sweet-sour; slightly acidic taste; pH 3.8;	food or refreshing beverage, consumed at religious ceremonies and for its curative properties;	homemade in rural and urban areas of southeast Mexico; small- and large-scale producers;	[85,97]
Shalgam	red colour; sour taste;	used as a medicine because of its antiseptic agents;	home-scale level; small scale producers;	[59,86]
Tobwa/Togwa	opaque and brownish colour; sweet, occasionally sour taste; pH~4;	consumed as a popular energy source;	industrially produced in Tanzania;	[87,88]

2.2. Processing Technologies and Their Outcome

The dietary attributes and sensory traits of cereal products can be at times viewed as inferior or deficient in comparison with those of milk and milk-based foods. Some of the reasons behind this include the smaller protein quantities, deficiency in certain amino acids (lysine), the presence of antinutrients (phytic acid, tannins, and polyphenols), and the coarse nature of grains [19].

The fermentation of starchy sources is more complex compared to that of low molecular sugars (glucose or sucrose) because, in general, the starch must at first be converted into fermentable sugars. To achieve an almost complete starch degradation, two main types of amylolytic enzymes are required (α-amylase and glucoamylase) [98].

Several methods have been engaged with the aim of enhancing the nutritional qualities of cereals. These include genetic improvement and amino acid supplementation with protein concentrates or other protein-rich sources, such as grain legumes or defatted oil seed meals of cereals. Additionally, several processing technologies, which include cooking, sprouting, milling, and fermentation, have been put into practice to improve the nutritional properties of cereals, although the best one is probably fermentation. In general, the spontaneous fermentation of cereals leads to a decrease in the level of carbohydrates, as well as some non-digestible poly- and oligosaccharides [35]. Certain amino acids may be synthesized, and the availability of B group vitamins may be improved [99]. Fermentation also provides optimum pH conditions for the enzymatic degradation of phytate, which is present in cereals in the form of complexes with polyvalent cations, such as iron, zinc, calcium, magnesium, and proteins. Such a reduction in phytate may result in a several fold increase in the amounts of soluble Fe, Zn, and Ca [19,100–102].

The current article advances one general processing technique, by compiling all the traditional recipes assessed and integrating germinated and non-germinated grains, to adapt it to an industrial scale with functional improvement. Several traditional processing steps can be applied in the production. The general outline of the process is essentially the same and that presented in Figure 1: the grains, after conditioning, are either soaked/wet milled or the grains are dry milled and the flour is extracted in water afterwards, the mix is boiled to gelatinize the starch, a source of enzymes to hydrolyse the gelatinized starch into fermentable sugars is added, and finally, spontaneous fermentation occurs [19,103].

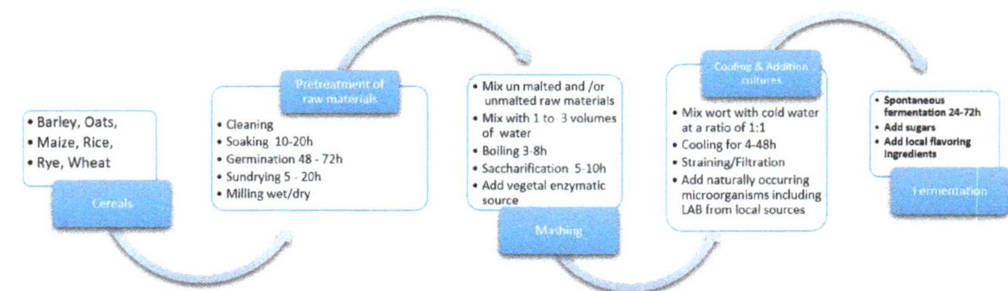

Figure 1. General flow diagram of traditional NFCB production.

2.2.1. Pre-Treatment of Raw Materials

Cereal processing is an essential component of the brewing production chain and the milling process is the main procedure. Before milling, the cleaning and conditioning of the grains is required. The cleaning process allows for the removal of various impurities, depending on the raw material types. The most common grain impurities are shrivelled grains, other cereals, grains damaged by pests, grains with discoloured germs, and sprouted grains. There are also miscellaneous impurities, including extraneous seeds, damaged grains, extraneous matter, husks, ergots, decayed grains, and insects. The conditioning or tempering of grains is performed using the monitored addition of water, which turns the endosperm softer and the bran harder. Doing this prevents the bran from breaking up, aids gradual separation throughout milling, and enhances sieving efficiency [104].

There are two milling categories, namely dry and wet milling, each having its own characteristics. Dry milling removes the germ and the outer fibrous materials of grains, as these by-products are not used in traditional ways [105]. Malting utilizes the power of natural germination when the grains, after absorbing water, germinate in the presence of oxygen to achieve a moisture content of up to 47% [106]. Another by-product is represented by floating kernels, which are unsuitable for malting. Throughout germination, the grain's embryo expands, and rooting begins. Moreover, the germination and steeping processes frequently overlap. It is recommended to keep the germination time and temperature at low values, given that long and warm germination processes lead to longer roots, resulting in larger malt yield losses [104,106].

During germination, grain enzymes start to break down the endosperm high-molecular-weight material into easily digestible components for the yeasts. Drying the malt stops the germination process [104].

In the case of traditional methods, the grains are usually superficially cleaned. A part of the cleaned grains (in variable percentages) is washed and soaked (10 to 20 h) at room temperature (26–35 °C). The soaked grains are drained and left for germination for 48 to 72 h with a frequent spraying of tap water. The germinated grains are sun dried (5 to 20 h), so the success of the process depends on the weather, mostly the intensity of the sunshine. Afterwards, the grains (malted and non-malted) are milled separately or in a mixture to obtain flour using rudimentary equipment [107].

Industrial cleaning processes aim at removing impurities and all other materials except for grains, using specific equipment such as magnetic separators, disc or sieve separators, aspirators, destoners, colour sorters, etc. The conditioning process ensures the complete hydration of grains, holding them in suitable containers for specific time intervals. Usually, depending on the grain varieties and initial moisture levels, the soaking time and temperature may be different [104].

For industrial-scale malting (Figure 2), the cereals are dried, and kilning is used to stop further transformations.

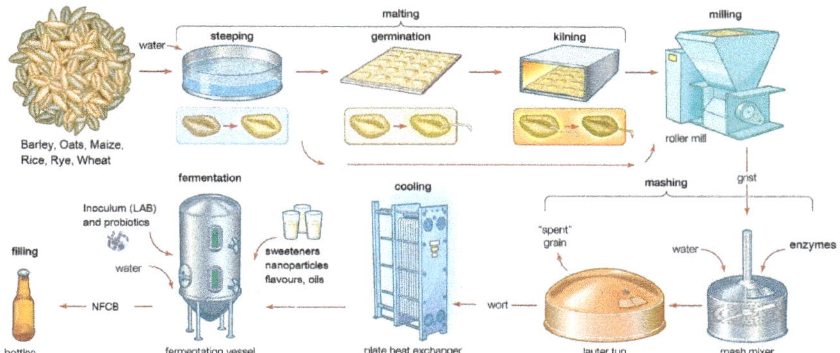

Figure 2. General flow diagram of industrial NFCB production (compilation from Encyclopaedia Britannica's brewing process).

During controlled drying, the water content should go under 5%, to stop the enzymatic activity whilst colour and flavour compounds are formed. A lot of by-products can additionally be recovered and further capitalized at an industrial scale. Dry milling can be extended to pearling, an abrasive procedure for gradually removing the testa and pericarp, aleurone and sub aleurone layers, and the germ. This results in polished grains and by-products with enhanced contents of bioactive compounds. Alternatively, wet milling, mostly used for producing starch and gluten, can increase the value of the cereals, as it is a source of by-products, with coproducts such as steep solids (abundant in compounds of pharmaceutical interest), germs (used in the oilseed-crushing industry), and bran [104].

2.2.2. Mashing

The traditional mashing process differs greatly, depending on the local culture, and has low efficiency. In Africa, either sorghum or finger millet malt is commonly used as a source of enzymes [22,108]. In the case of gowé production, a beverage based on maize and sorghum, the grains are milled to flour, which is optionally mixed/kneaded with tap or hot water or with a supernatant of a previous production and left to ferment at room temperature. Additionally, this processing technique can include a saccharification step, where a part of the malted sorghum or maize flour is kneaded with water and left for saccharification for 5 to 10 h [107]. For munkoyo, a maize-based beverage with a variable ABV, the specific feature is the usage of *Rhynchosia* roots or watery root extract as an enzymes source [22,77,78]. Starch gelatinization facilitates the activity of α- and β-amylases from the *Rhynchosia* roots for at least 4 h up to a maximum of 24 h, which hydrolyse the starch into fermentable sugars [22,103,109]. Mahewu is an example of a non-alcoholic sour beverage made from corn meal, consumed in Africa and some Arabian Gulf countries [74]. It is prepared from maize porridge further mixed with water. Sorghum, millet malt, or wheat flour is then added and left to ferment [19]. The production techniques of obtaining kvass, a cereal-based beverage produced from rye and barley malt, rye flour, and stale rye bread, uses as raw material either stale sourdough bread or malt rye malt and rye flour [110]. Boza is a colloid suspension, non-alcoholic beverage consumed in Bulgaria, Albania, Turkey, and Romania, made from wheat, rye, millet, maize, and other cereals mixed with sugar or saccharine [92]. Boza's preparation involves six stages: the preparation of raw materials, boiling, cooling, straining, the addition of sugar, and fermentation. Another option for its production is the use of previously fermented boza as an inoculum [59].

The industrial mashing process resembles the beer production process. Between one and three volumes of water are added per volume of milled grains, and during the cooking process, the mixture turns into a mash. The mixture is cooked at a normal pressure or in an autoclave for about 2 h at 4–5 atmospheres.

The macromolecular profile of cereal-based beverages is generally determined by polymeric compounds (proteins, polysaccharides, and polyphenols) and their progress in depolymerization during processing [111,112]. Given that yeasts or specific strains of lactic acid bacteria (LAB) cannot metabolize high-molar-mass substances [113], the macromolecules are solely depolymerized during the malting and mashing process by the malt's intrinsic enzymes. During malting, the enzymatic degradation of polymers is technologically controlled by the degree of steeping, germination time, and germination temperature [114,115]. Modern brewing barley varieties are bred to be balanced in malting performance and to meet the required brewing specifications. The degradation of starch into fermentable sugars (amylolysis) is the primary objective of mashing (substrate production for fermentation) [116], since during the subsequent fermentation process, only low-molar-mass compounds, fermentable sugars, and low-molar-mass proteinaceous compounds are metabolized by microorganisms (e.g., LAB and yeasts) [117,118].

Lautering is the next step in the large-scale process, through which solid and liquid fractions, respectively, the spent grain, composed of sugar-extracted grist or solids remaining in the mash, and the sweet wort with high contents of fermentable sugars are obtained. The spent grain is the major by-product of the brewing industry and represents a valuable source of bioactive ingredients and a potential ingredient for functional foods [119]. In small-scale production, malt extract can be used instead, thus skipping the use of grain malt (including milling, mashing, and lautering) [120].

2.2.3. Cooling and Addition of Yeast, LAB Cultures, and Other Ingredients

After boiling, the mash is gradually mixed with cold water in a 1:1 ratio. This slurry is often filtered or decanted to remove the grinding waste and insoluble plant material. In many traditional processes, where cereals are soaked in water for a few days, a succession of naturally occurring microorganisms will result in a population dominated by LAB. In such types of fermentations of endogenous grains, amylases generate fermentable sugars, which serve as energy sources for lactic acid bacteria. When no malted cereals are used, sucrose is added to the beverage to mimic the malt's sweet taste.

The bacterial flora formed in each fermented cereal drink is influenced by several factors, such as water activity, pH level, salt concentration, temperature, and the composition of the grain matrix, which must be considered in industrial processes. However, most fermented drinks, including the well-known products commonly met in the Western world, as well as those beverages of other origins which are less well studied and characterised, rely on lactic acid bacteria to mediate the fermentation process [19,121]. Lactic acid fermentation contributes towards the nutritional value, shelf life, safety status, and acceptability of a wide range of cereal-based foods [118]. Fermentation is often just one step in the process of fermented food preparation. Other operations, such as volume reduction, salting, or heating, also affect the final product characteristics [19,122]. Depending on the desired product, further steps can be applied afterwards, such as standardisations and the addition of other ingredients like flavourings, sugar, and stabilizers.

2.2.4. Fermentation Process

Despite the lack of process control, dealing with unstandardised microbial flora composition, delayed fermentation, and imperfect reproducibility of the fermentation process, spontaneous fermentation offers complex microbial diversity, providing higher levels of intrinsic stability to the microbial community [123]. This is achievable due to stabilizing interactions between species that prevent and inhibit the proliferation of unwanted microorganisms, including pathogenic ones. Recently, the existence of stability criteria for complex microbial communities was proven [124]. In terms of spontaneous fermentation, this can be explained by the resilience to small perturbations when there is a balance between the availability of resources, namely nutrients, and consumers (e.g., lactic acid bacteria). Commonly, the production of artisanal fermented beverages is conducted in successive batches [55], by using a natural starter from the previous fermentation. At all traditional sites, spontaneous fermentation proceeds after cooling. A 24 h fermentation period is sufficient for some traditional beverages to develop their characteristic sensory attributes, although in practice, fermentation time can go on for up to three days. Some brews, obtained through a 6 to 15 h fermentation interval of non-malted grain flour, are enhanced with commercial sugar (sucrose) to obtain the sweet taste. This process is a distortion of the original germination technique for gowé production [22]. A modern industrial process is different from the traditional process regarding the introduction and control of thermophilic LAB cultures, which only produce lactic acid, the extension of the product's shelf life by pasteurization and/or chemical preservation, and the inclusion of sugar and/or artificial sweeteners [125,126]. Controlled fermentation also leads to a general improvement in the shelf life, texture, taste, and aroma of the final product. During cereal fermentation, several volatile compounds are formed, which contribute to a complex blend of flavours [127]. Moreover, there is a good opportunity to apply colloidal dispersions in the form of nanoemulsions to deliver food grade nanoparticles, which contain water-insoluble molecules that were formerly unsuitable due to their poor soluble characteristics. Thus, there is a wide range of healthy foods which can be further designed, such as cereal-based fermented beverages enhanced with nanomolecules possessing beneficial health attributes [27].

2.3. Fermentation Microbiota and Safety of NFCBs

Originally, fermented beverages were only consumed in their native regions, however, due to increasing demand and interest, some of those traditional beverages are available to international markets. The attributes of traditional fermented beverages are influenced by several factors, such as the use of different raw materials, manufacturing methods, natural microbiota, and fermentation conditions. The microbiology of many traditional fermented drinks prepared from the most common types of cereals is quite complex.

There is a lot of diversity in the traditional processing techniques used for cereal-based fermented beverages all over the world, integrating single or multigrain cereals, germinated and/or non-germinated grains. Many types of cereal-based fermented beverages are produced in Africa, such as *togwa* in Tanzania, *mahewu* in Zimbabwe and South Africa, *mawè* in Benin, and *munkoyo* in Zambia and the Democratic Republic of the Congo [9,19,128,129]. Generally, the preparation of these products is a traditional family activity with an uncontrolled fermentation process by diverse microbial communities. The composition of the microbial community in a fermented food product largely determines the key product properties [130,131]. In other words, variations in microbial communities may result in differences in product quality, taste, acceptability, and microbial stability.

In one of our previous papers [55], we proved the impact of processing parameters, namely temperature and batch fermentation cycles, on the chemical composition of borș. Interestingly, the final composition of cereal-based beverages might not only be influenced by the physical process parameters. Processing practice variation affects the microbial composition of the fermenting microbial community. Despite the decreased pH caused by the lactic acid bacteria activity, the low pH of munkoyo also permitted the development of acidifying bacteria [22]. An important role is played by the initial microbial composition of the raw materials used in the process. The same study also referred to the fact that further investigations might be needed into the soil composition of the harvested raw materials. Along with the prevention of growth of most pathogenic strains [132], a low pH (of 2.5 to 3.5) improves food safety and expands the shelf life of this type of beverage [22]. The pathogenic microorganisms that are aerobes and facultative anaerobes and ferment simple sugars have an optimum pH for growth of 6.0 to 8.0. However, growth can occur at a pH as low as 4.3 and as high as 9, but with a combination of factors (pH, water activity), the control of foodborne pathogen growth can be done [133].

The fermentation of most cereals is natural and involves mixed cultures of yeasts, bacteria, and fungi [134]. Some microorganisms may participate in parallel, while others act in a sequential manner with a changing dominant flora during fermentation. The challenge, though, is the generally uncontrolled nature of the fermentation, which raises safety concerns, as well as the lack of standardization in the methods used, thus further research and development are needed to improve the traditional fermentation processes [80]. In this regard, the introduction of starter culture technology has led to greater consistency and safety and to better product quality [135].

Furthermore, the main difficulties encountered in the production of cereal-based beverages using traditional processes are linked to the high variability of unit operations and the unhygienic conditions of the processing environment. The soaking and germination parameters (temperature, duration, moisture) vary within and between processors. Moreover, during soaking and germination, the grains can be infested by fungi with the potential development of mycotoxins (aflatoxins) [136].

The continuous study of the fungi, yeasts, and bacteria strongly involved in ensuring a certain quality of the NFCB allows for the optimization of same-product delivery [85,137,138]. Such upgraded fermented beverages are sometimes the outcome of general efforts to enhance original recipes [139]. *Kunun-zaki* is an example of a traditional wheat and sorghum fermented beverage now also commercially available in the form of a powder [68], and there are several strategies proposed to upgrade and re-engineer the process of *gowé* production, a beverage obtained from sorghum and maize [66]. As seen so far, it is possible to prolong shelf life through the co-incubation of probiotic cultures, as seen in the development of some cereal-based fermented beverages [37,140] or to market to a group of consumers looking for healthy and functional foods by using oats, for example, as a main ingredient [141]. In-depth studies are still being conducted to ensure food safety in the processing technology of fermented foods through the keen selection of starter cultures and thorough examination of the specific microorganisms [142,143]. For example, it was recently shown that *Enterococcus faecium* YT52 isolated from *boza* is susceptible to clinically relevant antibiotics and contains low numbers of virulence factors and antibiotic resistance genes. Therefore, the enterocin-producing *E. faecium* YT52 strain poses a low risk to consumer health, and this strain may be used as a starter or a co-starter culture for improving the food safety of fermented products by acting against foodborne pathogens, such as *Listeria monocytogenes* and *Bacillus cereus* [144]. Moreover, rethinking the technological processes for obtaining various cereal-based fermented beverages helps to increase their functionality and overall therapeutic and nutritional properties. Such an example is *boza*, enhanced by fermenting cereals with *Lactobacillus acidophilus*, *Bifidobacterium bifidum*, and *Saccharomyces boulardii*, and by adding chickpea flour to the fermentable substrate for an elevated protein content [145]. Along with these interventions, narrowing the risks of product spoilage through the inoculation of specific strains of LAB and yeasts and processing the cereals in a certain manner repetitively allows the producers to deliver same-quality functional NFCBs.

Fermentation is one of the oldest known biotechnologies and the most economical procedure for producing new foods and ensuring product conservation. The yeasts responsible for fermentation in fermented drinks include species of *Saccharomyces*, *Saccharomycopsis*, *Schizosaccharomyces*, *Pichia*, *Candida*, *Torulopsis*, and *Zygosaccharomyces*. Brewer's yeast, *Saccharomyces cerevisiae*, metabolizes various sugars, mainly into alcohol, but also into other flavour-active substances. The most used microorganisms in fermentation processes belong to *Lactobacillus* species, which synthesize many flavour-active substances and lactic acid and are considered probiotic microorganisms known to support intestinal microbiota [146]. Fermented beverages are known to be rich in bioactive compounds, such as immune globulin peptides and the bioactive hormone cytokinin [19,71,82,147,148]. The use of a wide variety of microorganisms and yeasts is implied in the production of NFCBs by both traditional and modern means.

Largely responsible for the fermentation process are the indigenous microbiota present on the substrate or which can be added artificially as a culture. Basically, there are four main fermentation processes, which include alcohol production, lactic acid development, acetic acid production, and alkaline fermentation [19]. The modern industry for fermentative foods and beverages is innovative, given that it currently employs thermophilic fermentation, DNA technologies, molecular devices, designed starter cultures, genetic engineering, etc. Recombinant DNA technology has provided new insights for enhancing the product quality by designing tailor-made starter cultures that perform better than those found naturally [60]. For example, Basinskiene et al. used *Lactobacillus sakei*

KTU05-6 and *Pediococcus pentosaceus* KTU05-10 to ferment extruded rye for obtaining *kvass*, a traditional Lithuanian NFCB. They showed that the pH of the beverages fermented by LAB reached lower values compared to yeast fermentation and a they had a higher amount of organic acids. Innovative technology was applied, such as xylanolytic enzymes and antimicrobial active LAB, to improve the product's functional properties [71].

Functional cereal fermented beverages are becoming more attractive as they represent healthy alternatives for lactose intolerant consumers and for those who avoid certain allergens. For example, a study describes the production of kefir-like riboflavin-enriched beverages based on oat, maize, and barley flours. To obtain this beverage, riboflavin-producing Andean LAB were used, consisting of five *Lactobacillus plantarum* strains and two *Leuconostoc mesenteroides* strains [149].

As previously discussed, ensuring product safety, although difficult at times, is a crucial step in the production of traditional beverages where the fermentation process is spontaneous. Thus, microorganisms found on the brewers' skin, hair, and clothes can alter the product safety, therefore, high standards of hygiene are mandatory. Nevertheless, through the identification of bacteria strains and yeasts, the safety and quality status of the fermented beverages are guaranteed. A safety case study was conducted concerning "*obushera*", a Ugandan traditional fermented cereal beverage where important steps, such as pasteurization and ensuring water quality, are in the loop to ensure product safety and quality, as pathogens can also change the product's sensory characteristics [80].

Superior results can be obtained in the production of NFCBs by better understanding the interaction of microorganisms in the fermented substrates. The beneficial outcomes of controlling microorganisms are mirrored in product safety, increased shelf life, improved nutrient contents and availability, palatability, and enhanced sensory traits. For example, it was shown that *S. cerevisiae* improves LAB growth by transmitting essential metabolites, such as pyruvate, amino acids, and vitamins, while it uses some metabolites produced by LAB as carbon sources [150]. Moreover, Salari et al. concluded in their study that malt and *L. delbrueckii* were the best substrate and lactic strain for producing a functional beverage with the highest cell viability (1.2×10^6 cfu/mL after 4 weeks) [151].

Probiotics

Non-alcoholic fermented cereal-based beverages contain a wide range of diverse probiotics, depending on their cereal substrate and overall production methods. The processing method should assure the stability of the bacterial composition in order for the final product to possess probiotic functionality [152]. Still, the threshold to declare a beverage a probiotic one must be higher than 10^7 CFU/mL. Moreover, not all lactic acid bacteria possess probiotic activity. *L. rhamnosus*, also present in NFCBs (Table 1), was efficient in treatment and prevention of gastrointestinal disease [152]. The traditional Romanian NFCB, *borș*, has been consumed since ancient times as a gastric remedy. Several potentially probiotic bacteria—*L. casei*, *L. plantarum*, *L. brevis*, and *L. fermentum*—were isolated from borș, explaining the traditional consumption of this beverage for curative purposes [56].

The predominant microorganisms in the spontaneous fermentation of the African *mahewu*, a non-alcoholic sour beverage made from corn meal and consumed in Africa and some Arabian Gulf countries, belong to *Lactococcus lactis* subsp. *lactis*. On the other hand, the microbiota identification of Bulgarian boza shows that it mainly consists of lactic acid bacteria and yeasts, such as *Lactobacillus plantarum*, *L. acidophilus*, *L. fermentum*, *L. coprophilus*, *Leuconostoc raffinolactis*, *Ln. mesenteroides*, and *Ln. brevis* and *Saccharomyces cerevisiae*, *Candida tropicalis*, *C. glabrata*, *Geotrichum penicillatum*, and *G. candidum*, respectively [19].

In the case of Turkish *boza*, the use of LAB and yeast isolates as starter cultures allows for controlled fermentation studies to be carried out. The selection of proper strains with probiotic and antimicrobial properties enhances the functional properties of *boza* [59].

As seen in the case of African countries, the primary challenge for the development and use of fermented cereal-based probiotic beverages is the common lack of knowledge regarding the health and nutritional benefits of such foods and beverages. Insufficient common knowledge on probiotics and their benefits creates a sense of scepticism among consumers. Moreover, there is also an imperative need to ensure proper facilities for probiotic starter cultures, given that through spontaneous fermentation, the organoleptic and functional qualities of the resulting products are variable [153].

2.4. The Nutritional and Bioactive Composition of Commonly Consumed NFCBs

Consumers are aware of the importance of maintaining a strong immune system to prevent illnesses and they are actively looking for products which can help maintain their health status and alleviate health problems. It has been scientifically proven that probiotics isolated from functional beverages boost the immune system [154], and that, along with prebiotics, they are able to improve the intestinal homeostasis, immunomodulating ability, and general health of the host [27].

Looking at past publications, it is shown how fermented beverages have transitioned from traditional natural fermented products to beverages formulated with functional ingredients meant to offer cardiovascular health benefits, and then to functional fermented drinks which improve gastrointestinal health, which could then evolve into fermented products containing specific bioactive nanoparticles [134].

NFCBs are receiving increased attention from researchers and consumers more recently due to their proven probiotic characteristics and disease prevention perspectives [17]. The perceived health outcomes of fermented beverages are strongly related to the microbial content and implicit improvement of gastrointestinal health [38]. Moreover, non-alcoholic fermented beverages offer a sense of wellbeing, as they stimulate the metabolic system [126].

Fermented cereal-based foods, including NFCBs, are a potential source of new functional lactic acid bacteria species besides various nutrients and bioactive compounds, with beneficial effects on human health [56]. Furthermore, as briefly mentioned previously, NFCBs are healthy alternatives to the traditionally consumed food probiotics of dairy origin (e.g., yogurt, kefir, etc.), especially for people with lactose intolerance and milk protein allergies [27].

Given the functional components of fermented beverages and their bioactive compounds released through fermentation by cultures, NFCBs have been linked with many potential and some proven health benefits and actions on digestive, endocrine, cardiovascular, immune, and nervous levels [155]. They present beneficial actions for vital body functions and contribute to the prevention and reduction of risk factors for various diseases [19,23]. NFCBs are rich sources of minerals, vitamins, fibre, flavonoids, phenolic compounds, antioxidants, omega-3 fatty acids, plant extracts, sterols/stanols, amino acids, and biopeptides, among others, which could also protect from oxidative stress and inflammation diseases [17,134]. The presence of numerous valuable compounds in NFCBs grants several health benefits upon consumption, as presented in Figure 3.

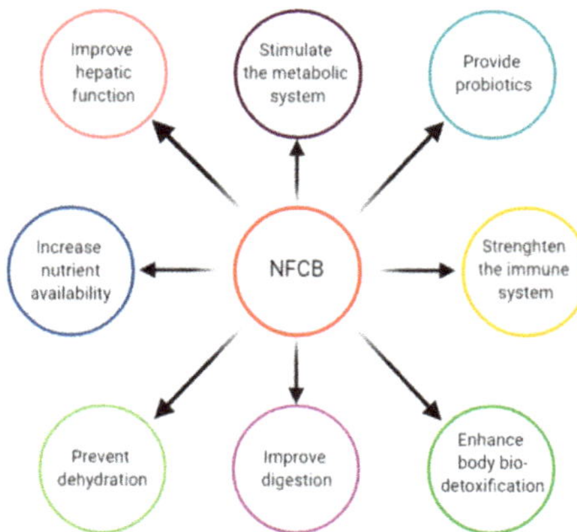

Figure 3. Health benefits associated with NFCB consumption.

Considering group B vitamins, these are unequally distributed in grain tissues. What counts the most in terms of their potential functionality in cereal-based food and beverages is their biological availability. During thermal food processing, these vitamins are destroyed almost completely, however, lactic acid fermentation represents a great tool for food industrialists interested in developing novel vitamin-fortified products. For example, Capozzi et al. (2012) pointed out the ability of many lactic acid bacteria strains, such as *Lactobacillus delbrueckii, Lactobacillus plantarum, Lactobacillus rhamnosus, Lactobacillus fermentum*, and *Pediococcus lactis*, reported in the composition of several NFCBs (Table 1), to biosynthesize riboflavin [156].

A new concept attributed to non-alcoholic cereal-based beverages is that they are healthy drinks with important impacts on human health. It has been shown that consuming NFCBs leads to liver function improvement, increased levels of lactobacilli and bifidobacteria in the intestinal microbiota [157], balanced gut microbiota, and the prevention of bacterial translocation with reduced incidences of nosocomial infections [27]. Furthermore, because non-alcoholic fermented beverages contain several LAB metabolites, their consumption confers bactericidal, bacteriolytic, and bacteriostatic properties, resulting in therapeutic effects at a digestive level. Antimicrobial compounds identified in functional beverages exhibited activities against several Gram-positive and Gram-negative bacteria and against yeasts and moulds [17]. The functional compounds and potential or proven health benefits of NFCBs are presented in Table 1.

As previously discussed, the potential benefits of NFCBs can be ensured in a safe and non-contaminated production environment to prevent the risk of product spoilage. Moreover, there are diverse nutritional and bioactive components of NFCBs, depending on the cereal substrate, microorganisms, and fermentation parameters, and on the technological process followed.

Currently, novel NFCBs are designed based on traditional recipes for delivering improved functional fermented beverages. Such an example is "*amahewu*" or "*mahewu*", a southern African LAB-fermented non-alcoholic maize-based beverage, which is deficient in vitamin A. In a study published in 2015, the traditional drink was redesigned using provitamin A-biofortified maize in exchange for the traditionally used white maize and it resulted in a functional beverage addressing vitamin A deficiency, a concerning health problem in sub-Saharan Africa [43].

2.4.1. Phenolic Compounds and Antioxidant Activity

Cereal grains and germs contain various major phytochemicals, such as phenolic acids, flavones, phytic acid, flavonoids, coumarins, and terpenes. The bran layers of cereal grains are relatively rich in water soluble and liposoluble antioxidants. Oats include phenolic acids, flavonoids, tocopherols, tocotrienols, and avenanthramides. On the other hand, insoluble fibre of wheat bran contains about 0.5%–1% phenolics and most phenolic compounds, like ferulic acid, are abundant in whole grains [158].

Depending on the cereal substrate, fermentation type, and other added ingredients, each NFCB can have quite different compositions. For example, the phenolic profile of brown beer vinegar indicates that such a fermented beverage is a rich source of phenolic compounds (661 mg GAE/L) and antioxidants, with 30% increased antioxidant activity after acetic fermentation. The same trend was observed regarding some of the drink's individual phenolic compounds. With cohumulone I (4.44 mg CE/L), cohumulone II (6.58 mg/L), 8-prenylnaringenin (2.33 mg CE/L), 6- prenylnaringenin (1.86 CE/L), humulone (5.62 CE/L), and isohumulone (4.14 CE/L) [159], this non-alcoholic fermented beverage can be considered an alternative source of valuable compounds, which could also be part of specific diets.

Fermentation parameters are of high importance in what concerns an NFCB's antioxidant profile. For example, a fermentation temperature of 24° C was considered a positive factor influencing the antioxidant activity of *borș*, a Romanian traditional wheat bran-based fermented beverage [55]. The antioxidant activity ranged from 3.05 to 8.52 mmol/L Trolox.

Some of the ancient traditional fermented beverages are currently produced at industrial scale. Still, the original recipes and the nutritional components are being altered because of thermal treatments applied to obtain product safety and stability. The addition of natural starters allowed for the increase in phenolic contents and the enhancement of final antioxidant activity. The same trend considering phenolic content between fermentation stages was found in the case of *borş* [56]. Depending on the processing operations applied, the phenolic compounds ranged considerably (4-hydroxybenzoic acid: 0.9–5.9 mg/L; vanillic acid: 0.6–3.2 mg/L; syringic acid: 0.5–2.5 mg/L; p–coumaric acid: 0.5–1.5 mg/L; sinapic acid: 0.6–2.7 mg/L; ferulic acid: 13.7–47.8 mg/L).

2.4.2. Amino Acids

Although cereals pose few challenges from a nutritional standpoint, especially with regard to the increased starch content upon cooking, the limited amino acid contents of their protein fractions, or the reduced bioavailability of their zinc and iron contents, there are already some solutions meant to correct and further enhance the nutritional status of cereal-based foods and beverages [160].

For example, in a study conducted on *borş*, an increase in amino acid content was found from one fermentation stage to another. The same study stated that the LAB fermentation of cereals improves the protein quality as well as the level of certain free amino acids by enhanced endogenous proteolysis and/or microbial action. Among the identified amino acids, there were isoleucine and threonine [56].

In general, the natural fermentation of cereals allows for amino acids to be synthesised, while it also helps to enhance the availability of group B vitamins. Fermentation processes also offer suitable pH conditions for phytate enzymatic degradation. Such a reduction in phytate may increase the amount of soluble iron, zinc, and calcium several fold, correcting the nutritional status of cereal-based beverages and foods [19].

3. Future Perspectives to Enhance the Functional Properties of NFCBs

3.1. Consumers' Preferences and Requirements

The way we perceive healthiness in foods varies between cultures and is reflected in the consumers' familiarity with health-related information [161]. Moreover, modern-day buyers might show increased interest in foods which can improve well-being [162], reduce the risk of developing illnesses, and satisfy nutrition-related conditions such as food intolerances and allergies. Currently, consumers approach the concept of wellness with a holistic view and are becoming more health conscious, particularly due to the growing incidence of diseases such as type 2 diabetes, coronary heart disease, cancer, and obesity [163]. There is a heightened demand for food products which are nutritious, functional, attractive, with clean labels, and which are ready to eat. Considering such attributes, the beverage sector seems to be one of the most suitable for addressing customers' demands and for increasing the rate of functional product consumption. Market data indicate that functional food products are among people's choices even when economic issues arise. The global functional beverage market has increased rapidly, and it is expected to grow further, due to their sensory appeal and health-promoting attributes [38].

Studies showed that providing information on labels regarding health-related claims associated with functional beverages determines consumers' preferences for a specific product [164]. Not only health-related claims but also the sensory properties determine consumers' preferences for a specific functional beverage [34]. It was suggested that the importance of specific testing conditions for functional beverages will help product developers to reformulate the product with the proper design of experiments, focusing on the consumers' needs [165], as they will determine the product acceptance [166]. The urgent need for educational campaigns designed with accessible and easily understood wording about the nutritional components of foods has already been pointed out [167]. Although the previously mentioned study was conducted in Brazil, insufficient knowledge about nutritional facts for food products is a worldwide issue strongly related to the level of education [168].

3.2. Possibilities of Improving the Appeal and Functionality of NFCBs

The functional beverage market is competitive and driven by product innovation and health awareness trends concerning optimum nutritional diets. The biological activities and the sensory properties of a beverage arise from individual components, along with chemical and physical interactions within the food matrix during processing, storage, ingestion, and digestion [169]. Beverages deliver nutrients, bioactive components, antioxidants, vitamins, minerals, fatty acids, plant extracts, probiotics, prebiotics, and micronutrients [13]. Cereal-based beverages include a complex mix of different polymers, such as proteins, polyphenols, and polysaccharides. These polymers affect the sensory perception of beverages in terms of mouthfeel, depending on their substance properties [112].

Several strategies have been set to intensify the production, availability, accessibility, and consumption of beverages rich in bioactive compounds. These include combining cereals with pseudocereals (e.g., quinoa, amaranth, etc.), legumes, vegetables extracts, fruits, aromatic plants, and herbs to improve the quality of the final food product [19,56,170,171]. Furthermore, mixing cereals with legumes could improve protein quality [172]. Fruit juices (e.g., apple, pineapple, mango, orange, lemon, peach, lychee, and strawberry) are considered as health-promoting foods and are an important basis of enrichment on which to append an extra functional constituent that can significantly augment the appeal to customers. Cereal beverages are based on grain suspensions. The viscosity, mouthfeel, and sweetness of the drink can be adjusted to the consumers' tastes using enzyme compounds. Wort can be mixed with various plant- and fruit-based juices to obtain a beverage rich in dietary fibre suitable even for athletes and sport amateurs. Moreover, innovative flavours can be obtained by fermenting the extract with specific microorganisms. As previously mentioned, fruits were also employed in the overall production scheme of fermented cereal beverages, as these are the added sugar necessary for initiating the fermentation process. Chemical analyses of ancient organics absorbed into pottery jars from the early Neolithic village of Jiahu, Henan Province, China have revealed that a mixed fermented beverage of rice, honey, and fruit (hawthorn fruit and or grape) was being produced as early as the seventh millennium B.C. [173]. Such findings inspire the development of new functional products, such as yogurt-like beverages made of a mixture of rice, barley, emmer wheat, oats, soy, and grape must [174] or of other milk-based functional drinks using fermented plant juices [175,176].

Plants are valued for their nutrients such as vitamins, dietary fibre, antioxidants, and flavonoids, which have shown nutritious and health-promoting properties [134]. Bioactive compounds derived from fruits and vegetables can be good vehicles for probiotics, prebiotics, and synbiotics (e.g., fermented cereal beverages) [57,58]. Spices such as tarhana herb, mint, and thyme are mixed with wheat flour, yogurt, vegetables, and herbs to produce a traditional Turkish fermented food called tarhana [177,178] and improve taste, aroma, and other profile characteristics. Moreover, it is suggested that adding herbs with proven beneficial compounds to otherwise traditional fermented soft drinks can augment the nutritional profile of fermented cereal beverages and add therapeutic potential [179]. Among the suggested herbs, we can find echinacea (*Echinacea angustifolia*) for antibiotic action and immune system support, ginkgo (*Ginkgo biloba*) for enhancing memory and alertness, guarana (*Paulina cupana*) for improved cognitive performance, kava (*Piper methysticum*) for stress reduction and mental balance, and St John's Wort (*Hypericum perforatum*) for anxiety reduction [180]. Vegetables such as beets, tomatoes, carrots, and cabbage can also be included in the production of some functional NFCBs, as they provide supplementary fermentable substrates and nutrients, and act as prebiotics in the final product [181]. *Shalgam* is a traditional Turkish NFCB made of turnip bulb, purple carrot, salt, sourdough, bulgur, or bulgur flour [182], and is thought to regulate the digestive system's pH and to act as an antiseptic agent. The results of a research study conducted on the co-culture probiotic fermentation of a protein-enriched cereal medium suggests that plant protein may be exploited for achieving protein supplementation of NFCBs. Legumes like chickpeas, individually or in combination with cereals, can provide a good substrate for probiotic microorganism production [145]. Additionally, the composition of legumes, vegetables, and fruits is known to be rich in protein, phytochemicals, dietary fibre, vitamins, and other micronutrients beneficial for human health [183].

Currently, concepts like "green consumerism" and "minimally processed foods" are on the rise, as consumers prefer food lacking synthetic additives. Natural compounds that can be added to functional beverages, such as essential oils and plant extracts including rosemary, peppermint, bay, basil, tea tree, celery seed, and fennel with antimicrobial activity, might represent an alternative to synthetic preservatives. The challenge is to define the optimal dosage of such compounds to have a positive impact on the product's final nutritional status, sensory properties, and consumer acceptance [12].

3.3. Perspectives for Future NFCBs

Due to the increasing prevalence of lactose intolerance and milk protein allergies, as well as the general aim of controlling cholesterol intake, and along with the growing interest in vegetarianism, research efforts have been focused on the feasibility of using cereals as fermentation substrates for the development of probiotic, prebiotic, and synbiotic beverages [27,113,184]. In the production of various non-alcoholic drinks, probiotic lactic acid bacteria are used to boost the product's functional value [143]. Although ensuring nutritive compounds within NFCB is a crucial step for product development, flavour improvement remains one of the main challenges for the development of LAB-fermented beverages obtained from cereal-based substrates [185].

It is a fact that improved and more appealing NFCB sensory properties would help consumers shift their interest intensively towards such healthy products. The efficacy of probiotics in the treatment of bowel disorders, the prevention of antibiotic-associated diarrhoea, and the improvement of lactose metabolism has been proven [186]. It has been also concluded that both fermentation and acidification with lactic acid have the potential to improve the nutritional quality of cereal-based foods as a method to combat protein malnutrition and iron and zinc deficiencies [187].

Biotechnological processes, such as malting and lactic acid fermentation, are recommended for producing functional beverages with increased contents of functional bioactive components [188]. There are numerous important benefits of enhancing NFCBs through biotechnological processes and it is worth mentioning the reduction of phytates through LAB-fermentation, which in turn leads to the increased absorption of Fe and other minerals [189]. Due to cereal fermentation, quantitative and qualitative changes take place in small molecules, ensuring the high bioavailability of macro- and micronutrients [190,191]. As mentioned earlier, prebiotics and dietary fibre can be added to fermented drinks to heighten their nutritive and functional contents. For example, it was shown that the addition of soy fibre not only improves *Lactococcus lactis* counts but also enhances the beverage characteristics regarding acidity, viscosity, and syneresis [192].

Currently, emerging techniques, such as nanotechnology, are in discussion to enhance the nutrient composition and functionality of NFCBs. However, continuous research and technological improvements are required to better design NFCBs and ensure product safety. It is also crucial to improve the quality of the main ingredients and it is imperative to integrate food safety management systems for industrial scale production. The proposed solutions include the development of new production technologies for obtaining functional NFCBs by extending the spectrum of raw materials used and applying new biotechnological resources (enzymes and lactic acid bacteria) [71].

4. Conclusions

Science and technology have the potential to produce superior functional NFCBs and to deliver consistent product quality, to improve shelf life, and to enhance nutritional values to finally meet the consumers' demands. Past research studies have proven that lactic [55] and acetic [159] fermentation enhance the nutrient content and their bioaccessibility in the case of cereal-based fermented beverages. Adding probiotics and prebiotics to a beverage is perhaps one of the most convenient ways of turning it into a functional beverage.

The research efforts on the enrichment of non-alcoholic fermented cereal beverages are still in their early stages but appear to be more promising than ever. Upcoming studies should focus on traditional non-alcoholic fermented beverages around the world to identify new potential probiotic microorganisms and starter cultures, new ingredients as potential substrates, and to elucidate the contributions of microorganisms and enzymes to the fermentation process.

Knowledge about processing applications and bacterial strains is essential to control the production methods and to design proper mixes of microorganisms, ensuring product safety. Afterwards, starter cultures with expected outcomes can be used for the industrial production of standard-quality fermented beverages with functional attributes.

The effect of advanced technologies on NFCB functional properties during processing requires additional studies to ensure that these technologies can prevent the loss of product quality and nutritive compounds. Moreover, for both scientific and industrial actors, the main challenge is to manage the large-scale production of fermented beverages without losing the unique flavours and other properties associated with the original products. Given this, it is highly recommended to explore the sensory properties of NFCBs when obtained from a combination of cereals, legumes, fruits, and plants.

The current review investigates some of the most scientifically documented traditional non-alcoholic beverages from all over the globe, highlighting their functional compounds and associated health benefits upon consumption. Moreover, processing technologies and their outcome were highlighted and discussed from a large-scale production standpoint. NFCBs were also reviewed, concerning fermentation microbiota and product safety, highlighting the need to apply starter cultures to ensure food safety and standard quality. The nutritional and bioactive compositions of commonly consumed NFCBs were reviewed, showing the functional compounds of NFCBs and their associated health benefits.

NFCBs are associated with functional benefits to one's health status, as discussed. However, new studies should be carried out to produce new NFCBs, combining probiotic fermented beverages and products such as fruit juices, vegetables, and cereals, meant to address specific health concerns. Future cereal-based fermented beverages need to be balanced regarding sensory properties, nutritional composition, alcohol content, and resource investments to be even more attractive, healthy, and affordable.

Author Contributions: Conceptualization, L.C.S. and M.V.I.; Writing—original draft preparation, M.V.I., L.C.S., C.R.P., O.L.P., T.E.C. and A.B.; Writing—review and editing, L.C.S., M.V.I., A.P. and T.E.C.; Supervision, L.C.S., M.T., A.P. and E.M.; Funding acquisition, L.C.S. All authors have read and agreed to the published version of the manuscript.

Funding: This research was funded by two grants, as follows: from the Ministry of Research and Innovation, CNCS-UEFISCDI, project number PN-III-P2-2.1-CI-2018-1462, within PNCDI III; and from the European Foundation for Alcohol Research (ERAB), Belgium, project number Ref EA 15 45.

Acknowledgments: The publication was supported by funds from the National Research Development Projects to finance excellence (PFE)-37/2018–2020 granted by the Romanian Ministry of Research and Innovation.

Conflicts of Interest: The authors declare no conflict of interest.

References

1. Pop, O.L.; Salanţă, L.C.; Pop, C.R.; Coldea, T.; Socaci, S.A.; Suharoschi, R.; Vodnar, D.C. *Prebiotics and Dairy Applications*; Academic Press: Cambridge, MA, USA, 2019; ISBN 9780128164952.
2. Peyer, L.C.; Zannini, E.; Arendt, E.K. Lactic acid bacteria as sensory biomodulators for fermented cereal-based beverages. *Trends Food Sci. Technol.* **2016**, *54*, 17–25. [CrossRef]
3. Salanță, L.C.; Uifălean, A.; Iuga, C.A.; Tofană, M.; Cropotova, J.; Pop, O.L.; Pop, C.R.; Rotar, A.M.; Bautista-Ávila, M.; Velázquez González, C. Valuable Food Molecules with Potential Benefits for Human Health. In *The Health Benefits of Foods—Current Knowledge and Further Development*; IntechOpen: London, UK, 2020. [CrossRef]
4. Phan, U.T.X.; Chambers, E. Application of an Eating Motivation Survey to Study Eating Occasions. *J. Sens. Stud.* **2016**, *31*, 114–123. [CrossRef]
5. Phan, U.T.X.; Chambers, E. Motivations for choosing various food groups based on individual foods. *Appetite* **2016**, *105*, 204–211. [CrossRef] [PubMed]

6. Phan, U.T.X.; Chambers, E. Motivations for meal and snack times: Three approaches reveal similar constructs. *Food Qual. Prefer.* **2018**, *68*, 267–275. [CrossRef]
7. Wang, Q.J.; Mielby, L.A.; Junge, J.Y.; Bertelsen, A.S.; Kidmose, U.; Spence, C.; Byrne, D.V. The role of intrinsic and extrinsic sensory factors in sweetness perception of food and beverages: A review. *Foods* **2019**, *8*, 211. [CrossRef]
8. Chambers, D.; Phan, U.; Chanadang, S.; Maughan, C.; Sanchez, K.; Di Donfrancesco, B.; Gomez, D.; Higa, F.; Li, H.; Chambers, E.; et al. Motivations for Food Consumption during Specific Eating Occasions in Turkey. *Foods* **2016**, *5*, 39. [CrossRef]
9. Fernandes, C.G.; Sonawane, S.K.; Arya, S.S. Cereal based functional beverages: A review. *J. Microbiol. Biotechnol. Food Sci.* **2018**, *8*, 914–919. [CrossRef]
10. Uifălean, A.; Schneider, S.; Ionescu, C.; Lalk, M.; Iuga, C.A. Soy Isoflavones and Breast Cancer Cell Lines: Molecular Mechanisms and Future Perspectives. *Molecules* **2015**, *21*, 13. [CrossRef]
11. Uifălean, A.; Schneider, S.; Gierok, P.; Ionescu, C.; Iuga, C.A.; Lalk, M. The impact of soy isoflavones on MCF-7 and MDA-MB-231 breast cancer cells using a global metabolomic approach. *Int. J. Mol. Sci.* **2016**, *17*, 1443. [CrossRef]
12. Corbo, M.R.; Bevilacqua, A.; Petruzzi, L.; Casanova, F.P.; Sinigaglia, M. Functional Beverages: The Emerging Side of Functional Foods: Commercial Trends, Research, and Health Implications. *Compr. Rev. Food Sci. Food Saf.* **2014**, *13*, 1192–1206. [CrossRef]
13. Ghoshal, G.; Kansal, S.K. The Emerging Trends in Functional and Medicinal Beverage Research and Its Health Implication. In *Functional and Medicinal Beverages*; Elsevier Inc.: Amsterdam, The Netherlands, 2019; ISBN 9780128163979.
14. Sethi, S.; Tyagi, S.K.; Anurag, R.K. Plant-based milk alternatives an emerging segment of functional beverages: A review. *J. Food Sci. Technol.* **2016**, *53*, 3408–3423. [CrossRef] [PubMed]
15. Şanlier, N.; Gökcen, B.B.; Sezgin, A.C. Health benefits of fermented foods. *Crit. Rev. Food Sci. Nutr.* **2019**, *59*, 506–527. [CrossRef] [PubMed]
16. Angelescu, I.R.; Zamfir, M.; Stancu, M.M.; Grosu-Tudor, S.S. Identification and probiotic properties of lactobacilli isolated from two different fermented beverages. *Ann. Microbiol.* **2019**, *69*, 1557–1565. [CrossRef]
17. Baschali, A.; Tsakalidou, E.; Kyriacou, A.; Karavasiloglou, N.; Matalas, A.L. Traditional low-alcoholic and non-alcoholic fermented beverages consumed in European countries: A neglected food group. *Nutr. Res. Rev.* **2017**, *30*, 1–24. [CrossRef]
18. Chaves-López, C.; Rossi, C.; Maggio, F.; Paparella, A.; Serio, A. Changes Occurring in Spontaneous Maize Fermentation: An Overview. *Fermentation* **2020**, *6*, 36. [CrossRef]
19. Blandino, A.; Al-Aseeri, M.E.; Pandiella, S.S.; Cantero, D.; Webb, C. Cereal-based fermented foods and beverages. *Food Res. Int.* **2003**, *36*, 527–543. [CrossRef]
20. Tolun, A.; Altintas, Z. Medicinal Properties and Functional Components of Beverages. In *Functional and Medicinal Beverages*; Elsevier Inc.: Amsterdam, The Netherlands, 2019; ISBN 9780128163979.
21. Anal, A.K. Quality ingredients and safety concerns for traditional fermented foods and beverages from Asia: A review. *Fermentation* **2019**, *5*, 8. [CrossRef]
22. Phiri, S.; Schoustra, S.E.; van den Heuvel, J.; Smid, E.J.; Shindano, J.; Linnemann, A. Fermented cereal-based Munkoyo beverage: Processing practices, microbial diversity and aroma compounds. *PLoS ONE* **2019**, *14*, e0223501. [CrossRef]
23. Achi, O.K.; Asamudo, N.U. Cereal-Based Fermented Foods of Africa as Functional Foods. *Int. J. Microbiol. Appl.* **2019**, 1527–1558. [CrossRef]
24. Alu'datt, M.H.; Rababah, T.; Alhamad, M.N.; Gammoh, S.; Alkhaldy, H.A.; Al-Mahasneh, M.A.; Tranchant, C.C.; Kubow, S.; Masadeh, N. *Fermented Malt Beverages and Their Biomedicinal Health Potential: Classification, Composition, Processing, and Bio-Functional Properties*; Elsevier Inc.: Amsterdam, The Netherlands, 2019; ISBN 9780128152713.
25. Kårlund, A.; Gómez-Gallego, C.; Korhonen, J.; Palo-Oja, O.M.; El-Nezami, H.; Kolehmainen, M. Harnessing microbes for sustainable development: Food fermentation as a tool for improving the nutritional quality of alternative protein sources. *Nutrients* **2020**, *12*, 1020. [CrossRef]
26. Menezes, A.G.T.; Ramos, C.L.; Dias, D.R.; Schwan, R.F. Combination of probiotic yeast and lactic acid bacteria as starter culture to produce maize-based beverages. *Food Res. Int.* **2018**, *111*, 187–197. [CrossRef] [PubMed]

27. Salmerón, I. Fermented cereal beverages: From probiotic, prebiotic and synbiotic towards Nanoscience designed healthy drinks. *Lett. Appl. Microbiol.* **2017**, *65*, 114–124. [CrossRef] [PubMed]
28. Schwan, R.F.; Ramos, C.L. Functional Beverages from Cereals. In *Functional and Medicinal Beverages;* Elsevier Inc.: Amsterdam, The Netherlands, 2019; ISBN 9780128163979.
29. Păucean, A.; Man, S.M.; Chiş, M.S.; Mureşan, V.; Pop, C.R.; Socaci, S.A.; Mureşan, C.C.; Muste, S. Use of pseudocereals preferment made with aromatic yeast strains for enhancing wheat bread quality. *Foods* **2019**, *8*, 1–14. [CrossRef] [PubMed]
30. Ripari, V. Techno-Functional Role of Exopolysaccharides in Cereal-Based, Yogurt-Like Beverages. *Beverages* **2019**, *5*, 16. [CrossRef]
31. Călinoiu, L.F.; Vodnar, D.C. Whole grains and phenolic acids: A review on bioactivity, functionality, health benefits and bioavailability. *Nutrients* **2018**, *10*, 1615. [CrossRef]
32. Bernardo, C.O.; Ascheri, J.L.R.; Carvalho, C.W.P.; Chávez, D.W.H.; Martins, I.B.A.; Deliza, R.; de Freitas, D.G.C.; Queiroz, V.A.V. Impact of extruded sorghum genotypes on the rehydration and sensory properties of soluble beverages and the Brazilian consumers' perception of sorghum and cereal beverage using word association. *J. Cereal Sci.* **2019**, *89*, 102793. [CrossRef]
33. Munekata, P.E.S.; Domínguez, R.; Budaraju, S.; Roselló-Soto, E.; Barba, F.J.; Mallikarjunan, K.; Roohinejad, S.; Lorenzo, J.M. Effect of innovative food processing technologies on the physicochemical and nutritional properties and quality of non-dairy plant-based beverages. *Foods* **2020**, *9*, 228. [CrossRef]
34. Yu, H.; Bogue, J. Concept optimisation of fermented functional cereal beverages. *Br. Food J.* **2013**, *115*, 541–563. [CrossRef]
35. Nyanzi, R.; Jooste, P.J. Cereal-Based Functional Foods, Probiotics, Everlon Cid Rigobelo. *IntechOpen* **2012**, 161–197. [CrossRef]
36. Granato, D.; Nunes, D.S.; Barba, F.J. An integrated strategy between food chemistry, biology, nutrition, pharmacology, and statistics in the development of functional foods: A proposal. *Trends Food Sci. Technol.* **2017**, *62*, 13–22. [CrossRef]
37. Granato, D.; Barba, F.J.; Lorenzo, J.M.; Cruz, A.G.; Putnik, P. Functional Foods: Product Development, Technological Trends, Efficacy Testing, and Safety. *Annu Rev Food Sci. T* **2020**, *11*, 93–118. [CrossRef] [PubMed]
38. Marsh, A.J.; Hill, C.; Ross, R.P.; Cotter, P.D. Fermented beverages with health-promoting potential: Past and future perspectives. *Trends Food Sci. Technol.* **2014**, *38*, 113–124. [CrossRef]
39. Müller, M.; Bellut, K.; Tippmann, J.; Becker, T. Physical Methods for Dealcoholization of Beverage Matrices and their Impact on Quality Attributes. *ChemBioEng Rev.* **2017**, *4*, 310–326. [CrossRef]
40. Kubo, M.; Nozu, Y.; Kataoka, C.; Kudo, M.; Taniguchi, S.; Sato, Y.; Nakayama, N.; Watanabe, M. Correlation between Non-Alcoholic Beverage Consumption and Alcohol Drinking Behavior among Japanese Youths. *Open J. Prev. Med.* **2015**, *5*, 31–37. [CrossRef]
41. Arici, M.; Daglioglu, O. Boza: A lactic acid fermented cereal beverage as a traditional Turkish food. *Food Rev. Int.* **2002**, *18*, 39–48. [CrossRef]
42. Ramashia, S.E.; Anyasi, T.A.; Gwata, E.T.; Meddows-Taylor, S.; Jideani, A.I.O. Processing, nutritional composition and health benefits of finger millet sub-Saharan Africa. *Food Sci. Technol.* **2019**, *39*, 253–266. [CrossRef]
43. Awobusuyi, T.D.; Siwela, M.; Kolanisi, U.; Amonsou, E.O. Provitamin A retention and sensory acceptability of amahewu, a non-alcoholic cereal-based beverage made with provitamin A-biofortified maize. *J. Sci. Food Agric.* **2016**, *96*, 1356–1361. [CrossRef]
44. Oguro, Y.; Nakamura, A.; Kurahashi, A. Effect of temperature on saccharification and oligosaccharide production efficiency in koji amazake. *J. Biosci. Bioeng.* **2019**, *127*, 570–574. [CrossRef]
45. Olusanya, R.N.; Kolanisi, U.; van Onselen, A.; Ngobese, N.Z.; Siwela, M. Nutritional composition and consumer acceptability of Moringa oleifera leaf powder (MOLP)-supplemented mahewu. *S. Afr. J. Bot.* **2020**, *129*, 175–180. [CrossRef]
46. Gernet, M.V.; Gribkova, I.N.; Kobelev, K.V.; Nurmukhanbetova, D.E.; Assembayeva, E.K. Biotechnological aspects of fermented drinks production on vegetable raw materials. *News Natl. Acad. Sci. Repub. Kazakhstan Ser. Geol. Tech. Sci.* **2019**, *1*, 223–230. [CrossRef]

47. Fallourd, M.J.; Viscione, L. *Ingredient Selection for Stabilisation and Texture Optimisation of Functional Beverages and the Inclusion of Dietary Fibre*; Woodhead Publishing Limited: Cambridge, UK; Sawston, UK, 2009; ISBN 9781845693428.
48. Vinicius De Melo Pereira, G.; De Carvalho Neto, D.P.; Junqueira, A.C.D.O.; Karp, S.G.; Letti, L.A.J.; Magalhães Júnior, A.I.; Soccol, C.R. A Review of Selection Criteria for Starter Culture Development in the Food Fermentation Industry. *Food Rev. Int.* **2020**, *36*, 135–167. [CrossRef]
49. Pontonio, E.; Rizzello, C.G. *Minor and Ancient Cereals: Exploitation of the Nutritional Potential Through the Use of Selected Starters and Sourdough Fermentation*, 2nd ed.; Elsevier Inc.: Amsterdam, The Netherlands, 2019; ISBN 9780128146392.
50. Oura, S.; Suzuki, S.; Hata, Y.; Kawato, A.; Abe, Y. Evaluation of physiological functionalities of amazake in mice. *J. Brew. Soc. Jpn.* **2007**, *102*, 781–788. [CrossRef]
51. Michio Sata, Y.N. Effect of a Late Evening Snack of Amazake in Patients with Liver Cirrhosis: A Pilot Study. *J. Nutr. Food Sci.* **2013**, *3*. [CrossRef]
52. Alauddin, M.; Kabir, Y. Functional and Molecular Role of Processed-Beverages toward Healthier Lifestyle. In *Nutrients in Beverages*; Elsevier Inc.: Amsterdam, The Netherlands, 2019; ISBN 9780128168424.
53. Maruki-Uchida, H.; Sai, M.; Yano, S.; Morita, M.; Maeda, K. Amazake made from sake cake and rice koji suppresses sebum content in differentiated hamster sebocytes and improves skin properties in humans. *Biosci. Biotechnol. Biochem.* **2020**, 1–7. [CrossRef]
54. Nicolau, A.I.; Gostin, A.I. *Safety of Borsh*; Elsevier Inc.: Amsterdam, The Netherlands, 2016; ISBN 9780128006207.
55. Pasqualone, A.; Summo, C.; Laddomada, B.; Mudura, E.; Coldea, T.E. Effect of processing variables on the physico-chemical characteristics and aroma of borş, a traditional beverage derived from wheat bran. *Food Chem.* **2018**, *265*, 242–252. [CrossRef]
56. Grosu-Tudor, S.-S.; Stefan, I.-R.; Stancu, M.-M.; Cornea, C.-P.; De Vuyst, L.; Zamfir, M. Microbial and nutritional characteristics of fermented wheat bran in traditional Romanian borş production. *Rom. Biotechnol. Lett.* **2019**, *24*, 440–447. [CrossRef]
57. Yeğin, S.; Üren, A. Biogenic amine content of boza: A traditional cereal-based, fermented Turkish beverage. *Food Chem.* **2008**, *111*, 983–987. [CrossRef]
58. Kedia, G.; Wang, R.; Patel, H.; Pandiella, S.S. Use of mixed cultures for the fermentation of cereal-based substrates with potential probiotic properties. *Process Biochem.* **2007**, *42*, 65–70. [CrossRef]
59. Altay, F.; Karbancioglu-Güler, F.; Daskaya-Dikmen, C.; Heperkan, D. A review on traditional Turkish fermented non-alcoholic beverages: Microbiota, fermentation process and quality characteristics. *Int. J. Food Microbiol.* **2013**, *167*, 44–56. [CrossRef]
60. Sahu, L.; Panda, S.K. Innovative Technologies and Implications in Fermented Food and Beverage Industries: An Overview. In *Innovations in Technologies for Fermented Food and Beverage Industries*; Springer: Berlin, Germany, 2018; pp. 1–23. [CrossRef]
61. Ramakrishnan, S.R.; Ravichandran, K.; Antony, U.M. *Whole Grains: Processing, Product Development, and Nutritional Aspects*; Mir, S.A., Manickavasagan, A., Shah, M.A., Eds.; CRC Press: Boca Raton, FL, USA, 2019; pp. 103–127.
62. Lee, Y. Characterization of Weissella kimchii PL9023 as a potential probiotic for women. *FEMS Microbiol. Lett.* **2005**, *250*, 157–162. [CrossRef]
63. Vieira-Dalodé, G.; Jespersen, L.; Hounhouigan, J.; Moller, P.L.; Nago, C.M.; Jakobsen, M. Lactic acid bacteria and yeasts associated with gowé production from sorghum in Bénin. *J. Appl. Microbiol.* **2007**, *103*, 342–349. [CrossRef] [PubMed]
64. Mandal, V.; Sen, S.K.; Mandal, N.C. Isolation and Characterization of Pediocin NV 5 Producing Pediococcus acidilactici LAB 5 from Vacuum-Packed Fermented Meat Product. *Indian J. Microbiol.* **2011**, *51*, 22–29. [CrossRef] [PubMed]
65. Adinsi, L.; Akissoé, N.H.; Dalodé-Vieira, G.; Anihouvi, V.B.; Fliedel, G.; Mestres, C.; Hounhouigan, J.D. Sensory evaluation and consumer acceptability of a beverage made from malted and fermented cereal: Case of gowe from Benin. *Food Sci. Nutr.* **2015**, *3*, 1–9. [CrossRef] [PubMed]
66. Adinsi, L.; Mestres, C.; Akissoé, N.; Vieira-Dalodé, G.; Anihouvi, V.; Durand, N.; Hounhouigan, D.J. Comprehensive quality and potential hazards of gowe, a malted and fermented cereal beverage from West Africa. A diagnostic for a future re-engineering. *Food Control* **2017**, *82*, 18–25. [CrossRef]

67. Agarry, O.O.; Nkama, I.; Akoma, O. Production of Kunun-zaki (A Nigerian fermented cereal beverage) using starter culture. *Int. Res. J. Microbiol.* **2010**, *1*, 18–25.
68. Nkama, I.; Agarry, O.O.; Akoma, O. Sensory and nutritional quality characteristics of powdered' Kunun-zaki': A Nigerian fermented cereal beverage. *Afr. J. Food Sci.* **2010**, *4*, 364–370.
69. Wartu, J.R.; Whong, C.M.Z.; Abdullahi, I.O.; Ameh, J.B.; Musa, B.J. Trends in total aflatoxins and nutritional impact of triticum spp during fermentation of kunun-zaki; a Nigeria sorghum bicolor based non-alcoholic beverage. *Int. J. Life Sci. Pharma Res.* **2015**, *5*, 26–35.
70. Olufunke, A.P. Assessment of Nutritional and Sensory Quality of Kunun Zaki—A Homemade Traditional Nigerian Beverage. *J. Nutr. Food Sci.* **2014**, *05*, 3–5. [CrossRef]
71. Basinskiene, L.; Juodeikiene, G.; Vidmantiene, D.; Tenkanen, M.; Makaravicius, T.; Bartkiene, E. Non-alcoholic beverages from fermented cereals with increased oligosaccharide content. *Food Technol. Biotechnol.* **2016**, *54*, 36–44. [CrossRef]
72. Gambuś, H.; Mickowska, B.; Bartoń, H.; Augustyn, G.; Zięć, G.; Litwinek, D.; Szary-Sworst, K.; Berski, W. Health benefits of kvass manufactured from rye wholemeal bread. *J. Microbiol. Biotechnol. Food Sci.* **2015**, *4*, 34–39. [CrossRef]
73. Bvochora, J.M.; Reed, J.D.; Read, J.S.; Zvauya, R. Effect of fermentation processes on proanthocyanidins in sorghum during preparation of Mahewu, a non-alcoholic beverage. *Process Biochem.* **1999**, *35*, 21–25. [CrossRef]
74. Idowu, O.O.; Fadahunsi, I.F.; Onabiyi, O.A. A Production and Nutritional Evaluation of Mahewu: A Non-Alcoholic Fermented Beaverage of South Africa. *Int. J. Res. Pharm. Biosci.* **2016**, *3*, 27.
75. Salvador, E.M.; McCrindle, C.M.E.; Buys, E.M.; Steenkamp, V. Standardization of cassava mahewu fermentation and assessment of the effects of iron sources used for fortification. *Afr. J. Food Agric. Nutr. Dev.* **2016**, *16*, 10898–10912. [CrossRef]
76. Fadahunsi, I.; Soremekun, O. Production, Nutritional and Microbiological Evaluation of Mahewu a South African Traditional Fermented Porridge. *J. Adv. Biol. Biotechnol.* **2017**, *14*, 1–10. [CrossRef]
77. Foma, R.K.; Destain, J.; Mobinzo, P.K.; Kayisu, K.; Thonart, P. Study of physicochemical parameters and spontaneous fermentation during traditional production of munkoyo, an indigenous beverage produced in Democratic Republic of Congo. *Food Control* **2012**, *25*, 334–341. [CrossRef]
78. Zulu, R.M.; Dillon, V.M.; Owens, J.D. Munkoyo beverage, a traditional Zambian fermented maize gruel using Rhynchosia root as amylase source. *Int. J. Food Microbiol.* **1997**, *34*, 249–258. [CrossRef]
79. Chileshe, J.; van den Heuvel, J.; Handema, R.; Zwaan, B.J.; Talsma, E.F.; Schoustra, S. Nutritional composition and microbial communities of two non-alcoholic traditional fermented beverages from Zambia: A study of mabisi and munkoyo. *Nutrients* **2020**, *12*, 1628. [CrossRef]
80. Byakika, S.; Mukisa, I.M.; Byaruhanga, Y.B.; Male, D.; Muyanja, C. Influence of food safety knowledge, attitudes and practices of processors on microbiological quality of commercially produced traditional fermented cereal beverages, a case of Obushera in Kampala. *Food Control* **2019**, *100*, 212–219. [CrossRef]
81. Misihairabgwi, J.M.; Ishola, A.; Quaye, I.; Sulyok, M.; Krska, R. Diversity and fate of fungal metabolites during the preparation of oshikundu, a Namibian traditional fermented beverage. *World Mycotoxin J.* **2018**, *11*, 471–481. [CrossRef]
82. Misihairabgwi, J.; Cheikhyoussef, A. Traditional fermented foods and beverages of Namibia. *J. Ethn. Foods* **2017**, *4*, 145–153. [CrossRef]
83. Embashu, W.; Nantanga, K.K.M. Pearl millet grain: A mini-review of the milling, fermentation and brewing of ontaku, a non-alcoholic traditional beverage in Namibia. *Trans. R. Soc. S. Afr.* **2019**, *74*, 276–282. [CrossRef]
84. Ben Omar, N.; Ampe, F. Microbial Community Dynamics during Production of the Mexican Fermented Maize Dough Pozol. *Appl. Environ. Microbiol.* **2000**, *66*, 3664–3673. [CrossRef] [PubMed]
85. Todorov, S.D.; Holzapfel, W.H. *Traditional Cereal Fermented Foods as Sources of Functional Microorganisms*; Elsevier Ltd.: Amsterdam, The Netherlands, 2015; ISBN 9781782420248.
86. Coskun, F. A Traditional Turkish Fermented Non-Alcoholic Beverage, "Shalgam". *Beverages* **2017**, *3*, 49. [CrossRef]
87. Gadaga, T.H.; Mutukumira, A.N.; Narvhus, J.A.; Feresu, S.B. A review of traditional fermented foods and beverages of Zimbabwe. *Int. J. Food Microbiol.* **1999**, *53*, 1–11. [CrossRef]
88. Oi, Y.; Kitabatake, N. Chemical Composition of an East African Traditional Beverage, Togwa. *J. Agric. Food Chem.* **2003**, *51*, 7024–7028. [CrossRef]

89. Waters, D.M.; Mauch, A.; Coffey, A.; Arendt, E.K.; Zannini, E. Lactic Acid Bacteria as a Cell Factory for the Delivery of Functional Biomolecules and Ingredients in Cereal-Based Beverages: A Review. *Crit. Rev. Food Sci. Nutr.* **2015**, *55*, 503–520. [CrossRef]
90. Ajiro, M.; Araki, M.; Ishikawa, M.; Kobayashi, K.; Ashida, I.; Miyaoka, Y. Analysis of Functional Relationships between Rice Particles and Oral Perception Using Amazake: A Traditional Japanese Beverage of Malted Rice. *Food Nutr. Sci.* **2017**, *08*, 901–911. [CrossRef]
91. Kawakami, S.; Ito, R.; Maruki-Uchida, H.; Kamei, A.; Yasuoka, A.; Toyoda, T.; Ishijima, T.; Nishimura, E.; Morita, M.; Sai, M.; et al. Intake of a mixture of sake cake and rice malt increases mucin levels and changes in intestinal microbiota in mice. *Nutrients* **2020**, *12*, 449. [CrossRef]
92. Petrova, P.; Petrov, K. Traditional cereal Beverage Boza: Fermentation technology, microbial content and healthy effects. In *Fermented Foods Part II Technological Interventions*; Ray, R.C., Montet, D., Eds.; CRC Press: Boca Raton, FL, USA, 2017; pp. 284–305. [CrossRef]
93. Oguro, Y.; Nishiwaki, T.; Shinada, R.; Kobayashi, K.; Kurahashi, A. Metabolite profile of koji amazake and its lactic acid fermentation product by Lactobacillus sakei UONUMA. *J. Biosci. Bioeng.* **2017**, *124*, 178–183. [CrossRef]
94. Akpinar-Bayizit, A.; Yilmaz-Ersan, L.; Ozcan, T. Determination of boza's organic acid composition as it is affected by raw material and fermentation. *Int. J. Food Prop.* **2010**, *13*, 648–656. [CrossRef]
95. Gotcheva, V.; Pandiella, S.S.; Angelov, A.; Roshkova, Z.; Webb, C. Monitoring the fermentation of the traditional Bulgarian beverage boza. *Int. J. Food Sci. Technol.* **2001**, *36*, 129–134. [CrossRef]
96. Mukisa, I.M.; Nsiimire, D.G.; Byaruhanga, Y.B.; Muyanja, C.M.B.K.; Langsrud, T.; Narvhus, J.A. Obushera: Descriptive sensory profiling and consumer acceptability. *J. Sens. Stud.* **2010**, *25*, 190–214. [CrossRef]
97. Jean-Pierre, G.; Serge, T.; Dolores, R.; Judith, E.; Dora, C.; Carmen, W. Pozol, a popular Mexican traditional beverage made from a fermented alkaline cooked maize dough. *Food Besed Approch. Health Nutr.* **2003**, *11*, 23–28.
98. Iskakova, J.; Smanalieva, J.; Methner, F.J. Investigation of changes in rheological properties during processing of fermented cereal beverages. *J. Food Sci. Technol.* **2019**, *56*, 3980–3987. [CrossRef] [PubMed]
99. Nkhata, S.G.; Ayua, E.; Kamau, E.H.; Shingiro, J.B. Fermentation and germination improve nutritional value of cereals and legumes through activation of endogenous enzymes. *Food Sci. Nutr.* **2018**, *6*, 2446–2458. [CrossRef] [PubMed]
100. Gillooly, M.; Bothwell, T.H.; Charlton, R.W.; Torrance, J.D.; Bezwoda, W.R.; MacPhail, A.P.; Derman, D.P.; Novelli, L.; Morrall, P.; Mayet, F. Factors affecting the absorption of iron from cereals. *Br. J. Nutr.* **1984**, *51*, 37–46. [CrossRef]
101. Khetarpaul, N.; Chauhan, B.M. Effect of fermentation by pure cultures of yeasts and lactobacilli on the available carbohydrate content of pearl millet. *Food Chem.* **1990**, *36*, 287–293. [CrossRef]
102. Nout, M.J.R.; Motarjemi, Y. Assessment of fermentation as a household technology for improving food safety: A joint FAO/WHO workshop. *Food Control* **1997**, *8*, 221–226. [CrossRef]
103. Kitabatake, N.; Gimbi, D.M.; Oi, Y. Traditional non-alcoholic beverage, Togwa, in East Africa, produced from maize flour and germinated finger millet. *Int. J. Food Sci. Nutr.* **2003**, *54*, 447–455. [CrossRef]
104. Papageorgiou, M.; Skendi, A. *Introduction to Cereal Processing and By-Products*; Elsevier: Amsterdam, The Netherlands, 2018; ISBN 9780081022146.
105. Hemery, Y.; Chaurand, M.; Holopainen, U.; Lampi, A.M.; Lehtinen, P.; Piironen, V.; Sadoudi, A.; Rouau, X. Potential of dry fractionation of wheat bran for the development of food ingredients, part I: Influence of ultra-fine grinding. *J. Cereal Sci.* **2011**, *53*, 1–8. [CrossRef]
106. Kunze, W. *Technology Brewing and Malting*; VLB Berlin: Berlin, Germany, 2004; Volume 949. [CrossRef]
107. Adinsi, L.; Vieira-Dalode, G.; Akissoe, N.; Anihouvi, V.; Mestres, C.; Jacobs, A.; Dlamini, N.; Pallet, D.; Hounhouigan, D.J. Processing and quality attributes of gowe: A malted and fermented cereal-based beverage from Benin. *Food Chain* **2014**, *4*, 171–183. [CrossRef]
108. Igyor, M.A.; Ogbonna, A.C.; Palmer, G.H. Effect of malting temperature and mashing methods on sorghum wort composition and beer flavour. *Process Biochem.* **2001**, *36*, 1039–1044. [CrossRef]
109. Bamforth, C.W. *Barley and Malt Starch in Brewing: A General Review*; Technical quarterly-Master Brewers Association of the Americas: Madison, WI, USA, 2003; Volume 40, pp. 89–97.
110. Dlusskaya, E.; Jänsch, A.; Schwab, C.; Gänzle, M.G. Microbial and chemical analysis of a kvass fermentation. *Eur. Food Res. Technol.* **2008**, *227*, 261–266. [CrossRef]

111. Krebs, G.; Becker, T.; Gastl, M. Characterization of polymeric substance classes in cereal-based beverages using asymmetrical flow field-flow fractionation with a multi-detection system. *Anal. Bioanal. Chem.* **2017**, *409*, 5723–5734. [CrossRef] [PubMed]
112. Krebs, G.; Müller, M.; Becker, T.; Gastl, M. Characterization of the macromolecular and sensory profile of non-alcoholic beers produced with various methods. *Food Res. Int.* **2019**, *116*, 508–517. [CrossRef] [PubMed]
113. Nsogning Dongmo, S.; Procopio, S.; Sacher, B.; Becker, T. Flavor of lactic acid fermented malt based beverages: Current status and perspectives. *Trends Food Sci. Technol.* **2016**, *54*, 37–51. [CrossRef]
114. Carvalho, D.O.; Gonçalves, L.M.; Guido, L.F. Overall Antioxidant Properties of Malt and How They Are Influenced by the Individual Constituents of Barley and the Malting Process. *Compr. Rev. Food Sci. Food Saf.* **2016**, *15*, 927–943. [CrossRef]
115. Wannenmacher, J.; Gastl, M.; Becker, T. Phenolic Substances in Beer: Structural Diversity, Reactive Potential and Relevance for Brewing Process and Beer Quality. *Compr. Rev. Food Sci. Food Saf.* **2018**, *17*, 953–988. [CrossRef]
116. Gupta, M.; Abu-Ghannam, N.; Gallaghar, E. Barley for brewing: Characteristic changes during malting, brewing and applications of its by-products. *Compr. Rev. Food Sci. Food Saf.* **2010**, *9*, 318–328. [CrossRef]
117. Krebs, G.; Becker, T.; Gastl, M. Influence of malt modification and the corresponding macromolecular profile on palate fullness in cereal-based beverages. *Eur. Food Res. Technol.* **2020**, *246*, 1219–1229. [CrossRef]
118. Liptáková, D.; Matejčeková, Z.; Valík, L. Lactic Acid Bacteria and Fermentation of Cereals and Pseudocereals. *Ferment. Process.* **2017**, 223–254. [CrossRef]
119. Fărcaş, A.; Tofană, M.; Socaci, S.; Mudura, E.; Scrob, S.; Salanţă, L.; Mureşan, V. Brewers' spent grain—A new potential ingredient for functional foods. *J. Agroaliment. Process. Technol.* **2014**, *20*, 137–141.
120. Nachel, M. *Homebrewing for Dummies*, 2nd ed.; Wiley Publishing, Inc.: Hoboken, NJ, USA, 2008; ISBN 9780470230626.
121. Rezac, S.; Kok, C.R.; Heermann, M.; Hutkins, R. Fermented foods as a dietary source of live organisms. *Front. Microbiol.* **2018**, *9*, 1785. [CrossRef] [PubMed]
122. Motarjemi, Y.; Nout, M.J.R. Food fermentation: A safety and nutritional assessment. *Bull. World Health Organ.* **1996**, *74*, 553–559.
123. Erkus, O.; De Jager, V.C.L.; Spus, M.; Van Alen-Boerrigter, I.J.; Van Rijswijck, I.M.H.; Hazelwood, L.; Janssen, P.W.M.; Van Hijum, S.A.F.T.; Kleerebezem, M.; Smid, E.J. Multifactorial diversity sustains microbial community stability. *ISME J.* **2013**, *7*, 2126–2136. [CrossRef]
124. Butler, S.; O'Dwyer, J.P. Stability criteria for complex microbial communities. *Nat. Commun.* **2018**, *9*, 2970. [CrossRef]
125. Taylor, J.R.N. *Fermentation: Foods and Nonalcoholic Beverages*, 2nd ed.; Elsevier Ltd.: Amsterdam, The Netherlands, 2015; Volume 3–4, ISBN 9780123947864.
126. Xiang, H.; Sun-Waterhouse, D.; Waterhouse, G.I.N.; Cui, C.; Ruan, Z. Fermentation-enabled wellness foods: A fresh perspective. *Food Sci. Hum. Wellness* **2019**, *8*, 203–243. [CrossRef]
127. Kohajdov, Z.; Karovi, J. Fermentation of cereals for specific purpose. *J. Food Nutr. Res.* **2007**, *46*, 51–57.
128. Aka, S.; Konan, G.; Fokou, G.; Dje Koffi, M.; Bonfoh, B. Review on African traditional cereal beverages. *Am. J. Res. Commun.* **2014**, *2*, 103–153.
129. Mugula, J.K.; Nnko, S.A.M.; Narvhus, J.A.; Sørhaug, T. Microbiological and fermentation characteristics of togwa, a Tanzanian fermented food. *Int. J. Food Microbiol.* **2003**, *80*, 187–199. [CrossRef]
130. Macori, G.; Cotter, P.D. Novel insights into the microbiology of fermented dairy foods. *Curr. Opin. Biotechnol.* **2018**, *49*, 172–178. [CrossRef] [PubMed]
131. Smid, E.J.; Hugenholtz, J. Functional Genomics for Food Fermentation Processes. *Annu. Rev. Food Sci. Technol.* **2010**, *1*, 497–519. [CrossRef] [PubMed]
132. Theron, M.; Lues, J.R. *Organic Acids and Food Preservation*; CRC Press: Boca Raton, FL, USA, 2010.
133. Bintsis, T. Foodborne pathogens. *AIMS Microbiol.* **2017**, *3*, 529–563. [CrossRef] [PubMed]
134. Srikaeo, K. *Biotechnological Tools in the Production of Functional Cereal-Based Beverages*; Elsevier Inc.: Amsterdam, The Netherlands, 2019; ISBN 9780128166789.
135. Tamang, J.P.; Cotter, P.D.; Endo, A.; Han, N.S.; Kort, R.; Liu, S.Q.; Mayo, B.; Westerik, N.; Hutkins, R. Fermented foods in a global age: East meets West. *Compr. Rev. Food Sci. Food Saf.* **2020**, *19*, 184–217. [CrossRef]

136. Gernah, D.I.; Ariahu, C.C.; Ingbian, E.K. Effects of malting and lactic fermentation on some chemical and functional properties of maize (Zea mays). *Am. J. Food Technol.* **2011**, *6*, 404–412. [CrossRef]
137. Sharaf, O.M.; El-Shafei, K.; Ibrahim, G.A.; El-Sayed, H.S.; Kassem, J.M.; Assem, F.M.; Tawfek, N.F.; Effat, B.A.; Abd El-Khalek, A.B.; Dabiza, N. Preparation, properties and evaluation of folate and riboflavin enriched six functional cereal-fermented milk beverages using encapsulated Lactobacillus plantarum or Streptococcus thermophiles. *Res. J. Pharm. Biol. Chem. Sci.* **2015**, *6*, 1724–1735.
138. Chavan, M.; Gat, Y.; Harmalkar, M.; Waghmare, R. Development of non-dairy fermented probiotic drink based on germinated and ungerminated cereals and legume. *LWT Food Sci. Technol.* **2018**, *91*, 339–344. [CrossRef]
139. Achi, O.K. The potential for upgrading traditional fermented foods through biotechnology. *Afr. J. Biotechnol.* **2005**, *4*, 375–380. [CrossRef]
140. Russo, P.; Arena, M.P.; Fiocco, D.; Capozzi, V.; Drider, D.; Spano, G. Lactobacillus plantarum with broad antifungal activity: A promising approach to increase safety and shelf-life of cereal-based products. *Int. J. Food Microbiol.* **2017**, *247*, 48–54. [CrossRef]
141. Angelov, A.; Yaneva-Marinova, T.; Gotcheva, V. Oats as a matrix of choice for developing fermented functional beverages. *J. Food Sci. Technol.* **2018**, *55*, 2351–2360. [CrossRef]
142. Bourdichon, F.; Casaregola, S.; Farrokh, C.; Frisvad, J.C.; Gerds, M.L.; Hammes, W.P.; Harnett, J.; Huys, G.; Laulund, S.; Ouwehand, A.; et al. Food fermentations: Microorganisms with technological beneficial use. *Int. J. Food Microbiol.* **2012**, *154*, 87–97. [CrossRef] [PubMed]
143. Ogunremi, O.R.; Banwo, K.; Sanni, A.I. Starter-culture to improve the quality of cereal-based fermented foods: Trends in selection and application. *Curr. Opin. Food Sci.* **2017**, *13*, 38–43. [CrossRef]
144. Gök Charyyev, M.; Özden Tuncer, B.; Akpınar Kankaya, D.; Tuncer, Y. Bacteriocinogenic properties and safety evaluation of Enterococcus faecium YT52 isolated from boza, a traditional cereal based fermented beverage. *J. fur Verbraucherschutz und Leb.* **2019**, *14*, 41–53. [CrossRef]
145. Arslan-Tontul, S.; Erbas, M. Co-Culture Probiotic Fermentation of Protein-Enriched Cereal Medium (Boza). *J. Am. Coll. Nutr.* **2020**, *39*, 72–81. [CrossRef] [PubMed]
146. Markowiak, P.; Ślizewska, K. Effects of probiotics, prebiotics, and synbiotics on human health. *Nutrients* **2017**, *9*, 1021. [CrossRef] [PubMed]
147. Kaur, P.; Ghoshal, G.; Banerjee, U.C. *Traditional Bio-Preservation in Beverages: Fermented Beverages*; Elsevier Inc.: Amsterdam, The Netherlands, 2019; ISBN 9780128166857.
148. Simango, C. Lactic acid fermentation of sour porridge and mahewu, a non-alcoholic fermented cereal beverage. *J. Appl. Sci. S. Afr.* **2002**, *08*, 89–98. [CrossRef]
149. Yépez, A.; Russo, P.; Spano, G.; Khomenko, I.; Biasioli, F.; Capozzi, V.; Aznar, R. In situ riboflavin fortification of different kefir-like cereal-based beverages using selected Andean LAB strains. *Food Microbiol.* **2019**, *77*, 61–68. [CrossRef]
150. Adesulu-Dahunsi, A.T.; Dahunsi, S.O.; Olayanju, A. Synergistic microbial interactions between lactic acid bacteria and yeasts during production of Nigerian indigenous fermented foods and beverages. *Food Control* **2020**, *110*, 106963. [CrossRef]
151. Salari, M.; Razavi, S.H.; Gharibzahedi, S.M.T. Characterising the synbiotic beverages based on barley and malt flours fermented by Lactobacillus delbrueckii and paracasei strains. *Qual. Assur. Saf. Crop. Foods* **2015**, *7*, 355–361. [CrossRef]
152. Kort, R.; Sybesma, W. Probiotics for every body. *Trends Biotechnol.* **2012**, *30*, 613–615. [CrossRef]
153. Setta, M.C.; Matemu, A.; Mbega, E.R. Potential of probiotics from fermented cereal-based beverages in improving health of poor people in Africa. *J. Food Sci. Technol.* **2020**. [CrossRef]
154. Azad, M.A.K.; Sarker, M.; Wan, D. Immunomodulatory Effects of Probiotics on Cytokine Profiles. *BioMed. Res. Int.* **2018**, *2018*. [CrossRef] [PubMed]
155. Mota de Carvalho, N.; Costa, E.M.; Silva, S.; Pimentel, L.; Fernandes, T.H.; Estevez Pintado, M. Fermented foods and beverages in human diet and their influence on gut microbiota and health. *Fermentation* **2018**, *4*, 1–13. [CrossRef]
156. Capozzi, V.; Russo, P.; Dueñas, M.T.; López, P.; Spano, G. Lactic acid bacteria producing B-group vitamins: A great potential for functional cereals products. *Appl. Microbiol. Biotechnol.* **2012**, *96*, 1383–1394. [CrossRef] [PubMed]

157. Tannock, G.W. Identification of lactobacilli and bifidobacteria. *Curr. Issues Mol. Biol.* **1999**, *1*, 53–64. [CrossRef] [PubMed]
158. Sidhu, J.S.; Kabir, Y.; Huffman, F.G. Functional foods from cereal grains. *Int. J. Food Prop.* **2007**, *10*, 231–244. [CrossRef]
159. Mudura, E.; Coldea, T.E.; Socaciu, C.; Ranga, F.; Pop, C.R.; Rotar, A.M.; Pasqualone, A. Brown beer vinegar: A potentially functional product based on its phenolic profile and antioxidant activity. *J. Serbian Chem. Soc.* **2018**, *83*, 19–30. [CrossRef]
160. Nout, M.J.R. Rich nutrition from the poorest—Cereal fermentations in Africa and Asia. *Food Microbiol.* **2009**, *26*, 685–692. [CrossRef]
161. Lähteenmäki, L. Consumer interpretation of nutrition and other information on food and beverage labels. In *Advances in Food and Beverage Labelling: Information and Regulations*; Elsevier Ltd.: Cambridge, UK, 2015; pp. 133–148. [CrossRef]
162. Salanţă, L.C.; Tofană, M.; Domokos, B.; Socaci, S.A.; Pop, C.R.; Fărcaş, A.C. Development of functional beverage from wheat grass juice. *Bull. UASVM Food Sci. Technol.* **2016**, *73*, 155–156. [CrossRef]
163. Mozaffarian, D. Dietary and Policy Priorities for Cardiovascular Disease, Diabetes, and Obesity. *Circulation* **2016**, *133*, 187–225. [CrossRef]
164. Klopčič, M.; Slokan, P.; Erjavec, K. Consumer preference for nutrition and health claims: A multi-methodological approach. *Food Qual. Prefer.* **2020**, *82*. [CrossRef]
165. Kim, M.K.; Kwak, H.S. Influence of functional information on consumer liking and consumer perception related to health claims for blueberry functional beverages. *Int. J. Food Sci. Technol.* **2015**, *50*, 70–76. [CrossRef]
166. Costell, E.; Tárrega, A.; Bayarri, S. Food acceptance: The role of consumer perception and attitudes. *Chemosens. Percept.* **2010**, *3*, 42–50. [CrossRef]
167. Viana, J.V.; Da Cruz, A.G.; Zoellner, S.S.; Silva, R.; Batista, A.L.D. Probiotic foods: Consumer perception and attitudes. *Int. J. Food Sci. Technol.* **2008**, *43*, 1577–1580. [CrossRef]
168. Sinclair, S.; Hammond, D.; Goodman, S. Sociodemographic differences in the comprehension of nutritional labels on food products. *J. Nutr. Educ. Behav.* **2013**, *45*, 767–772. [CrossRef] [PubMed]
169. Kasapoğlu, K.N.; Daşkaya-Dikmen, C.; Yavuz-Düzgün, M.; Karaça, A.C.; Özçelik, B. Enrichment of Beverages with Health Beneficial Ingredients. In *Value-Added Ingredients and Enrichments of Beverages*; Academic Press: Cambridge, MA, USA, 2019; ISBN 9780128166871.
170. Aadil, R.M.; Roobab, U.; Sahar, A.; ur Rahman, U.; Khalil, A.A. *Functionality of Bioactive Nutrients in Beverages*; Elsevier Inc.: Amsterdam, The Netherlands, 2019; ISBN 9780128168424.
171. Pop, A.; Muste, S.; Păucean, A.; Chiş, S.; Man, S.; Salanţă, L.; Marc, R.; Mureşan, A.M. Herbs and Spices in Terms of Food Preservation and Shelf Life. *Hop Med. Plants* **2019**, *27*, 57–65.
172. Campbell-Platt, G. Fermented foods—A world perspective. *Food Res. Int.* **1994**, *27*, 253–257. [CrossRef]
173. McGovern, P.E.; Zhang, J.; Tang, J.; Zhang, Z.; Hall, G.R.; Moreau, R.A.; Nuñez, A.; Butrym, E.D.; Richards, M.P.; Wang, C.S.; et al. Fermented beverages of pre- and proto-historic China. *Proc. Natl. Acad. Sci. USA* **2004**, *101*, 17593–17598. [CrossRef]
174. Coda, R.; Lanera, A.; Trani, A.; Gobbetti, M.; Di Cagno, R. Yogurt-like beverages made of a mixture of cereals, soy and grape must: Microbiology, texture, nutritional and sensory properties. *Int. J. Food Microbiol.* **2012**, *155*, 120–127. [CrossRef]
175. Cilla, A.; Garcia-Llatas, G.; Lagarda, M.J.; Barberá, R.; Alegría, A. Development of Functional Beverages: The Case of Plant Sterol-Enriched Milk-Based Fruit Beverages. In *Functional and Medicinal Beverages*; Elsevier Inc.: Amsterdam, The Netherlands, 2019; ISBN 9780128163979.
176. Bancalari, E.; Castellone, V.; Bottari, B.; Gatti, M. Wild Lactobacillus casei Group Strains: Potentiality to ferment plant derived juices. *Foods* **2020**, *9*, 314. [CrossRef]
177. Ozdemir, S.; Gocmen, D.; Kumral, A.Y. A traditional Turkish fermented cereal food: Tarhana. *Food Rev. Int.* **2007**, *23*, 107–121. [CrossRef]
178. Kivanç, M.; Funda, E.G. A functional food: A traditional tarhana fermentation. *Food Sci. Technol.* **2017**, *37*, 269–274. [CrossRef]
179. Kreisz, S.; Arendt, E.K.; Hübner, F.; Zarnkov, M. Cereal-based gluten-free functional drinks. In *Gluten-Free Cereal Products and Beverages*; Elsevier Inc.: Amsterdam, The Netherlands, 2008; pp. 373–392.
180. Katan, M.B.; De Roos, N.M. Promises and problems of functional foods. *Crit. Rev. Food Sci. Nutr.* **2004**, *44*, 369–377. [CrossRef] [PubMed]

181. Shori, A.B. Influence of food matrix on the viability of probiotic bacteria: A review based on dairy and non-dairy beverages. *Food Biosci.* **2016**, *13*, 1–8. [CrossRef]
182. Tanguler, H.; Selli, S.; Sen, K.; Cabaroglu, T.; Erten, H. Aroma composition of shalgam: A traditional Turkish lactic acid fermented beverage. *J. Food Sci. Technol.* **2017**, *54*, 2011–2019. [CrossRef]
183. Awobusuyi, T.D.; Siwela, M. Nutritional properties and consumer's acceptance of provitamin a-biofortified amahewu combined with bambara (Vigna subterranea) flour. *Nutrients* **2019**, *11*, 1476. [CrossRef]
184. Enujiugha, V.N.; Badejo, A.A. Probiotic potentials of cereal-based beverages. *Crit. Rev. Food Sci. Nutr.* **2017**, *57*, 790–804. [CrossRef]
185. Dongmo Nsogning, S.; Kollmannsberger, H.; Fischer, S.; Becker, T. Exploration of high-gravity fermentation to improve lactic acid bacteria performance and consumer's acceptance of malt wort-fermented beverages. *Int. J. Food Sci. Technol.* **2018**, *53*, 1753–1759. [CrossRef]
186. Kechagia, M.; Basoulis, D.; Konstantopoulou, S.; Dimitriadi, D.; Gyftopoulou, K.; Skarmoutsou, N.; Fakiri, E.M. Health Benefits of Probiotics: A Review. *Hindawi Publ. Corp.* **2013**, *2013*. [CrossRef]
187. Adeyanju, A.A.; Kruger, J.; Taylor, J.R.N.; Duodu, K.G. Effects of different souring methods on the protein quality and iron and zinc bioaccessibilities of non-alcoholic beverages from sorghum and amaranth. *Int. J. Food Sci. Technol.* **2019**, *54*, 798–809. [CrossRef]
188. Hassani, A.; Procopio, S.; Becker, T. Influence of malting and lactic acid fermentation on functional bioactive components in cereal-based raw materials: A review paper. *Int. J. Food Sci. Technol.* **2016**, *51*, 14–22. [CrossRef]
189. Towo, E.; Matuschek, E.; Svanberg, U. Fermentation and enzyme treatment of tannin sorghum gruels: Effects on phenolic compounds, phytate and in vitro accessible iron. *Food Chem.* **2006**, *94*, 369–376. [CrossRef]
190. Kwon, D.Y.; Daily, J.W.; Kim, H.J.; Park, S. Antidiabetic effects of fermented soybean products on type 2 diabetes. *Nutr. Res.* **2010**, *30*, 1–13. [CrossRef] [PubMed]
191. Kwon, D.Y.; Jang, J.S.; Hong, S.M.; Lee, J.E.; Sung, S.R.; Park, H.R.; Park, S. Long-term consumption of fermented soybean-derived Chungkookjang enhances insulinotropic action unlike soybeans in 90% pancreatectomized diabetic rats. *Eur. J. Nutr.* **2007**, *46*, 44–52. [CrossRef] [PubMed]
192. Baú, T.R.; Garcia, S.; Ida, E.I. Optimization of a fermented soy product formulation with a kefir culture and fiber using a simplex-centroid mixture design. *Int. J. Food Sci. Nutr.* **2013**, *64*, 929–935. [CrossRef] [PubMed]

Article

Nutritional Features and Bread-Making Performance of Wholewheat: Does the Milling System Matter?

Maria Ambrogina Pagani [1], Debora Giordano [2], Gaetano Cardone [1], Antonella Pasqualone [3], Maria Cristina Casiraghi [1], Daniela Erba [1], Massimo Blandino [2,*] and Alessandra Marti [1,*]

[1] Department of Food, Environmental and Nutritional Sciences (DeFENS), Università degli Studi di Milano, via G. Celoria 2, 20133 Milan, Italy; ambrogina.pagani@unimi.it (M.A.P.); gaetano.cardone@unimi.it (G.C.); maria.casiraghi@unimi.it (M.C.C.); daniela.erba@unimi.it (D.E.)

[2] Department of Agricultural, Forest and Food Sciences (DISAFA), Università degli Studi di Torino, Largo P. Braccini 2, 10095 Grugliasco (TO), Italy; debora.giordano@unito.it

[3] Food Science and Technology Unit, Department of Science of Soil, Università degli Studi di Bari, via Amendola, 165/A, 70126 Bari, Italy; antonella.pasqualone@uniba.it

* Correspondence: massimo.blandino@unito.it (M.B.); alessandra.marti@unimi.it (A.M.)

Received: 27 June 2020; Accepted: 29 July 2020; Published: 1 August 2020

Abstract: Despite the interest in stone-milling, there is no information on the potential advantages of using the resultant wholegrain flour (WF) in bread-making. Consequently, nutritional and technological properties of WFs obtained by both stone- (SWF) and roller-milling (RWF) were assessed on four wheat samples, differing in grain hardness and pigment richness. Regardless of the type of wheat, stone-milling led to WFs with a high number of particles ranging in size from 315 to 710 μm), whereas RWFs showed a bimodal distribution with large (>1000 μm) and fine (<250 μm) particles. On average, the milling system did not affect the proximate composition and the bioactive features of WFs. The gluten aggregation kinetics resulted in similar trends for all SWFs, with indices higher than for RWFs. The effect of milling on dough properties (i.e., mixing and leavening) was sample dependent. Overall, SWFs produced more gas, resulting in bread with higher specific volume. Bread crumb from SWF had higher lutein content in the wheat *cv* rich in xanthophylls, while bread from RWF of the blue-grained *cv* had a moderate but significantly higher content in esterified phenolic acids and total anthocyanins. In conclusion, there was no relevant advantage in using stone- as opposed to roller-milling (and vice versa).

Keywords: wholewheat flour; stone milling; roller milling; bioactive compounds; bread; pigmented wheat; dough rheology

---

1. Introduction

Whole grains have been (and still are in several countries) the most important energy source of mankind. They constitute a valuable and inexpensive source of numerous nutrients and phytochemicals, including fiber, phenolic compounds, minerals, and vitamins, mainly located in the germ and bran regions [1,2]. Although a universally accepted definition of whole grain has yet to be formulated [3], it is widely recognized that a whole grain product must contain bran, germ and endosperm in the same proportions as in the original, native grain [3]. Whatever the official definition is, the relationship between the consumption of whole grain foods and a lower incidence of diseases in the Occident has been suggested by numerous epidemiological studies [4–8].

Thanks to public institutions (governments and health organizations) and extensive promotional campaigns, consumers nowadays are fully aware of the numerous health benefits associated with the consumption of products made from whole grain flour (WF). Consequently, the demand for wholegrain products—in particular for staple foods such as bread and pasta—has been constantly growing [9]. This trend is recognizable even in countries where products made from refined flour have always been preferred for their undisputed superior sensory characteristics.

At present, two milling processes are mainly used to produce WF from wheat: (i) single–stream milling, and (ii) multiple-stream milling with recombination [10]. The former—where stone mills are used—is the world's oldest system for flour production. The stone mill is quite a simple machine, formed by two horizontal and overlapping grinding stones where the upper revolves while the other (bedstone) is stationary [11]. Grains are fed in the gap between the two stones and undergo shear, compression and abrasion forces. The ground fractions stay together during the entire milling process, and they are collected (with an extraction rate of 100%) at the bottom of the bedstone without separation according to particle size or composition [12].

Multiple-stream milling became established in the mid-1800s [11]. According to this process, wheat grains are progressively ground by means of a sequence of cast iron roller mills followed by sieving and sifting of the ground material [13]. Due to differences in composition, bran, germ and endosperm exhibit different friability and breakage patterns, thus facilitating the separation of bran and germ from flour, which derives from the endosperm. The separation of the three regions is enhanced by conditioning the kernels before milling [11,13]. The flour extraction rate is usually about 70% [12]. Even this milling process may lead to a WF if the different "fraction streams"—originating with the repeated grinding and sieving steps—are gathered at the end of the process. The extraction rate of the recombined WF should be about 100%, and the endosperm, germ and bran should be present in the same ratio as in the native whole grain [10]. Roller milling guarantees higher productivity, flexibility and more constant results over time; consequently, this process is highly preferred for industrial applications [13,14]. Moreover, the separation of bran and germ fractions during the process allows further treatments before recombination, such as mild heating and/or bran grinding, improving both technological performance and storage of the WF [13,15].

On the other hand, stone milling has been rediscovered in recent years by small farmers and bread-making artisans as it requires relatively low capital inputs [16]. Certainly, the simplicity and cheapness of the process (only one operation; grain tempering is not mandatory) also makes it suitable for household milling, favoring whole grain consumption for some segments of population [14]. Nevertheless, despite these advantages, stone milling not only is characterized by a lower yield but may also worsen the rheological properties of WF due to varying degrees of heat development [16,17] according to the type of small-scale mills used [14]. The negative effects of stone milling can be reduced by choosing suitable wheat *cultivars (cvs)* and/or farming procedures, as suggested by Gélinas et al. [18].

Despite the great interest in stone milling and increasing demands for WF-produced food, few studies have compared the effects of stone and roller milling on the same wheat varieties to identify and evaluate analogies and differences between the two kinds of WF [15,19]. Some research has found similarities in proximate chemical composition and phenolic profile of WF from stone milling (SWF) and roller milling (RWF) [17,20,21]. On the contrary, by promoting high heat development due to friction, compression and shear phenomena, stone milling leads to a significant loss in aminoacids and unsaturated fatty acids [17]. This worsening can affect not only the nutritional aspects of flour but also its technological properties, due to a partial denaturation of gluten proteins, relevant starch damage and/or differences in particle size distribution [15,17,20]. Specifically, SWFs exhibited lower pasting properties, higher water absorption, and lower stability compared to RWFs [14,21].

Regarding the relationship between milling process and bread quality, the data are not in agreement. Liu et al. [21] emphasized that RWFs exhibited the best steam-bread making performance, while according to Kihlberg et al. [22], bread produced using RWF was characterized by regular shape but higher compactness. Nevertheless, differences in bread volume in WFs obtained by the two types of milling can be resolved by adopting sourdough fermentation [14].

The few studies published so far on the effects of the milling process on WF and bread properties obtained by the same type of wheat presented no univocal findings and prompted our study to aim at giving a complete overview of behavior–from milling to baking–of WFs from four types of common wheat (*T. aestivum* L.), all belonging to the bread-making class and characterized by high protein content. Beyond these common traits, the four wheat samples differed in protein strength and physical structure of kernels–hard, semi-hard and soft endosperm–a property that highly impacts milling behavior and performance [23]. Moreover, two pigmented varieties were distinguished by their richness in bioactive compounds, differently located in the kernel (polyphenols and anthocyanins in the external layers and xanthophylls in the endosperm), whose retention has to be carefully ensured along the whole transformation chain.

2. Materials and Methods

2.1. Wheat Samples

The present study analyzed the SWF and RWF derived from four common wheat (*Triticum aestivum* L.) samples.

Table 1 compares the main information regarding the wheat samples. The two pigmented wheat *cvs* were chosen for their interesting bioactive compound content [24].

Table 1. Main information of common wheat samples.

Cultivar	Cultivation Region	Hardness	Color	TW kg/hL	TKW g
Bolero	Piedmont region, Italy	soft	white	78.0	33.3
CWRS	Manitoba region, Canada	hard	red	77.8	34.8
Bona Vita	Piedmont region, Italy	medium	yellow	74.6	36.5
Skorpion	Piedmont region, Italy	medium	blue	70.4	48.0

TW, test weight; TKW, thousand kernel weight.

2.2. Milling Procedures

An aliquot (70 kg) of each sample was stone-milled (Molino Tomatis; Niella Tanaro, Italy), without kernel conditioning through a single-stream process, without any sifting, to produce SWF. The stone mill used was made of two discs of natural French burrstones, from the district of La Ferté-sous-Jouarre (France). The stones, 1.3 m in diameter, are arranged on a vertical axis, with the upper stone rotating at 70 rpm. The distance between the two stones at their centers was about 1 mm. The opposing surfaces of the stones were subdivided into 10 harps and grooved from the center to the circumference. After the proper cleaning operation, 40 kg of each sample was milled and the flour discarded, in order to reach stable operative conditions (speed, temperature) of the stone mill before sampling. The remaining 30 kg aliquot was then milled, carefully mixed to favor fraction blending and subsampled for rheological and chemical analysis. Another aliquot (10 kg) of grains was conditioned (16 h at 20 °C) till reaching 16% moisture, and then submitted to multiple-stream milling by using a roller mill (Bona lab-scale mill, Labormill 4RB, Monza, Italy). This lab scale mill, equipped with 4 rollers, 0.07 m in diameter and 0.20 m wide were horizontally arranged can simulate the industrial milling process e by separating the different parts of the wheat kernel: the external coarse bran, the intermediate layers (middlings) and the inner endosperm (refined flour). The kernels are milled in three steps: a first break phase passing through fluted rolls and two reducing phases with smooth rolls. Gap settings of the break and reducing rolls were adjusted to 0.4 mm, 0.2 mm and 0.05 mm, respectively. The feed rate was adjusted to about 8 kg/h. The three milling fractions obtained (refined flour, middlings and coarse bran) were recombined to produce the RWF. WF yield was about 100% for both milling processes. Refined flours obtained from roller milling were produced for bread production and functional characterization (see Section 2.3.2).

Samples were stored in a polypropylene bag at 4 °C and under vacuum until used. Samples were used after two weeks of resting.

2.3. Methods

2.3.1. Particle Size Distribution

Particle size distribution of SWFs and RWFs was assessed in single by means of an automatic mechanical sifter (AS 200, Retsch GmbH, Düsseldorf, Germany) equipped with 8 sieves: 1000 μm, 800 μm, 710 μm, 500 μm, 425 μm, 315 μm, 250 μm, 160 μm. The test was carried out on 100 g of WF, setting 1.5 mm of oscillation for 5 min.

2.3.2. Chemical Analysis

The moisture content of WFs was determined by means of thermo-balance (MA 210.R, Radwag Wagi Elektroniczne, Radom, Poland), by drying the sample at 130 °C until its weight did not change by 1 mg in 10 s. Ash (AACC 08-01.01) and damaged starch (AACC 76-31.01) contents were evaluated according to AACC standard methods [25]. The amounts of protein (AOAC 34.01.05 No. 925.31), fat (AOAC 31.04.02 No. 963.15), total (AOAC 31.04.02 No. 985.29), soluble and insoluble dietary fiber contents were measured according to AOAC standard methods (AOAC 31.04.02 No. 991.43) [26]. Total and water soluble arabinoxylans were evaluated as previously reported by Manini et al. [27]. Total starch content was calculated as what remained after moisture, protein, ash, fat and total fiber determinations had been accounted for. α-amylase activity was determined according to the AACC method 22-02.01 [25].

The SWF and RWF from each *cv* were analysed for soluble phenolic acids (SPAs) and cell wall-bound phenolic acids (CWBPAs) and total antioxidant capacity. Flours from Bona Vita and Skorpion *cvs* were further analysed for xanthophylls and total anthocyanin content (TAC), respectively. Extraction of phenolic acids and xanthophylls and their quantification by means of RP-HPLC was performed as reported by Giordano et al. [28]. The antioxidant capacity was determined by means of FRAP (Ferric Reducing Antioxidant Power) and the ABTS [2,2′-Azino-bis(3-ethylbenzthiazoline-6-sulfonic acid)] assays adapted into QUENCHER method as described by Serpen et al. [29]. The TAC was determined on samples (1 g) extracted using 8 mL of ethanol acidified with 1 N HCl (85:15, *v/v*) for 30 min. The absorbance was measured after centrifugation at 20,800× *g* for 2 min at 540 nm as reported by Siebenhandl et al. [30]. TAC was expressed as mg cyanidin-3-*O*-glucoside (Cy-3-glc) equivalents/kg of sample (db).

Moisture, ash, protein, fat, soluble and insoluble dietary fiber, total and water soluble arabinoxylans, and amylase activity were evaluated in duplicate, whereas damaged starch content and bioactive compounds in triplicate.

2.3.3. Rheological Properties

Gluten aggregation properties were measured in triplicate by means of the GlutoPeak (Brabender GmbH and Co KG, Duisburg, Germany) device, by using distilled water as solvent; 9:10 flour:solvent ratio, 2750 rpm as paddle speed rate and 35 °C as temperature of water circulating bath. The main indices considered were: (i) maximum torque (the highest consistency value reached, evaluated in GlutoPeak Units, GPU); (ii) peak maximum time (time required to achieve the maximum torque, evaluated in seconds, s); and (iii) aggregation energy (the area under the curve between 15 s before and 15 s after the maximum torque; evaluated in GlutoPeak Equivalents, GPE).

Mixing properties were evaluated in duplicate by using the Farinograph-E (Brabender GmbH and Co KG, Duisburg, Germany) device equipped with a 50 g mixing bowl, according to the standard method ICC 115/1 [31].

Dough leavening properties were determined in single by means of the Rheofermentometer F4 (Chopin Technologies, Villeneuve La Garenne CEDEX, France) device on 315 g of dough at 30 °C for 4.5 h. Dough samples were prepared by mixing flour, in a spiral mixer (Artisan, KitchenAid®, Whirlpool, Benton Harbor, USA) with fresh yeast (2 g/100 g of flour; Carrefour, Annecy, France) and salt (NaCl; 1 g/100 g of flour; Candor®, Com-Sal s.r.l., Pesaro, Italy). The amount of tap water added, and the kneading time used were previously determined by means of the farinographic test (Table 2).

Table 2. Amount of water, kneading and leavening times used in bread-making for each sample.

	Bolero *cv*		**CWRS**		**BonaVita *cv***		**Skorpion *cv***	
	Stone-Milling	**Roller-Milling**	**Stone-Milling**	**Roller-Milling**	**Stone-Milling**	**Roller-Milling**	**Stone-Milling**	**Roller-Milling**
Amount of tap water (g/100 g of flour)	64.2	61.7	61.2	61.5	68.1	64.7	65.5	64.5
Kneading time (min)	5.0	5.5	6.7	6.0	5.6	5.9	4.3	4.9
Leavening time (h)	1.38	2.36	1.44	1.36	1.09	1.15	1.36	1.31

2.3.4. Bread Preparation and Characterization

Dough samples were prepared from either WFs or refined flours (obtained by roller milling) in the same conditions as those reported for the rheofermentographic test. After kneading, the samples were left to rest for 15 min, then divided into portions of 250 g, modelled in cylindrical shapes, put into aluminum pans (length: 12.5 cm; width: 6 cm; height: 5 cm) and leavened at 30 °C and 70% relative humidity in a combined proofer oven (Self-Cooking Center®, Rational International AG, Heerbrugg, Switzerland), until the dough exceeded the top of the baking pans by about 2.5 cm. Then, the leavened dough samples were baked (Self-Cooking Center®, Rational International AG, Heerbrugg, Switzerland) at 200 °C for 30 min (85% relative humidity). One baking test was performed, yielding two loaves for each sample.

Two hours after baking, the loaves were characterized in terms of specific volume, through the ratio between the bread volume—evaluated according to the standard method AACC 10-05.01 [25]– and its weight. Bread height was determined by measuring the maximum height of the slice by means of image analysis (Image ProPlus, v6; Media Cybernetics, Inc., Rockville, MD, USA). Loaf specific volume was determined on two loaves, while bread height was evaluated on three central slices of each bread, for a total of six replicates.

SPAs, CWBPAs, total antioxidant capacity, and xanthophylls (only for flour from Bona Vita *cv*) and TAC (only for flour from Skorpion *cv*) were carried out as described in Section 2.3.2, on bread samples obtained from SWF, RWF and refined flours of pigmented *cvs*, which provide a higher AC and the possibility to investigate the impact of milling method also on the bioactive compounds responsible for flour color. Before analysis, bread samples were ground to a fine powder (particle size <300 µm) with a Cyclotec 1093 sample mill (Foss, Padova, Italy). The same grinding procedure was carried out for bread crust (about 3.5 mm thick), and crumb after freeze-drying (−80 °C for 72 h; Alpha 1–2 LD plus; Deltek s.r.l., Naples, Italy). All samples were stored at −25 °C. Bioactive compounds in bread were evaluated in triplicate.

2.4. Statistical Analysis

Statistical analysis (t-test) was carried out in order to identify significant differences between SWF and RWF from the same *cv* by using Statgraphics Plus 5.1 (StatPoint Inc., Warrenton, CT, USA) at the 1% (* $p < 0.01$) significance level. Data obtained from the functional characterization of flours and breads were analyzed separately for each *cv* using analysis of variance (ANOVA), by comparing raw material (flour), bread crust and bread crumb obtained from refined flour, SWF and RWF. A 0.01 threshold was used to reject the null hypothesis. The REGW-F test was performed for multiple comparisons, by using SPSS for Windows statistical package, Version 24.0 (SPSS Inc., Chicago, IL, USA).

3. Results

3.1. Particle Size Distribution

Regardless of the type of wheat, stone milling led to WF with a higher number of particles from 315 to 710 μm (from 32.2% for CWRS to 45.9% for Bona Vita *cv*) (Figure 1). On the contrary, in WF obtained by roller milling (i.e., RWFs) such "intermediate" fractions extended from 7% for CWRS to 22% for Bolero *cv*. In addition, recombination of roller milling fractions led to a peculiar, bimodal distribution with large (>1000 μm, mainly composed by bran) and fine (<250 μm) particles. Indeed, the large particles extended from 15% to 25% (except for CWRS) in RWFs, whereas they did not exceed 8% in SWFs. On the other hand, fine particles represented more than 50% (*w/w*) in RWFs, reaching 75% for samples with the highest kernel hardness (i.e., CWRS); whereas this fraction was only 30% of the mass in SWFs.

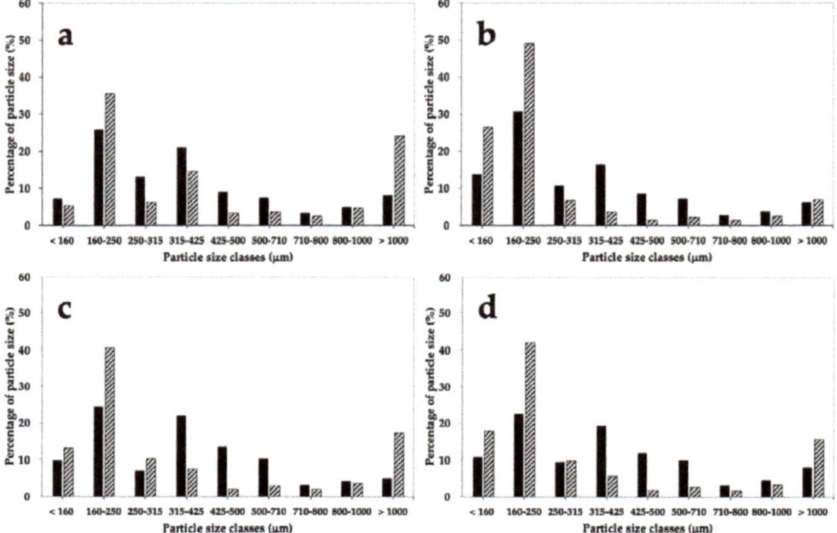

Figure 1. Effect of stone milling (black bars) and roller milling (dash bars) on the particle size distribution of whole grain flours from Bolero *cv* (**a**), CWRS (**b**), Bona Vita *cv* (**c**) and Skorpion *cv* (**d**). CWRS: commercial Canada Western Red Spring Wheat.

3.2. Chemical Composition and Bioactive Compounds in WFs

The milling system did not significantly affect the chemical composition of WFs, except for the moisture content that decreased by 17% and 11% when Bona Vita and Skorpion *cvs* were milled using the stone milling system (Table 3). Although the fiber content did not change, roller milling caused a slight but significant decrease in the total arabinoxylan content of both CWRS and Bona Vita *cv* (about 19% and 12%, respectively). In contrast, a significant increase in this parameter was found in RWF from Skorpion *cv* (about 70%). As regards the water soluble arabinoxylan fraction, the roller milling system resulted in an increase of about 63% for only Bona Vita *cv* (Table 3). The effect of milling on the bioactive compound concentration was sample and compound dependent. RWFs resulted in a significantly higher content of CWBPAs, CWB-ferulic acid only in Bona Vita and Skorpion *cvs*. A similar trend was observed for CWB-sinapic acid, with a significant difference observed in the CWRS. RWF also showed a significant higher TAC in the blue-grained Skorpion *cv* and a higher AC_{FRAP} in the yellow-grained Bona Vita *cv*. Conversely, no difference was observed for SPAs in any of the samples and for xanthophyll in Bona Vita *cv*.

Table 3. Chemical composition and bioactive compounds of whole flours obtained by stone-milling and roller-milling.

	Bolero *cv*		CWRS		Bona Vita *cv*		Skorpion *cv*		Average of All Samples	
	Stone-milling	Roller-milling	Stone-milling	Roller-milling	Stone-milling	Roller-milling	Stone-milling	Roller-milling	Stone-milling	Roller-milling
Moisture	11.8 ± 0.1	13.2 ± 0.7	14.8 ± 0.1	14.9 ± 0.7	11.8 ± 0.1	14.2 ± 0.1 *	12.4 ± 0.2	13.9 ± 0.1 *	12.7	14.05
Protein	16.7 ± 0.3	16.8 ± 0.2	16.9 ± 0.2	16.8 ± 0.2	15.4 ± 0.2	15.9 ± 0.2	15.3 ± 0.3	15.8 ± 0.8	16.1	16.3
Ash	1.5 ± 0.1	1.8 ± 0.1	1.5 ± 0.1	1.3 ± 0.1	1.7 ± 0.2	2.0 ± 0.2	1.7 ± 0.2	1.6 ± 0.1	1.6	1.7
Fat	1.8 ± 0.1	2.1 ± 0.01	1.9 ± 0.002	2.3 ± 0.1	2.4 ± 0.1	2.2 ± 0.04	1.9 ± 0.2	2.3 ± 0.3	2.0	2.2
Starch										
Total	66.5 ± 0.7	65.9 ± 0.3	67.9 ± 1.0	70.4 ± 0.4	66.8 ± 0.2	65.6 ± 0.8	68.5 ± 0.2	67.5 ± 0.5	67.4	67.4
Damaged	3.3 ± 0.1	2.2 ± 0.3	5.0 ± 0.5	6.2 ± 0.5	4.0 ± 0.3	3.5 ± 0.3	4.7 ± 1.0	4.3 ± 0.4	4.3	4.1
Dietary Fiber										
Total	13.4 ± 0.4	13.4 ± 0.1	11.7 ± 1.2	9.3 ± 0.2	13.6 ± 0.1	14.3 ± 0.4	12.5 ± 0.2	12.8 ± 0.2	12.8	12.5
Insoluble	11.2 ± 0.01	11.5 ± 0.2	9.8 ± 0.6	7.0 ± 0.5	11.8 ± 0.1	12.1 ± 0.2	10.6 ± 0.3	10.6 ± 0.3	10.9	10.3
Soluble	2.2 ± 0.4	1.9 ± 0.3	1.9 ± 0.6	2.3 ± 0.3	1.8 ± 0.2	2.2 ± 0.2	1.9 ± 0.1	2.2 ± 0.2	2.0	2.2
Arabinoxylan										
Total	4.0 ± 0.3	3.7 ± 0.12	3.74 ± 0.04	3.04 ± 0.05 *	5.44 ± 0.20	4.79 ± 0.01 *	4.01 ± 0.40	6.8 ± 0.2 *	4.3	4.6
Water extractable	0.23 ± 0.01	0.19 ± 0.05	0.33 ± 0.02	0.29 ± 0.02	0.32 ± 0.02	0.52 ± 0.01 *	0.50 ± 0.05	0.44 ± 0.01	0.35	0.36
CWBPAs	606 ± 42	626 ± 30	639 ± 54	684 ± 10	766 ± 16	886 ± 1 7 *	897 ± 19	991 ± 7 *	727	797
CWB-Ferulic acid	551 ± 39	570 ± 28	591 ± 52	629 ± 9	672 ± 16	783 ± 12 *	792 ± 17	874 ± 6 *	652	714
CWB-Sinapic acid	24.8 ± 2.3	25.5 ± 1.8	22.4 ± 0.9	26.4 ± 0.9 *	50.5 ± 2.1	53.1 ± 6.5	46.2 ± 0.4	48.8 ± 1.9	36.0	38.5
SPAs	52.8 ± 3.9	52.2 ± 3.4	59.0 ± 5.5	56.9 ± 1.5	62.1 ± 5.2	74.5 ± 1.0	84.7 ± 13.7	91.7 ± 2.6	64.7	68.8
S-Ferulic acid	13.9 ± 0.9	13.9 ± 1.0	15.8 ± 1.9	14.9 ± 0.6	15.7 ± 2.2	18.7 ± 1.4	20.8 ± 3.4	23.1 ± 1.0	16.6	17.7
S-Sinapic acid	24.1 ± 1.4	23.7 ± 1.6	29.5 ± 2.7	29.1 ± 0.7	32.2 ± 2.0	38.4 ± 2.6	43.4 ± 7.1	45.2 ± 1.4	32.3	34.1
Xanthophylls										
Lutein	n.d.	n.d.	n.d.	n.d.	3.3 ± 0.1	3.0 ± 0.2	n.d.	n.d.	-	-
Zeaxanthin	n.d.	n.d.	n.d.	n.d.	0.28 ± 0.02	0.24 ± 0.02	n.d.	n.d.	-	-
TAC	n.d.	n.d.	n.d.	n.d.	n.d.	n.d.	22.7 ± 0.4	26.3 ± 0.2 *	-	-
AC_{ABTS}	20.3 ± 0.3	20.2 ± 0.3	19.2 ± 0.3	19.1 ± 1.1	18.2 ± 0.8	18.0 ± 0.4	19.9 ± 0.7	18.6 ± 0.3	19.4	19.0
AC_{FRAP}	7.3 ± 0.5	6.7 ± 0.4	6.3 ± 0.4	6.4 ± 0.2	6.4 ± 0.1	8.1 ± 0.4 *	7.9 ± 0.6	7.3 ± 0.2	7.0	7.1

Protein, ash, fat, total dietary fiber and total arabinoxylans values are expressed as g/100 g dry basis. Damaged starch is expressed as g/100 g of total starch. Insoluble and soluble dietary fiber are reported as g/100 g dry basis of total dietary fiber. Water-extractable arabinoxylans are expressed as g/100 g dry basis the total arabinoxylans. Cell wall-bound phenolic acids (CWBPAs) and soluble (free and conjugated forms) phenolic acids (SPAs) are the sum of the single phenolic acids determined by means of RP-HPLC/DAD and are expressed as mg/kg dry basis. Xanthophyll (lutein and zeaxanthin) are expressed as mg/kg dry basis. Total anthocyanin content (TAC) is expressed as mg Cy-3-glc eq/kg dry basis. Antioxidant capacity (AC) measured by means of the ABTS and FRAP assays is expressed as mmol TE/kg dry basis. CWRS: commercial Canada Western Red Spring Wheat. Data are presented as mean ± standard deviation. The asterisks indicate significant differences between the means of stone- and roller-milled samples of each *cv* (* $p < 0.01$). The absence of asterisk indicates a not significant difference. n.d.: not determined.

3.3. Gluten Aggregation Properties

The gluten aggregation profiles of the WFs were consistent in showing a slower aggregation when the samples were milled by stone milling (Figure 2). Specifically, the peak maximum time was significantly higher in SWFs than RWFs (86 vs. 71 s, 86 vs. 64 s, 101 vs. 79 s, and 72 vs. 58 s for Bolero, CWRS, Bona Vita, and Skorpion *cvs*, respectively).

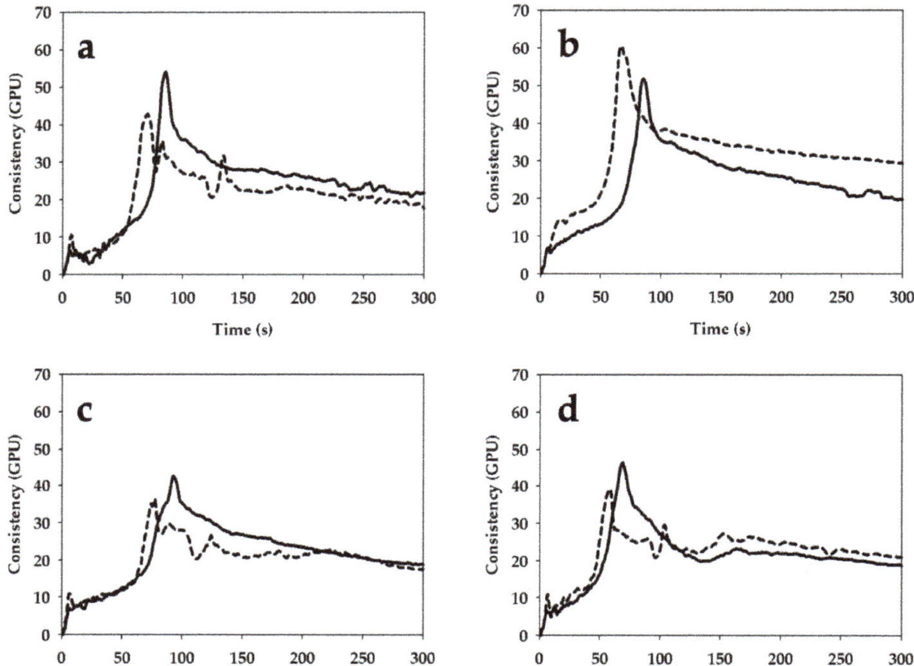

Figure 2. Effect of stone milling (solid line) and roller milling (dash line) on gluten aggregation properties, assessed by GlutoPeak®, of whole grain flours from Bolero *cv* (**a**), CWRS (**b**), Bona Vita *cv* (**c**) and Skorpion *cv* (**d**). CWRS: commercial Canada Western Red Spring Wheat; GPU: GlutoPeak Units.

In the case of maximum torque and aggregation energy, the effect of the milling system depended on the type of wheat. Specifically, SWFs showed a significantly higher maximum torque than RWFs (55.1 vs. 43.7 GPU and 46.2 vs. 37.3 GPU for Bolero and Bona Vita *cvs*, respectively), as well as significantly higher aggregation energy (1220 vs. 1001 GPE, 1099 vs. 864 GPE, and 1080 vs. 908 GPE, for Bolero, Bona Vita and Skorpion *cvs*, respectively) (Figure 2a,c,d). On the contrary, significantly lower values for both indices were found in the case of CWRS (52.2 vs. 59.7 GPU and 1216 vs. 1387 GPE for maximum torque and aggregation energy, respectively) (Figure 2b).

3.4. Mixing Properties

As regards the mixing properties (Figure 3), stone milling resulted in a significantly higher water absorption value for Bolero *cv* (61.7 vs. 64.2%, RWF vs. SWF) and Bona Vita *cv* (64.7 vs. 68.1% RWF vs. SWF). On the contrary, the milling system did not significantly affect the dough development time.

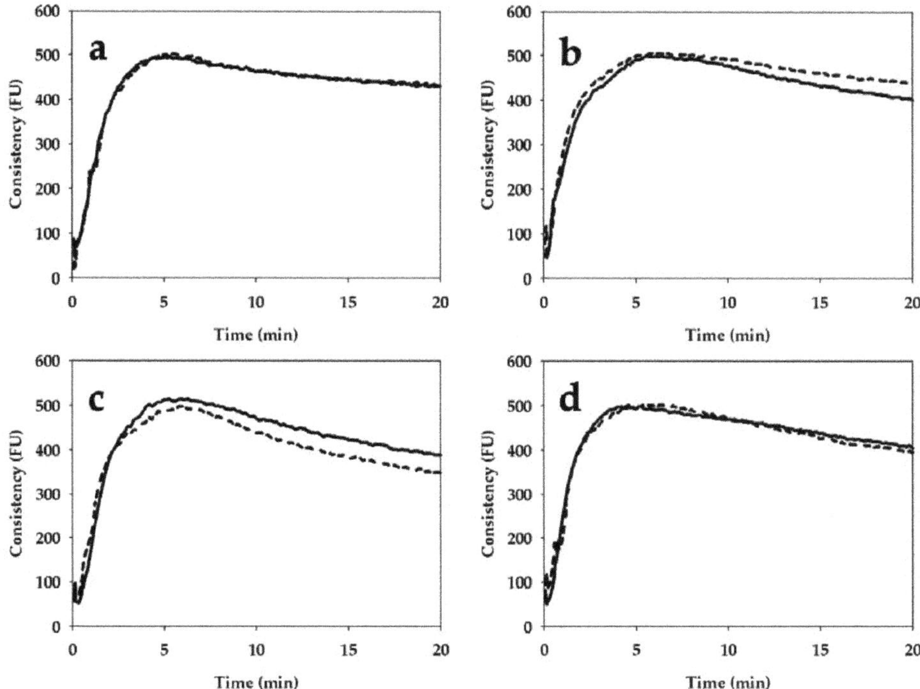

Figure 3. Effect of stone milling (solid line) and roller-milling (dash line) on the mixing properties, assessed by Farinograph®, of whole grain flours from Bolero *cv* (**a**), CWRS (**b**), Bona Vita *cv* (**c**) and Skorpion *cv* (**d**). CWRS: commercial Canada Western Red Spring Wheat; FU: Farinographic Units.

In addition, stone milling led to a significant increase in dough stability only in the case of Bolero *cv* (from 4.6 to 6.2 min for RWF and SWF). Finally, stone milling significantly influenced mixing resistance (evaluated as the degree of softening) only for CWRS and Bona Vita *cv*. Specifically, a higher degree of softening was observed in SWF (88 FU) compared to RWF (65 FU) in the case of CWRS, while this parameter decreased (from 138 to 119 FU, for RWF and SWF, respectively) in the case of Bona Vita *cv*.

3.5. Leavening Properties

The greatest impact of stone milling on dough leavening properties was observed for CWRS and Bolero cv, but each sample exhibited a different trend (Figure 4). Specifically, in the case of CWRS, stone milling caused a decrease in maximum dough development (from 40 to 33 mm), instead an opposite trend was found for Skorpion cv (from 37 to 42 mm). Moreover, stone milling resulted in increased dough height at the end of the test for CWRS (from 24 to 33 mm), as opposed to decreased height for the Bona Vita cv (from 34 to 20 mm). Moreover, the time required to reach maximum dough development was lower in stone milling for Bolero cv (2.36 and 1.38 h, for SWF and RWF respectively), as opposed to an increase in this index for CWRS (1.36 and 4.49 h, for SWF and RWF respectively), whereas no differences were observed for Bona Vita and Skorpion cvs (Figure 4).

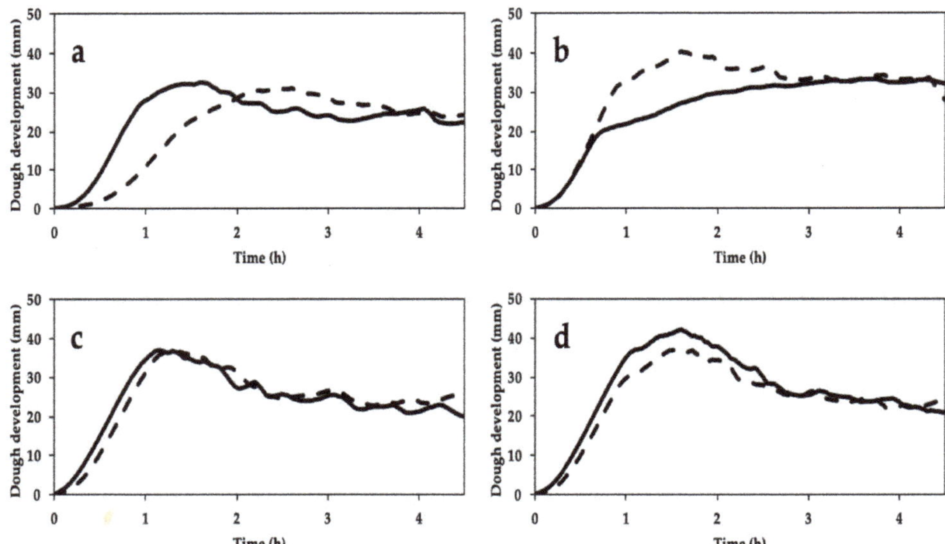

Figure 4. Effect of stone milling (solid line) and roller milling (dash line) on dough development, assessed by Rheofermentometer®, during leavening of whole grain flours from Bolero cv (**a**), CWRS (**b**), Bona Vita cv (**c**) and Skorpion cv (**d**). CWRS: commercial Canada Western Red Spring Wheat.

As regards the volume of CO_2 developed (Figure 5), stone milling resulted in an increase in this index for Bolero (from 1150 to 1398 mL) and Skorpion (from 1532 to 1659 mL) cvs. In addition, regardless of the type of wheat, stone milling led to an increase in the amount of CO_2 released (from 124 to 197 mL, from 195 to 211 mL, from 191 to 205 mL, and from 260 to 332 mL for Bolero, CWRS, Bona Vita and Skorpion cvs, respectively). Finally, no difference was observed in terms of the retention coefficient between the two milling approaches for the four samples.

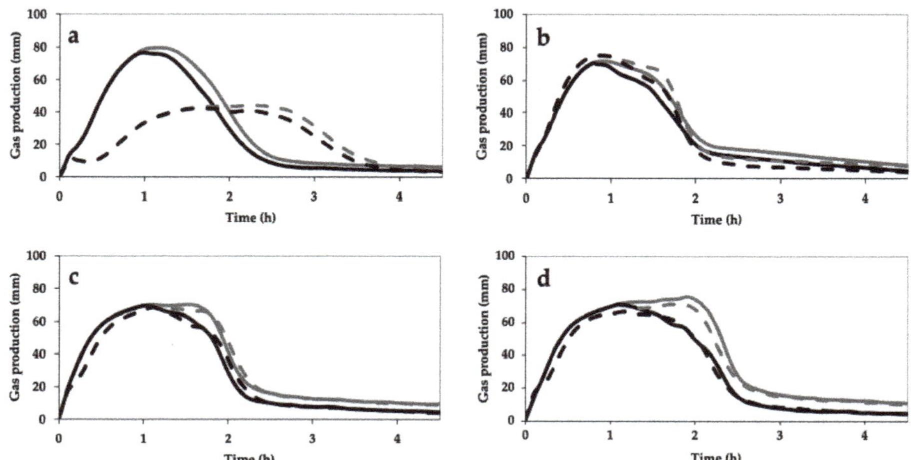

Figure 5. Effect of stone milling (solid line) and roller milling (dash line) on the total gas production (grey line) and on the retained gas (black line) in the dough, assessed by Rheofermentometer®, of whole grain flours from Bolero *cv* (**a**), CWRS (**b**), Bona Vita *cv* (**c**) and Skorpion *cv* (**d**).

3.6. Bread Properties

As for bread properties, stone milling resulted in higher bread height for Bolero *cv* and CWRS, instead no significant differences were observed for Bona Vita and Skorpion *cvs* (Figure 6). As regards volume and specific volume, bread from CWRS, Bona Vita and Skorpion *cvs* showed significant higher values when the stone milling system was used. However, bread from Bona Vita and Skorpion *cvs* exhibited a large bubble in the crumb layer under the upper surface of the bread, due to a collapse of the crumb structure.

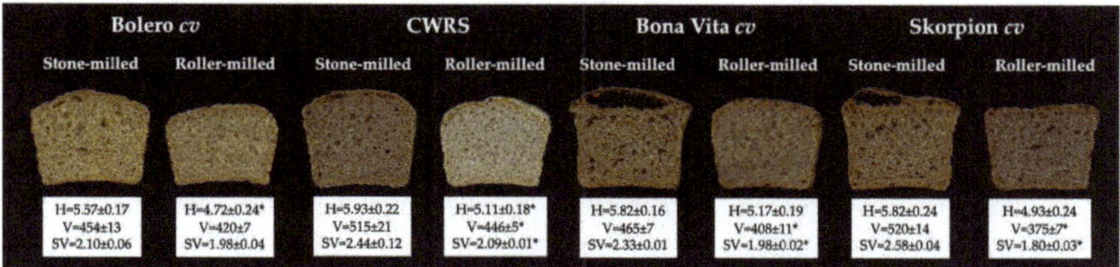

Figure 6. Effect of stone and roller milling on bread height (H; cm), volume (V; mL) specific volume (SV; mL/g). Data are presented as mean ± standard deviation. The asterisks indicate significant differences between the mean of the bread from stone and roller milled flours of each *cvs* (* $p < 0.01$). The absence of asterisk indicates a not significant difference. CWRS: commercial Canada Western Red Spring Wheat.

The analyses of bioactive compounds and AC were carried out on bread samples obtained from SWF, RWF and refined flours of pigmented *cvs*, chosen to provide more interesting phytochemical comparisons (Table 4). Compared to WFs, the removal of wheat bran led to a significant decrease in several antioxidant compounds both in refined flours and derived breads. On average, bread obtained from refined flours showed a lower content of CWBPA (−94%), SPA (−89%), AC_{ABTS} (−27%), AC_{FRAP} (−90%), zeaxanthin (−64%, Bona Vita *cv*) and TAC (−122 times, Skorpion *cv*). Conversely, lutein was higher (+29%) in bread from refined flour compared to WF samples. As highlighted in Table 4, bread samples obtained from the blue-grained *cv* showed levels of CWBPA (crumb and crust), SPA (crust), AC_{FRAP} (crumb) and TAC (crumb and crust) significantly higher for RWF as compared to SWF. RWF bread (crumb and crust) obtained from Bona Vita *cv* showed a significantly higher AC_{FRAP}, but a significantly lower content of lutein, than SWF. No difference in the relative abundance of CWBPAs was observed by comparing refined flour, SWF and RWF: ferulic acid was in both flour and bread the most representative (88%), followed by sinapic acid (5%) and *p*-cumaric acid (4%). Conversely, as far as the SPAs were concerned, a different behavior was observed in all compared flours: bread-making increased the relative abundance of soluble ferulic acid (on average from 27% to 46%), while reducing the percentage of soluble sinapic acid (from 41% to 32%).

Table 4. Cell wall-bound phenolic acids (CWBPAs), soluble phenolic acids (SPAs), antioxidant capacity (AC), xanthophylls (lutein and zeaxanthin) and total anthocyanin content (TAC) in raw material (flour), bread crust and bread crumb obtained from refined white flour, stone (SWF) and roller-milled (RWF) whole-grain flour of Bona Vita and Skorpion *cvs*.

cv	Flour	Product	CWBPAs	SPAs	Lutein	Zeaxantin	TAC	AC_{ABTS}	AC_{FRAP}
Bona Vita	refined	raw material	56.4 c	9.6 d	4.0 a	0.16 c	n.d.	13.5 b	0.2 g
		bread crumb	52.6 c	10.1 d	1.9 d	0.08 e	n.d.	12.8 bc	0.2 g
		bread crust	59.4 c	9.6 d	1.4 e	0.07 e	n.d.	11.3 c	1.8 f
	SWF	raw material	765.5 b	62.1 c	3.3 b	0.28 a	n.d.	18.2 a	6.4 e
		bread crumb	774.4 b	79.7 ab	1.5 e	0.14 cd	n.d.	17.5 a	7.5 d
		bread crust	770.2 b	76.6 b	1.0 f	0.12 d	n.d.	17.5 a	13.2 b
	RWF	raw material	885.7 a	74.5 b	3.0 c	0.24 b	n.d.	18.0 a	8.1 cd
		bread crumb	795.9 b	86.0 a	1.2 f	0.11 d	n.d.	17.2 a	8.8 c
		bread crust	808.6 ab	73.8 b	1.0 f	0.12 d	n.d.	17.6 a	15.9 a

Table 4. *Cont.*

cv	Flour	Product	CWBPAs	SPAs	Lutein	Zeaxantin	TAC	AC_{ABTS}	AC_{FRAP}
Skorpion	refined	raw material	50.3 d	8.4 e	n.d.	n.d.	0.1 e	14.8 d	0.2 e
		bread crumb	56.8 d	12.3 e	n.d.	n.d.	0.1 e	13.4 e	0.5 e
		bread crust	59.2 d	10.3 e	n.d.	n.d.	0.1 e	10.8 f	3.3 d
	SWF	raw material	897.4 b	84.7 cd	n.d.	n.d.	22.8 b	19.9 a	7.9 c
		bread crumb	781.4 c	95.5 abc	n.d.	n.d.	11.2 d	17.2 c	8.2 c
		bread crust	822.5 c	75.7 d	n.d.	n.d.	10.7 d	17.2 c	13.7 a
	RWF	raw material	990.5 a	91.7 bc	n.d.	n.d.	26.2 a	18.6 ab	7.3 c
		bread crumb	929.6 ab	107.2 a	n.d.	n.d.	13.3 c	18.1 bc	9.9 b
		bread crust	938.7 ab	98.9 ab	n.d.	n.d.	13.9 c	17.9 bc	13.6 a

Cell wall-bound phenolic acids (CWBPAs) and soluble (free and conjugated forms) phenolic acids (SPAs) are the sum of the single phenolic acids determined by means of RP-HPLC/DAD and are expressed as mg/kg dry basis. Xanthophylls (lutein and zeaxanthin) are expressed as mg/kg dry basis. Total anthocyanin content (TAC) is expressed as mg Cy-3-glc eq/kg dry basis. Antioxidant capacity (AC) measured by means of the ABTS and FRAP assays is expressed as mmol TE/kg dry basis. For each *cv*, value followed with different letters are significantly different (one-way ANOVA, $p < 0.01$), according to the REGW-F test.

4. Discussion

Despite the numerous websites and skilled marketing operations that declare the uniqueness and authenticity of SWF, the effects of stone milling—in comparison with those promoted by roller milling—on the chemical, rheological, and bread-making properties of the related flours have not yet been investigated systematically. The present study seeks to fill this gap. Moreover, since differences in kernel characteristics affect the milling process, four types of wheat were chosen for their variations in hardness, gluten strength, and richness in biocomponents. Specifically, the two wheat samples frequently used in the bread sector were Bolero (soft white winter wheat *cv*) and a commercial Canada Western Red Spring (CWRS). The other two wheat *cvs* were Bona Vita (medium-hard winter wheat with yellow endosperm for the high xanthophyll content) and Skorpion (medium-hard winter wheat with a blue external layer, rich in anthocyanins).

There is a widespread belief among consumers that, from a nutritional point of view, SWF are better than RWF, thus the label "made with stone-ground flour" is a powerful marketing tool for both producers and retailers [32]. In accordance with the AACC International definition of whole grain [33], neither milling process selected the anatomical regions, and endosperm, bran and germ have to be present in the same proportions as in the intact caryopsis. In agreement with other authors [17,20,22], no significant changes in the proximate composition were found regardless of the milling process used (Table 3). The few differences between SWFs and RWFs concerned moisture, which was significantly higher in RWFs from Bona Vita and Skorpion *cvs*. This result might firstly be related to the lack of conditioning before stone milling; moreover, a drop of moisture might be associated with heat development during this milling process, as mentioned by several authors [14,15,17].

The total dietary fiber content of WFs was included in the range 9.3%–14.3% (Table 3), similar to that observed for WFs examined within the European HEALTHGRAIN Project [10]. Regarding the potential effects of milling process on fiber fraction content, the SWF and RWF of each wheat sample exhibited no differences in total, insoluble and soluble fractions related to the milling process. Regarding the arabinoxylan fraction, the differences in the amounts of total and water-extractable arabinoxylans did not show a common trend regardless of the milling process used (Table 3). Nevertheless, in the two pigmented *cvs*, both of these parameters were considerably higher than in Bolero *cv* and CWRS, highlighting that Skorpion and Bona Vita *cvs* contain other interesting nutritional traits in addition to polyphenols or xanthophylls suitable for their exploitation. Although the occurrence of several macronutrients was unchanged in the compared WFs, RWFs of pigmented *cvs* resulted in a higher content of antioxidant compounds than SWF. Carcea et al. [20] reported no compositional difference regarding the total polyphenol and alkylresorcinol contents between stone milled or roller milled flours. In our study, in particular, a significantly higher content of CWBPAs and TAC were present in RWFs. Results could be related to the higher heat generated during stone milling due to friction [15].

Prabhasankar and Rao [17] observed that the higher temperature detected in SWF (85 °C), resulted in protein degradation, a reduction of amino acid content and a loss of some essential fatty acid compared to RWF (32 °C). Similarly, by comparing a stone milling process (60 °C) to a watermill process able to generate lower temperatures (30 °C), Di Silvestro et al. [34] observed a decrease in bound phenolic fraction, while no effect was detected for arabinoxylans and β-glucans.

In conventional roller milling, the importance of kernel tempering (or conditioning) to guarantee high yield and high quality of flour is widely recognized [11–13]. Indeed, particle size distribution after the first break and, consequently, the behavior of the "broken material" in all the remaining milling passages, is strictly influenced by kernel moisture [35]. On the contrary, no mention is made about the need to modify the native moisture of kernels before stone milling [11,16]. Such differences, and specifically, the increased moisture of the pericarp assured by tempering before roller milling could be the main reason for the differences in particle size distribution between SWFs and RWFs (Figure 1). Indeed, by lowering the native friability of the bran layers, moistening facilitates their separation from the starchy endosperm in large flakes during roller milling. At the same time, the increased endosperm moisture—in respect to the native kernel—induces the efficacious breakage of this region, yielding a high percentage of fine particles (Figure 1). This pattern is congruent with the results of Kihlberg et al. [22] and Ross and Kongraksawech [14], the latter investigating eight different small-scale mills. Although a common trend in all RWFs is recognizable, the moisture distribution inside the kernels could not be optimal irrespective of their hardness, as tempering conditions were the same for all wheat samples (moisture before milling equal to 16%). Bearing in mind the results obtained by Doblado-Maldonado et al. [36], CWRS could present, particularly in the external layers, lower moisture than required for optimal roller milling; this physical condition might account for the low percentage of large bran particles (Figure 1). Roller milling of medium-hard (i.e., Bona Vita and Skorpion *cvs*) and soft (Bolero *cv*) wheat, likely optimally moistened, yielded WFs with similar particle size distribution, characterized by a high percentage of both coarse and fine particles. On the contrary, when stone milling was applied, due to its different breakage system and lack of conditioning before milling, bran and endosperm regions exhibited a similar behavior during the breakage actions. Consequently, particles in SWFs were more homogeneously distributed in classes of different size, particularly in medium-sized classes (from about 300 to 700 μm) (Figure 1): the pattern was similar for all varieties, regardless of their hardness.

Evaluation of particle size distribution is important for understanding the rheological properties of dough: indeed, the particle size of bran and/or flour influences several features, including water absorption and gluten aggregation kinetics. Nevertheless, the literature has yet to indicate the 'optimal' particle size distribution for bread-making [37,38].

Although the milling system did not affect the protein content of the WFs (Table 3), some changes in protein properties—that are important for bread-making performance—were highlighted by the rheological tests. The analysis of gluten aggregation properties by means of a rapid shear-based method (i.e., GlutoPeak test) indicated that gluten proteins were able to aggregate and show a peak (Figure 2), which represents the maximum extent of gluten formation before its breaking due to the intense shear-stress [39]. Overall, RWFs exhibited faster gluten aggregation (lower peak maximum time), required less energy to aggregate and resulted in lower maximum consistency (except for CWRS) than SWFs, suggesting gluten weakening. A similar trend has been observed while comparing refined and whole flours due to the interference of fiber in network formation [40,41]. In the case of RWFs, the weakening of the gluten network could be due to depolymerization phenomena, favored by the presence of free-SH groups, particularly abundant in WFs with coarse particles (average particle size: 830 μm) [42]. Certainly, the presence of high amounts (more than 15% *w/w*) of large particles (>1000 μm size) in RWFs from Bolero, Bona Vita and Skorpion *cvs* might have negatively affected protein-protein interactions via physical mechanisms [42]. Similarly, the low percentage (only 7%) of the same size class in the RWF from CWRS might account for its opposite performance: both maximum torque and aggregation energy exhibited higher values than those determined in SWF.

Moving to the mixing properties evaluated by the farinographic test (Figure 3), the milling process did not seem to have a conclusive effect on such properties, that are greatly affected by the type of wheat. As emphasized by Ross and Kongraksawech [14], the farinographic indices were primarily influenced by *cv* and less by the milling process. In contrast to the gluten aggregation kinetics (evaluated by the GlutoPeak test on a slurry), the mixing properties were evaluated on a dough applying lower stress to the system (63 rpm vs. 2750 rpm). Thus, the apparent different findings could be attributed to the differences in the test conditions.

In general, the water absorption index was higher in SWFs (in Bolero and Bona Vita *cvs*), probably as the consequence of their higher (although not significant) amounts of damaged starch (Table 3), as the role of bran particle size, proposed as a valid explanation by Kihlberg et al. [22], was not highlighted. Stability was significantly affected only for Bolero *cv* (6.2 and 4.6 min for SWF and RWF, respectively). An important role might be played by the distribution of large/coarse bran particles (>1000 μm size) (Figure 1) which were three times higher in the RWF of Bolero *cv*. They were probably responsible for the weakness in its gluten network and, consequently, the significant decrease in dough stability. Moreover, the high percentage of large bran particles could impair not only dough properties during bread making but also the bioavailability of minerals, as indicated by Miller Jones et al. [10]. Nevertheless, the role of bran particle size on dough and bread characteristics needs to be further investigated as the results of works on this subject are still contradictory [38].

Regardless of the milling process, Skorpion and Bona Vita *cvs* showed similar leavening profiles (Figure 4), in agreement with their similar proximate composition (Table 3) and trends observed through the GlutoPeak test (Figure 2) and the farinographic test (Figure 3). Anyway, Skorpion WFs (both SWF and RWF) resulted in good dough development (Figure 4d) and gas production (Figure 5g,h), likely due to the high dough stability as shown by the farinographic test.

Differences in rheological properties associated with the milling systems were evident only for Bolero *cv* and CWRS. As considering the indications of the other rheological tests, Bolero *cv* and CWRS showed an opposite behavior according to the milling process; moreover, despite a similar protein content (Table 3), these samples were characterized by relevant differences in protein quality (Figure 2). Specifically, as for Bolero *cv*, SWF reached similar dough heights than RWF but faster, probably due to its higher—although not significant—damaged starch content as a quick source of simple sugars for yeast growth. In addition, RWF produced the least gas (Figure 5b) and the lowest bread volume (Figure 6). Both leavening properties and bread-making performance might be due to the high percentage of large particle size in RWF in Bolero *cv* (Figure 1). As expected, RWF of CWRS performed best during leavening in terms of dough development and time to reach it, indicative of the ability of this wheat type to withstand leavening stresses. Other reasons which might account for this result include good gluten aggregation (Figure 2) and mixing (Figure 3) properties, associated with the high damaged starch content and the low fiber percentage (Table 3) among the samples considered in this study.

Among the rheological tests used to predict the bread-making performance of samples, only the GlutoPeak tests (Figure 2) agreed for all samples as volume, specific volume and height of bread (Figure 6). Indeed, for all the wheat types, both the loaf height and the specific volume of bread samples produced with SWFs were higher than those obtained from RWFs, in agreement with the observations by Kihlberg et al. [22]. Our findings were also congruous with those of Gélinas et al. [18] which showed that dough mixing properties of WFs—in terms of farinographic absorption and stability—did not always relate to bread properties and, therefore, did not explain why some varieties performed better than others.

As the characteristics of bread are related not only to dough properties—generally evaluated by tests carried out at temperatures below 30 °C—but also to phenomena occurring during baking, we can hypothesize that, during baking, proteins in dough from SWFs might have retained extensibility for a longer time, assuring a higher bread volume. According to the literature [14,22,38,43,44], particle size distribution of WF might represent another trait able to influence bread volume. As previously discussed, stone milling produced a large amount of medium-coarse particles (from 300 to 700 μm) (Figure 1) that, according to Doblado-Maldonado et al. [15], could be considered the most advantageous for bread production. The particle size distribution observed in RWF (especially fine and large particles, simultaneously) (Figure 1) accounted for the low bread development. Indeed, small particles (<250 μm) could have a negative effect on bread characteristics as they promptly interfere in protein-protein interactions due to their high contact surface [43]. Also large particles might exert an undesirable action towards gluten development and gas cell stabilization [42] and bread appearance and texture [45].

Despite their high volume, in the case of Bona Vita and Skorpion *cvs*, stone milling resulted in a huge bubble just under the crust, together with the collapse of the underlying crumb (Figure 6). These behaviors cannot be attributed to differences in α-amylase activity (data not shown). The pattern of the corresponding rheofermentograph traces (Figure 4) allows us to hypothesize that, at the beginning of baking, the gas produced by yeasts in large amounts was not efficaciously held inside the gluten network and gathered in the upper part of loaf, causing the formation of a big bubble and a partial collapse of the underlying region.

The slight but significantly higher content of bioactive compounds in RWF flour compared to SWF (Table 3) was confirmed after bread-making. In particular, a higher CWBPA and TAC content found in both RWF bread crumb and crust for the blue-grained *cv*, resulting in a higher AC (FRAP assay, Table 4). As observed in previous studies [46,47], during bread-making significant changes occur in both bioactive compounds and AC. In the present study, bread-making caused a significant loss of the antioxidants responsible for the grain and flour pigmentation (xanthophylls and anthocyanins). Nevertheless, an increase in the AC was observed in the bread crust (Table 4). This could be due to the neo-formation of Maillard reaction products [46,48].

5. Conclusions

Most consumers believe that only the stone-milling process is able to preserve all the nutrients and bioactive compounds of wheat grains as, in this process, all kernel regions form a single stream. Indeed, the roller-milling process (a multiple-stream approach where the fractions must be recombined to obtain WF) is wrongly but commonly associated with a partial depletion of the native nutrients of the kernel. Our results proved that SWFs have neither a better proximate composition, nor a better bioactive compound concentration than RWFs. Only for blue-grained *cv* (Skorpion) RWFs resulted in a slight but significant higher content of CWBPAs and TAC compared to SWFs and this feature was observed also in bread.

The comparison of SWF and RWF properties highlighted a different particle size distribution. Indeed, during the grinding of caryopsis through the stone or the roller-milling, compression, shear, and cutting stresses exhibited different intensities and degrees (due to intrinsic kernel factors and process conditions), promoting the formation of large bran particles and very small flour particles in RWFs, while a more homogeneous particle size distribution was observed in SWFs. Although it can be assumed that these physical features could greatly affect the surface properties and the hydration properties of flour, only the GlutoPeak test, a quite recent rheological approach proposed for evaluating the protein-protein aggregation kinetics in wheat, highlighted significant differences in the gluten properties of WFs according to the milling process which were congruent with their bread-making performances.

The rheological differences between the WFs obtained from stone- or roller-milling, although significant, do not make one process clearly preferable to the other one. However, further information on the sensory profile of bread is worthy of interest. Nevertheless, the lower productivity of the former is acceptable for artisan or home-made processes, while the higher flexibility and versatility of multiple-stream roller-milling, and its fully automated management, can better satisfy industrial purposes. On the other hand, the effect of heat treatments (for stabilizing bran and germ) on the nutritional features of RWFs should be considered, as well as the effect of re-milling large bran particles on the technological performances.

Author Contributions: Conceptualization, M.A.P., A.M. and M.B.; methodology, G.C., M.C.C., D.E., D.G. and A.P.; validation, G.C., M.C.C. and D.G.; formal analysis, G.C., M.C.C., D.E. and D.G.; investigation, G.C., M.C.C. and D.G.; resources, M.A.P., A.M., M.C.C., A.P. and M.B.; data curation, G.C., M.C.C., D.E., D.G. and M.B.; writing—original draft preparation, M.A.P., G.C., M.C.C. and D.G.; writing—review and editing, A.M., M.B. and A.P., M.C.C., and D.E.; visualization, G.C. and D.G.; supervision, A.M. and M.B.; project administration, A.M., M.C.C. and M.B., funding acquisition, M.A.P., M.B, M.C.C., and D.E. All authors have read and agreed to the published version of the manuscript.

Funding: The authors acknowledge the financial support of the Italian Ministry of Education, University and Research (MIUR), program PRIN 2015 (Grant Number 2015SSEKFL) "Processing for healthy cereal foods".

Acknowledgments: The authors thank Simone Tomatis (Molino Tomatis S.N.C, Niella Tanaro, Italy), Francesca Vanara (Università degli Studi di Torino), Daniele Noé and Franca Criscuoli (Università degli Studi di Milano) for technical assistance.

Conflicts of Interest: The authors declare no conflict of interest

References

1. Rosa-Sibakov, N.; Poutanen, K.; Micard, V. How does wheat grain, bran and aleurone structure impact their nutritional and technological properties? *Trends Food Sci. Technol.* **2015**, *41*, 118–134. [CrossRef]
2. Gebruers, K.; Dornez, E.; Boros, D.; Dynkowska, W.; Bedo, Z.; Rakszegi, M.; Delcour, J.A.; Courtin, C.M. Variation in the content of dietary fiber and components thereof in wheats in the healthgrain diversity screen. *J. Agric. Food Chem.* **2008**, *56*, 9740–9749. [CrossRef] [PubMed]
3. Van Der Kamp, J.W.; Poutanen, K.; Seal, C.J.; Richardson, D.P. The healthgrain definition of "whole grain". *Food Nutr. Res.* **2014**, *58*, 1–8. [CrossRef] [PubMed]
4. Jonnalagadda, S.S.; Harnack, L.; Hai Liu, R.; McKeown, N.; Seal, C.; Liu, S.; Fahey, G.C. Putting the whole grain puzzle together: Health benefits associated with whole grains—Summary of american society for nutrition 2010 satellite symposium. *J. Nutr.* **2011**, *141*, 1011S–1022S. [CrossRef]
5. Miller Jones, J.; García, C.G.; Braun, H.J. Perspective: Whole and refined grains and health—Evidence supporting "Make Half Your Grains Whole". *Adv. Nutr.* **2020**, *11*, 492–506. [CrossRef] [PubMed]
6. Seal, C.J.; Brownlee, I.A. Whole-Grain foods and chronic disease: Evidence from epidemiological and intervention studies. *Proc. Nutr. Soc.* **2015**, *74*, 313–319. [CrossRef] [PubMed]
7. Willet, W. Food Planet Health. Available online: https://eatforum.org/content/uploads/2019/01/EAT-Lancet_Commission_Summary_Report.pdf (accessed on 18 May 2020).
8. Zhang, G.; Hamaker, B.R. The nutritional property of endosperm starch and its contribution to the health benefits of whole grain foods. *Crit. Rev. Food Sci. Nutr.* **2017**, *57*, 3807–3817. [CrossRef] [PubMed]
9. Nielsen Food Marker * Track. Prodotti Vincenti, Punti Vendita Esperienziali e Consumatori Smart: Ecco Il Largo Consumo del Futuro. Available online: https://www.nielsen.com/it/it/insights/article/2016/prodotti-vincenti-punti-vendita-esperienziali-e-consumatori-sma/ (accessed on 9 June 2020).
10. Miller Jones, J.; Adams, J.; Harriman, C.; Miller, C.; Van Der Kamp, J.W. Nutritional impacts of different whole grain milling techniques: A review of milling practices and existing data. *Cereal Foods World* **2015**, *60*, 130–139. [CrossRef]
11. Madureri, E. *Storia Della Macinazione dei Cereali*; Chiriotti Editori: Pinerolo (TO), Italy, 1995; ISBN 8885022545.
12. Pagani, M.A.; Marti, A.; Bottega, G. Wheat Milling and Flour Quality Evaluation. In *Bakery Products Science and Technology*; Zhou, W., Hui, Y.H., De Leyn, I., Pagani, M.A., Rosell, C.M., Selman, J.D., Therdthai, N., Eds.; John Wiley & Sons, Ltd.: Chichester, UK, 2014; pp. 17–53.
13. Posner, E.S.; Hibbs, A.N. *Wheat Flour Milling*; American Association of Cereal Chemists, Inc.: St. Paul, MN, USA, 2005.
14. Ross, A.S.; Kongraksawech, T. Characterizing whole-wheat flours produced using a commercial stone mill, laboratory mills, and household single-stream flour mills. *Cereal Chem.* **2018**, *95*, 239–252. [CrossRef]
15. Doblado-Maldonado, A.F.; Pike, O.A.; Sweley, J.C.; Rose, D.J. Key issues and challenges in whole wheat flour milling and storage. *J. Cereal Sci.* **2012**, *56*, 119–126. [CrossRef]
16. Gélinas, P.; Dessureault, K.; Beauchemin, R. Stones adjustment and the quality of stone-ground wheat flour. *Int. J. Food Sci. Technol.* **2004**, *39*, 459–463. [CrossRef]
17. Prabhasankar, P.; Haridas Rao, P. Effect of different milling methods on chemical composition of whole wheat flour. *Eur. Food Res. Technol.* **2001**, *213*, 465–469. [CrossRef]

18. Gélinas, P.; Morin, C.; Reid, J.F.; Lachance, P. Wheat cultivars grown under organic agriculture and the bread making performance of stone-ground whole wheat flour. *Int. J. Food Sci. Technol.* **2009**, *44*, 525–530. [CrossRef]
19. Cappelli, A.; Oliva, N.; Cini, E. Stone milling versus roller milling: A systematic review of the effects on wheat flour quality, dough rheology, and bread characteristics. *Trends Food Sci. Technol.* **2020**, *97*, 147–155. [CrossRef]
20. Carcea, M.; Turfani, V.; Narducci, V.; Melloni, S.; Galli, V.; Tullio, V. Stone milling versus roller milling in soft wheat: Influence on products composition. *Foods* **2019**, *9*, 3. [CrossRef]
21. Liu, C.; Liu, L.; Li, L.; Hao, C.; Zheng, X.; Bian, K.; Zhang, J.; Wang, X. Effects of different milling processes on whole wheat flour quality and performance in steamed bread making. *LWT Food Sci. Technol.* **2015**, *62*, 310–318. [CrossRef]
22. Kihlberg, I.; Johansson, L.; Kohler, A.; Risvik, E. Sensory qualities of whole wheat pan bread—Influence of farming system, milling and baking technique. *J. Cereal Sci.* **2004**, *39*, 67–84. [CrossRef]
23. Turnbull, K.-M.; Rahman, S. Endosperm texture in wheat. *J. Cereal Sci.* **2002**, *36*, 327–337. [CrossRef]
24. Giordano, D.; Locatelli, M.; Travaglia, F.; Bordiga, M.; Reyneri, A.; Coïsson, J.D.; Blandino, M. Bioactive compound and antioxidant activity distribution in roller-milled and pearled fractions of conventional and pigmented wheat varieties. *Food Chem.* **2017**, *233*, 483–491. [CrossRef]
25. AACC International. *AACC Approved Methods of Analysis*, 11th ed.; Cereals & Grains Association: St. Paul, MN, USA, 2011.
26. *Official Methods of Analysis of AOAC International*, 19th ed.; AOAC International: Gaithersburg, MD, USA, 2012.
27. Manini, F.; Brasca, M.; Plumed-Ferrer, C.; Morandi, S.; Erba, D.; Casiraghi, M.C. Study of the chemical changes and evolution of microbiota during sourdoughlike fermentation of wheat bran. *Cereal Chem.* **2014**, *91*, 342–349. [CrossRef]
28. Giordano, D.; Reyneri, A.; Locatelli, M.; Coïsson, J.D.; Blandino, M. Distribution of bioactive compounds in pearled fractions of tritordeum. *Food Chem.* **2019**, *301*, 125228. [CrossRef] [PubMed]
29. Serpen, A.; Gökmen, V.; Fogliano, V. Solvent effects on total antioxidant capacity of foods measured by direct quencher procedure. *J. Food Compos. Anal.* **2012**, *26*, 52–57. [CrossRef]
30. Siebenhandl, S.; Grausgruber, H.; Pellegrini, N.; Del Rio, D.; Fogliano, V.; Pernice, R.; Berghofer, E. Phytochemical profile of main antioxidants in different fractions of purple and blue wheat, and black barley. *J. Agric. Food Chem.* **2007**, *55*, 8541–8547. [CrossRef] [PubMed]
31. International Association for Cereal Science and Technology. *Standard No. 115/1, Method for Using the Brabender Farinograph*; International Association for Cereal Science and Technology: Vienna, Austria, 1992.
32. Guerrini, L.; Parenti, O.; Angeloni, G.; Zanoni, B. The bread making process of ancient wheat: A semi-structured interview to bakers. *J. Cereal Sci.* **2019**, *87*, 9–17. [CrossRef]
33. AACC International Defines Whole Grain. Available online: http://aaccnet.org/definitions/wholegrain.asp (accessed on 4 June 2020).
34. Di Silvestro, R.; Di Loreto, A.; Marotti, I.; Bosi, S.; Bregola, V.; Gianotti, A.; Quinn, R.; Dinelli, G. Effects of flour storage and heat generated during milling on starch, dietary fibre and polyphenols in stoneground flours from two durum-type wheats. *Int. J. Food Sci. Technol.* **2014**, *49*, 2230–2236. [CrossRef]
35. Fuh, K.F.; Coate, J.M.; Campbell, G.M. Effects of roll gap, kernel shape, and moisture on wheat breakage modeled using the double normalized kumaraswamy breakage function. *Cereal Chem.* **2014**, *91*, 8–17. [CrossRef]
36. Doblado-Maldonado, A.F.; Flores, R.A.; Rose, D.J. Low moisture milling of wheat for quality testing of wholegrain flour. *J. Cereal Sci.* **2013**, *58*, 420–423. [CrossRef]
37. Deroover, L.; Tie, Y.; Verspreet, J.; Courtin, C.M.; Verbeke, K. Modifying wheat bran to improve its health benefits. *Crit. Rev. Food Sci. Nutr.* **2020**, *60*, 1104–1122. [CrossRef]
38. Heiniö, R.L.; Noort, M.W.J.; Katina, K.; Alam, S.A.; Sozer, N.; de Kock, H.L.; Hersleth, M.; Poutanen, K. Sensory characteristics of wholegrain and bran-rich cereal foods—A review. *Trends Food Sci. Technol.* **2016**, *47*, 25–38. [CrossRef]
39. Marti, A.; Cecchini, C.; D'Egidio, M.; Dreisoerner, J.; Pagani, M. Characterization of durum wheat semolina by means of a rapid shear-based method. *Cereal Chem.* **2014**, *91*, 542–547. [CrossRef]
40. Cardone, G.; D'Incecco, P.; Pagani, M.A.; Marti, A. Sprouting improves the bread-making performance of whole wheat flour (*Triticum aestivum* L.). *J. Sci. Food Agric.* **2020**, *100*, 2453–2459. [CrossRef] [PubMed]

41. Malegori, C.; Grassi, S.; Ohm, J.-B.; Anderson, J.; Marti, A. GlutoPeak profile analysis for wheat classification: Skipping the refinement process. *J. Cereal Sci.* **2018**, *79*, 73–79. [CrossRef]
42. Bressiani, J.; Oro, T.; Santetti, G.S.; Almeida, J.L.; Bertolin, T.E.; Gómez, M.; Gutkoski, L.C. Properties of whole grain wheat flour and performance in bakery products as a function of particle size. *J. Cereal Sci.* **2017**, *75*, 269–277. [CrossRef]
43. Noort, M.W.J.; van Haaster, D.; Hemery, Y.; Schols, H.A.; Hamer, R.J. The effect of particle size of wheat bran fractions on bread quality—Evidence for fibre-protein interactions. *J. Cereal Sci.* **2010**, *52*, 59–64. [CrossRef]
44. Pasqualone, A.; Laddomada, B.; Centomani, I.; Paradiso, V.M.; Minervini, D.; Caponio, F.; Summo, C. Bread making aptitude of mixtures of re-milled semolina and selected durum wheat milling by-products. *LWT Food Sci. Technol.* **2017**, *78*, 151–159. [CrossRef]
45. Zhang, D.; Moore, W.R. Effect of wheat bran particle size on dough rheological properties. *J. Sci. Food Agric.* **1997**, *74*, 490–496. [CrossRef]
46. Yu, L.; Beta, T. Identification and antioxidant properties of phenolic compounds during production of bread from purple wheat grains. *Molecules* **2015**, *20*, 15525–15549. [CrossRef]
47. Çelik, E.E.; Gökmen, V. Effects of fermentation and heat treatments on bound-ferulic acid content and total antioxidant capacity of bread crust-like systems made of different whole grain flours. *J. Cereal Sci.* **2020**, *93*, 102978. [CrossRef]
48. Yu, L.; Nanguet, A.-L.; Beta, T. Comparison of antioxidant properties of refined and whole wheat flour and bread. *Antioxidants* **2013**, *2*, 370–383. [CrossRef]

Article

Bio-Functional and Structural Properties of Pasta Enriched with a Debranning Fraction from Purple Wheat

Parisa Abbasi Parizad [1], Mauro Marengo [1], Francesco Bonomi [1], Alessio Scarafoni [1], Cristina Cecchini [2], Maria Ambrogina Pagani [1], Alessandra Marti [1,*] and Stefania Iametti [1,*]

1 Department of Food, Environmental, and Nutritional Sciences (DeFENS), Università degli Studi di Milano, Via G. Celoria 2, 20133 Milan, Italy; parisa.abbasi@unimi.it (P.A.P.); mauro.marengo@unito.it (M.M.); francesco.bonomi@unimi.it (F.B.); alessio.scarafoni@unimi.it (A.S.); ambrogina.pagani@unimi.it (M.A.P.)

2 Consiglio per la Ricerca in Agricoltura e L'analisi Dell'economia Agraria (CREA), Centro di Ricerca Ingegneria e Trasformazioni Agroalimentari, Via Manziana 30, 00189 Roma, Italy; cristina.cecchini@crea.gov.it

* Correspondence: alessandra.marti@unimi.it (A.M.); stefania.iametti@unimi.it (S.I.); Tel.: +39-02-5031-6656 (A.M.); +39-02-5031-6819 (S.I.)

Received: 20 January 2020; Accepted: 6 February 2020; Published: 8 February 2020

Abstract: A colored and fiber-rich fraction from the debranning of purple wheat was incorporated at 25% into semolina- and flour-based pasta produced on a pilot-plant scale, with the aim of increasing anthocyanin and total phenolic content with respect to pasta obtained from whole pigmented grains. The debranning fraction impaired the formation of disulfide-stabilized protein networks in semolina-based systems. Recovery of phenolics was impaired by the pasta making process, and cooking decreased the phenolic content in both enriched samples. Cooking-related losses in anthocyanins and total phenolics were similar, but anthocyanins in the cooked semolina-based pasta were around 20% of what was expected from the formulation. HPLC (High Performance Liquid Chromatography) profiling of phenolics was carried out on extracts from either type of enriched pasta both before and after cooking and indicate possible preferential retention of specific compounds in each type of enriched pasta. Extracts from cooked samples of either enriched pasta were tested as inhibitors of enzymes involved in glucose metabolism and uptake, as well as for their capacity of suppressing the response to inflammatory stimuli. Results of both biological tests indicate that the phenolics in extracts from both cooked pasta samples had inhibitory capacities higher than extracts of the original debranning fraction at identical concentrations of total bioactives.

Keywords: pigmented wheat; anthocyanins; polyphenols; alpha-amylase inhibition; anti-inflammatory activity

1. Introduction

Many studies have shown that anthocyanins and other polyphenolic compounds have anti-inflammatory and antioxidant properties that may play a positive effect in preventing chronic diseases ranging from cardiovascular diseases to metabolic syndrome. The relationship between specific anthocyanins and their biological activities, such as anti-inflammatory, anti-obesity, anti-diabetes, has been the subject of a number of recent reviews [1,2]. Anthocyanins and polyphenols have also been reported to control intracellular signaling cascades as the process of inflammation progresses within the cells [3–5]. Numerous studies have demonstrated that anthocyanins can exert the beneficial effects in diabetes by acting on various molecular targets and regulate different signaling pathways in multiple organs and tissues such as liver, pancreas, kidney, adipose, skeletal muscle, and brain [6].

A critical aspect of most of the reported bioactivities for this class of chemical species relates to the poor absorption of several phenolics that impairs their presence at high enough concentration in biological fluids [7,8]. However, many of the health benefits associated with anthocyanins bioactivity relate to the effects these molecules reportedly exert on proteins in the intestine. Among the significant targets that do not require a transit of phenolics across the gastrointestinal epithelia are enzymes involved in glucose metabolism, such as pancreatic amylases involved in the starch enzymatic breakdown and the brush-border alpha-glucosidase relevant to glucose uptake [9–14].

One possible strategy to ensure adequate uptake of anthocyanins and phenolics is their incorporation into staple foods [8,15]. In this frame, pigmented grains have received particular attention as they may represent the starting ingredients to produce staple foods such as pasta or bread. High concentrations of phenolic compounds are present in the outer layers of a number of varieties of common grains such as wheat, corn, and rice [16]. In the case of colored grains, anthocyanins are responsible for their purple, blue, or red color [17]. These compounds are present in various amounts in bran layers as either the glycosylated or aglycone form [18].

The growing interest for new products and the visual appeal of naturally colorful products has led to the introduction of "colored" bread and pasta to the market. Most of the available products are prepared by using whole grains as the source of anthocyanins [16,19]. Recently, an innovative milling process has been developed in order to separate bran components to be used in specific processes [20,21]. In the case of colored grains, the usually discarded outermost part of the kernels is a suitable ingredient for the production of enriched food. According to Zanoletti et al., [20], it was possible to produce pasta with a high amount (up to 15% (*w/w*)) of bran fraction, with fiber and polyphenol content sensibly higher than in products prepared from the corresponding whole grains. The enriched pasta samples were characterized mainly in terms of chemical composition and cooking behavior, and only limited data are available for the retention of the bioactive properties of the incorporated materials.

This study attempts to fill evident gaps in the studies mentioned above by studying the effects of processing (and cooking) on retention of the bioactives' functionality in pasta samples prepared by incorporating at least 25% of anthocyanin-rich bran fractions from purple wheat debranning into pasta made from semolina or common wheat flour. This amount was chosen to provide nutritionally relevant amounts of both dietary fibers and phenolics in a product with attractive color features, avoiding in the meantime excessive changes in the pasta-making process and allowing the retention of the structural integrity of the cooked product. The focus in this study was on the in vitro inhibitory capacity of phenolics toward specific enzymes involved in the carbohydrate metabolism, as well as on their anti-inflammatory properties.

2. Materials and Methods

2.1. Chemicals and Enzymes

Unless otherwise specified, all chemicals and enzymes (namely, rat intestinal acetone powders of alpha-glucosidase (EC 3.2.1.20) and porcine pancreatic alpha-amylase (EC 3.2.1.1)) were from Sigma-Aldrich (Milan, Italy).

2.2. Raw Materials

An anthocyanin-rich bran fraction (henceforth, DF) was obtained by the debranning of commercial purple wheat, essentially as reported in [20]. The DF used in the studies reported below corresponds to an abrasion level of 3.7%, with respect to the whole grain and had a particle size in the 500–700 μm range. Refined common wheat flour (12.5% protein, dw; ashes <0.5%, dw) and durum wheat semolina (14.4% protein, dw; ashes 0.85%, dw) were provided by Molino Quaglia (Vighizzolo d'Este, Padua, Italy) and by F.lli De Cecco (Fara San Martino, Chieti, Italy), respectively.

2.3. Pasta

Pasta was produced from either commercial flour or semolina that was enriched by dry mixing with 25% (*w*/*w*) of the bran fraction obtained by the debranning of purple wheat (DF, see above). Macaroni-shaped pasta samples were produced using the DeFENS pilot plant, essentially as described in [22]. In particular, moisture in the dough was adjusted to a final level of 31.8%, and mixing and extrusion were completed in 20 min. Drying was carried out at 60 °C for 12 h. Each sample of pasta was cooked in tap water (1 L water per 100 g pasta) at the optimum cooking time, according to the AACC method 16–50 [23]. The sensory acceptability of the cooked pasta was assessed by ten untrained panelists. Prior to further characterization, samples of dried uncooked pasta were ground to <250 μm in a laboratory mill, whereas samples of cooked pasta were frozen in a deep freezer at −80 °C and lyophilized (Alpha 2-4 LD freeze dryer, Martin Christ, Osterhode am Harz, Germany) prior to further characterization.

2.4. Protein Solubility and Thiol Accessibility

The solubility of proteins in pasta samples was determined in triplicate using buffers of increasing dissociating ability, as described elsewhere [24,25]. Results are expressed as (mg soluble protein)/(g total proteins) to account for the protein content of individual samples, assessed through a dye-binding method [26]. Accessible thiol groups were determined (in triplicate) by using the spectrophotometric thiol reagent 5,5′-dithiobis-(2-nitrobenzoate) (DTNB) as also described in [24,25]. Results are expressed as μmol thiols/g total protein, to account for the protein content of individual samples.

2.5. Phenolics Extraction

Each sample (2 g) was defatted with petroleum ether and extracted twice with 15 mL of an ethanol/HCl mixture (15 mL of a mixture made up of 65 volumes of 95% ethanol and 35 volumes of aqueous 0.3 M HCl). The pooled ethanol/HCl extracts were used for analytical measurements and HPLC profiling. Extracts to be used with Caco-2 cells were by using water in place of dilute HCl, in order to avoid interference with the cellular assays and cell viability issues [13]. Assays were carried out in triplicate for each sample.

2.6. Determination of the Total Polyphenols and the Total Anthocyanins Content

Total polyphenols (as gallic acid equivalents) and total anthocyanins (as cyanidin-3-*O*-glucoside) were measured as in [27] on individual extracts from both cooked and uncooked pasta samples.

2.7. Anthocyanin and Phenolics Profiling by RP-HPLC

RP-HPLC profiling was performed on a C18 column (5 μm, 4.6 mm × 250 mm, Waters, Milan, IT) fitted to a Waters 600 E HPLC, equipped with a 996 PDA (Photo Diode Array) detector, by adapting procedures reported elsewhere [9]. Extracts (0.1–0.2 mL) were loaded on the column, and components eluted at a solvent flow of 0.8 mL min^{-1}, using a gradient from 100% A (0.1% trifluoroacetic acid in water) to 100% B (0.1% trifluoroacetic acid in acetonitrile): 0% to 5% B in 5 min; 5% to 40% B from 5 to 40 min; 40% to 70% B from 40 to 48 min. The eluate was monitored at 520 nm (for anthocyanins), 350 nm (for rutin and quercetin), 320 nm (for ferulic acid), and 280 nm (for catechin and epicatechin). Calibration was carried out by using suitable standards, and results are expressed as content of individual species in the original sample, on a dry matter basis.

2.8. Enzyme Inhibition Studies

Alpha-Glucosidase: Rat intestinal acetone powder was used in these assays, following established procedures [10–14] and testing inhibition by ethanol/HCl extracts from individual samples, diluted as appropriate in ethanol/HCl. Blanks were prepared in the absence of enzyme and of the ethanol/HCl mixture, whereas controls were complete reaction mixtures containing the appropriate volumes of the ethanol/HCl mixture used for extraction, but no bioactives. Acarbose (from a stock solution in ethanol/HCl) was used as the reference inhibitor. Tests were carried out in triplicate.

α-Amylase: Enzymatic activity was measured according to established procedures [10–14], and inhibition was tested by using ethanol/HCl extracts from individual samples, diluted as appropriate in ethanol/HCl. Blanks were prepared in the absence of enzymes, whereas controls were complete reaction mixtures, containing the appropriate volumes of the ethanol/HCl mixture used for extraction, but no bioactives. Acarbose (from a stock solution in ethanol/HCl) was used as the reference inhibitor. Tests were carried out in triplicate.

2.9. Immunomodulatory Properties of Extracts

Human intestinal epithelial Caco-2 cells were provided by Maria Rosa Lovati (Dipartimento di Scienze Farmacologiche e BioMolecolari, University of Milan, Milan, Italy). All experiments involving cultured Caco-2 cells have been carried out in the DeFENS Cell Culture Laboratory, a core facility of the Department of Food, Environmental and Nutritional Sciences (University of Milan, Milan, Italy). The experimental setup allows us to monitor the effects of extracts for inhibiting IL1β-stimulated synthesis of NF-κB in the transfected Caco-2 cells. In short, the test measures inhibition of luciferase expression in Caco-2 cells transiently transfected with the pNiFty2-Luc plasmid (InvivoGen, Rho, Italy), that combines five NF-κB binding sites with the luciferase reporter gene *luc*, allowing expression of the luciferase gene in the presence of NF-κB upon stimulation with interleukin 1β (IL1β) at a final concentration of 20 ng mL^{-1} in the presence/absence of extracts from individual samples. It should be noted that the extracts used in this study were prepared with aqueous ethanol in the absence of HCl, as noted above (Section 2.5). The experimental setup for these experiments has been reported in detail elsewhere [13]. The activity of the expressed luciferase was measured in cellular extracts by adding ATP and D-luciferin, followed by monitoring bioluminescence in a VICTOR3 1420 Multilabel Counter (PerkinElmer, Waltham, MA, USA) Each assay was carried out at least in triplicate.

2.10. Statistical Analysis

Analysis of variance (one-way ANOVA) was carried out by using Statgraphic Plus v. 5.1 (StatPoint Inc., Warrenton, VA, USA). The addition of DF to each pasta samples was considered as a factor for ANOVA. Results are reported as averages ± SD. The number of replicates for individual measurements is given under the Materials and Methods section for individual measurements, and in the legend to individual tables.

3. Results and Discussion

3.1. Molecular Organization of Proteins in Enriched Pasta

The possible effects of incorporating high levels of a bran-derived fraction into pasta on the overall structure of the protein network in the product were investigated by taking into account the conditional solubility of proteins and the accessibility of cysteine thiols to water-soluble reagents in the uncooked products [25,28]. The conditional solubility approach offers a simple way for estimating the nature of interprotein bonds, due to their different sensitivity to the action of chaotropes in the presence/absence of disulfide-breaking agents. Measuring the accessibility of residual thiols in the presence/absence of chaotropes provides information on the overall compactness of the protein network and complements protein solubility data.

Results are presented in Table 1, which also include values of these parameters for reference semolina-only pasta made in the same pilot plant under very similar processing conditions. The protein solubility data in Table 1 indicate the prevalence of urea-sensitive hydrophobic interactions in protein networks formed from wheat flour, as well as their lower compactness with respect to those formed by proteins in semolina. However, a comparison with the reference semolina pasta highlights the well-known "destructive" effect of the addition of a bran-derived fraction on the formation of a disulfide-stabilized protein network in semolina-based systems [29]. This is evident when considering that the DTT-dependent increase in soluble proteins is about 25% in the enriched pasta, whereas, in the reference semolina-based pasta, the breakdown of disulfide bonds by DTT results in a 4-fold increase of the amount of solubilized proteins.

Table 1. Properties of the protein network in uncooked pasta.

Measured Parameter	Solvent	Enriched Pasta, Wheat Flour	Enriched Pasta, Semolina	Reference Semolina Pasta
soluble proteins, mg/g protein	buffer + NaCl	28.8 ± 1.8 [a]	21.4 ± 1.6 [b]	25.0 ± 4.2 [a]
	+ 8 M urea	90.6 ± 7.2 [b]	70.1 ± 5.7 [a]	59.1 ± 9.0 [a]
	+ 10 mM DTT	129.0 ± 7.8 [b]	88.5 ± 4.7 [c]	295.3 ± 29.8 [a]
accessible thiols, µmol/g protein	buffer + NaCl	0.426 ± 0.060 [b]	0.213 ± 0.002 [c]	1.527 ± 0.242 [a]
	+ 8 M urea	1.320 ± 0.240 [b]	0.614 ± 0.119 [c]	6.319 ± 1.346 [a]

Addition of anthocyanin-rich bran fraction (DF) was considered as a factor for ANOVA. Different letters in a row indicate a significant difference at $p < 0.05$ (Tukey test; $n = 4$). DTT

A comparison of data on thiol accessibility is also informative, as the number of accessible thiols in either enriched pasta sample is by far lower than what expected from the properties of the starting material or—in the case of semolina-based enriched pasta—from the figures obtained for non-enriched samples. Freely accessible thiols (i.e., those accessible in the absence of chaotropes) in wheat flour or semolina proteins are in the 3–4 micromolar range when expressed on a protein basis [24,25]. The addition of chaotropes results in a 3-fold increase of accessible thiols for wheat flour-based pasta, and almost doubles the amount of accessible thiols in semolina.

A detailed molecular-level investigation of these observations is beyond the scope of this report. However, it seems reasonable to hypothesize that the observed decrease in accessible thiols—regardless of the presence/absence of chaotropes—may relate to some physical features of the enriched pasta that do not necessarily involve the properties of the protein network itself and may involve protein/polysaccharide interactions, as observed, for instance, in a rice-based pasta [30].

3.2. *Total Anthocyanins and Total Polyphenol Incorporation in Enriched Pasta*

The total anthocyanins content (TAC) and the total polyphenol content (TPC) in both cooked and uncooked pasta samples—prepared from either common wheat flour or semolina and 25% (*w*/*w*) of the polyphenol-rich fraction obtained from the debranning of purple wheat grains (DF)—are reported in Table 2. The values are compared with those expected from the content of each subclass of phenolics in the original bran fraction used to enrich both pasta samples (TAC, 690 mg Cya-3-O-glycoside eq/kg; TPC, 47,400 mg GAE/kg [13,20]). Separate analyses indicated that the contribution of phenolics from either the wheat flour or the semolina used in this study was negligible.

The data in Table 2 indicate an apparent decrease in both TAC and TPC in uncooked pasta samples with respect to what expected from the mixing formula. Losses in either family of phenolics may have occurred in the pasta-making process (most likely as a consequence of the drying steps in pasta production [7,31,32]), but the observed decrease may also be attributed to difficulties in recovering either species from the highly structured protein matrix formed upon kneading and drying.

The fact that apparent recovery figures in uncooked pasta are lower for semolina-based samples than for the flour-based ones offers circumstantial support for the latter hypothesis. In other words, the stiff protein network likely present in pasta could make the ethanol/HCl extraction procedure less effective, an effect most evident in semolina-based pasta because of the high protein content of semolina and the relevance of covalent disulfide bonds in the stabilization of interprotein networks formed by *T. durum* (*Triticum durum*) proteins in semolina-based dough in contrast with the prevalence of non-covalent hydrophobic interactions in the stabilization of interprotein networks formed by *T. aestivum* proteins in wheat flour dough [33,34].

Table 2. Efficiency and stability of phenolics incorporation.

Title	Expected *	Wheat Flour + DF		Semolina + DF	
		Uncooked	Cooked	Uncooked	Cooked
TAC content (mg Cyn-3-O-Glc/kg)	138	122 ± 9 [a]	51 ± 2 [c]	89 ± 12 [b]	31 ± 4 [d]
TAC retention, % of expected	100	88.4	36.9	64.5	22.4
TAC loss upon cooking, % of uncooked		100	58.2	100	65.2
TPC content (mg GAE/kg)	9480	3110 ± 188 [a]	1254 ± 98 [b]	2997 ± 302 [a]	1467 ± 111 [b]
TPC retention, % of expected	100	32.8	13.2	31.6	15.5
TPC loss upon cooking, % of uncooked		100	59.7	100	51.1

Addition of DF was considered as a factor for ANOVA. Different letters in a row indicate a significant difference at $p < 0.05$ (Tukey test; $n = 3$). * As calculated from formulation and previous or current analytical data [20].

As also evident from Table 2, cooking induced a further decrease of the phenolics content in both pasta samples. At this stage, we are unable to discriminate between this decrease being due to the release of the bioactives in the cooking water or to their sensitivity to temperature. In general, it seems safe to assume that there should be no issues for the yield of the ethanol/HCl extraction procedure when applied to cooked and lyophilized pasta samples, given the highly porous structure of the lyophilized samples.

It may be noted that the losses observed upon cooking flour-based enriched pasta are similar when considering anthocyanins alone or total phenolics. Conversely, cooking semolina-based enriched pasta gave loss in total phenolics slightly lower than those observed in common wheat-flour enriched pasta, as expected from the higher tenacity of the protein network in semolina-based pasta according to plentiful literature reports and to the protein network stability data reported in the previous section of this report. In spite of this, losses in anthocyanins upon cooking semolina-based enriched pasta were the highest, resulting in the final content of these species not exceeding 22% of what was calculated for the untreated formulation, based on proportion among individual ingredients and on their content in the various bioactives.

HPLC profiling was used to address whether any particular chemical species was involved in the compositional changes reported in Table 2 before and after cooking. To avoid possible sources of confusion [9], the HPLC profiling in this study takes into account only the aglycones of the most abundant species present in ethanol/HCl extracts of DF, namely, cyanidin and delphinidin (as representative of anthocyanins), and ferulic acid, quercetin and rutin (as representative of not-colored phenolics). In this general frame, it is worth mentioning that glycosylated and non-glycosylated anthocyanins are present in almost equivalent amounts in purple wheat [13], whereas most of the non-pigmented phenolics are present in their glycosylated forms. Also, the amount of ferulic acid derivatives in the debranning fractions from pigmented wheat varieties is reportedly much lower than in bran fractions from non-pigmented wheat varieties, where ferulic acid and its glycosylated forms may account for more than 80% of the total phenolics [35].

The results of HPLC profiling carried out on ethanol/HCl extracts from the various samples are reported in Table 3, which also provides—as a reference—the expected content in individual aglycones, calculated from quantitative HPLC profiling of ethanol/HCl extracts from DF. As already observed for the data presented in Table 2, the content of individual anthocyanin aglycones is lower than expected even in the uncooked samples, suggesting that the yield of the extraction procedure may be impaired by processing. This hypothesis is circumstantially supported by the apparent (although not statistically significant) increase in the content of each of the two anthocyanin aglycones in the cooked and lyophilized pasta samples.

A comparison among the data in Table 2; Table 3 also makes it evident that recovery figures for representative anthocyanin aglycones in uncooked pasta (30–50% of the expected, Table 3) are much lower than recovery figures for total anthocyanins (90–65%, Table 2), suggesting that the glycosylated forms of anthocyanins are less sensitive to matrix-related recovery issues. Extending the above comparison to the cooked samples, it appears evident that the abundant glycosylated anthocyanins are much more prone to being released upon cooking than their aglycones, and that anthocyanin aglycones are retained by the matrix in cooked pasta independently of whether the enriched pasta was based on semolina or wheat-flour (see Table 3), at difference with what observed with total anthocyanins (see Table 2).

Table 3. Content of representative aglycones in enriched pasta.

Content in Individual Species, µg aglycone/g Pasta		Expected *	Common Wheat Flour + DF		Semolina + DF	
			Uncooked	Cooked	Uncooked	Cooked
Anthocyanins	*Cyanidin*	9.05	3.11 ± 1.10 [a]	4.89 ± 1.21[a]	4.02 ± 0.81 [a]	4.92 ± 1.08 [a]
	Delphinidin	13.11	6.07 ± 0.91 [a]	6.02 ± 1.11 [a]	6.15 ± 0.81 [a]	8.78 ± 2.11 [a]
Phenolics	*Ferulic acid*	4.51	4.10 ± 0.61 [a]	1.84 ± 0.12 [c]	4.73 ± 0.39 [a]	2.38 ± 0.11 [b]
	Rutin	5.00	1.85 ± 0.41 [a]	1.65 ± 0.08 [a]	1.73 ± 0.28 [a]	1.45 ± 0.10 [a]
	Quercetin	4.52	0.48 ± 0.11 [a]	0.25 ± 0.06 [b]	0.43 ± 0.07 [a]	0.38 ± 0.02 [a]

Different letters in a row indicate a significant difference at $p < 0.05$ (Tukey test; $n = 4$). * As calculated from formulation and previous analytical data on DF [20].

Figures for the aglycone forms of representative non-pigmented phenolics indicate relevant differences in the efficiency of their incorporation in non-cooked pasta. Whereas ferulic acid was incorporated almost completely regardless of the use of flour or semolina in the formulation, rutin levels in pasta were roughly 30% of what expected, and quercetin fared even worse, showing a 10% incorporation, again regardless of the use of flour or semolina. This is in keeping with the data in Table 2, indicating a 30% retention of total phenolics.

Conversely, bioactive losses upon cooking were highest for the aglycone form of ferulic acid (30–50%) and statistically negligible for the rutin aglycone. In contrast, about 50% of the quercetin aglycone was lost upon cooking flour-based pasta, but no significant losses were observed in the case of semolina-based pasta. In this frame, it should be noted that losses in total phenolics upon cooking were ranging from 50 to 60 percent (see Table 2), again with the lowest losses being recorded for semolina-based pasta.

3.3. *Inhibition of Enzymes Relevant to Glucose Metabolism and Uptake*

Ethanol/HCl extracts from cooked pasta were tested for their capability of inhibiting pancreatic α-amylase and brush border alpha-glucosidase. Cooked pasta was used for these studies, in order to evaluate the possible effects of the bioactives in both types of enriched pasta after undergoing all the required steps prior to their consumption.

The data in the upper panel of Figure 1 indicate that extracts from either type of pasta had an inhibitory activity towards pancreatic alpha-amylase remarkably higher than the reference drug acarbose or extracts from the debranning fraction (DF) used in the formulation of either type of pasta (on a weight basis, as anthocyanin equivalents). Extracts recovered from semolina-based pasta were found to be less effective inhibitors than those from flour-based pasta, in particular at low anthocyanin concentrations.

A comparison with the profile of individual classes of phenolics in the extracts (see Table 3) does not offer valuable hints for a straightforward interpretation of these results. Further investigation is required to assess whether the difference between extracts from the two pasta samples could depend on the absence of synergistic effects among individual classes of phenolics (and individual chemical species, as reported in previous studies on alpha-amylase inhibition with various classes of phenolics of different origin [11,13,14], possibly as a consequence of losses or alterations of specific individual molecules during the processing and cooking steps.

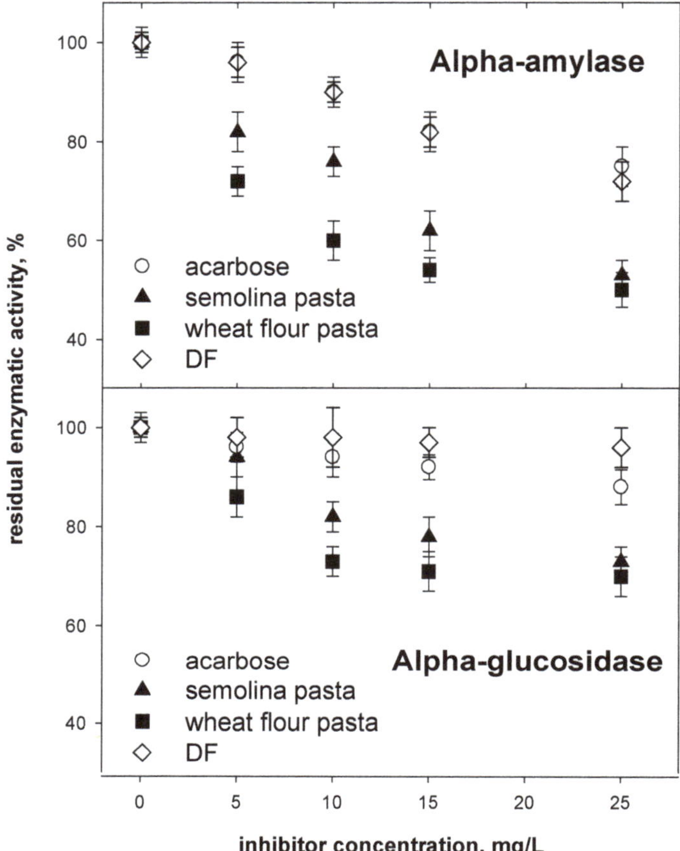

Figure 1. Inhibition of pancreatic alpha-amylase (upper panel) and brush-border alpha-glucosidase (lower panel) by HCl-ethanol extracts from cooked pasta, the original debranning fraction (DF), and by the reference drug, acarbose. Concentration of bioactives in extracts from either type of cooked pasta and from the debranning fraction (DF) is given as cyanidin equivalents. Results are reported as averages ± SD (n = 3).

The data on glucosidase inhibition by extracts from either type of cooked pasta are reported in the lower panel of Figure 1 and indicate that brush border alpha glucosidase is less sensitive to phenolics in the HCl-ethanol extracts (and to the reference drug, acarbose) than pancreatic alpha-amylase, confirming a number of previous reports [13,14]. Also, in the case of alpha-glucosidase, extracts from flour-based pasta appear more active than those from semolina-based pasta at identical concentrations of anthocyanins. As already pointed out for alpha-amylase inhibition, extracts from either type of pasta were much more efficacious in alpha-glucosidase inhibition than extracts from the debranning fraction (DF) used in the formulation of either type of pasta (on a weight basis, as anthocyanin equivalents).

As in the case of alpha-amylase, the differences in alpha-glycosidase inhibitory activity between extracts from the two pasta samples are most evident at low concentrations (<15 mg/L) of bioactives. Again, comparison with data reported in previous work [13] indicates that the inhibitory effects of anthocyanins extracted from cooked pasta samples towards alpha-glycosidase were higher (about 50% inhibition at 25 mg/L total anthocyanin, almost regardless of the type of pasta, see Figure 1) than those for extracts from the same debranning fraction used in this study, which gave approximately 30% inhibition at the same concentrations.

The results discussed above may be interpreted by taking into account the different phenolics profile in the extracts used in this study (see Table 3). These figures suggest that some of the phenolics originally present in DF are better retained in the cooked products than others or, conversely, that the retained ones are more powerful inhibitors than those lost in any of the steps leading to the final cooked pasta. As for the differences observed at low concentrations of anthocyanins in the inhibition assays, a working hypothesis could ascribe them to a different interplay among the individual species that are retained in each system and are responsible for specific inhibition of either enzyme. Indeed, a number of studies have pointed out synergistic inhibitory effects among various molecules in this general class of compounds [10,13,14].

3.4. *In Vitro Study of Anti-Inflammatory Activity of Cooked Pasta Extracts on Caco-2 Cells*

The anti-inflammatory activity of extracts from either type of enriched pasta was tested on the same model used in a number of previous studies [3–5]. In short, the model allows us to estimate the effects of extracts for inhibiting IL1β-stimulated synthesis of NF-κB in suitably transfected Caco-2 cells. Anthocyanidins are among the phenolics reportedly able to suppress the expression of inflammatory mediators such as cyclooxygenase (COX-2) by attenuating various forms of cellular signaling, including pathways involving NF-κB and MAPK [1,4–7,36].

From a methodological standpoint, it has to be underscored that the extracts from cooked pasta samples used in the experiments involving cells were prepared in aqueous ethanol to avoid cell viability issues related to residual traces of HCl [13]. This experimental detail prevents any immediate comparison with the ethanol/HCl extracts discussed above but allows straightforward comparison with the acid-free extracts from the same debranning fraction used in the formulation of the pasta samples used here and already tested for their anti-inflammatory activity in previous studies [13].

The data in Figure 2 highlight the ability of extracts from both types of cooked pasta to suppress NF-κB expression in the cellular model at concentrations much lower than those required for inhibiting enzymes relevant to glucose metabolism. As observed for enzyme inhibitory activities, the higher efficacy of extracts from flour-based pasta in repressing response to IL-1β with respect to the semolina-based one seems to level off at a high concentration of phenolics (as anthocyanin equivalents).

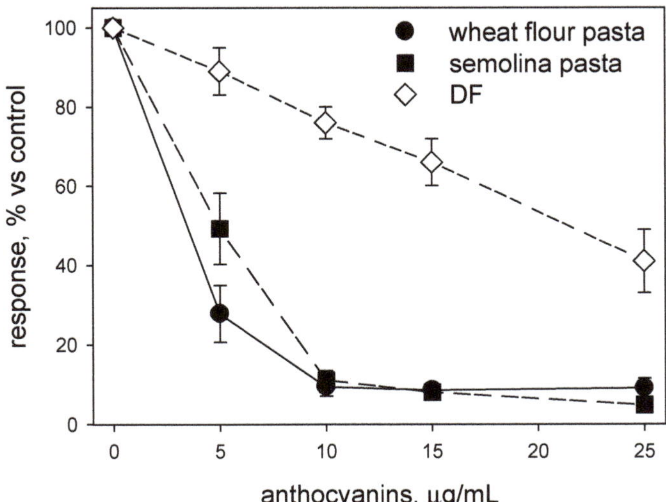

Figure 2. Immunosuppressive effects of the aqueous ethanol extracts from both types of cooked pasta and the original debranning fraction (DF) Data are presented as percent inhibition of IL1β-stimulated expression of NF-κB. Acid-free extracts were used in all these assays. Results are reported as averages ± SD (n = 3).

Figure 2 makes also evident that acid-free extracts from DF were sensibly less efficacious in suppressing NF-κB expression in the cellular model used here than that of similar extracts from either type of cooked pasta all over the concentration of anthocyanins tested here. As discussed above, these differences likely stem from the different phenolics profile in the various extracts (see Table 3). Again, as in the case of enzyme inhibition, some of the phenolics in DF could be better retained in the cooked products than others, and the retained ones may include species that are particularly efficient in suppressing inflammatory response either as individual compounds or through synergistic effects [10,13,14]. Work currently in progress will hopefully contribute to elucidating in sufficient detail the molecular determinants of the observed differences and offer some useful clues as to the suspected synergies among individual components in the extracts.

4. Conclusions

Pilot-plant scale incorporation of a phenolics- and fiber-rich fraction (from the debranning of purple wheat) into both semolina- and flour-based pasta was tested at 25% content of the debranning fraction, an amount suitable for avoiding excessive changes in the pasta-making process itself, as well as (1) preserving structural integrity of the product, (2) ensuring some attractive color characters in pasta, and (3) providing added value by incorporating nutritionally relevant amounts of both dietary fibers and phenolics.

Protein solubility and thiol accessibility approaches indicated that protein networks in flour-based enriched pasta were mostly stabilized by hydrophobic interactions and had lower compactness than those in the semolina-based enriched sample. However, a comparison with non-enriched semolina-based pasta indicated that the addition of the bran-derived fraction in the enriched pasta resulted in a sensible destabilization of disulfide-stabilized protein networks in semolina-based systems.

Either type of enriched pasta had a content in anthocyanins and total phenolics much higher than what reported in literature for products obtained from whole pigmented grains, but lower than what was expected from what was calculated from their formulation, at least on the basis of previous analytical data on the debranning fraction of purple wheat used in this study [13,20,21]. Losses in either anthocyanins or phenolics may have occurred as a consequence of thermal treatments in the pasta-making process (most likely as a consequence of the drying steps in pasta production) [7,8,10,15,31]. However, the decrease observed for both anthocyanins and phenolics before cooking could also stem by analytical recovery issues, as suggested by the observation that retention figures in uncooked pasta were lower in semolina-based samples than in flour-based ones.

Cooking induced a further decrease of the phenolics content in both pasta samples. In the case of flour-based enriched pasta, similar cooking losses were observed for anthocyanins alone and total phenolics. Conversely, cooking semolina-based enriched pasta gave a loss in total phenolics slightly lower than those observed in flour-based enriched pasta, but losses in anthocyanins upon cooking semolina-based enriched pasta were the highest, resulting in a final content of these species never exceeding 20% of what was expected on a formulation basis.

HPLC profiling of phenolics was carried out on extracts from either type of enriched pasta both before and after cooking. Although the characterization reported here may hardly be seen as complete, results are suggesting the possible "preferential retention" of specific compounds in each type of enriched pasta, both before and after cooking. The relevance of specific/preferential retention was evident from measurements of biological activities, which were carried out using suitable extracts from cooked samples of either type of enriched pasta.

Extracts from cooked pasta samples were tested for inhibitory capacity towards enzymes involved in glucose metabolism and uptake, and for the ability to suppress the cellular response to inflammatory stimuli. Results of both types of test (i.e., "in vitro" inhibition of enzymes and suppression of response to inflammatory stimuli in a widely used cellular model [1,3,13,36]) indicated that the phenolics retained in both samples of cooked, enriched pasta may be regarded as the "good" ones, at least in terms of biological activity. Indeed, extracts from either type of enriched pasta had inhibitory capacities higher than extracts of the original debranning fraction used for formulating these products at identical concentrations of bioactives.

A coarse estimate from data in Table 2 indicates that a 60 g serving (dry weight) of pasta enriched with suitable fractions from purple wheat debranning could provide between 1.8 and 2.4 mg of total anthocyanins in the cooked product. This should result in an estimated duodenal total concentration of anthocyanins in the range of 4 to 6 mg/L. In this concentration span, the activity of enzymes involved in glucose metabolism is decreased (decrease is in the 20% to 25% range, see Figure 1), as is the response of epithelial cells to inflammatory stimuli (decrease is in the 75% to 60% range, see Figure 2).

The findings reported here may be seen as exciting from the viewpoint of obtaining products with possible physiological impact and of general appeal to the consumer (with the bonus of added nutritional value as a consequence of their content in dietary fibers) by using established production processes. On the other hand, from the standpoint of the food chemist or biochemist, these results bring forward a number of daunting challenges. Indeed, providing a molecular-based rationale for the observations reported above will require substantial efforts, including a more thorough characterization of components in the various samples and, possibly, the use of specific mixtures of individual components to verify, in vitro at least, the observed effects and the occurrence of possible synergies among specific compounds.

Author Contributions: Conceptualization, M.A.P., A.M., S.I.; Methodology, P.A.P., M.M., A.S.; C.C.; Formal analysis A.M., S.I., F.B.; Writing—original draft preparation, P.A.P., M.M., S.I.; Writing—review and editing, F.B., M.A.P., A.M. All authors have read and agreed to the published version of the manuscript.

Funding: The authors acknowledge the financial support of the Italian Ministry of Education, University and Research (MIUR), program PRIN 2015 (Grant Number 2015SSEKFL) "Processing for healthy cereal foods". M.M. is the grateful recipient of a post-doctoral fellowship from the University of Milan.

Conflicts of Interest: The authors declare no conflict of interest.

References

1. Serra, D.; Almeida, L.M.; Dinis, T.C. Anti-inflammatory protection afforded by cyanidin-3-glucoside and resveratrol in human intestinal cells via Nrf2 and PPAR-γ: Comparison with 5-aminosalicylic acid. *Chem. Biol. Interact.* **2016**, *260*, 102–109. [CrossRef] [PubMed]
2. Iametti, S.; Abbasi, P.; Bonomi, F.; Marengo, M. Pigmented grains as a source of bioactives. In *Encyclopedia of Food Security and Sustainability*; Ferranti, P., Berry, E.M., Anderson, J.R., Eds.; Elsevier: Amsterdam, The Netherlands, 2019; Volume 1, pp. 261–270.
3. Taverniti, V.; Fracassetti, D.; Del Bò, C.; Lanti, C.; Minuzzo, M.; Klimis-Zacas, D.; Riso, P.; Guglielmetti, S. Immunomodulatory effect of a wild blueberry anthocyanin-rich extract in human Caco-2 intestinal cells. *J. Agric. Food Chem.* **2014**, *62*, 8346–8351. [CrossRef] [PubMed]
4. Sogo, T.; Terahara, N.; Hisanaga, A.; Kumamoto, T.; Yamashiro, T.; Wu, S.; Sakao, K.; Hou, D.X. Anti-inflammatory activity and molecular mechanism of delphinidin 3-sambubioside, a Hibiscus anthocyanin. *Biofactors* **2015**, *41*, 58–65. [CrossRef] [PubMed]
5. Vendrame, S.; Klimis-Zacas, D. Anti-inflammatory effect of anthocyanins via modulation of nuclear factor-κB and mitogen-activated protein kinase signaling cascades. *Nutr. Rev.* **2015**, *73*, 348–358. [CrossRef]
6. Gowd, V.Z.; Jia, Z.; Chen, W. Anthocyanins as promising molecules and dietary bioactive components against diabetes—A review of recent advances. *Trends Food Sci. Technol.* **2017**, *68*, 1–13. [CrossRef]
7. Karakaya, S.; Simsek, S.; Eker, A.T.; Pineda-Vadillo, C.; Dupont, D.; Perez, B.; Viadel, B.; Sanz-Buenhombre, M.; Rodriguez, A.G.; Kertész, Z. Stability and bioaccessibility of anthocyanins in bakery products enriched with anthocyanins. *Food Funct.* **2016**, *7*, 3488–3496. [CrossRef]
8. Angelino, D.; Cossu, M.; Marti, A.; Zanoletti, M.; Chiavaroli, L.; Brighenti, F.; Del Rio, D.; Martini, D. Bioaccessibility and bioavailability of phenolic compounds in bread: A review. *Food Funct.* **2017**, *8*, 2368–2393. [CrossRef]
9. Tadera, K.; Minami, Y.; Takamatsu, K.; Matsuoka, T. Inhibition of alpha-glucosidase and alpha-amylase by flavonoids. *J. Nutr. Sci. Vitamin.* **2006**, *52*, 149–153. [CrossRef]

10. Akkarachiyasit, S.; Charoenlertkul, P.; Yibchok-anun, S.; Adisakwattana, A. Inhibitory activities of cyanidin and its glycosides and synergistic effect with acarbose against intestinal alpha-glucosidase and pancreatic alpha-amylase. *Int. J. Mol. Sci.* **2010**, *11*, 3387–3396. [CrossRef]
11. Lavelli, V.; Harsha, P.S.C.S.; Ferranti, P.; Scarafoni, A.; Iametti, S. Grape skin phenolics as inhibitors of mammalian alpha-glucosidase and alpha-amylase—Effect of food matrix and processing on efficacy. *Food Funct.* **2016**, *7*, 1655–1663. [CrossRef]
12. Malunga, L.N.; Eck, P.; Beta, T. Inhibition of intestinal α-glucosidase and glucose absorption by feruloylated arabinoxylan mono- and oligosaccharides from corn bran and wheat aleurone. *J. Nutr. Metab.* **2016**. [CrossRef] [PubMed]
13. Abbasi, P.; Capraro, J.; Scarafoni, A.; Bonomi, F.; Blandino, M.; Marengo, M.; Giordano, D.; Carpen, A.; Iametti, S. The bio-functional properties of pigmented cereals may involve synergies among different bioactive species. *Plant Foods Hum. Nutr.* **2019**, *74*, 128–134.
14. Mattio, L.M.; Marengo, M.; Parravicini, C.; Eberini, I.; Dallavalle, S.; Bonomi, F.; Iametti, S.; Pinto, A. Inhibition of pancreatic alpha-amylase by resveratrol derivatives: Biological activity and molecular modelling evidence for cooperativity between viniferin enantiomers. *Molecules* **2019**, *24*, 3225. [CrossRef] [PubMed]
15. Ficco, D.B.M.; De Simone, V.; De Leonardis, A.M.; Giovanniello, V.; Del Nobile, M.A.; Padalino, L.; Lecce, L.; Borrelli, G.M.; De Vita, P. Use of purple durum wheat to produce naturally functional fresh and dry pasta. *Food Chem.* **2016**, *205*, 187–195. [CrossRef] [PubMed]
16. Abdel-Aal, E.S.M.; Young, J.C.; Rabalski, I. Anthocyanin composition in black, blue, pink, purple and red cereal grains. *J. Agric. Food Chem.* **2006**, *54*, 4696–4704. [CrossRef] [PubMed]
17. Liu, Q.; Qiu, Y.; Beta, T. Comparison of antioxidant activities of different colored wheat grains and analysis of phenolic compounds. *J. Agric. Food Chem.* **2010**, *58*, 9235–9241. [CrossRef] [PubMed]
18. Chen, C.; Somavat, P.; Singh, V.; de Mejia, E.G. Chemical characterization of proanthocyanidins in purple, blue, and red maize coproducts from different milling processes and their anti-inflammatory properties. *Ind. Crops Prod.* **2017**, *109*, 464–475. [CrossRef]
19. Pasqualone, A.; Gambacorta, G.; Summo, C.; Caponio, F.; Di Miceli, G.; Flagella, Z.; Marrese, P.P.; Piro, G.; Perrotta, C.; De Bellis, L. Functional, textural and sensory properties of dry pasta supplemented with lyophilized tomato matrix or with durum wheat bran extracts produced by supercritical carbon dioxide or ultrasound. *Food Chem.* **2016**, *213*, 545–553. [CrossRef]
20. Zanoletti, M.; Abbasi Parizad, P.; Lavelli, V.; Cecchini, C.; Menesatti, P.; Marti, A.; Pagani, M.A. Debranning of purple wheat: Recovery of anthocyanin-rich fractions and their use in pasta production. *LWT—Food Sci. Technol.* **2017**, *75*, 663–669. [CrossRef]
21. Delcour, J.A.; Rouau, X.; Courtin, C.M.; Poutanen, K.; Ranieri, R. Technologies for enhanced exploitation of the health-promoting potential of cereals. *Trends Food Sci. Technol.* **2012**, *25*, 78–86. [CrossRef]
22. Marti, A.; Seetharaman, K.; Pagani, M.A. Rheological approaches suitable for investigating starch and protein properties related to cooking quality of durum wheat pasta. *J. Food Qual.* **2013**, *36*, 133–138. [CrossRef]
23. *AACC (1995) Approved methods of the American Association of Cereal Chemists*; AACC International: St Paul, MN, USA, 1995.
24. Iametti, S.; Bonomi, F.; Pagani, M.A.; Zardi, M.; Cecchini, C.; D'Egidio, M.G. Properties of the protein and carbohydrate fractions in immature wheat kernels. *J. Agric. Food Chem.* **2006**, *54*, 10239–10244. [CrossRef]
25. Bonomi, F.; D'Egidio, M.G.; Iametti, S.; Marengo, M.; Marti, A.; Pagani, M.A.; Ragg, E.M. Structure-quality relationship in commercial pasta: A molecular glimpse. *Food Chem.* **2012**, *135*, 348–355. [CrossRef]
26. Bradford, M.M. A rapid and sensitive method for the quantitation of microgram quantities of protein utilizing the principle of protein-dye binding. *Anal. Biochem.* **1976**, *72*, 248–254. [CrossRef]
27. Hosseinian, F.S.; Li, W.; Beta, T. Measurement of anthocyanins and other phytochemicals in purple wheat. *Food Chem.* **2008**, *109*, 916–924. [CrossRef]
28. Iametti, S.; Marengo, M.; Miriani, M.; Pagani, M.A.; Marti, A.; Bonomi, F. Integrating the information from proteomic approaches: A "thiolomics" approach to assess the role of thiols in protein-based networks. *Food Res. Int.* **2013**, *54*, 980–987. [CrossRef]
29. Kaur, G.; Sharma, S.; Nagi, H.; Dar, B.N. Functional properties of pasta enriched with variable cereal brans. *J. Food Sci. Technol.* **2012**, *49*, 467–474. [CrossRef]
30. Barbiroli, A.; Bonomi, F.; Casiraghi, M.C.; Iametti, S.; Pagani, M.A.; Marti, A. Process conditions affect starch structure and its interactions with proteins in rice pasta. *Carbohydr. Polym.* **2013**, *92*, 1865–1872. [CrossRef]

31. Li, W.; Pickard, M.D.; Beta, T. Effect of thermal processing on antioxidant properties of purple wheat bran. *Food Chem.* **2007**, *104*, 1080–1086. [CrossRef]
32. Moreno, C.R.; Fernández, P.C.R.; Rodríguez, E.O.C.; Carrillo, J.M.; Rochín, S.M. Changes in nutritional properties and bioactive compounds in cereals during extrusion cooking. *IntechOpen* **2017**. [CrossRef]
33. Bonomi, F.; Iametti, S.; Mamone, G.; Ferranti, P. The performing protein: Beyond wheat proteomics? *Cereal Chem.* **2013**, *90*, 358–366. [CrossRef]
34. Jazaeri, S.; Bock, J.E.; Bagagli, M.P.; Iametti, S.; Bonomi, F.; Seetharaman, K. Structural modifications of gluten proteins in strong and weak wheat dough during mixing. *Cereal Chem.* **2015**, *92*, 105–113. [CrossRef]
35. Siebenhandl, S.; Grausgruber, H.; Pellegrini, N.; Del Rio, D.; Fogliano, V.; Pernice, R.; Berghofer, E. Phytochemical profile of main antioxidants in different fractions of purple and blue wheat, and black barley. *J. Agric. Food Chem.* **2007**, *55*, 8541–8547. [CrossRef]
36. Lee, S.G.; Kim, B.; Yang, Y.; Pham, T.X.; Park, Y.-K.; Manatou, J.E.; Koo, S.I.; Chuna, O.K.; Lee, J.Y. Berry anthocyanins suppress the expression and secretion of proinflammatory mediators in macrophages by inhibiting nuclear translocation of NF-κB independent of Nrf2-mediated mechanism. *J. Nutr. Biochem.* **2014**, *25*, 404–411. [CrossRef]

Article

Fresh Pasta Manufactured with Fermented Whole Wheat Semolina: Physicochemical, Sensorial, and Nutritional Properties

Simonetta Fois, Marco Campus, Piero Pasqualino Piu, Silvia Siliani, Manuela Sanna, Tonina Roggio and Pasquale Catzeddu *

Porto Conte Ricerche Srl, Località Tramariglio, 07041 Alghero (SS), Italy; fois@portocontericerche.it (S.F.); campus@portocontericerche.it (M.C.); piu@portocontericerche.it (P.P.P.); siliani@portocontericerche.it (S.S.); sanna@portocontericerche.it (M.S.); roggio@portocontericerche.it (T.R.)

* Correspondence: catzeddu@portocontericerche.it; Tel.: +39-(0)79-998448

Received: 30 August 2019; Accepted: 16 September 2019; Published: 18 September 2019

Abstract: Fresh pasta (SP) was prepared by mixing semolina with liquid sourdough, whole wheat semolina based, and the effects of sourdough inclusion were evaluated against a control sample (CP) prepared using semolina and whole wheat semolina. Physicochemical, nutritional, and sensorial analyses were performed on pasteurized fresh pasta, before and after cooking. The optimum cooking time was not affected by whole wheat sourdough, whereas differences were found in color, firmness, and cooking loss. Changes of in vitro digested starch fractions in SP pasta were affected by a higher cooking loss. Overall, SP samples were characterized by improved nutraceutical features, namely higher content of free essential amino acids and phenolic compounds, lower phytic acid content, and higher antioxidant activity. Sensory analyses (acceptability and check-all-that-apply (CATA) tests) showed significantly higher scores for the SP, and the differences were enhanced when the consumers were informed about the product composition and how it was manufactured. Consumers checked for more positive sensory parameters for the SP than the CP.

Keywords: sourdough; fiber; amino acids; phenolic compounds; phytic acid

---

1. Introduction

Pasta is a staple food in the Mediterranean area, and it is produced and consumed worldwide. It is a good source of carbohydrates and proteins, with interesting nutritional properties, i.e., a low glycemic index. In recent years there has been a trend towards the production of whole wheat pasta, which represents a good source of fiber. The intake of dietary fiber exerts beneficial effects on human health. In fact, bran, a by-product of wheat milling, obtained from the outer layers of wheat kernel, contains fibers, vitamins, and minerals. Many advantages have been associated with the consumption of bran fiber, in terms of risk reduction of hypertension, breast cancer, and type 2 diabetes, and in terms of prevention of colon disease and gastric cancer [1,2]. Bran fiber is resistant to digestion and absorption at the small intestine level, and thus reaches the colon where bacterial fermentation occurs and produces short chain fatty acids [3]. On the other hand, wheat bran contains phytic acid, recognized as an anti-nutritional compound [4], which reduces the nutritional value by chelating ions (such as Ca^{2+}, Fe^{2+}, Mg^{2+}, and Zn^{2+}). Mineral deficiency could lead to decreased function of the immune system and reduced body growth and development [5]. With regards to bread making technology, the use of bran has some drawbacks which negatively affects volume, texture, and sensory acceptance [6]. Pasta prepared with the addition of bran has an inferior technological quality as compared with pasta prepared with semolina [7] or wheat flour [8], because bran interferes with gluten development, especially when bran presents inappropriately sized particles [9].

The general recognition of the positive nutritional effects of fiber consumption increase the demand for technological solutions to overcome the negative effects of bran supplementation. In bread making technology, fermentation of bran by microbial strains has been suggested as a method to reduce the negative effects of phytic acid [10] and to improve the volume and sensory properties of bread containing bran [11].

Few scientific papers have reported on the use of fermentation technology in pasta making. A gluten-free pasta was produced by [12] using buckwheat flour and 24 h fermented semolina, a vitamin B2-enriched pasta was produced by [13] using a 16 h prefermented semolina, and fresh pasta was produced by [14] using semolina and semolina-based liquid sourdough. In this study, fermented whole wheat semolina was used as a functional ingredient in pasta making, in order to ameliorate the detrimental effects of bran fraction over the structure and sensory features, while retaining the advantageous effects of bran on human health. Physical, chemical, sensorial, and nutritional characteristics were evaluated.

2. Materials and Methods

2.1. Raw Materials

Commercial whole semolina (Integrale, Selezione Casillo S.r.l., Corato, Bari, Italy), and commercial semolina (Extra Arancio, Selezione Casillo S.r.l., Corato, Bari, Italy) were used in this study. The whole semolina had the following percent composition, as is or on the basis of dry matter (DM): moisture 14.1%, ash 1.6% DM, protein 12.5% DM, fiber 7.8% DM, dry gluten 8.5% DM, gluten index 60, alveographic W 199 ($J \times 10^{-4}$) and P to L ratio 5.12. The composition of semolina was the following: moisture 14.0%, ash 0.75% DM, protein 13% DM, fiber 2.7% DM, dry gluten 11% DM, gluten index 88, alveographic W 176 ($J \times 10^{-4}$) and P to L ratio 1.31.

2.2. Preparation and Maintenance of Liquid Whole Wheat Semolina-Based Sourdough (LWS)

A liquid whole wheat semolina-based sourdough (LWS) was prepared starting from the semolina-based liquid sourdough used by Fois et al. [14], which was refreshed for several days by back-slopping using whole wheat semolina and water, at a ratio of 1:1.5:1.5, in order to obtain a dough yield of 200. Back-slopping was done in the bioreactor GL MINI 25 (Esmach S.p.A., Grisignano di Zocco, Italy). The fermentation process was conducted at 26 °C for 5 h, and then LWS was kept at 5 °C until the subsequent daily back-slopping. The product was monitored daily in order to achieve stable values of pH and total titratable acidity (TTA).

The ripe sourdough had a pH value of 4.3, a TTA of 12.4 mL NaOH (0.1 mol/L) in 10 g, and a viable cell number of approximately 10^7 cfu/g for yeast and 10^5 cfu/g for lactobacilli. Yeast cells were enumerated on Rose Bengal Chloramphenicol agar, and lactobacilli on de Man, Rogosa and Sharpe (MRS) modified agar medium [14]. Media and supplements were purchased from Oxoid (Basingstoke, England).

2.3. Physicochemical Analyses of Pasta and LWS

A thermogravimetric analyzer Thermostep (Eltra GmbH, Haan, Germany) was used for moisture and ashes content determination, at 105 °C and at 600 °C respectively. Both TTA and pH were determined with an automatic titrator (Crison, Hach Lange, Barcelona, Spain), after homogenization of 10 g of sample in 90 mL of distilled water. After 30 min of gentle stirring for sourdough, and 60 min for pasta, the pH was determined, and the samples were titrated to a pH of 8.5 with NaOH 0.1 mol/L. The TTA was reported as mL of NaOH per 10 g of sample. The total dietary fiber (TDF) content of pasta was measured using the Total Dietary Fiber assay kit (Megazyme, Wicklow, Ireland). The color was determined on raw spaghetti placed side by side without any gap, using a CM-700d spectrophotometer (Konica Minolta, Osaka, Japan), using Standard Illuminant D65/10°. Prior to measurements, the instrument was calibrated against the white tile. CIE L*a*b* color space coordinates, lightness (L*), color in the red/green field (a*) and color in the blue/yellow field (b*), were computed. The Euclidean distance between colors, calculated as ΔE_{76}, was used to estimate the range of perceived difference between samples of close chroma [15]:

$$\Delta E_{76} = \sqrt{(L_2^* - L_1^*)^2 + (a_2^* - a_1^*)^2 + (b_2^* + b_1^*)^2} \quad (1)$$

0 < Δ76 < 1 the difference is unnoticeable; 1 < Δ76 < 2 the difference is only noticed by an experienced observer; 2 < Δ76 < 3.5 the difference is also noticed by an unexperienced observer; 3.5 < Δ76 < 5 the difference is clearly noticeable; 5 < Δ76 gives the impression that these are two different colors

Protein content ($N \times 5.27$) was determined in 200 mg of pasta by the crude protein AACC (American Association of Cereal Chemists) combustion method 46–30 [16] using a Rapid N Cube analyzer (Elementar Analysensysteme GmbH, Langenselbold, Germany).

2.4. Pasta Making

The pasta was manufactured using the pasta maker "La Monferrina Dolly" (Moncalieri, Italy) equipped with a bronze die. The control pasta (CP) was prepared by mixing semolina (700 g) and whole wheat semolina (280 g) with 318 mL of water. Pasta with LWS, hereafter called sourdough pasta (SP), was prepared by mixing 38 mL of water, 700 g of semolina, and 560 g of LWS (consisting of 280 g of whole meal semolina and 280 mL of water). The dough was mixed for 20 min before extrusion into "spaghetti" of 3 mm diameter. After production, the CP and the SP were immediately pasteurized (at 97 °C for 3 min) and packaged under modified atmosphere (CO_2:N_2 = 30:70) [14]. For the analysis of total starch, protein digestibility, free amino acids, total phenolic compounds, antioxidant activity, and phytic acid, the samples were homogenized using a cryogenic mill (SpexSamplePrep, Stanmore, UK) and stored at −80 °C. All the analyses were replicated three times (n = 3).

2.5. Cooking Quality and Texture Analysis

The optimum cooking time (OCT), cooking loss (grams of solids in cooking water per 100 g of pasta as is), and swelling index (SI), i.e., grams of absorbed water per gram of pasta DM were measured for the CP and the SP according to the AACC Approved method 66–50 [16].

The texture of pasta, cooked to OCT, was evaluated according to the AACC Approved method 66–50 [16], using a TA.XTPlus Texture Analyzer (Stable Microsystems, Godalming, UK). The spaghetti strands were rinsed in cool water (4 °C) for 30 s to avoid overcooking. Tests were performed on 5 spaghetti strands, cut crosswise by the plexiglass blade probe A/LKB-F, at a test speed of 0.17 mm/s and a distance of 4.5 mm. The maximum force (N) of the curve, referred to as "firmness", was computed. The software Texture Expert Exceed (v1.21) (Stable Microsystems, Godalming, UK) was used for texture data processing.

2.6. In Vitro Starch Digestibility

In vitro digestion of starch was performed on the CP and the SP, in order to quantify rapidly digestible starch (RDS), slowly digestible starch (SDS), and inaccessible digestible starch (IDS). RDS is the glucose released after 20 min of in vitro digestion. SDS and IDS are defined as the glucose released in the time frame between 20 and 120 min and between 120 and 180 min, respectively. IDS is defined as "inaccessible digestible starch" since it is not actually digestion-resistant starch, but just physically inaccessible to the digestive enzymes, and it was made accessible by homogenization of the sample after 120 min of in vitro digestion (Sanna et al., 2019).

Samples were cut 2 cm long and cooked to OCT, and then processed as in Sanna et al., [17].

2.7. Protein Digestibility and Free Amino Acid Analysis

The raw and cooked CP and SP samples were analyzed for protein digestibility [18,19] and free amino acid content. Free amino acid extract was prepared mixing 10 mL of 0.1 M hydrochloric acid solution and 1.5 g of homogenized sample, then, the mixture was vortexed for 10 s and centrifuged at 18,000× *g* at 4 °C for 15 min. Finally, the supernatant was filtered through a 0.22 µm PTFE syringe filter (Phenomenex, Macclesfield Cheshire, UK). The amino acid analysis was performed using an Agilent 1200 series HPLC system (Santa Clara, California, CA, USA), equipped with a binary pump with integrated vacuum degasser, an autosampler, a thermostated column compartment, and a diode array detector (DAD). The system was controlled by the Agilent Chemstation chromatography manager. The pre-column derivatization, gradient eluent method, and injection program were performed according to Agilent Application Note [20], using an Agilent Poroshell HPH-C18 column (4.6 × 100 mm, 2.7 µm pore size; Agilent Technologies, Santa Clara, California, USA) with HPH-C18 fast guard (Agilent Technologies, Santa Clara, California, USA), at a temperature of 40 °C. The pre-column derivatization was performed using *o*-phthalaldehyde reagent (OPA), 9-fluorenylmethyl chloroformate reagent (FMOC), and borate buffer supplied by Agilent Technologies. Standard solutions were purchased from Agilent Technologies, whereas GABA (γ-aminobutyric acid) was purchased from Sigma-Aldrich S.r.l. (Milano, Italy). Results were expressed as mg/100 g DM.

2.8. Total Phenolic Content and Antioxidant Activity

The total phenolic content (TPC) and 2,2-diphenyl-1-picrylhydrazyl (DPPH) radical scavenging activity were determined on raw and cooked CP and SP, by suspending 1 g of sample in 10 mL of an 80% aqueous methanol solution (20:80, *v*/*v*). The mixture was shaken for 2 h at 750 rpm in a thermomixer (Thermomixer Comfort, Eppendorf), then centrifuged at 800× *g* for 10 min. Supernatant was filtered through a 0.22 µm PTFE (Polytetrafluoroethylene) syringe filter (Phenomenex, Macclesfield Cheshire, UK) and stored at −20 °C until analyses.

The TPC of sample extracts was determined using the Folin–Ciocalteau method [21] with the following modifications: 0.2 mL of the sample extract (or 80% methanol for the blank) was mixed with 1.5 mL of Folin–Ciocalteau reagent, previously diluted with water (1:10 *v*/*v*), and 1.5 mL of saturated sodium carbonate solution (7.5% *w*/*v*). The mixture was allowed to stand in the dark at room temperature for 1 h, then the absorbance was read at 735 nm, against the blank. The gallic acid calibration curve was built in the range 25–600 mg L^{-1} (y = 0.0051x + 0.071 R^2 = 0.999) and results were expressed in terms of Gallic acid equivalents (GAE mg mL^{-1}).

The antioxidant activity of pasta was determined through the evaluation of free radical scavenging effect on the DPPH radical, according to [22] with some modifications: 1.4 mL of DPPH solution (0.10 mM in an 80% aqueous methanol) was mixed with 0.1 mL of sample extract, in a 1.5 mL centrifuge tube, vortexed, and then allowed to stand for 30 min in the dark. The discoloration of DPPH against the 80% aqueous methanol was monitored after 30 min, measuring the absorbance at 517 nm. The antioxidant activity of the sample was expressed as the percentage discoloration of DPPH solution, by the following equation:

$$\%\ \text{discoloration} = [(\text{AbsDPPH} - \text{AbsSample})/\text{AbsDPPH}] \times 100 \quad (2)$$

where, Abs_{DPPH} is the absorbance of the DPPH solution without extract, and Abs_{Sample} is the absorbance of the sample solution after 30 min of reaction.

All reagents were analytical grade and purchased from Sigma-Aldrich S.r.l. (Milano, Italy).

2.9. Phytic Acid Determination

Phytic acid was determined on the raw and cooked CP and SP using the Phytic Acid (Phytate)/Total Phosphorus assay kit (Megazyme, Wicklow, Ireland).

2.10. Consumer Testing

An acceptability test and a check-all-that-apply (CATA) method [23] were performed on the CP and SP samples. The test was carried out by 54 consumers, 26 women and 28 men, most of them recruited on the basis of interest and willingness. They were regular pasta consumers, aged between 32 and 60 years, not trained on sensory analysis of pasta products. Approximately, 40 g of packed sample were supplied to each consumer and the test was performed at their own home, under real conditions of use and consumption [24], so that consumers could season the pasta with the preferred sauce. The optimal cooking time, and the avoidance of spice and chili were suggested.

The evaluation of the SP and CP samples was carried out in two separate conditions, at two different sessions. The first condition was a blind test and the second condition was an informed test, in which consumers, before evaluating pasta, were informed on how the SP and CP pasta were produced.

Consumers were asked to give a judgment for acceptability by scoring for the following attributes: flavor, taste, texture, and overall acceptability. A nine-point structured hedonic scale ranging from 1 (extremely disliked it) to 9 (extremely liked it) was used, and a sample was considered acceptable when it scored above 5 (neither like nor dislike). Then, consumers answered the CATA questionnaire, containing 12 phrases related to sensory, hedonic, and functional properties of the samples, which were the following: Do you believe that it is a wholesome food? Do you believe that it is ideal for a balanced diet? Do you feel it satiating? Do you sense a distinctive odor? Does it remind you of home-made pasta? Does it feel hard to chew? Do you feel it is tasty? Do you feel any unusual taste? Do you feel it "al dente" (cooked to OCT)? Does it absorb the sauce well? Do you feel it is sticky? Is it pleasantly sour?

The consumers were forced to answer "yes" or "no", checking all phrases that applied as suitable to describe the product.

2.11. Statistical Analysis

Standard ANOVA procedure (randomized complete design with three replicates and two treatments) was applied on the dataset. The means were separated by the LSD test at $p = 0.05$ significance level, using the Statgraphics Centurion software package (version16.1.11, Statpont Technologies Inc, Warrenton, VA, USA).

Hedonic scores collected in the acceptability test were analyzed by analysis of variance (ANOVA). The Cochran's Q test at $p = 0.05$ was used to analyze the CATA data.

3. Results

3.1. Physicochemical Characteristics and Cooking Properties

Physicochemical characteristics of samples are reported in Table 1. The fermentation process did not affect ashes, moisture, and protein content, which were similar in the SP and CP samples, nor the total dietary fiber content, which was measured prior to cooking. Moisture increased after cooking, similarly in the CP and SP, and this was obviously due to the water absorbed during cooking. Ashes decreased after cooking, probably because mineral salts were lost in the boiling water. After cooking, protein percentage increased in the CP, whereas it did not vary in the SP and probably this was an effect of cooking loss, as discussed later and displayed in Table 2. Cooking caused an increase in the pH of the SP and a decrease of the TTA values in the CP and SP, indicating that organic acids diffused into the cooking water. The lightness (L*) and the yellow color (b*) were significantly higher in the SP than in the CP, as showed in Table 1, suggesting a greater retention of the pigments which could contribute to a higher antioxidant activity [14,25]. The Euclidean color distance, ΔE_{76}, was 4.26, depicting a clearly noticeable difference at human sight [15].

Table 2 shows the results of cooking quality analysis and texture of cooked pasta. The optimal cooking time was seven minutes for both the SP and the CP. The CP and SP showed the same swelling index, indicating the same amount of water absorbed, whereas differences were found in the cooking loss, which was significantly higher for the SP samples. Analysis of solids leached into the cooking water, indicate that the SP released more proteinaceous substances than the CP. Moreover, the SP showed a lower firmness than the CP.

Table 1. Physicochemical parameters of pasta. CP, control pasta; SP, pasta with sourdough. Mean values are reported.

	Ashes	Moisture	Protein Content	Dietary Fiber	pH	TTA	L	a	b	ΔE_{76}
	g/100 g DM [1]	g/100 g	g/100 g DM	g/100 g DM		mL NaOH 0.1*N*				
fermentation	ns	ns	ns	ns	***	***				
cooking	***	***	ns		***	***				
fermentation*cooking	ns	ns	**		***	***				
CP raw	0.97 ^{a}	30.51 ^{b}	13.75 ^{b}	8.01	6.50 ^{a}	2.48 ^{b}	63.96 ^{a}	2.45 ^{a}	14.16 ^{a}	
CP cooked	0.70 ^{b}	58.24 ^{a}	14.64 ^{a}		6.65 ^{a}	1.25 ^{c}				4.26
SP raw	0.97 ^{a}	31.01 ^{b}	14.32 ^{b}	8.16	5.11 ^{c}	5.98 ^{a}	68.01 ^{b}	2.50 ^{a}	15.46 ^{b}	
SP cooked	0.61 ^{b}	58.39 ^{a}	13.72 ^{b}		5.67 ^{b}	2.45 ^{b}				

Significance of the F-test after ANOVA: ns, not significant; **, significant for $p \leq 0.01$; ***, significant for $p \leq 0.001$. Different superscript letters for the same treatment denote a statistically significant difference at $p \leq 0.05$. [1] DM, dry matter; TTA, total titratable acidity; L, lightness; a, color in the red/green field; b, color in the blue/yellow field; CP, control pasta; SP, sourdough pasta. ^{a,b,c} Means with different letters for each parameter indicate significant differences ($p < 0.05$).

Table 2. Cooking quality parameters, texture properties, and protein digestibility and availability of cooked pasta. CP, control pasta; SP, pasta with sourdough. Mean values are reported.

Cooked Samples	Cooking Loss * (%)	Swelling Index ** ($g \cdot g^{-1}$)	Protein Loss in Water *** (%)	Firmness (*N*)	Protein Digestibility (%)	Protein Availability (%)
CP	4.61 ^{a}	1.34 ^{a}	0.36 ^{a}	6.88 ^{b}	86.6 ^{b}	12.6 ^{b}
SP	5.27 ^{b}	1.36 ^{a}	0.45 ^{b}	5.64 ^{a}	82.9 ^{a}	11.9 ^{a}

Different superscript letters for the same treatment denote a statistically significant difference at $p \leq 0.05$. *, grams of solids in cooking water per 100 g of pasta (dry matter); **, grams of absorbed water per gram of pasta (dry matter); ***, g of proteins lost in water after cooking per 100 g of pasta (dry matter). ^{a,b} Means with different letters for each parameter indicate significant differences ($p < 0.05$).

During the cooking of pasta, starch is normally leached into the water, together with soluble proteinaceous material. Kordonowy and Youngs [26] reported that the cooking loss was higher in bran-containing food, as a consequence of water-soluble components of bran and gluten dilution. In the SP the protein loss in the cooking water was higher than in the CP, probably because proteins have been partially hydrolyzed to peptides and amino acids by microbial proteases and, to a greater extent, by endogenous proteases and peptidases, which are active at a low pH. The proteolytic activity on the gluten proteins also explains the lower firmness of cooked pasta.

3.2. *In Vitro Starch Digestibility*

The results of glucose release after in vitro starch digestion and total starch values are reported in Figure 1. No significant differences were found between samples for the rapidly digestible starch (RDS), slowly digestible starch (SDS), and total digestible starch (TDS), whereas a significant difference was found for the inaccessible digestible starch (IDS), which resulted in being significantly lower in the SP sample. The differences observed for the TDS (44 g/100 g DM for the SP against 48 g/100 g DM for the CP), nevertheless not significant at $p = 0.05$, are fairly high. The acidic conditions of the SP before the heat treatment (pasteurization) should have been responsible for a stricter interaction between starch and gluten [27], and thus an increase of IDS, the starch fraction indigestible because of the food structure, was expected in the SP. Moreover, several studies have investigated the effects of dietary fiber in food and pasta on in vitro digestibility [28,29] showing that dietary fiber might have been responsible for the formation of a polysaccharide network that could encapsulate the starch granules during processing. The entrapment of starch reduces accessibility to enzymatic degradation, and therefore reduces the sugars released in the blood [30]. Sourdough technology applied to whole wheat bread already has been proved to retard postprandial glucose and insulin response of bread, with respect to yeast leavened whole wheat bread [31]. The decrease of IDS in the SP could be explained taking into account the higher cooking loss observed in the SP samples, which was a consequence of a greater disruption of food structure during cooking, thus altering the relative percentages of starch fractions. Sourdough determines the partial hydrolysis of proteins, causing a higher loss of peptides and amino acids in the cooking water and, at the same time, a weakening of the gluten matrix. A weaker gluten matrix is less able to entrap swollen starch granules during cooking, resulting in higher cooking loss of starch, which explains the lower content of some starch fractions in the SP.

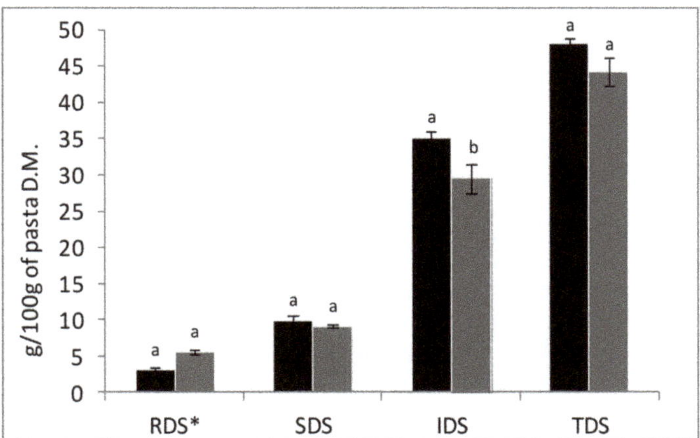

Figure 1. Results of in vitro digestion of pasta samples. RDS, rapidly digestible starch; SDS, slowly digestible starch; IDS, inaccessible digestible starch; TDS, total digestible starch. Black bars, CP and grey bars, SP. Different letters (a, b) for the same starch fraction denote a statistically significant difference between samples at $p \leq 0.05$.

3.3. Protein Digestibility and Amino Acid Content

The digestibility of protein and the total protein availability (i.e., protein content*digestibility) are reported in Table 2, expressed as a percentage of the average of triplicate runs. The digestibility is significantly lower in the SP (83%) as compared with the CP (86%), and consequently the value of protein availability is lower in the SP. The method used to estimate protein digestibility [18] measures the pH drop, after 10 min of hydrolysis, of an aqueous protein suspension that has been adjusted to a pH of 8.0 using NaOH. During enzymatic hydrolysis, carboxyl (-COO^-) and amino (-$NH3^+$) groups are released from proteins. Protons, (H^+), released into the surrounding reaction medium give rise to a decrease of the pH. The lower protein digestibility found in the SP is, therefore, due to the fact that part of the accessible peptidic bonds, which are the source of protons measured during the procedure, had already been broken by protease of the liquid sourdough. Moreover, as observed by Desai et al. [19] phenolic compounds (which are present at higher amount in the SP sample, as discussed later on) can contribute to a decrease in the protein digestibility. Therefore, pH drop method results must be carefully interpreted when they come out from the analysis of fermented products.

Table 3 shows the quantity of the 20 individual free amino acids (FAAs) measured in raw and cooked CP and SP. Aspartic acid, glutamic acid, asparagine, arginine, alanine, GABA and tryptophan had the highest concentrations in all samples, whereas the least abundant FAAs were glutamine, histidine, methionine and isoleucine. In the SP, the mean value of total FAAs was 197.41 mg/100 g and 171.91 mg/100 g, raw and cooked, respectively. For the CP it was 159.65 and 153.39 mg/100 g, raw and cooked, respectively. The SP had the highest total FAA content, likely due to the proteolytic activity of the sourdough, and significant ($p < 0.05$) differences could be found between the SP and CP samples for all amino acids, except for glycine, alanine, tryptophan, lysine, and proline. Commonly, the amount of each FAA was higher in the SP than the CP, except for GABA, asparagine, and glutamine which were higher in CP. SP also had a significantly higher content of total free essential amino acids (EAAs).

Table 3. Free amino acid content in pasta samples (mg/100 g of pasta dry matter).

Aminoacids	Fermentation	Cooking	Fermentation*Cooking	CP		SP	
				Raw	Cooked	Raw	Cooked
Aspartic acid	**	***	ns	23.34	24.32	33.27	29.70
Glutamic acid	***	**	ns	14.86	15.19	29.48	26.36
Asparagine	***	***	*	21.63	20.62	14.86	13.41
Glutamine	***	***	***	4.84	5.24	1.24	1.34
Serine	***	***	*	3.03	2.77	5.62	5.41
Histidine	*	**	ns	2.22	2.13	2.84	2.46
Glycine	ns	***	ns	3.81	4.10	4.39	4.52
Threonine	***	***	ns	2.20	2.27	3.75	3.48
Arginine	***	**	ns	10.86	8.55	17.56	13.58
Alanine	ns	***	ns	11.15	11.43	12.11	11.52
GABA	***	***	ns	17.88	17.19	13.44	12.40
Tyrosine	*	**	ns	5.06	4.37	6.59	5.14
Valine	**	***	ns	7.78	7.39	10.86	9.82
Methionine	***	**	ns	0.68	0.70	1.31	1.23
Tryptophan	ns	**	ns	14.60	11.37	13.97	10.02
Phenylalanine	***	**	ns	3.36	2.96	6.26	4.81
Isoleucine	**	***	**	1.91	1.83	1.83	1.28
Leucine	***	**	ns	3.09	3.04	9.24	7.70
Lysine	ns	ns	ns	3.52	2.86	3.86	2.95
Proline	ns	***	ns	3.81	5.05	4.92	4.80
FAA [1]	*	***	ns	159.65	153.39	197.41	171.91
EAA [2]	**	**	ns	39.38	34.55	53.92	43.74

Significance of the *F*-test after ANOVA: ns, not significant; *, significant for $p \leq 0.05$; **, significant for $p \leq 0.01$; ***, significant for $p \leq 0.001$. [1] FAA, free amino acids and [2] EAA, essential amino acids.

These results show the important role of the microbial fermentation in order to obtain a higher content of FAAs and EAAs, since the acidic conditions of sourdough activated the cereal proteinases. Furthermore, microbial peptidases released small peptides and FAAs into the food matrix [32]. Such a phenomenon already has been observed in bread. Lappi et al., [31] found more solubilized and smaller molecular weight proteins and peptides in sourdough than in yeast bread.

The high level of glutamic acid in the SP was an effect of microbial deamidation of glutamine [33]. This metabolic pathway plays an important role in pH homeostasis and acid resistance of microorganisms. As reported by Zhao et al. [34], a high level of glutamic acid can also result in a more pronounced food taste. Glutamic acid is a metabolic precursor of GABA, a non-protein amino acid naturally present in cereals in small quantities and having different health benefits. Numerous studies describe an increase of GABA in sourdough produced with selected lactic acid bacteria strains [35]. In this study a lower amount of GABA was found in the SP with respect to the CP, likely due to a consumption by the yeast [36]. Moreover, the decrease of asparagine observed in the SP as compared to the CP, also can be due to the yeast metabolism. These results confirm that microbial activity can affect nutritional properties and the taste of fresh pasta.

3.4. Total Phenolic Content, Antioxidant Activity, and Phytic Acid Content

The total phenolic content (TPC) and the antioxidant activity, measured as a percentage of discoloration of DPPH radical with respect to a blank sample, are reported in Table 4. In raw pasta, both the TPC and the antioxidant activity values were higher in the SP samples (37.0 mg/100 g pasta DM and 6.3%) than the CP samples (23.5 mg/100g pasta DM and 3.6%). It is known that fermentation improves the bioavailability of phenolic compounds and induces enhancement of antioxidant activity, through a mechanism reviewed by Hur et al. [37]. In addition, cooking had a positive effect on the TPC and antioxidant activity, which significantly increased in both the CP and SP. The increase of the TPC in pasta after cooking has already been reported in pasta enriched with bran fractions [38], and it was mainly ascribable to the increase of ferulic acid [39,40]. The increase of the antioxidant activity is consistent with the observed increase in the TPC, in agreement with Fares et al. [39], who stated that ferulic acid esters that are linked to cell walls are released by the pasta matrix during cooking and that they do not lose their antioxidant capacity, even after the hydrothermal treatment. Note that the combined effect of fermentation and cooking made the level of the TPC in the SP to be about three times higher than that of the CP sample, but the concomitant increase in antioxidant activity (from 6.33% to 7.83%) did not have such a high extent. This is probably because ferulic acid shows a relatively weak antiradical effect [41].

Table 4. Phytic acid and antioxidant activity in raw and cooked pasta. CP, control pasta and SP, pasta with sourdough.

Samples	Total Phenolic Content [1] (mg/100 g DM)	Antioxidant Activity [2] (%)	Phytic Acid (g/100 g DM)
fermentation	***	***	**
cooking	***	***	**
fermentation*cooking	***	**	ns
CP raw	23.53 ^{c}	3.57 ^{c}	0.26 ^{a}
CP cooked	36.13 ^{b}	6.53 ^{b}	0.40 ^{b}
SP raw	37.02 ^{b}	6.33 ^{b}	0.19 ^{c}
SP cooked	104.87 ^{a}	7.83 ^{a}	0.25 ^{d}

Significance of the F-test after ANOVA: ns, not significant; **, significant for $p \leq 0.01$; ***, significant for $p \leq 0.001$. Different superscript letters for the same treatment denote a statistically significant difference at $p \leq 0.05$. [1] mg of gallic acid per 100 g of pasta (dry matter); [2] percentage of discoloration referred to blank sample. ^{a,b,c} Means with different letters for each parameter indicate significant differences ($p < 0.05$).

The amount of phytic acid was different between samples (Table 4), being lower in the SP as compared with the CP, both in raw and cooked samples. The phytic acid increased significantly after cooking, in both SP and in CP.

Bioavailability of essential nutrients, such as minerals and amino acids, is strongly reduced by the chelating properties of phytic acid in food stuff containing bran. Kordonowy and Youngs [26] found that the addition of bran in pasta increased the phytic acid content, and also observed the loss of phytic acid into the cooking water. The chelating properties of phytic acid can be inactivated at acidic conditions by the endogenous phytase of wheat [42]. In our work the significant reduction of phytic acid in the SP with respect to the CP was due to the acidic conditions generated by fermentation, and this is an important consequence from a nutritional point of view.

3.5. Consumer Testing

The main purpose of this study was to analyze how the use of fermented whole wheat semolina affected sensory properties of fresh pasta, taking into account that other authors reported lower sensory scores for high fiber pasta than for semolina pasta [7,26]. Figure 2 shows the results of the acceptability test performed on the CP and the SP, which was planned in two subsequent steps, the first one aiming to know the responses of consumers not informed on the SP properties (blind test), and the second one to know the responses of consumers informed on the preparation of the sample (informed test). As indicated in Figure 2, the scores of the CP sample were significantly lower for most of the sensory parameters as compared with those of the SP samples, except for the flavor which was similar to the value of the SP sample in the blind test but not in the informed test. Significant differences between the blind and informed tests, for SP samples, were found for flavor and overall acceptability, indicating the significant influence of information on food acceptability. Previous researches demonstrated that information on food stuff (brand, manufacturing, nutritional properties, etc.) might affect its hedonic rating [43,44].

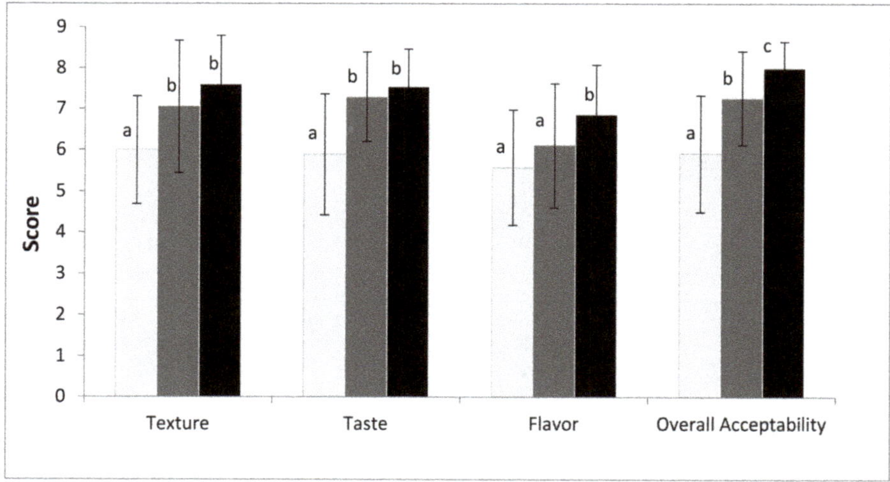

Figure 2. Results of the acceptability sensory test performed on the CP sample (white bars) and on the SP samples, the latter as blind (grey bars) and informed tests (black bars). Different letters (a, b, c) for the same sensory parameter denote a statistically significant difference between samples at $p \leq 0.05$.

Table 5 reports the relative frequencies of positive answers of consumers subjected to the CATA test. For the most part, the SP samples collect a greater number of positive responses than CP sample. Only the questions "Do you feel it hard to chew" and "Do you feel it sticky" collected a lower score and CP sample was found to be harder and more sticky than SP samples. To note that in the informed test the SP sample collected a greater number of positive responses than in the blind test, confirming that the information on the food preparation (i.e., the fermentation of semolina) had a positive impact on consumers' expected food quality. The information provided on pasta influenced positively the "satiating" sensation, the "al dente" property and the "absorption of sauce", which obtained an higher score than the SP blind sample, whereas the "unusual taste", observed in the blind sample, disappeared in the informed one.

Table 5. Check-all-that-apply (CATA) test results. The reported values indicate the relative frequency of positive answers. In the first column the main attributes representing the questions addressed to the consumers, as follows: Do you believe that it is a wholesome food? Do you believe that it is ideal for a balanced diet? Do you feel it is satiating? Do you sense a distinctive odor? Does it remind you of home-made pasta? Does it feel hard to chew? Do you feel it is tasty? Do you feel any unusual taste? Do you feel it "al dente" (cooked to OCT)? Does it absorb the sauce well? Do you feel it is sticky? Is it pleasantly sour?

Attributes	Significance	CP	SP Blind	SP Informed
wholesome	***	0.29 ^{a}	0.42 ^{a}	0.90 ^{b}
balanced diet	***	0.29 ^{a}	0.42 ^{a}	0.81 ^{b}
satiating	***	0.46 ^{a}	0.54 ^{a}	0.77 ^{b}
distinctive odor	***	0.29 ^{a}	0.79 ^{b}	0.98 ^{b}
home made	***	0.37 ^{a}	0.79 ^{b}	0.98 ^{b}
hard to chew	***	0.79 ^{b}	0.21 ^{a}	0.21 ^{a}
tasteful	***	0.21 ^{a}	0.81 ^{b}	1.00 ^{b}
unusual taste	***	0.04 ^{a}	0.29 ^{b}	0.034 ^{a}
"al dente"	***	0.19 ^{a}	0.50 ^{b}	1.00 ^{c}
adsorbs sauce	***	0.56 ^{a}	0.77 ^{b}	0.92 ^{b}
sticky	***	0.33 ^{b}	0.19 ^{a,b}	0.00 ^{a}
gently sour	***	0.00 ^{a}	0.48 ^{b}	0.71 ^{c}

Significance of the Cochran *Q*-test ***, significant for $p \leq 0.001$. Different superscript letters for the same treatment denote a statistically significant difference at $p \leq 0.05$. ^{a,b,c} Means with different letters for each parameter indicate significant differences ($p < 0.05$).

4. Conclusions

The addition of fermented whole wheat semolina affected many quality features of fresh pasta. Differences were found in color, firmness and cooking loss, while the optimal cooking time was the same for both samples. Notably, the SP samples were characterized by improved nutraceutical characteristics, showing a higher content in total and essential free amino acids, phenolic compounds, and resulting DPPH scavenging activity, and a decreased content of phytic acid. The results of sensorial analysis indicate an increase in the overall quality of pasta obtained using fermented whole wheat semolina, suggesting a new way for the sensorial improvement of high fiber pasta. The results of the acceptability test highlighted the differences between the CP and SP, with the latter having higher scores for all sensory parameters. The highest overall acceptability score was obtained from the SP sample after consumers were informed that the SP contained sourdough, indicating consumer interest in the addition of the functional ingredient. This study demonstrated that whole wheat sourdough is a valuable functional ingredient in fresh pasta making. Studies are in progress with in vivo trials, to investigate the nutraceutical properties of this innovative fresh pasta.

Author Contributions: Conceptualization, P.C; formal analysis, S.F, M.C, P.P.P, S.S, and M.S; funding acquisition, T.R; investigation, S.F, M.C, P.P.P, S.S, and M.S; resources, T.R; supervision, P.C; writing—original draft, S.F and M.C; writing—review and editing, P.C.

Funding: This work was supported by Sardegna Ricerche, Science and Technology Park of Sardinia, and Regione Autonoma della Sardegna, under grant program art.9 LR 20/2016 (2017) to Porto Conte Ricerche.

Acknowledgments: We gratefully acknowledge Paola Conte, PhD, for technical assistance in fiber analysis.

Conflicts of Interest: The authors declare that there are no conflicts of interest in this research article.

References

1. Aune, D.; Chan, D.S.M.; Lau, R.; Vieira, R.; Greenwood, D.C.; Kampman, E.; Norat, T. Dietary fibre, whole grains, and risk of colorectal cancer: Systematic review and dose-response meta-analysis of prospective studies. *BMJ* **2011**, *343*, d6617. [CrossRef] [PubMed]
2. Kumar, P.; Yadava, R.K.; Gollen, B.; Kumar, S.; Verma, R.K.; Yadav, S. Nutritional Contents and Medicinal Properties of Wheat: A Review. *Life Sci. Med. Res.* **2011**, *22*, 1–10.
3. Mongeau, R.; Brooks, S.P.J. Dietary fiber—properties and sources. In *Encyclopedia of Food Sciences and Nutrition*, 2nd ed.; Academic Press: Cambridge, MA, USA, 2003; pp. 1813–1823.
4. Hemery, Y.; Rouau, X.; Lullien-Pellerin, V.; Barron, C.; Abecassis, J. Dry processes to develop wheat fractions and products with enhanced nutritional quality. *J. Cereal Sci.* **2007**, *46*, 327–347. [CrossRef]
5. Cheryan, M.; Anderson, F.W.; Grynspan, F. Magnesium-phytate complexes: Effect of pH and molar ratio on solubility characteristics. *Cereal Chem.* **1983**, *60*, 235–237.
6. Hemdane, S.; Jacobs, P.J.; Dornez, E.; Verspreet, J.; Delcour, J.A.; Courtin, C.M. Wheat (*Triticum aestivum L.*) bran in bread making: A critical review. *Compr. Rev. Food Sci. Food Saf.* **2016**, *15*, 28–42. [CrossRef]
7. Aravind, N.; Sissons, M.; Egan, N.; Fellows, C. Effect of insoluble dietary fibre addition on technological, sensory, and structural properties of durum wheat spaghetti. *Food Chem.* **2012**, *130*, 299–309. [CrossRef]
8. Vignola, M.B.; Bustos, M.C.; Pérez, G.T. Comparison of quality attributes of refined and whole wheat extruded pasta. *LWT Food Sci. Technol.* **2018**, *89*, 329–335. [CrossRef]
9. Delcour, J.; Poutanen, K. *Fibre-Enriched and Whole Grain Breads*, 1st ed.; Martin Woodhead Publishing: Cambridge, UK, 2013.
10. Servi, S.; Özkaya, H.; Colakoglu, A.S. Dephytinization of wheat bran by fermentation with bakers' yeast, incubation with barley malt flour and autoclaving at different pH levels. *J. Cereal Sci.* **2008**, *48*, 471–476. [CrossRef]
11. Coda, R.; Kärki, I.; Nordlund, E.; Heiniö, R.; Poutanen, K.; Katina, K. Influence of particle size on bioprocess induced changes on technological functionality of wheat bran. *Food Microbiol.* **2014**, *37*, 69–77. [CrossRef]
12. Di Cagno, R.; De Angelis, M.; Alfonsi, G.; De Vincenzi, M.; Silano, M.; Vincentini, O.; Gobbetti, M. Pasta made from durum wheat semolina fermented with selected lactobacilli as a tool for potential decrease of the gluten intolerance. *J. Agric. Food Chem.* **2005**, *53*, 4393–4402. [CrossRef]
13. Capozzi, V.; Menga, V.; Digesu, A.M.; De Vita, P.; van Sinderen, D.; Cattivelli, L.; Fares, C.; Spano, G. Biotechnological production of vitamin B2-enriched bread and pasta. *J. Agric. Food Chem.* **2011**, *59*, 8013–8020. [CrossRef] [PubMed]
14. Fois, S.; Piu, P.P.; Sanna, M.; Roggio, T.; Catzeddu, P. Starch digestibility and properties of fresh pasta made with semolina-based liquid sourdough. *LWT Food Sci. Technol.* **2018**, *89*, 496–502. [CrossRef]
15. Mokrzycki, W.S.; Tatol, M. Color difference Delta E—A survey. *Mach. Graph. Vis.* **2001**, *20*, 383–411.
16. AACC. *Approved Methods of the American Association of Cereal Chemist*, 10th ed.; AACC Inc.: St. Paul, MN, USA, 2000.
17. Sanna, M.; Fois, S.; Falchi, G.; Campus, M.; Roggio, T.; Catzeddu, P. Effect of liquid sourdough technology on the pre-biotic, texture, and sensory properties of a crispy flatbread. *Food Sci. Biotechnol.* **2019**, *28*, 721–730. [CrossRef] [PubMed]
18. Hsu, H.W.; Vavak, D.L.; Satterlee, L.D.; Miller, G.A. A multienzyme technique for estimating protein digestibility. *J. Food Sci.* **1977**, *42*, 1269–1273. [CrossRef]
19. Desai, A.S.; Brennan, M.A.; Brennan, C.S. Effect of fortification with fish (*Pseudophycis bachus*) powder on nutritional quality of durum wheat pasta. *Foods* **2018**, *7*, 62. [CrossRef] [PubMed]
20. Long, W. Automated Amino Acid Analysis Using an Agilent Poroshell HPH-C18 Column. 2015. Available online: http://www.agilent.com/cs/library/applications/5991-5571EN.pdf (accessed on 25 February 2019).
21. Singleton, V.L.; Rossi, J.A. Colorimetry of total phenolics with phosphomolybdic-phosphotungstic acid reagents. *Am. J. Enol. Vitic.* **1965**, *16*, 144–158.
22. Boroski, M.; de Aguiar, A.C.; Boeing, J.S.; Rotta, E.M.; Wibby, C.L.; Bonafé, E.G.; de Souza, N.E.; Visentainer, J.V. Enhancement of pasta antioxidant activity with oregano and carrot leaf. *Food Chem.* **2011**, *125*, 696–700. [CrossRef]
23. Henrique, N.A.; Deliza, R.; Rosenthal, A. Consumer sensory characterization of cooked ham using the *Check-All-That-Apply* (CATA) methodology. *Food Eng. Rev.* **2015**, *7*, 265–273. [CrossRef]

24. Moskowitz, H.R.; Beckley, H.; Resurreccion, A.V.A. *Sensory and Consumer Research in Food Product Design and Development*; John Wiley & Sons: Hoboken, NJ, USA, 2009.
25. Coda, R.; Rizzello, C.G.; Pinto, D.; Gobbetti, M. Selected lactic acid bacteria synthesize antioxidant peptides during sourdough fermentation of cereal flours. *Appl. Environ. Microbiol.* **2012**, *78*, 1087–1096. [CrossRef]
26. Kordonowy, R.K.; Youngs, V.L. Utilization of durum bran and its effect on spaghetti. *Cereal Chem.* **1985**, *62*, 301–308.
27. Östman, E. Fermentation as a Means of Optimizing the Glycaemic Index—Food Mechanisms and Metabolic Merits with Emphasis on Lactic Acid in Cereal Products. Ph.D. Thesis, Lund University, Lund, Sweden, 2003.
28. Brennan, C.S.; Tudorica, C.M. Evaluation of potential mechanisms by which dietary fibre additions reduce the predicted glycaemic index of fresh pastas. *Int. J. Food Sci. Technol.* **2008**, *43*, 2151–2162. [CrossRef]
29. Dhingra, D.; Michael, M.; Rajput, H.; Patil, R.T. Dietary fibre in foods: A review. *J. Food Sci. Technol.* **2012**, *49*, 255–266. [CrossRef] [PubMed]
30. Foschia, M.; Peressini, D.; Sensidoni, A.; Brennan, M.A.; Brennan, C.S. Synergistic effect of different dietary fibres in pasta on in vitro starch digestion? *Food Chem.* **2015**, *172*, 245–250. [CrossRef] [PubMed]
31. Lappi, J.; Selinheimo, E.; Schwab, U.; Katina, K.; Lehtinen, P.; Mykkänen, H.; Kolehmainen, M.; Poutanen, K. Sourdough fermentation of wholemeal wheat bread increases solubility of arabinoxylan and protein and decreases postprandial glucose and insulin responses. *J. Cereal Sci.* **2010**, *51*, 152–158. [CrossRef]
32. Gänzle, M.; Gobbetti, M. Physiology and Biochemistry of Lactic Acid Bacteria. In *Handbook on Sourdough Biotechnology*; Gänzle, M., Gobetti, M., Eds.; Springer: New York, NY, USA, 2013; pp. 183–216.
33. Teixeira, J.S.; Seeras, A.; Sanchez-Maldonado, A.F.; Zhang, C.; Su, M.S.; Gänzle, M.G. Glutamine, glutamate, and arginine-based acid resistance in *Lactobacillus reuteri*. *Food Microbiol.* **2014**, *42*, 172–180. [CrossRef]
34. Zhao, C.J.; Kinner, M.; Wismer, W.; Gänzle, M.G. Effect of glutamate accumulation during sourdough fermentation with *Lactobacillus reuteri* on the taste of bread and sodium-reduced bread. *Cereal Chem.* **2015**, *92*, 224–230. [CrossRef]
35. Diana, M.; Quílez, J.; Rafecas, M. Gamma-aminobutyric acid as a bioactive compound in foods: A review. *J. Funct. Food.* **2014**, *10*, 407–420. [CrossRef]
36. Lamberts, L.; Joye, I.J.; Beliën, T.; Delcour, J.A. Dynamics of γ-aminobutyric acid in wheat flour bread making. *Food Chem.* **2012**, *130*, 896–901. [CrossRef]
37. Hur, A.J.; Lee, S.Y.; Kim, Y.; Choi, I.; Kim, G. Effect of fermentation on the antioxidant activity in plant-based foods. *Food Chem.* **2014**, *160*, 346–356. [CrossRef]
38. Ciccoritti, R.; Taddei, F.; Nicoletti, I.; Gazza, L.; Corradini, D.; D'Egidio, M.G.; Martini, D. Use of bran fractions and debranned kernels for the development of pasta with high nutritional and healthy potential. *Food Chem.* **2017**, *225*, 77–86. [CrossRef] [PubMed]
39. Fares, C.; Platani, C.; Baiano, A.; Menga, V. Effect of processing and cooking on phenolic acid profile and antioxidant capacity of durum wheat pasta enriched with debranning fractions of wheat. *Food Chem.* **2010**, *119*, 1023–1029. [CrossRef]
40. Zielinski, H.; Kozlowska, H.; Lewczuk, B. Bioactive compounds in the cereal grains before and after hydrothermal processing. *Innov. Food Sci. Emerg. Technol.* **2001**, *2*, 159–169. [CrossRef]
41. Dordević, T.M.; Šiler-Marinković, S.S.; Dimitrijević-Branković, S.I. Effect of fermentation on antioxidant properties of some cereals and pseudo cereals. *Food Chem.* **2010**, *119*, 957–963. [CrossRef]
42. Gobbetti, M.; Rizzello, C.G.; Di Cagno, R.; De Angelis, M. How the sourdough may affect the functional features of leavened baked goods. *Food Microbiol.* **2014**, *37*, 30–40. [CrossRef]
43. Resano, H.; Sanjuán, A.I.; Albisu, L.M. Consumers' acceptability of cured ham in Spain and the influence of information. *Food Qual. Prefer.* **2007**, *18*, 1064–1076. [CrossRef]
44. Laureati, M.; Conte, A.; Padalino, L.; Del Nobile, M.A.; Pagliarini, E. Effect of fiber information on consumer's expectation and liking of wheat bran enriched pasta. *J. Sens. Stud.* **2016**, *31*, 348–359. [CrossRef]

Article

The Effect of *Moringa oleifera* Leaf Powder on the Physical Quality, Nutritional Composition and Consumer Acceptability of White and Brown Breads

Laurencia Govender and Muthulisi Siwela *

Discipline of Dietetics and Human Nutrition, School of Agricultural, Earth and Environmental Sciences, University of KwaZulu-Natal, Private Bag X01, Scottsville 3209, Pietermaritzburg 3201, South Africa; GovenderL3@ukzn.ac.za

* Correspondence: Siwelam@ukzn.ac.za; Tel.: +27-33-260-5459

Received: 30 October 2020; Accepted: 3 December 2020; Published: 21 December 2020

Abstract: Fortifying popular, relatively affordable, but nutrient-limited staple foods, such as bread, with *Moringa oleifera* leaf powder (MOLP), could contribute significantly to addressing under nutrition, especially protein and mineral deficiencies, which are particularly prevalent among a large proportion of populations in sub-Saharan African countries. The current study aimed to determine the effect of MOLP on the physical quality, nutritional composition and consumer acceptability of white and brown breads. The texture, colour and nutritional composition of white and brown bread samples substituted with 5% and 10% (*w*/*w*) MOLPs were analysed using standard methods and compared with the control (0% MOLP). A consumer panel evaluated the acceptability of the bread samples using a nine-point hedonic scale. Bread samples became darker as the concentration of MOLP was increased, whilst nutrient levels increased. The overall consumer acceptability of the bread samples decreased with increasing concentrations of MOLP. However, brown bread samples were significantly more acceptable compared with corresponding white bread samples ($p < 0.05$). Under the experiment conditions of the current study, it seems that the bread containing 5% MOLP can be used to contribute significantly to addressing malnutrition, with respect to protein deficiency.

Keywords: *Moringa oleifera* leaf powder (MOLP); bread; fortification; nutritional composition; consumer acceptability

---

1. Introduction

Food insecurity and malnutrition are significant problems globally, specifically undernutrition (wasting, stunting, underweight and micronutrient deficiencies) and hunger [1]. Moderate and severe hunger affects approximately two billion individuals worldwide [1], which contributes significantly to the high rates of malnutrition seen in the world. Approximately 21.3%, 6.9% and 5.6% of children globally are stunted, wasted and overweight, respectively. Additionally, the 2020 nutrition global report indicates that one in nine and one in three individuals are hungry or malnourished and overweight or obese, respectively [2]. The COVID-19 pandemic has contributed to the increase in undernutrition, especially in countries where people are facing financial difficulties [1]. Monotonous diets consisting of largely starchy staple foods are mostly consumed in low- to middle-income countries, such as South Africa (SA). Further, the majority of communities from these countries consume limited amounts of fruits and vegetables and animal source foods [1]. Animal source foods are high in quality protein, but are less affordable to many impoverished households in SA and other developing countries, compared to plant-based protein sources [3]. This type of diet also lacks dietary diversity and could also lead to micronutrient deficiencies. Micronutrient deficiencies are a public health concern, especially in developing countries such as SA. Vitamin A, iron and zinc deficiencies are particularly problematic [4]. One of the main reasons for the trends of an increase in poor diets is the rising food costs. A basic 28-item food basket in SA in 2020 costs approximately USD 53.69 [3], which can be unaffordable to many, especially those that are hard hit by the current economic situation and global pandemic. The poorest are the most affected and are at significant risk for food insecurity [5]. Over the years, the consumption of indigenous crops such as *Moringa oleifera* has decreased due to more Westernised cultures being adopted [6]. This has resulted in less dietary diversity as foods from the formal markets (supermarkets, etc.) are expensive. Indigenous crops, such as *Moringa oleifera*, are known to be nutrient-rich as well as have many health beneficial properties. *Moringa oleifera* is one of 13 species of the *Moringaceae* family of plants and is widely researched. This plant originated in India and Africa but is now widely grown in other parts of the world [7]. Not only can *Moringa oleifera* thrive under different climatic conditions—i.e., in tropical and subtropical countries—it also has nutritional, antioxidant and phytochemical benefits [8].

Furthermore, Moringa is a good source of iron, which is generally deficient in most leading staple plant-based diets such as the starchy staples [9]. Iron is an important micronutrient, especially during pregnancy as it contributes to foetal growth. A pregnant woman who has iron-deficiency anaemia is at great risk for perinatal and maternal mortality, premature delivery and having a low birth weight infant [10,11]. *Moringa oliefera* is also rich in vitamins such as the provitamin A beta-carotene, folic acid, pyridoxine and nicotinic acids and vitamins C, D and E [12]. Vitamin A deficiency is also common in SA and most other developing countries, especially in the sub-Saharan African region. The human body needs vitamin A, and its deficiency affects vision, growth, development, protein synthesis and could result in a child not being able to reach their full potential, both physically and mentally [13]. *Moringa oleifera* is also a good source of protein and contains 16–19 amino acids. Ten of these amino acids are essential [14]. Wasting can present in the form of protein-energy malnutrition (PEM) and is caused by a deficiency of good quality protein in the diet [15]. When an individual is malnourished, their body goes into a state of starvation and negatively affects the immune system, kidneys, cardiac muscle, liver and gastrointestinal tract [13,15]. Micronutrient deficiencies and PEM are commonly seen in vulnerable population groups, such as of a woman of childbearing age and children under five years. These groups are thus targeted for nutritional interventions.

Staple foods such as bread are commonly consumed food items in SA [16], and since October 2003, wheat flour fortification was made mandatory [17]. However, access to fortified foods still remains challenging to many impoverished individuals, as many of these individuals rely on social grants to purchase food [3,18,19]. Baked bread contains high amounts of energy, carbohydrates and fat, but is limited in other nutrients such as protein, minerals and vitamins [20]. To increase the nutritional composition of bread, *Moringa oleifera* leaf powder (MOLP) could be used as it is rich in proteins and several micronutrients that are deficient in bread. Bread is an affordable source of energy (in the form of starch) and, therefore, would be a suitable candidate for supplementation with MOLP [21]. Several studies have investigated the nutritional composition and consumer acceptability of bread fortified with MOLP (Table 1). Consumers living in different geographical locations (e.g., countries) may show different preference and acceptability levels for the same innovative food product developed from the same conventional product. Thus, there may be differences in consumer acceptability of a MOLP fortified bread across different countries. While studies conducted in other countries found that the dark colour and bitter taste of MOLP impacted negatively on consumer acceptability of bread [22,23], different results may be obtained with consumers living in SA. To the best of the researchers' knowledge, this study is first to investigate consumer acceptability of a MOLP fortified bread in SA. Brown bread is dark in colour, so it may be a suitable food item for supplementing with MOLP as its dark colour might not significantly change due to the addition of MOLP, which is also dark in colour. It seems none of the previous studies listed in Table 1 compared consumer acceptable of corresponding samples of white and brown breads supplemented with MOLP. Furthermore, the other authors from earlier studies [23–28] (Table 1) did not compare the nutritional composition obtained from their respective study to the estimated average intake values. Consequently, the authors did not determine the exact amount of nutrients that would be obtained from the consumption of usual portions of MOLP fortified bread. Therefore, this study determined the effect of MOLP on the physical quality, nutritional composition and consumer acceptability of white and brown breads in SA.

Table 1. Studies conducted on the addition of *Moringa* to bread.

Authors	Study Methods	Location	Participants	Findings
Bolarinwa, Aruna and Raji (2019) [24]	Bread was fortified with 0%, 5%, 10%, 15% and 20% Moringa seed powders. The proximate, mineral and vitamin A contents were determined. A sensory evaluation was conducted using a seven-point hedonic scale to assess consumer acceptability for the sensory attributes—colour, shape, texture, sweetness, flavour, mouthfeel and overall acceptability.	Nigeria	Twenty randomly selected judges.	Study results indicated the as Moringa seed powder was added to bread, there was a significant increase in protein, ash, fat, fibre, phosphorus, potassium, calcium, iron and vitamin A contents. However, there was a decrease in moisture and carbohydrates. The sensory evaluation results indicated that there was no difference between the control bread and the 5% Moringa fortified bread. Further, the bread containing 5% Moringa seed powder was rated the best for the sensory attributes investigated.
Obichili and Ifediba (2019) [28]	Experimental research study. Two samples were prepared. The authors prepared a whole wheat bread (control) and a whole wheat bread fortified with Moringa leaf powder (MLP) using the ratio 1:4 (*v*/*v*). A nine-point hedonic scale was used to assess the sensory attributes—colour, taste, flavour, texture and general acceptability.	Nigeria	Thirty-seven evaluators categorised into two groups of seven lecturers and 30 registered postgraduate students in the Department of Home Economics, University of Nigeria Nsukka.	The whole wheat bread containing MLP was rated better for the sensory attributes flavour, taste and general acceptability. In contrast, the whole wheat bread (control) was rated better for colour and texture.
Bourekoua, Różyło, Gawlik-Dziki et al. (2018) [25]	Gluten-free bread (made with rice semolina) was fortified with 2.5%, 5%, 7.5% and 10% MLP. Antioxidant activity was determined. The sensory acceptability was determined using a nine-point hedonic scale. The following attributes were assessed—taste, aroma, texture, appearance and overall acceptability.	Poland	Fifty-two untrained consumers (23–48 years, 28 females and 24 males)	There was a significant decrease in the volume of the bread samples as MLP was added with the exception of 2.5% MLP. With the addition of 2.5% and 10% MLPs, there was a slight decrease in hardiness and chewiness. The lightness of the bread decreased when MLP was added. As MOLP was increased from 0% to 10%, so too was the antioxidant activity. In comparison to the control, the most acceptable MLP-containing bread was with the addition of 2.5% MLP.

Table 1. *Cont.*

Authors	Study Methods	Location	Participants	Findings
El-Gammal, Ghoneim and ElShehawy (2016) [27]	Pan bread was fortified with 5%, 10%, 15% and 20% MLPs. The moisture, fat, ash, crude fibre, magnesium, calcium, copper, zinc and iron contents were determined using standard methods. Sensory evaluation was conducted using the AACC method to access the smoothness, crust colour, crumb colour, taste and overall acceptability. Texture and stalling of the bread were also determined.	Egypt	Fifteen staff members from the Food Industries Department, Faculty of Agriculture, Mansoura University	The Moringa leaf powder contained a high protein, crude fibre, calcium, magnesium, phosphorous and iron content. The pan bread containing MLP had a high protein content but lower carbohydrate content in comparison to the control. Further, the 10% MLP pan bread had higher calcium, magnesium and iron contents compared to the control pan bread. The sensory acceptability decreased as the MLP concentration increased, especially with the 15% and 20% MLPs pan bread. When MLP was increased, there was a decrease in gumminess, chewiness and springiness and a gradual increase in freshness.
Sengev, Abu and Gernah (2013) [23]	Bread was prepared using 0%, 1%, 2%, 3%, 4% and 5% MLP. Moisture, crude protein, crude fat, crude fibre and ash content were determined according to standardised methods. The volume and weight of the baked loaves were determined. Meilgaard's procedure was used to determine the sensory evaluation after 24 h. The sensory attributes of the bread that were evaluated were crust colour, crumb colour, crumb texture, flavour and overall acceptability.	Nigeria	Was not provided	There was a significant increase in the protein, fibre, ash, magnesium, calcium and beta-carotene content and a decrease in iron and copper content as the MLP concentration increased in the bread samples. The volume and loaf height decreased while the weight of the loaf containing MLP increased. There was a negative effect on the sensory evaluation as the MLP concentration increased.
Ogunsina, Radha and Indrani (2011) [26]	The bread samples were prepared using 0%, 5%, 10% and 15% dry Moringa oliefera seed flours. The nutritional contents (moisture, crude fat, as, crude protein, crude fibre, iron and calcium) were determined. A quantitative descriptive analysis was used. The sensory evaluation was conducted using a quantified structure scale (crust colour (10); shape (15) and symmetry (15); crumb colour (10); grain (20); mouthfeel (20) and taste (10)).	India	Suitably trained panellists. The total number of panellist was not provided.	This study found that bread fortified with 10% Moringa was acceptable. Additionally, the 10% Moringa fortified bread had higher protein, iron and calcium contents compared to the other samples.

2. Methodology

2.1. Study Design

This study was a cross-sectional experimental design. Figure 1 presents a conceptual framework of the methodology.

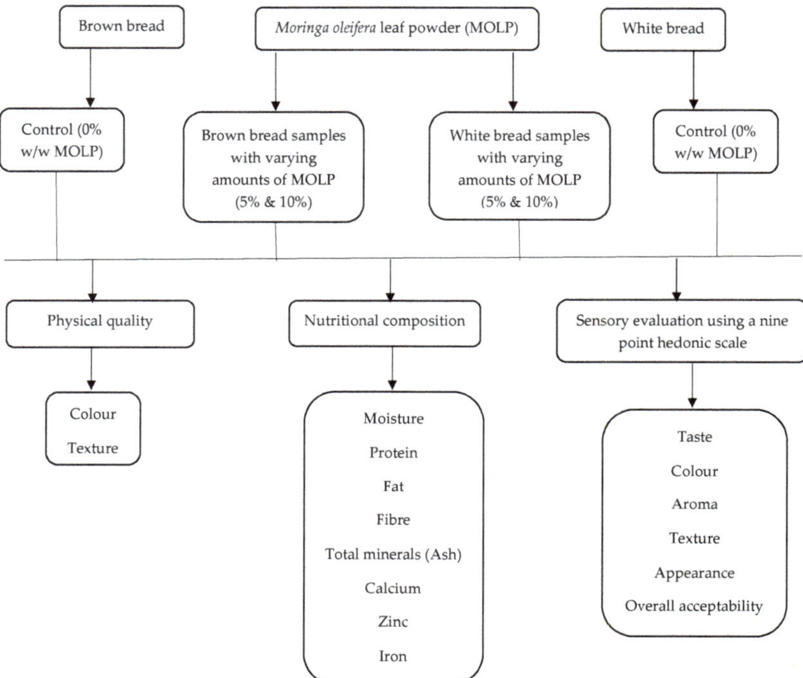

Figure 1. Conceptual framework of the methodology.

2.2. Preparation of Bread

Moringa oleifera leaf powder was purchased from a local pharmacy in Pietermaritzburg, SA. The bread (both white and brown) was prepared using a standardised bread making recipe [29]. *Moringa oleifera* leaf powder partially replaced wheat flour at 5% and 10% substitution levels of MOLP in both white and brown breads, respectively. Standard white and brown bread types served as corresponding controls (0% MOLP) for the white and brown bread samples, respectively. Fresh bread samples were baked on the day of data collection.

2.2.1. Ingredients

In this experiment, the ingredients listed below were used.

180–250 mL lukewarm water;
300 g brown/white bread flour;
10 g dry yeast;
3.8 g salt;
15 mL melted butter;
15 g or 30 g MOLP for substitution (This will be added after the flour and salt is sifted).

2.2.2. Method

The method for making bread involved the sequential steps described next. Sift flour and salt into a large bowl. Make a well in the centre of the flour in the large bowl. In a small bowl, mix the dry yeast with 60 mL water and stir until dissolved. Pour yeast mixture into the well of flour and add warm melted butter and 180 mL of water. Mix the ingredients with your fingers, then beat with your hands, adding a little more water if necessary, to make a firm dough. Using your hands, fold and slap the dough against the side of the bowl until it starts to feel elastic and leaves the sides of the bowl. Place the dough onto a floured work surface and knead by folding the far edge towards you, then pushing it firmly away with the heel of the hand. Turn the dough a little and repeat. Continue kneading until the dough is smooth and elastic and springs back when you make a dent with your finger. Place the dough in a clean, warm, oiled bowl, turn it over so that the dough is slightly oiled all over, then cover it with oiled plastic wrap and a cloth. Leave to rise in a warm place for 1 h. After the allocated time, test by pushing a finger into the dough—if the indent remains, it is ready to knock back the dough by punching with your fist several times, squeezing out any large bubbles. Place onto a lightly floured surface and knead three to four times. Pat the dough until round, then fold sides under to make a neat oblong. Press together and seal, then place in a lightly oiled 23 × 12 cm bread tin. Cover loosely with oiled plastic wrap and a cloth, leave in a warm place until it rises to the top of the tin. Preheat the Defy Thermofan Stove (Model 731 MF) oven to 230 °C. Bake the bread at 230 °C for 15 min, then turn the tins around, reduce to moderate heat, i.e., 180 °C, and bake for a further 20–25 min. The bread is cooked if it sounds hollow when the underside is knocked with the knuckles. Take the bread out of tins and leave to cool on a wire rack.

2.3. Physical Quality

A Hunter Lab colourimeter was used to measure the colour in all six bread samples. The L measured lightness, a measured redness/greenness and b measured yellowness/blueness. A TA-XT2 Plus texture analyser was used to determine the texture of the bread samples [30]. The texture probe analyser (TPA) applied a force to the bread sample at a speed of 2.0 mm/s, and the force compression was recorded.

2.4. Nutritional Composition

The nutrition analysis was conducted on all six bread samples in duplicate for moisture, protein, fat, fibre (NDF), total mineral (ash), calcium, iron and zinc using standard methods [31].

2.4.1. Moisture

The moisture content of all six bread samples was determined using the Association of Official Analytical Chemists (AOAC) Official Method 934.01 [31]. The bread samples were dried in an air circulated oven at 90 °C for 72 h. The moisture content was then calculated using the weight loss content of the samples.

2.4.2. Protein

The protein content was measured with a LECO Truspec Nitrogen Analyser, using the AOAC Official Method 990.03 [31]. The six bread samples were individually placed in a combustion chamber with an autoloader at 950 °C. The percentage protein content was calculated.

2.4.3. Fat

The fat content of the six bread samples was determined according to the AOAC Official Method 920.39 [31], following the Soxhlet procedure. A Büchi 810 Soxhlett Fat Extractor was used to extract the fat in petroleum ether which was then used to calculate the percentage fat.

2.4.4. Fibre (NDF)

The fibre content of the bread samples was measured as Neutral Detergent Fibre (NDF). The NDF was determined according to the AOAC Official Method 978.10, using the Dosi-fibre system [31].

2.4.5. Total Mineral Content (Ash)

The ash content, otherwise known as the total mineral content, was determined using the AOAC Official Method 942.05 [31]. The bread samples were placed in a furnace and heated at 550 °C for 72 h.

2.4.6. Calcium, Zinc and Iron

The calcium (Ca), zinc (Zn) and iron (Fe) contents were determined according to the AOAC Official Method. The bread samples were dried at 105 °C for 2 h, then ashed in a furnace at 550 °C for 4 h. Deionized water and hydrochloric acid (HCl) were added to the ash and boiled in a water bath until dry; this was repeated twice to allow minerals to be drawn into the solution. After 24 h, an atomic absorption spectrophotometer was used to measure the Ca, Zn and Fe contents.

2.5. Sensory Evaluation

Fifty-four students and staff members from the agricultural campus of the University of KwaZulu-Natal (UKZN) were recruited to participate in the study. A pilot study was conducted before the main study using 10 participants to test the acceptability of the recipes and adjust the methods accordingly. In the pilot study, 0%, 10% and 20% (*w*/*w*) substitutions were used; however, the 20% substitution was not well accepted. Therefore, the MOLP substitution levels were adjusted to 0%, 5% and 10% (*w*/*w*) for the main study. The pilot study participants were not allowed to participate in the main study. To prevent participants from communicating with one another, they were placed in separate cubicles. Each of the six samples was assigned a unique three-digit code obtained by a table of random numbers, and the tables of random permutations of nine were used to determine the serving order [32]. Each participant was given about a quarter of a slice of each of the six bread samples on a polystyrene plate. Each panellist was given a cup of water so that they could rinse their palate between tasting each bread sample and the sensory evaluation questionnaire. The questionnaire provided was in English, as this is the language of instruction at UKZN. The questionnaire made use of the nine-point hedonic scale (1 = dislike extremely to 9 = like extremely) together with the sensory attributes (taste, colour, aroma, texture, appearance and overall acceptability), which was explained to the panellist before the commencement of the sensory evaluation. Participants were given limited information on MOLP so that bias was prevented. Research assistants helped panellists, if necessary, during the sensory evaluation.

2.6. Ethical Consideration

Ethical approval was obtained from the UKZN Humanities and Social Science Ethics Committee (HSS/1244/015D). The gatekeeper's permission was obtained from the registrar of UKZN. All panellists were required to sign a written consent form before participating in the study.

2.7. Statistical Analysis

Data from the physical quality, nutritional composition and sensory evaluation questionnaires were entered into an Excel spreadsheet and cross-checked for accuracy. Thereafter, data were transferred to the Statistical Package for Social Science® (SPSS) version 25 (IBM Corp., Armonk, NY, USA) for analysis. Appropriate statistical techniques, including the Bonferroni and Tukey tests, were used to analyse the data. A p-value of <0.05 was considered to be statistically significant.

3. Results

3.1. The Effect of Moringa oleifera Leaf Powder on the Physical Quality of White and Brown Breads

Figure 2 shows the six bread samples, and Table 2 presents the effect of MOLP on the colour and texture of the bread.

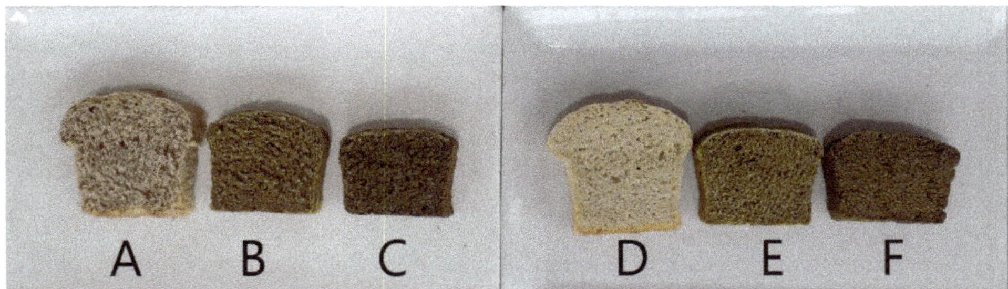

Figure 2. Depicts the different bread samples ((**A**): Brown bread (control, 0%); (**B**): Brown bread (5%) (**C**) *Moringa oleifera* leaf powder [MOLP]); Brown bread (10% MOLP); (**D**): White bread (Control, 0%); (**E**): White bread (5% MOLP); (**F**): White bread (10% MOLP)).

Table 2. The effect of *Moringa oleifera* leaf powder fortification on colour and texture of white and brown breads.

Bread Samples (% *w/w* MOLP Added)	Colour and Texture (Mean ± SD)			
	L (Lightness)	a (Redness)	b (Yellowness)	Texture (g)
White bread				
0% (control)	67.14 ± 0.77 ^{e}	1.56 ± 0.05 ^{a}	24.1 ± 0.1 ^{b}	0.09 ± 0.02 ^{a}
5%	54.02 ± 0.14 ^{b,c}	0.86 ± 0.02 ^{a}	27.4 ± 0.1 ^{d}	0.08 ± 0.01 ^{a}
10%	45.97 ± 0.29 ^{a}	1.31 ± 0.05 ^{a}	25.4 ± 0.2 ^{c}	0.13 ± 0.07 ^{a}
Brown bread				
0% (control)	63.1 ± 0.6 ^{d}	3.02 ± 0.07 ^{b}	21.5 ± 0.3 ^{a}	0.09 ± 0.02 ^{a}
5%	55.8 ± 0.4 ^{c}	2.67 ± 0.19 ^{b}	24.9 ± 0.2 ^{c}	0.08 ± 0.02 ^{a}
10%	51.7 ± 2.0 ^{b}	2.79 ± 0.87 ^{b}	27.5 ± 0.2 ^{d}	0.08 ± 0.03 ^{a}

Different letters in columns show significant difference according to the Bonferroni test ($p < 0.05$).

The Bonferroni test indicated that there was no significant difference between both control bread and bread containing 5% and 10% MOLPs. However, there was a significant decrease in the lightness of bread with the addition of MOLP. White bread containing 10% MOLP had a darker colour in comparison to brown bread containing 10% MOLP (Figure 2 and Table 2). There was no effect on the texture of the bread samples with the addition of MOLP at different substitution levels.

3.2. The Effect of MOLP on the Nutritional Composition of White and Brown Breads

Table 3 presents the nutritional composition of white and brown breads containing MOLP.

The Tukey test indicated that the total mineral content in white bread significantly increased when MOLP was added. There was a slightly significant increase in the total mineral content when 5% MOLP was added to brown bread, and there was a significant increase in brown bread with 10% MOLP ($p < 0.05$). A significant increase in protein content was seen in white bread containing 5% and 10% MOLPs. Further, as the MOLP concentration increased in white bread, so too did the protein content ($p < 0.05$). There was a slightly significant increase in the protein content when 5% MOLP was added to brown bread, and there was a significant increase in brown bread with 10% MOLP ($p < 0.05$). White bread containing MOLP has a higher protein content than brown bread containing

MOLP. Addition of MOLP to a concentration of 10% in white bread and 5% in brown bread would result in a significant increase in the iron contents of the two bread types, respectively.

Nutritional Composition of White and Brown Breads Compared to the Estimated Average Requirement

The nutritional composition of the bread was compared to the estimated average requirement (EAR) for protein and iron for vulnerable population groups (children under five years and women of childbearing age). The EAR value is one of the four dietary reference intake values and presents the daily average nutrient intake for particular nutrients for specific gender and age groups [33]. In the current study, an estimated weight for a child aged 1–3 years was 13 kg, 4–5 years was 24 kg, 14–18 years was 55 kg and 19–50 years was 62 kg. A standard size of bread is 30 g. A child aged 1–3 years consumes $\frac{1}{2}$ a slice of bread three times a day, a child 4–6 years consumes approximately one slice of bread three times a day, a female aged 14–18 years consumes two slices of bread three times a day and a female adult aged 19–50 years consumes three slices of bread three times a day. Tables 4 and 5 present the percentage of the EAR for protein and iron, respectively, that would be met from the consumption of estimated bread portions by the respective vulnerable population groups.

3.3. The Effect of MOLP on the Sensory Acceptability of White and Brown Breads

Fifty-four students and staff members from the UKZN agricultural campus were recruited to participate in the study. Table 6 indicates the effect of MOLP on the overall acceptability of white and brown breads, as indicated by the percentage distribution of the sensory evaluation scores.

Table 6 shows that as MOLP was increased in either white or brown bread at different substitution levels (5% and 10%), the overall acceptability decreased. However, in terms of the addition of MOLP to bread, there was a less negative effect when MOLP was added to brown bread at different substitution levels in comparison to white bread. Table 7 indicates the effect of MOLP on taste, colour, aroma, texture, appearance and overall acceptability of both white and brown breads.

The Tukey test indicated that there was a significant decrease in taste acceptability when MOLP was added to both white and brown breads, respectively ($p < 0.05$). However, unlike white bread which showed no significant difference in taste acceptability with the addition of either 5% or 10% MOLP, there was a slightly significant decrease in taste acceptability when MOLP was increased from 5% to 10% in brown bread. In terms of taste acceptability, brown bread containing 5% MOLP has significantly higher taste acceptability in comparison to white bread containing 5% MOLP ($p < 0.05$). The colour acceptability significantly decreased when MOLP was added to white bread. There was no significant effect when 5% MOLP was added to brown bread ($p > 0.05$). However, there was a significant decrease in colour acceptability when 10% MOLP was added to brown bread. Brown bread containing 5% MOLP had a significantly higher colour acceptability compared to white bread containing either 5% or 10% MOLP and brown bread containing 10% MOLP.

As MOLP was added to both white and brown breads, respectively, there was a significant decrease in aroma acceptability ($p < 0.05$) and there was a slightly significant decrease when MOLP was increased from 5% to 10% in both white and brown breads. The texture acceptability significantly decreased when MOLP was added to white bread; however, there was no significant difference between white bread containing 5% and 10% MOLPs. The appearance acceptability decreased when MOLP was added to white and brown breads. Additionally, when MOLP was increased to 10%, there was a further significant decrease in appearance acceptability seen for white bread. The overall acceptability significantly decreased when MOLP was increased ($p < 0.05$)—this was especially seen for brown bread. With this being said, brown bread containing 5% MOLP had a significantly higher overall acceptability in comparison to the other MOLP-containing bread.

Table 3. Nutritional composition of white and brown breads containing different concentrations of *Moringa oleifera* leaf powder (MOLP) on a dry weight basis (DW)].

Bread Samples (% MOLP Added)	Moisture g/100 g	Ash g/100 g	Fat g/100 g	NDF g/100 g	Protein g/100 g	Calcium mg/kg	Zinc mg/kg	Iron mg/kg
White bread								
0% (control)	35.18 ± 1.32 [a]	1.97 ± 0.06 [a]	3.12 ± 0.11 [a]	13.75 ± 0.10 [a]	13.68 ± 0.01 [c]	0.04 ± 0.00	3.00 ± 0.00 [a]	6.50 ± 2.83 [a]
5%	31.84 ± 0.25 [a]	2.40 ± 0.00 [b,c]	3.58 ± 0.06 [a]	19.02 ± 2.73 [a]	13.96 ± 0.00 [d]	0.10 ± 0.00	3.50 ± 9.90 [a]	7.60 ± 2.83 [a]
10%	32.39 ± 0.40 [a]	2.54 ± 0.15 [b,c]	3.33 ± 0.32 [a]	14.77 ± 2.94 [a]	14.59 ± 0.01 [e]	0.15 ± 0.00	2.95 ± 0.71 [a]	16.55 ± 2.12 [b]
Brown bread								
0% (control)	33.56 ± 0.40 [a]	2.28 ± 0.01 [a,b]	2.98 ± 0.42 [a]	23.87 ± 4.86 [a]	13.07 ± 0.06 [a]	0.03 ± 0.00	3.20 ± 0.00 [a]	16.50 ± 2.83 [b]
5%	38.93 ± 6.93 [a]	2.46 ± 0.06 [b,c]	2.78 ± 0.02 [a]	20.60 ± 1.20 [a]	13.16 ± 0.01 [ab]	0.05 ± 0.00	3.05 ± 0.71 [a]	23.20 ± 2.83 [c]
10%	33.60 ± 0.10 [a]	2.62 ± 0.04 [c]	3.36 ± 0.13 [a]	17.75 ± 1.14 [a]	13.34 ± 0.10 [b]	0.09 ± 0.00	3.10 ± 0.00 [a]	30.95 ± 9.19 [d]

Data reported as Mean ± SD of at least two replicates. NDF: Neutral detergent fibre; Different letters in columns show significant difference according to the Tukey test ($p < 0.05$).

Table 4. Percentage of the estimated average requirement met for protein for vulnerable population groups using the estimated portions of bread.

Bread Samples (% MOLP [a] Added)	1–3 Years			4–5 Years			14–18 Years (Female)			19–50 Years (Female)		
	Protein (g)	EAR [b] g/day	% EAR Met	Protein (g)	EAR g/day	% EAR Met	Protein (g)	EAR g/day	% EAR Met	Protein (g)	EAR g/day	% EAR Met
White bread												
0% (control)	6.15	11.31	54.4	12.30	18.24	67.4	24.60	39.05	63.0	36.90	40.92	90.2
5%	6.29	11.31	55.6	12.57	18.24	68.9	25.14	39.05	64.4	37.71	40.92	92.2
10%	6.57	11.31	58.1	13.14	18.24	72.0	26.28	39.05	67.3	39.42	40.92	96.3
Brown bread												
0%	5.88	11.31	52.0	11.76	18.24	64.8	23.52	39.05	60.2	35.28	40.92	86.2
5%	5.93	11.31	52.4	11.85	18.24	65.0	23.70	39.05	60.7	35.55	40.92	86.9
10%	6.00	11.31	53.1	12.00	18.24	65.8	24.00	39.05	61.5	36.00	40.92	88.0

[a] *Moringa oleifera* leaf powder; [b] Estimated average requirement (EAR) [33].

Table 5. Percentage of the estimated average requirement met for iron for vulnerable population groups using the estimated portions of bread.

Bread Samples (% MOLP [a] Added)	1–3 Years			4–5 Years			14–18 Years (Female)			19–50 Years (Female)		
	Iron (mg)	EAR [b] mg/day	% EAR Met	Iron (mg)	EAR mg/day	% EAR met	Iron (mg)	EAR mg/day	% EAR met	Iron (mg)	EAR mg/day	% EAR Met
White bread												
0% (control)	2.93	3.00	98.3	5.85	4.10	142.7	11.7	7.90	148.1	17.6	8.10	217.3
5%	3.42	3.00	114.0	6.84	4.10	166.8	13.7	7.90	173.4	20.5	8.10	253.1
10%	7.46	3.00	248.7	14.91	4.10	363.7	29.8	7.90	235.4	44.7	8.10	551.9
Brown bread												
0%	7.43	3.00	247.7	14.85	4.10	362.2	29.7	7.90	375.9	44.6	8.10	550.6
5%	10.44	3.00	348.0	20.88	4.10	509.3	41.8	7.90	529.1	62.6	8.10	772.8
10%	13.91	3.00	463.7	27.81	4.10	678.3	55.6	7.90	703.8	83.4	8.10	1029.6

[a] *Moringa oleifera* leaf powder; [b] EAR (Estimated average requirement) [33].

Table 6. The effect of *Moringa oleifera* leaf powder (MOLP) on the overall acceptability of white and brown breads as shown by percentage distribution of sensory evaluation scores.

Bread Samples (% MOLP Added)	Overall Acceptability										
	Dislike Extremely (1)	Dislike Very Much (2)	Dislike Moderately (3)	Dislike Slightly (4)	Neither Like Nor Dislike (5)	Like Slightly (6)	Like Moderately (7)	Like Very Much (8)	Like Extremely (9)	Overall Like (6–9)	Overall Dislike (1–4)
White bread											
0 (Control)	1 [a] (1.9) [b]	1 (1.9)	0 (0)	1 (1.9)	7 (13.0)	10 (18.5)	9 (16.7)	13 (24.1)	12 (22.2)	44 (81.5)	3 (5.6)
5	3 (5.6)	4 (7.4)	4 (7.4)	8 (14.8)	10 (18.5)	9 (16.7)	9 (16.7)	5 (9.3)	2 (3.7)	25 (46.3)	19 (35.2)
10	7 (13.0)	6 (11.1)	8 (14.8)	12 (22.2)	7 (13.0)	5 (9.3)	4 (7.4)	2 (3.7)	3 (5.6)	14 (25.9)	33 (61.1)
Brown bread											
0 (Control)	0 (0)	0 (0)	0 (0)	0 (0)	3 (5.6)	9 (16.7)	10 (18.5)	16 (29.6)	16 (29.6)	51 (94.4)	0 (0)
5	1 (1.9)	2 (3.7)	1 (1.9)	3 (5.6)	10 (18.5)	7 (13.0)	13 (24.1)	8 (14.8)	9 (16.7)	37 (68.5)	7 (13.0)
10	2 (3.7)	4 (7.4)	2 (3.7)	7 (13.0)	14 (25.9)	9 (16.7)	10 (18.5)	3 (5.6)	3 (5.6)	25 (46.3)	15 (27.8)

[a] number of panellists who gave the score; [b] percentage of the consumer panel (N = 54) that gave the score.

Table 7. The effect of *Moringa oleifera* leaf powder (MOLP) on sensory acceptability of white and brown breads.

Bread Samples (% MOLP Added)	Sensory Acceptability (Mean ± SD)					
	Taste	Colour	Aroma	Texture	Appearance	Overall Acceptability
White bread						
0 (Control)	$6.28 \pm 2.16^{c,d}$	7.46 ± 1.40^{c}	7.07 ± 1.52^{c}	6.32 ± 2.04^{b}	7.2 ± 1.59^{c}	6.98 ± 1.79^{cd}
5%	4.57 ± 2.07^{a}	$5.06 \pm 2.29^{a,b}$	$4.70 \pm 2.31^{a,b}$	5.11 ± 2.34^{a}	5.4 ± 2.31^{b}	5.19 ± 2.07^{ab}
10%	3.83 ± 2.25^{a}	4.09 ± 2.48^{a}	4.07 ± 2.26^{a}	5.11 ± 2.12^{a}	4.1 ± 2.70^{a}	4.20 ± 2.23^{a}
Brown bread						
0 (Control)	7.07 ± 1.46^{d}	7.57 ± 1.11^{c}	7.24 ± 1.36^{c}	6.85 ± 1.89^{b}	7.5 ± 1.33^{c}	7.61 ± 1.23^{d}
5%	$5.91 \pm 2.10^{b,c}$	6.56 ± 1.86^{c}	5.74 ± 1.99^{b}	6.06 ± 2.01^{ab}	$6.4 \pm 2.07^{b,c}$	6.44 ± 1.96^{c}
10%	$4.78 \pm 2.06^{a,b}$	5.44 ± 2.07^{b}	$4.85 \pm 1.99^{a,b}$	$5.87 \pm 1.82^{a,b}$	5.7 ± 2.06^{b}	5.35 ± 1.94^{b}

Means marked with different letters in the same column are significantly different at $p < 0.05$ according to the Tukey test.

4. Discussion

The bread samples became a darker colour when MOLP was included in the dough; this was particularly prominent in white bread samples (Table 2, Figure 2). This was an expected result as *Moringa oleifera* leaves are naturally a dark green colour due to the high chlorophyll content [22] and are thus responsible for the undesirable colour change. The brown bread prepared in this study became a darker colour but was not as noticeable as white bread, and this could be due to the fact that brown bread is a darker colour to start with due to the chocolate colour of the bran [28], masking the undesirable darker colour seen in white bread at the same concentration of MOLP. These results were consistent with a study conducted by Bourekoua et al. (2018), which found that, as the MOLP concentration increased, the lightness of bread crumb and crust decreased [25]. This dark colour may negatively affect consumer acceptability of bread fortified with MOLP as consumers are more accustomed to bread being a golden-brown colour. With this being said, more individuals are will to try a product that they are familiar with if it was beneficial to their health, thus bread containing MOLP may be acceptable despite being a darker colour. White bread is more commonly consumed than brown bread, therefore another solution to this could be to use a lightening agent to mask the dark colour [23], thus making the bread containing MOLP more acceptable. The use of lightening agents was not investigated in this study. Brown bread containing 5% MOLP had a similar colour to the control thus implies that fortifying with a concentration lower than 5% may result in lighter bread with all the nutritional benefits.

Although MOLP has an undesirable physical attribute, as mentioned earlier, it has many nutritional benefits. Table 3 indicates that the protein concentration increased when MOLP was added to white and brown breads. White bread containing 10% MOLP had the highest protein content (15.5 g/100 g). This was expected as Moringa leaves have a high protein content [34]. The study results were similar to another study which found unfortified bread had the lowest protein content (8.5%) and bread fortified with MOLP contained the highest protein content (13.5%) [24]. Similarly, other authors found that there was a gradual increase in protein content as MOLP was added at different concentration levels [27]. Furthermore, said study had a higher protein content than the current study when 5% MOLP was added to the bread (17.72%).

Table 4 shows results of assessing the potential contribution of MOLP-supplemented bread to the EAR for protein for vulnerable population groups if the amount equivalent to the usual portion size of standard was consumed. It was found that all bread samples containing MOLP would contribute to meeting more than 50% of the EAR for protein for each of the vulnerable groups. Moreover, white bread containing MOLP would contribute more to meeting the EAR for protein for all vulnerable population groups. To fully meet the EAR for protein, a 1–3-year-old would need to consume three slices of the MOLP-containing bread/day. A 4–5-year-old would need to consume 4.5 slices of white bread containing MOLP/day or five slices of brown bread containing MOLP/day. The 14–18 year female group would have to consume 9.5 slices of white bread containing 5% MOLP/day and nine slices of white bread containing 10% MOLP/day and ten slices of brown bread containing MOLP/day. Lastly, the 19–50 year female group would need to consume ten slices of bread containing 5% MOLP/daily and 9.5 slices of white bread containing 10% MOLP. In contrast, 10.5 slices of brown bread containing MOLP would need to be consumed daily. The fact that MOLP increases the protein content is encouraging as animal sources of protein are good but expensive and not affordable to many. The high protein content found in bread containing MOLP could assist in reducing PEM. However, in order not to promote a monotonous diet focused on MOLP bread, MOLP should be incorporated in other commonly consumed food items that are also deficient in protein.

The iron concentration increased in white bread when 10% MOLP was added, and when 5% MOLP was added to brown bread (Table 3). This was an expected result as MOLP generally contains a high iron content. The iron values obtained in the current study agree with the results obtained from previous studies, which showed an increase in iron content as MOLP was added [23,24,27].

Analysis of the percentage contribution of MOLP-containing bread samples to the EAR value for iron for each of the vulnerable population groups is presented in Table 5. It was found that consumption of MOLP-containing bread, up to an amount equivalent of the usual portion size of standard bread, would result in the achievement of more than 100% of the EAR value for iron for each of the vulnerable groups. The iron intake, of each age group, that would result from consumption of MOLP-containing bread, up to an amount equivalent to the usual portion size of standard bread, was also compared with the Tolerable Upper Intake Level (UL). UL refers to the highest amount of a nutrient that should be consumed from food without any adverse effects [33]. The results show that the iron intake of 1–5-year-old children would be way below the UL (40 mg/d) if they consumed MOLP-containing bread up to an amount equivalent to the usual portion size of standard bread. For the 14–18 year and 19–50 year female groups, consumption of all MOLP-containing bread types (except for the brown bread containing 10% MOLP), up to an amount equivalent to the usual portion size of standard bread, would result in an iron intake below the UL value. It is noted that, even for the MOLP-containing bread types that would result in an iron intake above the UL value, it is unlikely to be of health concern because of the limited bioavailability of divalent metal minerals such as iron in plant-based foods. Overall, the results suggest that bread containing MOLP could be a cheaper alternative to improve the iron intake in vulnerable population groups.

Overall, consumer acceptability was higher in the control white and brown breads (0%) compared with their respective bread samples containing MOLP (Tables 6 and 7). This could be lower due to the bitter taste [7] and the fact that it causes the colour of the product to become darker. These results concurred with other studies which found a decrease in overall acceptability with an increase in MOLP [25,27]. There was no significant difference in the instrument texture values of the bread samples. However, the texture acceptability of the bread samples decreased significantly in white bread with the addition of MOLP ($p < 0.05$), whilst there was a marginal difference in the texture acceptability of the control brown bread and bread samples containing MOLP. Thus, it appears that MOLP had a negative effect on the texture of white bread, as perceived by the consumers. MOLP increases the hardness of bread, which is likely due to its high fibre content, but the increase in hardness was perceived in white bread and not in brown bread, probably due to the fact that brown bread already has a high concentration of fibre compared to white bread.

The overall acceptability of brown bread containing 5% MOLP was higher than that of the white bread containing 5% MOLP, indicating that brown bread would be more suitable for fortification using MOLP compared to white bread. For consumers who prefer white bread to brown bread, the darkness imparted to the white bread by MOLP, which was disliked by the consumers, could be resolved by adding a lightening agent together with the MOLP to the white bread dough.

5. Conclusions

Protein-energy malnutrition and iron deficiency anaemia are public health concerns. A food-based intervention such as fortification of bread with MOLP could improve the protein and iron contents of bread. The results of the current study indicate that bread supplemented with about 5% MOLP could be used to complement existing strategies for addressing malnutrition, especially PEM. However, further research needs to be conducted in order to improve the physical attributes of bread containing MOLP. Further research involving incorporating MOLP in other popular, but nutrient-deficient foods, needs to be conducted to determine the most suitable foods for fortifying with MOLP.

Author Contributions: Conceptualization, L.G. and M.S.; methodology, L.G. and M.S.; investigation, L.G. and M.S.; resources, M.S.; data curation, L.G.; writing—original draft preparation, L.G.; writing—review and editing, L.G. and M.S.; visualization, L.G. and M.S.; supervision, L.G. and M.S.; project administration, L.G.; funding acquisition, M.S. All authors have read and agreed to the published version of the manuscript.

Funding: The Halley Stott Foundation for financial contribution for payment of the statistician.

Acknowledgments: The authors would like to thank the volunteers for participating herein. The authors would also like to thank Kaylé Higgs, Masana Makhubele, Nelisiwe Hlope and Raine Booy for assisting with data collection.

Conflicts of Interest: The authors declare no conflict of interest.

References

1. FAO; IFAD; UNICEF; WHO. *The State of Food Security and Nutrition in the World 2020. Transforming Food Systems for Affordable Healthy Diets*; FAO: Rome, Italy, 2020.
2. DevelopmentInitiatives. *2020 Global Nutrition Report: Action on Equity to End Malnutrition*; Development Initiatives: Bristol, UK, 2020.
3. NAMC. Food Basket Price Monthly. Available online: https://www.namc.co.za/wp-content/uploads/2020/02/Food-Basket-Feb-2020.pdf (accessed on 20 October 2020).
4. Bailey, R.L.; West, K.P., Jr.; Black, R.E. The epidemiology of global micronutrient deficiencies. *Ann. Nutr. Metab.* **2015**, *66*, 22–33. [CrossRef] [PubMed]
5. Laborde, D.; Martin, W.; Vos, R. *Poverty and Food Insecurity Could Grow Dramatically as Covid-19 Spreads*; Internationla Food Policy Research Institute (IFPRI): Washington, DC, USA, 2020; pp. 16–19.
6. Van der Hoeven, M.; Osei, J.; Greeff, M.; Kruger, A.; Faber, M.; Smuts, C.M. Indigenous and traditional plants: South african parents' knowledge, perceptions and uses and their children's sensory acceptance. *J. Ethnobiol. Ethnomed.* **2013**, *9*, 78. [CrossRef] [PubMed]
7. Method, A.; Tulchinsky, T.H. Commentary: Food Fortification: African Countries Can Make more Progress. *Adv. Food Technol. Nutr. Sci.* **2015**, *SE*, S22–S28. [CrossRef]
8. Shah, S.K.; Jhade, D.; Chouksey, R. Moringa oleifera lam. A study of ethnobotany, nutrients and pharmacological profile. *Res. J. Pharm. Biol. Chem. Sci.* **2016**, *7*, 2158–2165.
9. Mutiara, T.; Titi, E.; Estiasih, W. Effect lactagogue moringa leaves (Moringa oleifera lam) powder in rats. *J. Basic Appl. Sci. Res.* **2013**, *3*, 430–434.
10. Pasricha, S.-R.; Drakesmith, H.; Black, J.; Hipgrave, D.; Biggs, B.-A. Control of iron deficiency anemia in low-and middle-income countries. *Blood* **2013**, *121*, 2607–2617. [CrossRef] [PubMed]
11. Allen, L.H. Anemia and iron deficiency: Effects on pregnancy outcome. *Am. J. Clin. Nutr.* **2000**, *71*, 1280S–1284S. [CrossRef]
12. Mbikay, M. Therapeutic potential of Moringa oleifera leaves in chronic hyperglycemia and dyslipidemia: A review. *Front. Pharmacol.* **2012**, *3*, 24. [CrossRef]
13. Mahan, L.; Raymond, J. *Krauses Food & the Nutrition Care Process. 14a*; Elsevier Inc.: St. Louis, MA, USA, 2017; p. 1063.
14. Falowo, A.B.; Mukumbo, F.E.; Idamokoro, E.M.; Lorenzo, J.M.; Afolayan, A.J.; Muchenje, V. Multi-functional application of Moringa oleifera lam. In nutrition and animal food products: A review. *Food Res. Int.* **2018**, *106*, 317–334. [CrossRef]
15. Manary, M.J.; Sandige, H.L. Management of acute moderate and severe childhood malnutrition. *BMJ* **2008**, *337*, a2180. [CrossRef]
16. Labadarios, D.; Moodie, I.; Rensburg, A. *Vitamin a Status: In: Labadarios d. National Food Consumption Survey: Fortification Baseline, Chapter 9b: South africa 2005*; Department of Health: Stellenbosch, South Africa, 2007.
17. Department of Health (DoH); United Nations Children's Fund (UNICEF). *A Reflection of the South African Maize Meal and Wheat Flour Fortification Programme (2004 to 2007)*; Department of Health: Pretoria, South Africa, 2007.
18. Labadarios, D.; Swart, R.; Maunder, E.; Kruger, H.; Gericke, G.; Kuzwayo, P.; Ntsie, P.; Steyn, N.; Schloss, I.; Dhansay, M.; et al. The national food consumption survey fortification baseline (nfcs-fb). *S. Afr. J. Clin. Nutr.* **2008**, *21*, 246–271.
19. Statistics South Africa General Household Survey. 2018. Available online: http://www.statssa.gov.za/publications/P0318/P03182018.pdf (accessed on 20 September 2020).
20. Ameh, M.O.; Gernah, D.I.; Igbabul, B.D. Physico-chemical and sensory evaluation of wheat bread supplemented with stabilized undefatted rice bran. *Food Nutr. Sci.* **2013**, *4*, 43. [CrossRef]
21. Olushola, A. *Achieve Vibrant Health with Nature*; University of Jos Consultancy Ltd. Press: Jos, Nigeria, 2006.

22. Karim, O.; Kayode, R.; Oyeyinka, S.; Oyeyinka, A. Physicochemical properties of stiff dough 'amala'prepared from plantain (musa paradisca) flour and moringa (Moringa oleifera) leaf powder. *Hrana Zdr. Boles. Znan. Strucni Cas. Nutr. Dijetetiku* **2015**, *4*, 48–58.
23. Sengev, A.I.; Abu, J.O.; Gernah, D.I. Effect of Moringa oleifera leaf powder supplementation on some quality characteristics of wheat bread. *Food Nutr. Sci.* **2013**, *4*, 270.
24. Bolarinwa, I.F.; Aruna, T.E.; Raji, A.O. Nutritive value and acceptability of bread fortified with moringa seed powder. *J. Saudi Soc. Agric. Sci.* **2019**, *18*, 195–200. [CrossRef]
25. Bourekoua, H.; Różyło, R.; Gawlik-Dziki, U.; Benatallah, L.; Zidoune, M.N.; Dziki, D. Evaluation of physical, sensorial, and antioxidant properties of gluten-free bread enriched with Moringa oleifera leaf powder. *Eur. Food Res. Technol.* **2018**, *244*, 189–195. [CrossRef]
26. Ogunsina, B.; Radha, C.; Indrani, D. Quality characteristics of bread and cookies enriched with debittered Moringa oleifera seed flour. *Int. J. Food Sci. Nutr.* **2011**, *62*, 185–194. [CrossRef]
27. El-Gammal, R.E.; Ghoneim, G.A.; ElShehawy, S.M. Effect of moringa leaves powder (Moringa oleifera) on some chemical and physical properties of pan bread. *J. Food Dairy Sci.* **2016**, *7*, 307–314. [CrossRef]
28. Obichili, O.; Ifediba, D. Sensory acceptability studies of whole wheat bread fortified with moringa leaf powder. *South Asian J. Biol. Res.* **2019**, *2*, 58–67.
29. Smit, S.M.F. *The A–Z of Food & Cookery in South Africa*, 3rd ed.; Struik Lifestyle: Cape Town, South Africa, 2011.
30. Afoakwa, E.O.; Yenyi, S.E. Application of response surface methodology for studying the influence of soaking, blanching and sodium hexametaphosphate salt concentration on some biochemical and physical characteristics of cowpeas (vigna unguiculata) during canning. *J. Food Eng.* **2006**, *77*, 713–724. [CrossRef]
31. AOAC. *Official Methods of Analysis of the Association of Official Analytical Chemists International*, 17th ed.; AOAC International: Gaithersburg, ML, USA, 2003.
32. Heymann, H. *Three-Day Advanced Sensory Analysis Workshop*; Agricultural Research Council (ARC): Pretoria, South Africa, 1995.
33. Medicine, I.O. *Dietary dri Refernce Intakes: The Essential Guide to Nutrient Requirement*; National Academy Press: Washington, DC, USA, 2006.
34. Moyo, B.; Masika, P.J.; Hugo, A.; Muchenje, V. Nutritional characterization of moringa (*Moringa oleifera lam.*) leaves. *Afr. J. Biotechnol.* **2011**, *10*, 12925–12933.

Article

Effect of *Moringa oleifera* L. Leaf Powder Addition on the Phenolic Bioaccessibility and on In Vitro Starch Digestibility of Durum Wheat Fresh Pasta

Gabriele Rocchetti [1], Corrado Rizzi [2], Gabriella Pasini [3], Luigi Lucini [1], Gianluca Giuberti [1,*] and Barbara Simonato [2]

1 Department for Sustainable Food Process, Università Cattolica del Sacro Cuore, Via Emilia Parmense 84, 29122 Piacenza, Italy; gabriele.rocchetti@unicatt.it (G.R.); luigi.lucini@unicatt.it (L.L.)
2 Department of Biothechnology, University of Verona, Strada Le Grazie 15, 37134 Verona, Italy; corrado.rizzi@univr.it (C.R.); barbara.simonato@univr.it (B.S.)
3 Department of Agronomy, Food, Natural Resources, Animals and Environment (DAFNAE), University of Padova, Viale dell'Università 16, 35020 Legnaro (Padova), Italy; gabriella.pasini@unipd.it
* Correspondence: gianluca.giuberti@unicatt.it; Tel.: +39-0523-599-181

Received: 31 March 2020; Accepted: 13 May 2020; Published: 14 May 2020

Abstract: Fresh pasta was formulated by replacing wheat semolina with 0, 5, 10, and 15 g/100 g (*w/w*) of *Moringa oleifera* L. leaf powder (MOLP). The samples (i.e., M0, M5, M10, and M15 as a function of the substitution level) were cooked by boiling. The changes in the phenolic bioaccessibility and the in vitro starch digestibility were considered. On the cooked-to-optimum samples, by means of ultra-high-performance liquid chromatography-quadrupole time-of-flight (UHPLC-QTOF) mass spectrometry, 152 polyphenols were putatively annotated with the greatest content recorded for M15 pasta, being 2.19 mg/g dry matter ($p < 0.05$). Multivariate statistics showed that stigmastanol ferulate (VIP score = 1.22) followed by isomeric forms of kaempferol (VIP scores = 1.19) and other phenolic acids (i.e., schottenol/sitosterol ferulate and 24-methylcholestanol ferulate) were the most affected compounds through the in vitro static digestion process. The inclusion of different levels of MOLP in the recipe increased the slowly digestible starch fractions and decreased the rapidly digestible starch fractions and the starch hydrolysis index of the cooked-to-optimum samples. The present results showed that MOLP could be considered a promising ingredient in fresh pasta formulation.

Keywords: *Moringa oleifera*; phenolic bioaccessibility; starch digestion; slowly digestible starch; resistant starch

---

1. Introduction

Nowadays, one of the most applied strategies to increase the nutritional properties of a certain food and provide consumers with physiological functions is the incorporation of different functional ingredients during formulation [1]. This strategy would be useful to extend health benefits to the maximum number of consumers, contributing to the reduction of nutrient deficiencies, without impairing the eating habits of the population [2]. In this context, durum wheat semolina pasta, a widely consumed product, can be an excellent staple food for the addition of different bioactive compounds [3]. Indeed, pasta formulated with different sources of dietary fiber, proteins, omega-3 fatty acids, and/or bioactive compounds has been produced [4,5]. In this framework, the use of *Moringa oleifera* L. leaf powder (MOLP) in durum wheat semolina pasta formulation could be considered a promising strategy aiming to improve the overall nutritional quality of this food product.

The *Moringa oleifera* L. plant is native to India and is cultivated worldwide for its characteristic nutritional properties and for its variety of end-uses. Every part of the *Moringa oleifera* plant contains important nutrients and phytochemicals, such as vitamins, minerals, essential amino acids, bioactive compounds, and dietary fiber [6]. The leaves of Moringa are considered a valuable source of distinctive classes of polyphenols, including flavonoids, phenolic acids, and lignans [6,7]. Polyphenols have been studied for their potential health-promoting properties, including their antioxidant capacity [8,9]. However, these benefits are not only related to the content of polyphenols in a certain food, but also to their bioaccessibility, bioavailability, and bioefficacy in humans [10,11]. Therefore, MOLP polyphenols' bioaccessibility studies seem to be essential for a first-step investigation on the potential health benefits of this plant ingredient. However, no information is available on the changes in the phenolic profiles following in vitro digestion (i.e., bioaccessibility) for MOLP-enriched cooked fresh pasta. Besides, although the inclusion of MOLP has been reported to substantially improve the nutritional value of cereal-based foods, by increasing both the protein and dietary fiber contents, none of the studies have determined if the incorporation of MOLP could also contribute to modifying the starch digestibility, at least in vitro, in real food systems (i.e., after cooking) [6].

Considering the growing interest in MOLP in food formulation [6], due to its nutrient composition and the bioactive compound profile [7,12], in this work we produced durum wheat semolina fresh pasta with different substitution levels of MOLP, being 0, 5, 10, and 15 g/100 g (*w/w*), respectively. The MOLP substitution level up to 15 g/100 g (*w/w*) was selected considering that greater levels of MOLP in the recipe could impair the food sensory as well as the technological properties [6].

To better explore the nutritional role of MOLP in fresh pasta production, the present study aimed to evaluate the effect of increasing levels of MOLP in durum wheat semolina fresh pasta by focusing on (i) the phenolic bioaccessibility and (ii) the in vitro starch digestion of cooked pasta.

2. Materials and Methods

2.1. Materials and Fresh Pasta Sample Preparation

Durum wheat semolina and dried MOLP were acquired in a local market. As reported on the label, durum wheat semolina's nutritional composition was as follows (g/100 g product): total starch: 70.8 g; total protein: 11.0 g; total fat: 1.8 g; total dietary fiber: 3.0 g. For the dried MOLP (g/100 g product): total starch: 15.1 g; total sugars: 3.1 g; total protein: 29.9 g; total fat: 8.2 g; total dietary fiber: 30.7 g. The MOLP and durum wheat semolina had a particle size smaller than 0.2 mm.

Fresh durum wheat semolina pasta samples with 100% durum wheat semolina (control: M0) and by replacing semolina with 5, 10, and 15 g/100 g MOLP (*w/w*), obtaining the M5, M10, and M15 pasta samples were produced, respectively. The dough was made with the addition of 35% *v/w* of tap water (37 °C) to the pure semolina or the blend semolina–MOLP by using a pasta machine (Mod. Lillodue, Bottene, Italy). The mixing time was 15 min. The resulting dough was extruded through a bronze die for a spaghetti shape (0.22 cm diameter, approximately 25.0 cm length). For each recipe, three pasta production batches were produced on the same day.

2.2. Moisture Content, Water Activity and Pasta Cooking Properties

The moisture content of the fresh pasta samples was measured with the method 44-15A [13]. Water activity (a_w) was measured using a Hygropalm HC2-AW-meter (Rotronic Italia, Milano) at 23 °C. The AOAC approved method 66-50 was applied for the optimum cooking time (OCT) determination [13]. In particular, samples were cooked in distilled boiling water (ratio of 1:10, *w/v*). At 30 seconds intervals, spaghetti strands were picked from the boiling water and squeezed between 2 glass slides. The OCT for each pasta sample, by definition, is the time for disappearing the white central core of the spaghetti after being squeezed between 2 glass plates.

2.3. Cooking Process and Experimental Details

Prior to in vitro investigations, the spaghetti (5.0 g) were cooked in boiling water (1:10 *w/v*) according to the individual OCT, drained up for 1 min, chopped with a manual meat mincer to simulate mastication, and analyzed "as eaten". Three separate in vitro evaluations were conducted, as detailed below.

2.3.1. In Vitro Static Digestion of Cooked Samples for the Evaluation of the Fate of Polyphenols

The protocol involved an oral, a gastric, and an intestinal stage as reported by Minekus et al. [14]. The cooked-to-optimum pasta samples (i.e., 5.0 g) were sequentially hydrolyzed at 37 °C through (i) an oral phase, (5 mL of salivary fluid at pH = 7.0 plus human salivary α-amylase (A1031; Sigma-Aldrich; Milan, Italy; 75 U/mL) for 2 min; (ii) a gastric phase (10 mL of a simulated gastric fluid at pH 3.0 plus pepsin (P7012; Sigma-Aldrich; 2.000 U/mL) for 120 min; and (iii) an intestinal phase (20 mL of simulated intestinal fluid at pH = 7.0 plus pancreatin (P7545; Sigma-Aldrich; Milan, Italy; 100 U/mL) and bile salts (B8631; Sigma-Aldrich; Milan, Italy; 10 mM) for a further 120 min. Appropriate amounts of HCl (1 M) and NaOH (1 M) were added for the pH adjustment. Liquid aliquots were carefully removed from each hydrolyzed sample after each hydrolysis phase and stored at −20 °C.

2.3.2. Nutritional Starch Fractions Determination

The rapidly digestible starch (RDS) and slowly digestible starch (SDS) were measured with the method of Englyst et al. [15], with minor modifications as detailed by Simonato et al. [5]. The RDS and SDS contents were calculated considering the glucose released after 20 min and 120 min of incubation [15] by measuring the amount of glucose spectrophotometrically using a D-Glucose assay kit (GOPOD, Megazyme, Wicklow, Ireland). The resistant starch (RS) was quantified by a K-RSTAR assay kit (Megazyme, Wicklow, Ireland). The total starch content was calculated as the sum of non-resistant starch and RS following the K-RSTAR assay kit's instructions.

2.3.3. Starch Hydrolysis Index

The cooked-to-optimum spaghetti samples (100 mg) were dispersed in 4 mL of maleic buffer (pH 6), containing an enzyme mixture composed of amyloglucosidase (AMG; 4 μL; 300 U/mL; Megazyme, Wicklow, Ireland) and pancreatic α-amylase (40 mg; 3000 U/mg; Megazyme, Wicklow, Ireland). Samples were incubated in a shaking water bath at 37 °C. At selected time intervals (i.e., 0, 30, 60, 120, and 180 min) the reaction was stopped by adding absolute ethanol. Samples were then centrifuged at 2500× *g* for 10 min. The amount of glucose was quantified as previously detailed, after the correction for glucose present in the AMG solution. Values were plotted on a graph vs. time, and the area under the hydrolysis curve (AUHC; 0–180 min) was measured by using the trapezoid rule. A starch hydrolysis index (HI) value was calculated as the AUHC with the product as a percentage of the corresponding area with white wheat bread [16].

2.4. Extraction and Characterization of Untargeted Phenolic Profile by UHPLC-ESI/QTOF Mass Spectrometry

Three replicates (1.0 g) for each cooked-to-optimum pasta batch were extracted in 10 mL of a methanol/water 80:20 (*v/v*) solution, by using a homogenizer-assisted extraction with an Ultra-Turrax (Ika T25, Staufen, Germany; 5000× *g*; 3 min) [7]. The extracts were centrifuged (10,000× *g*; 10 min; 4 °C), filtered (0.22 μm cellulose syringe filters), and collected [7]. The bound phenolic fraction was extracted from the remaining solid residue [17]. After the alkaline hydrolysis (3 mL of 2 M sodium hydroxide; 1 h; room temperature), the pH was adjusted to 3 with 3 M citric acid and the bound phenolics were extracted with 8 mL of ethyl acetate. After 15 min at 6500 rpm centrifugation, 4 mL of the supernatant was dried under a nitrogen flow at 55 °C and the residue was dissolved in 1 mL of 1% formic acid in 80% methanol, vortexed, and centrifuged (10,000× *g* for 10 min). The resulting solution was filtered (0.22 μm cellulose syringe filters) and 200 μL aliquot was transferred to amber vials for analysis.

The modifications in the polyphenol profile after subjecting the cooked samples through the in vitro static digestion method (i.e., Section 2.3.1) were evaluated by ultra-high-performance liquid chromatography-quadrupole time-of-flight (UHPLC-ESI/QTOF) mass spectrometry [7]. Liquid aliquots collected after the oral, the gastric, and the pancreatic in vitro digestion phases were centrifuged at 7000× g for 10 min and then filtered (0.22 μm cellulose syringe filters). A mixture of water and acetonitrile (VWR, Milan, Italy; both acidified with 0.1% formic acid) as a mobile phase and an Agilent Zorbax Eclipse-plus C18 column (100 mm × 2.1 mm, 1.8 μm) were used. The gradient was from 6% acetonitrile to 94% acetonitrile in 30 min and the flow rate was 0.220 mL/min. The mass spectrometer worked in the positive scan mode (100–1200 *m/z*), injecting 6 μL and source conditions were: sheath gas nitrogen 10 L min^{-1} at 350 °C; drying gas 10 L min^{-1} at 280 °C; nebulizer pressure 60 psig, nozzle voltage 300 V, capillary voltage 3.5 kV. Three technical replicates were done for each pasta batch

The software Agilent Profinder B.06 was used to elaborate the raw features. Features were aligned, and the monoisotopic accurate mass was combined with the isotopic profile for the compounds' annotation, thus reaching a level 2 of confidence in annotation (i.e., putatively annotated compounds) [18]. The database Phenol-Explorer 3.6 was used. The mass accuracy tolerance was set to 5 ppm. Phenolic compounds passing the frequency of the detection thresholds (100% of replications within at least one condition) were classified and then quantitative information was produced using calibration curves (in the range 0.05–500 mg/L) from standard solutions of the single phenolic compounds (purchased from Extrasynthese; Genay; France, purity >98%). Selected representative compounds were as follows: cyanidin, quercetin, luteolin, catechin, tyrosol, and ferulic acid. Results were expressed as mg phenolic equivalents/g dry matter (DM). The polyphenols' bioaccessibility was calculated [19]:

$$\text{Bioaccessibility} = (\text{PCA}/\text{PCB}) \times 100$$

where PCA is the total phenolic subclass content in the samples (mg/g DM) collected after each individual in vitro digestion incubation phase, and PCB is the total phenolic subclass (free plus bound polyphenols) content in the cooked samples before the in vitro digestion.

2.5. Statistical Analysis

Analyses were run in triplicate on each batch and data were expressed as mean values ± standard deviation. The data were evaluated using a one-way analysis of variance (ANOVA). Differences among means were evaluated by Tukey's HSD tests ($p < 0.05$). The statistical software was R project (version 3.2.3, December 2015). Metabolomic data were pre-processed using the software Agilent Mass Profiler Professional B.12.06 (Agilent Technologies, Santa Clara, CA, USA). Compounds were aligned and filtered by abundance (peak area >5000 counts), normalized at the 75th percentile, and baselined against the median [7]. The metabolomics-based dataset was then exported into SIMCA 13 (Umetrics, Malmo, Sweden) to produce a supervised orthogonal projection to latent structures discriminant analysis (OPLS-DA) model. Hotelling's T2 together with CV-ANOVA ($p < 0.01$) and permutation testing were checked to cross-validate the model. Model parameters (i.e., R^2Y and Q^2Y) were recorded. The variable selection method variable importance in projection (VIP) was used to point out the phenolic compounds with the highest discrimination ability (VIP score >1) during the in vitro digestion.

3. Results and Discussion

3.1. Moisture Content, Water Activity and Optimal Cooking Time of Samples

The moisture content and the a_w values of the fresh pasta samples were on average 30.8 g water/100 g of fresh pasta and 0.96, respectively, and were not influenced by the inclusion of MOLP ($p > 0.05$; data not reported). The gradual substitution of semolina flour with MOLP induced a progressive reduction in the OCT, ranging from 5 min for the M0 to 2.5 min for the M15 pasta samples (i.e., 4 min and 3.5 min for M5 and M10, respectively). The progressive decrease in the OCT as a function of the MOLP inclusion level could be due to the great presence of fiber (30.7 g/100 g product) in the MOLP along with the reduction in the total starch content of the samples. For instance, fiber inclusion in wheat pasta can affect the starch–gluten structure, allowing a faster cooking water entrance in the core of the pasta and a resultant faster starch granule gelatinization, thus reducing the OCT [5,20].

3.2. Free and Bound Phenolic Profiles of Cooked-To-Optimum Samples

On the cooked pasta samples, 152 phenolic compounds were putatively annotated, being 38 flavone equivalents (mainly flavones and flavanones), 30 flavonols, 4 flavan-3-ols, 27 anthocyanins, 36 phenolic acids, and 17 remaining compounds. A comprehensive list of each phenolic compound annotated is provided in File S1, considering both the mass abundances and composite spectra. The most abundant compounds detected were tetramethylscutellarein (a flavone), glycosidic and isomeric forms of quercetin and kaempferol (belonging to flavonols), pyrogallol (a low molecular weight phenolic, characterizing mainly the bound phenolic fraction), and malvidin 3-*O*-galactoside (an anthocyanin).

Thereafter, the main phenolic classes characterizing the different cooked samples were targeted. The results are reported in Figure 1, considering both the free (A) and the bound (B) phenolic contents.

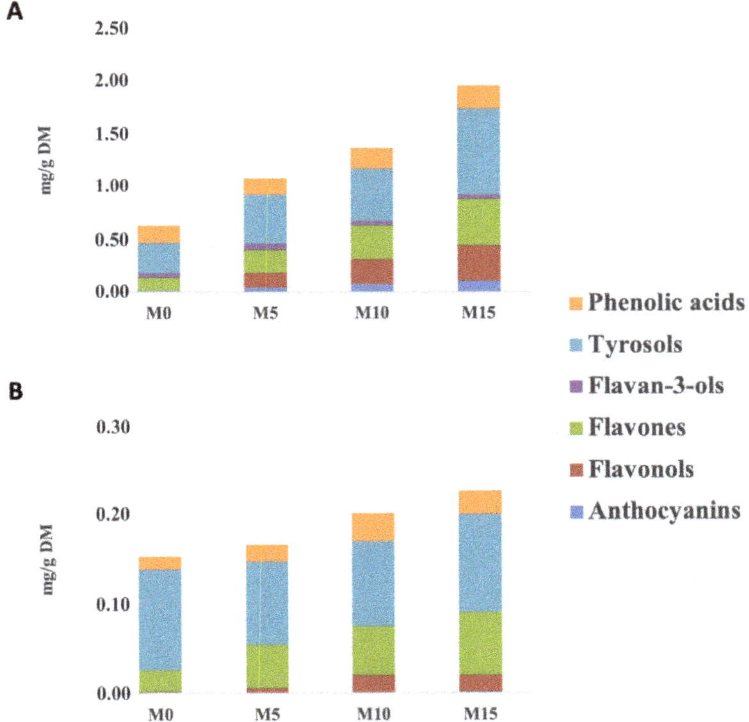

Figure 1. Cumulative phenolic composition (as mg phenolic equivalents/g dry matter) in the cooked pasta samples, considering the free (**A**) and bound (**B**) phenolic fractions. M0: 100% durum wheat semolina fresh pasta. M5, M10, and M15 = Fresh pasta produced with 5, 10, and 15 g/100 g *w/w Moringa oleifera* L. leaf powder, respectively.

Overall, the greatest ($p < 0.05$) total phenolic content (i.e., sum of the different phenolic classes) was found in the M15 sample, being 2.19 mg/g DM, followed by the M10 (1.58 mg/g DM), M5 (1.24 mg/g) and M0 pasta samples (0.78 mg/g DM), as a function of the increasing inclusion level of MOLP in the formulation. Interestingly, the highest inclusion level of MOLP (i.e., 15% *w/w*) was also found to impact the bound phenolic content of the cooked pasta samples (Figure 1B). When considering the specific phenolic composition of the different cooked samples, it was evident that the most represented phenolic classes (in terms of semi-quantitative contents) were low-molecular weight phenolic compounds grouped as tyrosol equivalents (according to the Phenol-Explorer Database), followed by flavonoids (mainly flavonol and flavone equivalents) and phenolic acids. In addition, the highest increase of polyphenols was observed in the M15 sample when considering the total flavonol equivalents; in fact, this class of compounds moved from 0.19 mg/100 g DM for M0 to 35.73 mg/100 g DM for the M15 sample.

Another phenolic class affecting the phenolic profile of the different cooked pasta samples was tyrosol equivalents. These low-molecular weight phenolics were greater in MOLP-substituted spaghetti when compared to M0. These differences may be related to the inherent phenolic profile of MOLP, along with the specific inclusion level in each recipe. Moringa leaves have been reported as a great source of polyphenols, such as flavonoids [7]. Likewise, an abundance of glycosidic forms of quercetin and kaempferol equivalents (i.e., flavonols), followed by hydroxycinnamic/hydroxybenzoic acids and low-molecular-weight phenolics (i.e., phlorin and protocatechuic aldehyde) has been previously indicated [7]. In addition, previous studies reported that cooking by boiling can cause substantial water-losses and/or oxidative degradation of several antioxidant components [21]. According to the literature, whole cereal grains (such as wheat) are reported to be abundant in bound phenolic compounds, such as phenolic acids (i.e., ferulic acid) and lignans. In fact, these compounds are particularly concentrated in the external bran tissues [22]. However, in this work, we found that the M0-cooked sample was characterized by a lower phenolic content, also when considering the bound phenolic composition (Figure 1). Overall, the trends observed for the M0 sample could be explained by considering the different variables such as (a) the milling process conditions, widely reported as one of the major factors able to affect the phenolic profile of durum wheat semolina [23]; (b) the cooking-by-boiling process used; and (c) the rupture of the plant cell structures as promoted by the extraction method, based on a homogenizer-assisted extraction [7].

3.3. Changes of Phenolic Profiles during In Vitro Static Digestion

The cooked-to-optimum samples were hydrolyzed through a standardized static digestion method, with the aim to describe the changes in the phenolic profile. In particular, 102 phenolic compounds were found. Flavonoids were the most represented (56 compounds), followed by hydroxycinnamic acids (24 compounds) and tyrosol equivalents. Overall, the phenolic compounds exhibited different bioaccessibility behaviors, mainly imposed by the presence of different food components (e.g., dietary fiber) in each pasta sample, in line with previous findings [8,10,21]. As can be observed in Table 1, lower percentage bioaccessibility values were detected during the entire gastrointestinal process for specific subclasses of compounds, namely anthocyanins, flavanols, and flavonols.

On the other hand, flavones, tyrosols, and phenolic acids had a moderate bioaccessibility during the in vitro digestion process (mainly after 120 min of the pancreatic step). In fact, the higher percentage bioaccessibility values were measured for the tyrosol equivalents in the M0 sample (i.e., 29.21%), followed by flavones characterizing the M5 sample (i.e., 28.26%). However, a linear trend between the polyphenols' bioaccessibility and MOLP increasing levels in the recipe was not observed. The similarities in percentage of bioaccessibility are expected because the polyphenol content was increased but the in vitro digestive conditions were the same.

Recent studies showed that several bioactivities including antioxidant, antiproliferative, immuneregulatory, hormonal regulation abilities, and neuro-/hepato-/cardioprotective effects can be related to the consumption of phenolic-rich foods [21]. However, these health benefits are greatly dependent on the bioaccessibility potential within the human digestive tract. Present findings corroborated the fact that food components–polyphenols interactions should be considered when studying the changes in the bioaccessibility values during the in vitro digestion in a real food system (i.e., cooked matrix) [8]. Another important factor is the impact of the cooking process, that can modify the bioaccessibility of several phenolic compounds [24,25]. Therefore, the relatively high-percentage bioaccessibility values observed in the pancreatic phase for some phenolic classes (such as phenolic acids, flavones, and tyrosol equivalents) are not surprising and could promote an antioxidant environment in the digestive tract [26]. According to the literature [21,27], the detected bioaccessibility trends may be related to the simulated gastrointestinal digestion conditions used. These latter are not only responsible to break down food matrices and thus release bound phenolic compounds but may also modify the phenolic hydroxyl group (major functional group) of the released phenolics, thus leading to a decrease or increase in the phenolic content in the digestion fluids. In addition, according to the phenolic profile reported for wheat flour, we found a greater bioaccessibility of alkylphenols (quantified as tyrosol equivalents) for the M0 sample during the intestinal step (File S1). In particular, greater percentage bioaccessibility values were measured for three wheat compounds, namely 5-heneicosenylresorcinol, 5-heneicosylresorcinol, and 5-tricosenylresorcinol, which are widely reported as the most abundant in wheat flour [28]. Present findings are difficult to compare with the literature, due to the lack of similar works. To the best of our knowledge, only the work by Caicedo-Lopez and co-authors [12] investigated the changes in the bioaccessibility, intestinal permeability, and antioxidant capacity of the free-phenolic fraction of MOLP after an in vitro gastrointestinal digestion. The authors showed that the greatest content of bioactive compounds was retained in the non-digestible fraction, with higher bioaccessibility values recorded for some phenolics acids, morin, and kaempferol, in line with our findings.

Table 1. Bioaccessibility values (expressed as % of phenolic equivalents) for the different phenolic subclasses during the in vitro static digestion of the cooked-to-optimum pasta samples formulated with different substitution levels of *Moringa oleifera* L. leaf powder (MOLP), considering the oral, gastric, and pancreatic phases.

Phenolic Subclasses	Pasta Samples	TPC Cooked Samples (mg Eq./100 g)	% Bioaccessibility		
			Oral	Gastric	Pancreatic
Anthocyanins	M0	0.87 ± 0.04 [a]	0.31	0.45	nd
	M5	4.47 ± 0.22 [b]	0.14	0.13	0.07
	M10	7.43 ± 0.36 [c]	0.17	0.19	0.37
	M15	10.41 ± 0.50 [d]	0.22	0.20	0.27
Flavonols	M0	0.19 ± 0.01 [a]	nd	nd	nd
	M5	14.74 ± 0.72 [b]	0.04	0.22	1.01
	M10	26.06 ± 1.20 [c]	0.09	0.26	1.13
	M15	35.73 ± 1.80 [d]	0.16	0.44	1.14
Flavones	M0	15.19 ± 0.71 [a]	13.08	9.31	5.53
	M5	26.59 ± 1.33 [b]	7.48	4.69	28.26
	M10	36.70 ± 1.82 [c]	5.01	3.10	17.27
	M15	50.28 ± 2.41 [d]	3.23	2.13	15.16
Flavan-3-ols	M0	4.36 ± 0.18	14.45	nd	nd
	M5	5.20 ± 0.21	7.05	nd	nd
	M10	3.91 ± 0.12	5.52	nd	nd
	M15	4.01 ± 0.13	8.50	nd	nd
Tyrosols	M0	39.25 ± 1.90 [a]	1.92	1.89	29.21
	M5	55.11 ± 2.71 [b]	1.41	1.93	13.80
	M10	60.11 ± 2.98 [b]	1.30	1.73	8.32
	M15	93.53 ± 4.65 [c]	0.87	1.17	3.01
Phenolic acids	M0	17.96 ± 0.86 [a]	3.30	3.10	12.45
	M5	17.95 ± 0.89 [a]	2.42	4.10	12.36
	M10	23.50 ± 1.18 [b]	1.81	3.09	8.97
	M15	24.86 ± 1.25 [b]	1.86	3.22	8.28

M0: wheat semolina fresh pasta. M5, M10, and M15: fresh pasta produced with 5, 10, and 15 g/100 g *w/w* MOLP, respectively. nd = not detected. Within each subclass, means within a column with different superscript letters for the total phenolic content (TPC) of the cooked samples differed at $p < 0.05$.

Multivariate statistics based on OPLS-DA-supervised modelling was then applied to the metabolomics-based dataset. The OPLS-DA score plot built considering each hydrolyzed cooked sample is reported in Figure 2.

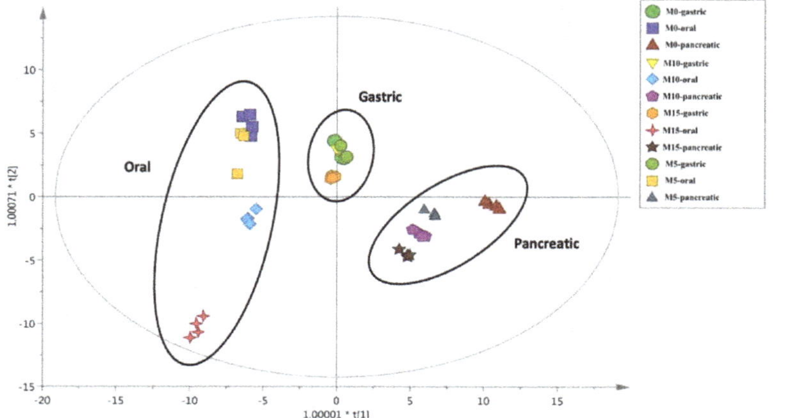

Figure 2. Orthogonal projections to latent structures discriminant analysis (OPLS-DA) score plot on the cooked pasta samples' phenolic profile after oral, gastric, and pancreatic phases of in vitro static digestion. M0: wheat semolina fresh pasta. M5, M10, and M15: fresh pasta produced with 5, 10, and 15 g/100 g *w/w* MOLP, respectively.

Most of the sample variability characterized the oral and the pancreatic phases of the in vitro static digestion, whilst the gastric phase clustered together. The oral-digested M15 sample clustered differently when compared with the others, likely due to the greater content of anthocyanins and flavonol equivalents (i.e., 0.02 and 0.06 mg/100 g DM, respectively). Interestingly, the OPLS-DA score plot confirmed a characteristic phenolic profile for the M0 spaghetti after the pancreatic phase, likely due to its inherent phytochemical composition when compared with the MOLP-substituted counterparts. The OPLS-DA score plot was also inspected for model accuracy parameters recording the acceptable goodness of fit/prediction values (i.e., $R^2X = 0.92$; $R^2Y = 0.66$; $Q^2cum = 0.50$). Finally, the VIP method was used to rank those phenolic compounds most affected during the in vitro digestion. A list containing these VIP markers is reported in File S1, together with the corresponding VIP score (i.e., their discrimination potential) and standard error. Overall, 41 compounds were detected, including flavonoids (such as glycosidic forms of flavonols and flavones, followed by anthocyanins) and phenolic acids (i.e., hydroxycinnamic acids). The highest VIP scores (i.e., representing those compounds most affected by the in vitro static digestion process) were recorded for stigmastanol ferulate (1.23), isomeric and glycosidic forms of kaempferol (1.19), and other flavones (such as luteolin 7-*O*-rutinoside, apigenin 6,8-di-*C*-glucoside, and chrysoeriol 7-*O*-apiosyl-glucoside). Interestingly, the VIP selection method revealed the presence of several anthocyanins, including acetyl-glycosidic forms of peonidin, malvidin, and petunidin. Regarding the other VIP markers discriminating the in vitro digestion process, we found a good distribution of flavones. In this regard, luteolin (VIP score = 1.16) was clearly related to the presence of MOLP in the recipe, being only detected in the M5, M10, and M15 samples (File S1). Similar findings were obtained for quercetin (presenting a VIP score = 1.08), which was found to be abundant in the MOLP-substituted samples during the pancreatic phase.

3.4. In Vitro Starch Digestion of Cooked Samples

The nutritional starch fraction contents, along with the HI of the cooked samples, are reported in Table 2.

Table 2. Starch fractions (g/100 g dry matter), total starch (g/100 g dry matter), and in vitro hydrolysis index (HI) of the cooked-to-optimum pasta samples formulated with different substitution levels of *Moringa oleifera* L. leaf powder (MOLP). Results are expressed as mean ± standard deviation ($n = 3$).

	Substitution with MOLP			
	M0	**M5**	**M10**	**M15**
Rapidly digestible starch	44.3 ± 0.31 [a]	43.8 ± 0.79 [a]	38.1 ± 1.76 [b]	34.1 ± 3.49 [b]
Slowly digestible starch	16.8 ± 0.70 [a]	16.8 ± 0.67 [a]	18.1 ± 0.20 [b]	20.8 ± 0.67 [c]
Resistant starch	2.1 ± 0.26 [a]	1.4 ± 0.04 [b]	1.3 ± 0.01 [c]	1.1 ± 0.04 [d]
Total starch	63.1 ± 1.33 [b]	62.1 ± 1.77 [b]	57.7 ± 1.21 [a]	55.9 ± 1.44 [a]
HI [1]	47.4 ± 1.05 [a]	45.4 ± 1.32 [ab]	43.9 ± 1.21 [ab]	41.8 ± 0.81 [b]

Values in the same row with different superscripts are significantly different ($p < 0.05$). M0: wheat semolina fresh pasta. M5, M10, and M15: fresh pasta produced with 5, 10, and 15 g/100 g *w/w* MOLP, respectively. [1] Calculated using commercial white wheat bread as a reference.

An increase in the SDS and a decrease in the RDS fractions were reported considering the gradual substitution of durum wheat semolina with MOLP. The M10 and M15 cooked pasta samples exhibited the lowest RDS value (i.e., 38.1 and 34.1 g/100 g DM; $p < 0.05$) along with the greatest SDS content (i.e., 18.1 and 20.8 g/100 g DM; $p < 0.05$), when compared with the other samples. From a nutritional standpoint, the RDS fraction was found to be responsible for a rapid increment in the blood glucose levels in humans, while the SDS fraction, being characterized by slow digestion properties, can provide a prolonged release of glucose over time [29]. The mechanism by which the MOLP addition affected the nutritional starch fraction contents of the samples may depend on the interactions among the major constituents (i.e., protein, starch, and fibre polysaccharides) and other minor compounds (i.e., certain classes of dietary polyphenols) [5,8,11,20]. It has been reported that the inclusion of fiber-rich ingredients could modulate the in vitro starch digestion to a certain extent, by changing both the physicochemical properties of the food system [5,30]. For instance, a reduction in the RDS fraction, along with an increment in the SDS fraction exerted by the addition of olive pomace has been reported in wheat-based spaghetti [5]. Lastly, the RS represents, by definition, a certain fraction of starch not degraded in the human small intestine but fermented in the large intestine, with a series of physiological benefits [31]. The RS content of the M0 pasta (i.e., 2.1 g/100 g DM) appeared in line with previous findings for similar food products [32]. However, the RS content of the samples decreased with the increasing inclusion level of MOLP in the recipe, with the lowest value recorded for M15 (i.e., 1.1 g/100 g DM; $p < 0.05$) (Table 2). A possible explanation is that the added MOLP could have contributed to undermine the compact microstructure of wheat pasta, by allowing water and heat to more easily penetrate the pasta during cooking, thus contributing to gelatinize the resistant starch granules present in the core region of the pasta strand to a greater extent [20,33]. In addition, the RS fraction in durum wheat pasta is mainly formed during the pasta extrusion at a high temperature and during the drying process, which in our case was not made [34]. Lastly, the gradual substitution of wheat semolina with MOLP decreased the total starch content of the samples ($p < 0.05$), probably due to a dilution effect of starch exerted by the addition of MOLP, as a consequence of the individual chemical compositions of the selected ingredients.

The starch HI can be used as a predictor of the in vivo glycemic response for a certain starch-based food product [35]. In addition, from the HI values, it is possible to calculate the glycemic index of starch-based foods by applying predictive equations [16]. As reported in Table 2, using white wheat bread as a reference, the HI of the M0 pasta was 47.4. The substitution of a part of the durum wheat semolina with increasing levels of MOLP decreased ($p < 0.05$) the HI of the cooked pasta, the lowest values being recorded for M15 (i.e., 41.8; $p < 0.05$). The decrease in the HI values as a function of the substitution level of MOLP in the recipe could be related to the decrease in the starch content for the replacement of semolina with different quantities of MOLP, as also described in the literature [5], or to the interplay of several factors related to the inherent chemical compositions of the samples. In particular, MOLP contains greater amounts of dietary fiber (about 30.7 g/100 g) and protein (about 29.9 g/100 g) than durum wheat semolina. This may have contributed to entrap starch granules into a non-starchy network with a limited enzyme accessibility [35,36]. Accordingly, cookies enriched with increasing amounts of alfalfa seed (*Medicago sativa* L.) flour showed a reduction in the in vitro starch digestibility compared with the control [37]. Furthermore, the specific phenolic composition characterizing the MOLP-substituted cooked pasta samples (Figure 1; File S1) could have contributed to modulate, at least in part, the in vitro starch digestion of the samples. In particular, certain classes of phenolic compounds may have a role in modulating the in vitro starch digestion, via the inhibition of the starch digestive enzymes (i.e., α-amylase and α-glucosidase enzymes) and/or through the non-covalent interactions with starch on cooking, thus contributing to the formation of inclusion and non-inclusion starch-complexes with a limited enzyme accessibility [11,38,39]. For instance, secondary metabolites characterizing MOLP-substituted pasta (e.g., flavones, flavonols, and hydroxycinnamic acids; File S1) have already been reported to inhibit both α-glucosidase and α-amylase during in vitro activities [40–42].

4. Conclusions

Fresh pasta was formulated by replacing durum wheat semolina with 0, 5, 10, and 15 g/100 g *w/w* of MOLP. After cooking and following an in vitro digestion process, the phenolic compounds exhibited different bioaccessibility behaviors, with an increase in bioaccessibility observed for flavonols characterizing the digested pasta sample formulated with the greatest inclusion level of MOLP in the recipe. Multivariate statistics highlighted a general abundance of flavonoids and phenolic acids among the discriminant markers. Additionally, the inclusion of MOLP in the pasta influenced the rate of in vitro starch digestion in the cooked samples, showing an increase in the SDS fraction, and a decrease in the RDS fraction and HI values. Taken together, the present findings support the fact that MOLP may represent a valuable ingredient to produce a functional pasta. Future investigations considering technological and sensorial aspects are needed to expand the knowledge on the effect of an MOLP addition in fresh pasta formulation.

Supplementary Materials: The following are available online at http://www.mdpi.com/2304-8158/9/5/628/s1, File S1: Metabolomics dataset containing polyphenols identified by UHPLC-QTOF mass spectrometry, together with semi-quantitative values for each phenolic class and VIP markers following OPLS-DA multivariate model.

Author Contributions: Conceptualization, C.R. and B.S.; methodology, G.R., L.L., G.P. and B.S.; software, L.L. and B.S.; validation, C.R., G.G., and B.S.; formal analysis, G.R. and G.G..; investigation, C.R. and B.S.; resources, G.G., B.S.; data curation, L.L., G.G. and B.S.; writing—original draft preparation, G.R., C.R., G.G., B.S.; writing—review and editing, G.R., C.R., L.L., G.G. and B.S.; supervision, B.S.; project administration, B.S.; funding acquisition, G.G. and B.S. All authors have read and agreed to the published version of the manuscript.

Funding: This work was supported by the Cremona FoodLAB (Fondazione Cariplo e Regione Lombardia, Italy).

Acknowledgments: The authors wish to thank the "Enrica e Romeo Invernizzi" Foundation for kindly supporting the metabolomic platform.

Conflicts of Interest: The authors declare no conflict of interest.

References

1. Sparvoli, F.; Laureati, M.; Pilu, R.; Pagliarini, E.; Toschi, I.; Giuberti, G.; Fortunati, P.; Daminati, M.G.; Cominelli, E.; Bollini, R. Exploitation of common bean flours with low antinutrient content for making nutritionally enhanced biscuits. *Front. Plant Sci.* **2016**, *7*, 00928. [CrossRef]
2. Neufeld, L.M.; Friesen, V.M. Chapter 32—Impact Evaluation of Food Fortification Programs: Review of Methodological Approaches Used and Opportunities to Strengthen Them. In *Food Fortification in a Globalized World*; Academic Press: Cambridge, MA, USA, 2018; pp. 305–315.
3. Spinelli, S.; Padalino, L.; Costa, C.; Del Nobile, M.A.; Conte, A. Food by-products to fortified pasta: A new approach for optimization. *J. Clean. Prod.* **2019**, *215*, 985–991. [CrossRef]

4. Mercier, S.; Moresoli, C.; Mondor, M.; Villeneuve, S.; Marcos, B. A meta-analysis of enriched pasta: What are the effects of enrichment and process specifications on the quality attributes of pasta? *Compr. Rev. Food Sci. F* **2016**, *15*, 685–704. [CrossRef]
5. Simonato, B.; Trevisan, S.; Tolve, R.; Favati, F.; Pasini, G. Pasta fortification with olive pomace: Effects on the technological characteristics and nutritional properties. *LWT* **2019**, *114*, 108368. [CrossRef]
6. Oyeyinka, A.T.; Oyeyinka, S.A. *Moringa oleifera* as a food fortificant: Recent trends and prospects. *J. Saudi Soc. Agric. Sci.* **2018**, *17*, 127–136. [CrossRef]
7. Rocchetti, G.; Blasi, F.; Montesano, D.; Gisoni, S.; Marcotullio, M.C.; Sabatini, S.; Cossignani, L.; Lucini, L. Impact of conventional/non-conventional extraction methods on the untargeted phenolic profile of *Moringa oleifera* leaves. *Food Res. Int.* **2019**, *115*, 319–327. [CrossRef]
8. Kardum, N.; Glibetic, M. Chapter Three—Polyphenols and their interactions with other dietary compounds: Implications for human health. *Adv. Food Nutr. Res.* **2018**, *84*, 103–144.
9. Acosta-Estrada, B.A.; Gutiérrez-Uribe, J.A.; Serna-Saldívar, S.O. Bound phenolics in foods, a review. *Food Chem.* **2014**, *152*, 46–55. [CrossRef]
10. Rocchetti, G.; Giuberti, G.; Lucini, L. Gluten-free cereal-based food products: The potential of metabolomics to investigate changes in phenolics profile and their in vitro bioaccessibility. *Curr. Opin. Food Sci.* **2018**, *22*, 1–8. [CrossRef]
11. Sun, L.; Miao, M. Dietary polyphenols modulate starch digestion and glycaemic level: A review. *Crit. Rev. Food Sci.* **2019**, *23*, 1–15. [CrossRef]
12. Caicedo-Lopez, L.H.; Luzardo-Ocampo, I.; Cuellar-Nunez, M.L.; Campos-Vega, R.; Mendoza, S.; Loarca-Pina, G. Effect of the in vitro gastrointestinal digestion on free-phenolic compounds and mono/oligosaccharides from *Moringa oleifera* leaves: Bioaccessibility, intestinal permeability and antioxidant capacity. *Food Res. Int.* **2019**, *120*, 631–642. [CrossRef] [PubMed]
13. Association of Official Analytical Chemists. *Official Methods of Analysis*, 17th ed.; AOAC, Inc.: Arlington, VA, USA, 2000.
14. Minekus, M.; Alminger, M.; Alvito, P.; Ballance, S.; Bohn, T.; Bourlieu, C.; Carrière, F.; Boutrou, R.; Corredig, M.; Dupont, D.; et al. A standardised static in vitro digestion method suitable for food - an international consensus. *Food Funct.* **2014**, *5*, 1113–1124. [CrossRef] [PubMed]
15. Englyst, H.N.; Kingman, S.M.; Cummings, J.H. Classification and measurement of nutritionally important starch fractions. *Eur. J. Clin. Nutr.* **1992**, *46*, 33–50.
16. Granfeldt, Y.; Björck, I.; Drews, A.; Tovar, J. An in vitro procedure based on chewing to predict metabolic response to starch in cereal and legume products. *Eur. J. Clin. Nutr.* **1992**, *46*, 649–660. [CrossRef] [PubMed]
17. Rocchetti, G.; Lucini, L.; Giuberti, G.; Bhumireddy, S.R.; Mandal, R.; Trevisan, M.; Wishart, D. Transformation of polyphenols found in pigmented gluten-free flours during in vitro large intestinal fermentation. *Food Chem.* **2019**, *298*, 125068. [CrossRef] [PubMed]
18. Salek, R.M.; Steinbeck, C.; Viant, M.R.; Goodacre, R.; Dunn, W.B. The role of reporting standards for metabolite annotation and identification in metabolomic studies. *Gigascience* **2013**, *2*, 13. [CrossRef]
19. Rocchetti, G.; Chiodelli, G.; Giuberti, G.; Lucini, L. Bioaccessibility of phenolic compounds following in vitro large intestine fermentation of nuts for human consumption. *Food Chem.* **2018**, *245*, 633–640. [CrossRef]
20. Rakhesh, N.; Fellows, C.M.; Sissons, M. Evaluation of the technological and sensory properties of durum wheat spaghetti enriched with different dietary fibres. *J. Sci. Food Agric.* **2015**, *95*, 2–11. [CrossRef]
21. Shahidi, F.; Peng, H. Bioaccessibility and bioavailability of phenolic compounds. *J. Food Bioact.* **2018**, *4*, 11–68. [CrossRef]
22. Fares, C.; Platani, C.; Baiano, A.; Menga, V. Effect of processing and cooking on phenolic acid profile and antioxidant capacity of durum wheat pasta enriched with debranning fractions of wheat. *Food Chem.* **2010**, *119*, 1023–1029. [CrossRef]
23. Žilić, S. Phenolic compounds of wheat: Their content, antioxidant capacity and bioaccessibility. *MOJ Food Process. Technol.* **2016**, *2*, 85–89. [CrossRef]
24. Sun-Waterhouse, D.; Jin, D.; Waterhouse, G.I.N. Effect of adding elderberry juice concentrate on the quality attributes, polyphenol contents and antioxidant activity of three fibre-enriched pastas. *Food Res. Int.* **2013**, *54*, 781–789. [CrossRef]

25. Rocchetti, G.; Lucini, L.; Chiodelli, G.; Giuberti, G.; Gallo, A.; Masoero, F.; Trevisan, M. Phenolic profile and fermentation patterns of different commercial gluten-free pasta during in vitro large intestine fermentation. *Food Res. Int.* **2017**, *197*, 78–86. [CrossRef]
26. Jara-Palacios, M.J.; Goncalves, S.; Heranz, D.; Heredia, F.J.; Romano, A. Effects of in vitro gastrointestinal digestion on phenolic compounds and antioxidant activity of different white winemaking byproducts extracts. *Food Res. Int.* **2018**, *109*, 433–439. [CrossRef] [PubMed]
27. Ti, H.; Zhang, R.; Li, Q.; Wei, Z.; Zhang, M. Effects of cooking and in vitro digestion of rice on phenolic profiles and antioxidant activity. *Food Res. Int.* **2015**, *76*, 813–820. [CrossRef] [PubMed]
28. Kruk, J.; Aboul-Enein, B.; Bernstein, J.; Marchlewicz, M. Dietary alkylresorcinols and cancer prevention: A systematic review. *Eur. Food Res. Technol.* **2017**, *243*, 1693–1710. [CrossRef]
29. Sajilata, M.G.; Singhal, R.S.; Kulkarni, P.R. Resistant starch—A review. *Compr. Rev. Food Sci. F* **2006**, *5*, 1–17. [CrossRef]
30. Jia, M.; Yu, Q.; Chen, J.; He, Z.; Chen, Y.; Xie, J.; Nie, S.; Xie, M. Physical quality and in vitro starch digestibility of biscuits as affected by addition of soluble dietary fiber from defatted rice bran. *Food Hydrocoll.* **2020**, *99*, 105349. [CrossRef]
31. Fuentes-Zaragoza, E.; Sánchez-Zapata, E.; Sendra, E.; Sayas, E.; Navarro, C.; Fernández-López, J.; Pérez-Alvarez, J.A. Resistant starch as prebiotic: A review. *Starch/Stärke* **2011**, *63*, 406–415. [CrossRef]
32. Espinosa-Solis, W.; Zamudio-Flores, P.B.; Tirado-Gallegos, J.M.; Ramírez-Mancinas, S.; Olivas-Orozco, G.I.; Espino-Díaz, M.; Hernández-González, M.; García-Cano, V.G.; Sánchez-Ortíz, O.; Buenrostro-Figueroa, J.J.; et al. Evaluation of cooking quality, nutritional and texture characteristics of pasta added with oat bran and apple flour. *Foods* **2019**, *8*, 299. [CrossRef]
33. Xou, W.; Sissons, M.; Gidley, M.J.; Gilbert, R.G.; Warren, F.J. Combined techniques for characterising pasta structure reveals how the gluten network slows enzymic digestion rate. *Food Chem.* **2015**, *188*, 559–568.
34. Yue, P.; Rayas-Duarte, P.; Elias, E. Effect of drying temperature on physicochemical properties of starch isolated from pasta. *Cereal. Chem.* **1999**, *76*, 541–547. [CrossRef]
35. Singh, J.; Dartois, A.; Kaur, L. Starch digestibility in food matrix: A review. *Trends Food Sci. Technol.* **2010**, *21*, 168–180. [CrossRef]
36. Dachana, K.B.; Rajiv, J.; Indrani, D.; Prakash, J. Effect of dried Moringa (Moringa Oleifera Lam) leaves on rheological, microstructural, nutritional, textural and organoleptic characteristics of cookies. *J. Food Qual.* **2010**, *33*, 660–677. [CrossRef]
37. Giuberti, G.; Rocchetti, G.; Sigolo, S.; Fortunati, P.; Lucini, L.; Gallo, A. Exploitation of alfalfa seed (Medicago sativa L.) flour into gluten-free rice cookies: Nutritional, antioxidant and quality characteristics. *Food Chem.* **2018**, *239*, 679–687. [CrossRef]
38. Zhu, F. Interactions between starch and phenolic compound. *Trends Food Sci. Technol.* **2015**, *43*, 129–143. [CrossRef]
39. Rocchetti, G.; Giuberti, G.; Gallo, A.; Bernardi, J.; Marocco, A.; Lucini, L. Effect of dietary polyphenols on the in vitro starch digestibility of pigmented maize varieties under cooking conditions. *Food Res. Int.* **2018**, *108*, 183–191. [CrossRef]
40. Khan, W.; Parveen, R.; Chester, K.; Parveen, S.; Ahmad, S. Hypoglycemic potential of aqueous extract of *Moringa oleifera* leaf and in vivo GC-MS metabolomics. *Front. Pharmacol.* **2017**, *8*, 577. [CrossRef]
41. Leone, A.; Bertoli, S.; Di Lello, S.; Bassoli, A.; Ravasenghi, S.; Borgonovo, G.; Forlani, F.; Battezzati, A. Effect of *Moringa oleifera* leaf powder on postprandial blood glucose response: In vivo study on Saharawi people living in refugee camps. *Nutrients* **2018**, *10*, 1494. [CrossRef]
42. Ademiluyi, A.O.; Aladeselu, O.H.; Oboh, G.; Boligon, A.A. Drying alters the phenolic constituents, antioxidant properties, α-amylase, and α-glucosidase inhibitory properties of Moringa (*Moringa oleifera*) leaf. *Food Sci. Nutr.* **2018**, *6*, 2123–2133. [CrossRef]

Article

The Effect of the Addition of *Apulian black* Chickpea Flour on the Nutritional and Qualitative Properties of Durum Wheat-Based Bakery Products

Antonella Pasqualone *, Davide De Angelis, Giacomo Squeo, Graziana Difonzo, Francesco Caponio and Carmine Summo

Department of Soil, Plant and Food Science (DISSPA), University of Bari Aldo Moro, Via Amendola, 165/a, I-70126 Bari, Italy; davide.deangelis@uniba.it (D.D.A.); giacomo.squeo@uniba.it (G.S.); graziana.difonzo@uniba.it (G.D.); francesco.caponio@uniba.it (F.C.); carmine.summo@uniba.it (C.S.)

* Correspondence: antonella.pasqualone@uniba.it

Received: 2 October 2019; Accepted: 14 October 2019; Published: 16 October 2019

Abstract: Historically cultivated in Apulia (Southern Italy), *Apulian black* chickpeas are rich in bioactive compounds such as anthocyanins. This type of chickpea is being replaced by modern cultivars and is at risk of genetic erosion; therefore, it is important to explore its potential for new food applications. The aim of this work was to assess the effect of the addition of *Apulian black* chickpea wholemeal flour on the nutritional and qualitative properties of durum wheat-based bakery products; namely bread, "focaccia" (an Italian traditional bakery product similar to pizza), and pizza crust. Composite meals were prepared by mixing *Apulian black* chickpea wholemeal flour with re-milled semolina at 10:90, 20:80, 30:70, and 40:60. The rheological properties, evaluated by farinograph, alveograph, and rheofermentograph, showed a progressive worsening of the bread-making attitude when increasing amounts of chickpea flour were added. The end-products expanded less during baking, and were harder and darker than the corresponding conventional products, as assessed both instrumentally and by sensory analysis. However, these negative features were balanced by higher contents of fibre, proteins, and bioactive compounds, as well as higher antioxidant activity.

Keywords: pulses; re-milled semolina; bread; pizza; focaccia; rheological properties; reofermentograph; bioactive compounds; texture; sensory profile

---

1. Introduction

The chickpea (*Cicer arietinum* L.) plays an important role in a healthy and environmentally sustainable diet. In fact, as with other pulses, the chickpea fixes atmospheric nitrogen, thus increasing soil fertility [1]. This species is rich in proteins, and particularly if consumed in combination with cereals to compensate for amino acid deficiencies, can help decrease the dietary intake of meat.

Chickpea is commonly classified in two main types: *kabuli*, characterized by large seeds with beige coats, and *desi*, characterized by small and rough seeds with a black or brown coat [2]. However, there is another type of chickpea, historically cultivated in Apulia (South of Italy), apparently similar to *desi* because of its black coat, but different from the genetic point of view [3]. This *Apulian black* type, which is being replaced by modern cultivars, and is therefore, at risk of genetic erosion [3], has an interesting potential for further commercial development due to its high content of antioxidant compounds, such as anthocyanins and carotenoids [4,5].

Both traditional and new food uses of pulses have been proposed in recent years [6]. Chickpea flour has been used as an ingredient in the production of vegetable beverages [7], extruded snacks [8], canned purée [9], pasta [10], and gluten-free bread [11]. Several attempts have been made to also use chickpea flour in conventional bread-making, rediscovering an ancient tradition of Albania and Turkey, where chickpea bread was commonly prepared in the past [12]. The addition of 10–30% chickpea wholemeal flour to common wheat flour was proposed, but a negative effect on bread quality in terms of volume, internal structure, texture, and sensory acceptability was observed [13–15]. Blends of different pulses (chickpeas, lentils, and beans, with chickpeas accounting for only 5% of the total flour) was then proposed [16] for sourdough, the latter being helpful to reduce antinutrients such as phytates. The use of dried chickpea sourdough in bread-making was studied, pointing out that 10% was the optimal amount [17]. Together with the addition of 5–30% fermented chickpea flour, the use of 0.5–3% xanthan gum was found to improve the viscoelastic properties of dough [18]. Alternatively, some improvers, such as ascorbic acid and sodium stearoyl-2-lactylate, were helpful when using 25–35% chickpea flour in combination with wheat wholemeal flour [19]. In this frame, however, only one study proposed to enrich durum wheat re-milled semolina with flour of pulses, namely yellow pea, for preparing bread [20], and no research was made to employ chickpea flour in durum wheat bread-making. Moreover, no study considered the use of pigmented chickpeas.

Durum wheat (*Triticum turgidum* var. *durum*) is largely used in the production of pasta and cous cous, and is characterized by tenacious gluten and by the presence of carotenoid pigments. The latter are important from the nutritional point of view due to provitamin A activity, and confer the typical yellowish colour to the end-products, much appreciated by consumers. In the Mediterranean area, part of durum wheat production is used to prepare traditional bread and other bakery products, with a typically dense crumb [21]. In particular, durum wheat re-milled semolina has to be used, which has a particle size similar to that of bread wheat flour (i.e., smaller than that of semolina used for pasta-making) [22], ensuring high hydration rate.

Pizza, originated in Italy, has become a popular food worldwide. The demand of pizza has continued to grow in recent decades, so that food industrial companies have shown growing interest in its production [23]. Pizza is usually prepared with bread wheat flour, but other types of flour have been proposed in the recent years, including durum wheat re-milled semolina [24]. Pizza crust is a convenient food that can be seasoned at home.

"Focaccia" is another Italian traditional bakery product, widely consumed as street food, similar to pizza but containing higher amounts of oil [25]. It may be defined as a leavened greasy flat bread varyingly seasoned, the most typical topping being cherry tomatoes and olives, accompanied by the rosemary and the potato and onion variants. Additionally, in the preparation of focaccia, the use of durum wheat re-milled semolina is quite common [25].

To the best of our knowledge, no papers have considered the enrichment of durum wheat bread, pizza, and focaccia with black chickpea flour. The aim of this work was, therefore, to assess the effect of the addition of *Apulian black* chickpea flour on the nutritional and qualitative properties of durum wheat based bakery products; namely, bread, focaccia, and pizza crust. In particular, a wholemeal flour of chickpeas was used, for accomplishing the current dietary guidelines that highlight the need of increasing fibre intake.

2. Materials and Methods

2.1. Materials

Apulian black chickpeas (*C. arietinum* L.) were supplied by CerealPuglia s.r.l. (Altamura, Italy). Durum wheat (*T. turgidum* var. *durum*) re-milled semolina was supplied by the milling company Industria Molitoria Mininni s.r.l. (Altamura, Italy). Extra virgin olive oil was supplied by Agridè (Bitonto, Italy). Kastalia stabilized liquid yeast, composed of *Saccharomyces cerevisiae*, salt, and xanthan gum as its stabilizer, and having fermentative power >120 mL CO_2 at 20 °C and 1 atm, was provided by Lesaffre Italia (Trecasali, Parma, Italy).

2.2. Preparation of the Composite Flours

Apulian black chickpeas were ground at the Food Science laboratory of the University of Bari using a laboratory-scale mill (ETA, Vercella Giuseppe, Mercenasco, Italy) equipped with a sieve of 0.6 mm, to obtain wholemeal flour. Durum wheat re-milled semolina was then used to prepare composite meals containing 10, 20, 30, and 40/100 g of *Apulian black* chickpea wholemeal flour.

2.3. Formulation of the Bakery Products

Three types of bakery products, namely bread, focaccia, and pizza crust were produced using a composite meal made of re-milled semolina (60/100 g) and *Apulian black* chickpea wholemeal flour (40/100 g). For each product, a control made of pure re-milled semolina was prepared. The formulation, reported in Table 1, was the same for the three types of products, except for the amount of oil, which was not used for preparing bread according to the traditional formulation of Italian durum wheat bread [21]. The quantity of water was added to flour in quantities sufficient to reach a dough consistency of 500 BU, assessed by preliminary farinograph analyses (Brabender instrument, Duisburg, Germany). The preparation of the experimental bakery products was carried out at the bakery laboratory "Buéne" of the Industria Molitoria Mininni s.r.l. (Altamura, Italy).

Table 1. Formulation of bread, focaccia, and pizza crust, given for 100 g of flour. DW = product prepared by using durum wheat re-milled semolina; BC = product prepared by using a composite meal containing 60/100 g of durum wheat re-milled semolina and 40/100 g of *Apulian black* chickpea wholemeal flour.

Product Type	Ingredient					
	Durum Wheat Re-milled Semolina (g)	*Apulian Black* Chickpea Wholemeal Flour (g)	Water (g)	Extra Virgin Olive Oil (g)	Salt (g)	Yeast (g)
Bread (BC)	60	40	67	-	2	1
Bread (DW)	100	-	62	-	2	1
Focaccia (BC)	60	40	67	20 [1]	2	1
Focaccia (DW)	100	-	62	20 [1]	2	1
Pizza crust (BC)	60	40	67	10	2	1
Pizza crust (DW)	100	-	62	10	2	1

[1] This amount is the sum of 10 g in the dough and 10 g used to oil the pans.

2.4. Preparation of Bread

Flour, water, and yeast were kneaded for 6 min by a spiral mixer (Mecnosud, Flumeri, Italy). Then, salt was added and kneading was continued for other 6 min. The formulation was as in Table 1. The homogeneous dough obtained was left to rise for 1.5 h at 35 °C, RH = 20% (Electric oven-leavening combo EKL 1264 TCR, Tecnoeka S.r.l., Borgoricco, Italy), then was divided into 110 g portions manually shaped at 14 cm length, 6 cm width, and 0.5 cm thickness ("ciabatta" bread type), and left to rise again for 30 min at 35 °C, RH = 20% (EKL 1264 proofer, Tecnoeka S.r.l., Borgoricco, Italy). Bread was finally baked at 220 °C for 20 min in an electric oven (Smeg SI 850 RA-5 oven, Smeg S.p.A., Guastalla, Italy).

2.5. Preparation of Focaccia

Flour, water, and yeast were kneaded for 6 min by a spiral mixer (Mecnosud, Flumeri, Italy). Then, salt and oil were added and kneading was continued for other 6 min. The formulation was as in Table 1. The homogeneous dough obtained was left to rise for 1 h and 15 min at 35 °C, RH = 20% (EKL 1264 proofer, Tecnoeka S.r.l., Borgoricco, Italy), then was divided into 110 g portions of spherical shape, which were manually flattened at 1.5 cm thickness and about 13 cm diameter, put in disposable aluminum pans, previously oiled with other 10 g of extra virgin olive oil, and left to rise again for 1 h and 15 min at 35 °C, RH = 20% (EKL 1264 proofer, Tecnoeka S.r.l., Borgoricco, Italy). Focaccia was finally baked at 220 °C for 15 min in an electric oven (Smeg SI 850 RA-5 oven, Smeg S.p.A., Guastalla, Italy).

2.6. Preparation of Pizza Crust

Flour, water, and yeast were kneaded for 6 min by a spiral mixer (Mecnosud, Flumeri, Italy). Then, salt and oil were added and kneading was continued for other 6 min. The formulation was as in Table 1. The homogeneous dough obtained was divided into 220 g portions of spherical shape which were left to rise for 1 h and 45 min at 35 °C, RH = 20% (EKL 1264 proofer, Tecnoeka S.r.l., Borgoricco, Italy). Subsequently, the dough portions were manually flattened to the thickness of 0.5 cm and diameter of about 28 cm and were immediately baked at 380 °C for 3 min in an electric oven for pizza, equipped with a refractory cooking stone (G3 pizza oven, Ferrari, Rimini, Italy).

2.7. Chemical Analyses

Protein (N × 5.7), ash, and moisture contents were determined according to the American Association of Cereal Chemists (AACC) methods 46-11.02, 44–19, and 08–01, respectively [26]. Fat was extracted and determined by Soxhlet apparatus using diethyl ether as solvent. Total dietary fibre was determined by the enzymatic-gravimetric procedure [27]. Carbohydrates were calculated by difference: 100 − (moisture + proteins + lipids + fibre + ash). Energy value (kJ) was calculated by Atwater general conversion factors, by considering the contribution of 8 kJ/g from total dietary fibre also, in accordance with the Annex XIV of the Regulation (EC) number 1169/2011 [28]. Total anthocyanins, total phenolic compounds and antioxidant activity by DPPH method were determined as in [29]. The antioxidant activity by ABTS method was assessed as in [30]. Total carotenoid pigments were determined according to AACC method 14–50.01 [26]. All analyses were carried out in triplicate.

2.8. Determination of the Rheological Properties and Fermentative Attitude of Flours and Composite Meals

The farinograph indices were determined according to the AACC 54–21 method [26] by a farinograph (Brabender instrument, Duisburg, Germany), equipped with the software Farinograph (Brabender instrument, Duisburg, Germany). Alveograph trials were performed according to the AACC method 54–30A [24] using an alveoconsistograph, equipped with the software Alveolink NG (Tripette et Renaud, Villeneuve-la-Garenne, France). The α-amylase activity was determined by using the Falling Number 1500 apparatus (Perten Instruments AB, Huddinge, Sweden), according to the ISO 3093:2009 method [31]. Rheofermentometer analysis (F3 rheofermentometer, Tripette et Renaud, Chopin Technologies, Villeneuve-la-Garenne, France) was carried out according to the AACC 89-01 method [26] at 28.5 °C for 3 h, with a 2000 g weight. All analyses were carried out in triplicate.

2.9. Physical Determinations of Bakery Products

Texture profile analysis (TPA) was performed on bread and focaccia. Pizza crust was not analyzed due to its very low thickness (0.7–1.0 cm). A Z1.0 TN texture analyzer (Zwick Roell, Ulm, Germany) was used, equipped with a stainless steel square probe (4 cm side) and a 50 N load cell. Data were acquired by means of the TestXPertII version 3.41 software (Zwick Roell, Ulm, Germany). Two centimeter thick slices (3.5 cm × 3.5 cm) were prepared and analyzed. The TPA conditions in the cyclic compression test were: 1 mm/s probe compression rate; 40% sample deformation in both the compressions; and a 5 s pause before second compression. The analyses were carried out in triplicate.

The color indices L^*, a^*, and b^* were measured by using a Chromameter CM-600d (Konica Minolta, Tokyo, Japan). The brown index was calculated as 100 − L^*. Five replicated analyses were made.

The respective diameter (D), length (L), width (W), and thickness (T) values of bread, focaccia, and pizza crust before and after baking were determined by a caliper and used to calculate the percentage variation due to baking as follows:

% of variation of D (or L, W, T) = [D (or L, W, T) after baking − D (or L, W, T) before baking]/D (or L, W, T) before baking × 100. The analyses were carried out in triplicate.

2.10. Sensory Analysis

Quantitative descriptive analysis (QDA) of bread, focaccia, and pizza crust respectively, was performed by a sensory panel consisting of 8 trained members, as described in [32]. The sensory panelists (4 males; 4 females; age range 35 to 52) were recruited based on their previous experience in the sensory evaluation of cereal-based foods among technicians and researchers of the laboratory of the Food Science and Technology unit of the Department of Plant, Soil, and Food Sciences of the University of Bari, Italy. All panel members had neither food allergies nor intolerances and were regular consumers of bread, baked goods, and chickpeas. Pre-test sessions were carried out: (i) to define the list of descriptors to be evaluated in the samples object of the study; (ii) to define the intensity range of each descriptor; (iii) to fix the scale anchors of each descriptor; (iv) to verify reliability, consistency, and discriminating ability of panelists when testing bread, focaccia, and pizza crust. The study protocol followed the ethical guidelines of the laboratory. Panelists were given information about study aims, and individually written informed consent was obtained from each participant. All tested samples were food-grade. A total number of 11 sensory descriptors of appearance, smell, texture, and taste were considered. Seven of them, i.e. external color, chickpea odor, crumb elasticity, crumb consistency, crumb moisture, saltiness, and sweetness were evaluated for all three products; namely bread, focaccia, and pizza crust. Another two descriptors, inner color and crumb porosity, were evaluated only in bread and focaccia due to difficulties in separating the crumb from the surface related to the reduced thickness of pizza crust. Greasiness was evaluated only in focaccia, which contained more oil than the other products, whereas the presence of surface bubbles was evaluated only in pizza crust, where it represents a typical feature. The descriptors were rated on an anchored line scale that provided a 0–9 score range (0 = minimum and 9 = maximum intensity). The analyses were carried out in triplicate.

2.11. Statistical Analysis

Statistical analysis was carried out using XLSTAT software (Addinsoft SARL, New York, NY, USA). Significant differences were determined at $p < 0.05$ by one-way analysis of variance (ANOVA) followed by Tukey's HSD test.

3. Results and Discussion

3.1. Nutritional Characteristics of the Starting Flours

Apulian black chickpea wholemeal flour showed a significantly higher ($p < 0.05$) content of proteins, lipids, and fibre than re-milled semolina (Table 2). The protein content of chickpea flour was slightly higher than those observed in previous works [4,5], whereas the value observed in re-milled semolina agreed with previous quality surveys [22,33]. The fibre content of chickpea flour was within the range observed in other studies [4,5].

The two types of flour exhibited a similar content of phenolic compounds. However, black chickpea flour was characterized by a significantly ($p < 0.05$) higher content of anthocyanins and carotenoids than re-milled semolina. Consequently, the antioxidant activity was also stronger.

These positive features of *Apulian black* chickpea flour confirmed the results of previous studies [4,5].

Table 2. Nutritional characteristics, bioactive compounds, and antioxidant activities of flours used in the production of experimental bread, pizza crust, and focaccia (values are expressed on fresh weight bases).

Parameter	Durum Wheat Re-milled Semolina	*Apulian Black* Chickpea Wholemeal Flour
Moisture (g/100 g)	14.7 ± 0.1 [a]	9.3 ± 0.2 [b]
Carbohydrates (g/100 g)	68.5 ± 1.9 [a]	47.4 ± 1.3 [b]
Proteins (g/100 g)	12.3 ± 0.1 [b]	21.4 ± 0.4 [a]
Lipids (g/100 g)	1.5 ± 0.1 [b]	4.3 ± 0.3 [a]
Fibre (g/100 g)	2.1 ± 0.3 [b]	14.9 ± 1.6 [a]
Ash (g/100 g)	0.88 ± 0.01 [b]	2.70 ± 0.01 [a]
Total anthocyanins (mg/kg cyanidin 3-O-glucoside)	n.d. [1]	69.5 ± 2.6
Total carotenoids (mg/kg β-carotene)	5.75 ± 0.11 [b]	32.7 ± 2.4 [a]
Total phenolic compounds (mg/g ferulic acid)	0.97 ± 0.03 [a]	1.07 ± 0.05 a
Antioxidant activity-ABTS method (μmol Trolox/g)	0.86 ± 0.04 [b]	1.89 ± 0.07 [a]
Antioxidant activity-DPPH method (μmol Trolox/g)	2.25 ± 0.29 [b]	4.04 ± 0.01 [a]

[1] n.d. = not detected. Different letters in a row indicate significant differences ($p < 0.05$).

3.2. Rheological and Fermentative Characteristics of Flours and Composite Meals

Composite meals were prepared by mixing *Apulian black* chickpea wholemeal flour with re-milled semolina at 10:90, 20:80, 30:70, and 40:60. The rheological properties of the obtained blends were then evaluated by farinograph, alveograph, and rheofermentograph, in order to assess the suitability to the production of fermented bakery products.

The farinograph parameters measured in pure re-milled semolina agreed with previous works [22]. The addition of black chickpea flour significantly ($p < 0.05$) influenced all farinograph parameters (Table 3). In particular, water absorption, dough development time, and loss of consistency progressively increased, whereas the dough stability decreased. The increase of water absorption was due to the presence of fibre in chickpea wholemeal flour, able to absorb high amounts of water. The dilution of gluten, chickpea flour being gluten free, and the presence of fibre able to interfer with a gluten network, were responsible for the decrease of dough stability, the increase of time needed to develop gluten, and the increase of consistency loss. These results highlight that the addition of chickpea flour worsened the bread-making attitude of wheat flour, with a more evident effect at higher doses. Similar negative effects on farinograph were reported when re-milled semolina was added of other fibre-rich ingredients, such as powdered almond skins [34], or yellow pea wholemeal flour [20]. Moreover, the same effects have been reported for blends of chickpea wholemeal flour with bread wheat flour [14,15] or with semolina for pasta-making [10], evidencing that the bread-making attitude of any gluten-containing flour is depressed by the addition of chickpea flour.

As for the alveograph strength (W), it progressively decreased by the addition of increasing amounts of chickpea flour (Table 3). Moreover, that result was imputable to the increasing content of fibre, contributed by chickpea wholemeal flour, and to gluten's dilution. The values of the alveograph tenacity/extensibility ratio (P/L) ratio, instead, remained almost constant after the addition of chickpea flour. The value of P/L observed for pure re-milled semolina was in the range observed in previous works, whereas W was particularly high, indicating a very good quality level [22].

The fermentative attitude of flour blends was evaluated by measuring the falling number, related to the amylase activity, and by performing the rheofermentograph analysis. The values of falling number showed a significant difference ($p < 0.05$) only when comparing pure re-milled semolina (with the highest amylase activity) to the composite flour containing the highest chickpea amount (with the lowest amylase activity) (Table 4). The decrease of amylase activity with the addition of 40/100 g of chickpea flour reflected the presence of α-amylase inhibitors, reported in chickpeas and in several other pulses [35]. These inhibitors are slightly more active in *desi* than in *kabuli* chickpea cultivars [36], but can be inactivated by heat treatment [35].

Table 3. Farinograph and alveograph data of pure durum wheat re-milled semolina and of blends with increasing amounts of *Apulian black* chickpea wholemeal flour.

Parameter	Amount of Black Chickpea Flour (g/100 g)				
	0	10	20	30	40
Farinograph					
Water absorption at 500 B.U. [1] (g/100 g)	61.8 ± 0.1 [e]	63.2 ± 0.1 [d]	64.8 ± 0.1 [c]	66.3 ± 0.1 [b]	67.2 ± 0.2 [a]
Development time (min)	2.0 ± 0.3 [d]	2.1 ± 0.3 [d]	4.1 ± 0.3 [c]	4.9 ± 0.2 [b]	5.7 ± 0.1 [a]
Dough stability (min)	8.5 ± 0.5 [a]	6.6 ± 0.6 [b]	5.7 ± 0.2 [c]	3.6 ± 0.2 [d]	2.6 ± 0.3 [e]
Loss of consistency at 12 min (B.U.)	49.7 ± 3.2 [d]	66.3 ± 5.1 [c]	76.1 ± 3.1 [b]	83.3 ± 3.8 [ab]	86.3 ± 1.5 [a]
Alveograph					
W (10^{-4} J)	282 ± 4 [a]	211 ± 12 [b]	150 ± 10 [c]	108 ± 7 [d]	97 ± 11 [e]
P/L	2.7 ± 0.1 [a]	2.6 ± 0.3 [a]	2.6 ± 0.1 [a]	2.9 ± 0.3 [a]	3.0 ± 0.3 [a]

[1] B.U. = Brabender units. Different letters in a row indicate significant differences ($p < 0.05$).

Table 4. The fermentative attitude of pure durum wheat re-milled semolina and of blends with increasing amounts of *Apulian black* chickpea wholemeal flour.

Parameter	Amount of Black Chickpea Flour (g/100 g)				
	0	10	20	30	40
Amylase activity					
Falling Number (s)	532 ± 8 [b]	537 ± 6 [ab]	541 ± 6 [ab]	539 ± 3 [ab]	549 ± 5 [a]
Rheofermentograph – Curve of gas production and retention					
Volume of gas produced (V_T) (mL)	2098 ± 8 [a]	2077 ± 7 [b]	2037 ± 3 [c]	1994 ± 4 [d]	1972 ± 8 [d]
Volume of gas retained (V_R) (mL)	1419 ± 11 [a]	1388 ± 13 [b]	1325 ± 15 [c]	1302 ± 21 [c]	1266 ± 6 [d]
Volume of gas lost (V_L) (mL)	679 ± 8 [b]	689 ± 6 [b]	712 ± 11 [a]	692 ± 14 [ab]	706 ± 8 [a]
Coefficient of gas retention (V_R/V_T) (%)	68 ± 1 [a]	67 ± 1 [ab]	65 ± 1 [bc]	64 ± 1 [c]	64 ± 1 [c]
Maximum height of gas production curve (H'_m) (mm)	84 ± 1 [a]	82 ± 1 [ab]	81 ± 1 [b]	74 ± 1 [c]	73 ± 1 [c]
Time needed to start losing gas (T_x) (min)	59 ± 2 [a]	57 ± 4 [ab]	56 ± 2 [ab]	53 ± 2 [b]	48 ± 2 [c]
Rheofermentograph – Curve of dough development					
Maximum dough height (*Hm*) (mm)	49 ± 2 [a]	44 ± 1 [b]	41 ± 1 [c]	35 ± 1 [d]	23 ± 1 [e]
Dough height after 3 h (*h*) (mm)	43 ± 2 [a]	38 ± 1 [b]	33 ± 1 [c]	28 ± 1 [d]	15 ± 1 [e]

Different letters in a row indicate significant differences ($p < 0.05$).

Rheofermentograph data of pure re-milled semolina agreed with other works [37]. Increasing amounts of chickpea flour significantly reduced ($p < 0.05$) the amount of gas produced (V_T) during the rheofermentograph analysis, in agreement with the decrease of amylase activity. Moreover, a greater amount of gas was lost by the dough (V_L) when composite meals were analyzed. Consequently, a progressively lower quantity of gas was retained (V_R) as the addition of chickpea flour increased, reflecting in a significantly lower coefficient of gas retention. Besides, the loss of gas appeared sooner (T_x) in chickpea-added dough than in case of pure re-milled semolina.

The volumetric increase of leavened bakery products depends on both the amount of CO_2 developed and the rheological properties of dough, in terms of quality and strength of the gluten network, which allows one to effectively retain the gas developed during fermentation. Therefore, these results were due to the diminished fermentative attitude of chickpea-added dough, as shown by the lower amylase activity, and to its weaker gluten network, evidenced by the alveograph and farinograph analyses.

As for the dough development curve, its maximum height (*Hm*), as well as the height after 3 h (*h*), decreased significantly with the increase of chickpea wholemeal flour added, indicating a lower inflation of the dough.

Overall, the bread-making attitude of re-milled semolina worsened by increasing the level of enrichment with black chickpea wholemeal flour. However, studies showed that consumer behavior is changing, driven by awareness of the relationship between nutrition and health. Consequently, whole and fibre-enriched bakery products are accepted better than in the past, despite the lower quality characteristics, particularly if information on the high fibre content is shown on the label [38]. Therefore, the search for the best rheological parameters should no longer be the exclusive criterion for selecting the optimal level of enrichment.

By calculating the theoretical fibre content of the end-products, it was found that the addition of 40/100 g black chickpea wholemeal flour would make it possible to claim the "fibre source" on the label, which requires a content of at least 3 g of fibre per 100 g of product [39]. Lower levels of addition would not reach the conditions for the inclusion of this statement in the label. Considering that information about the presence of fibre influences the acceptance of the modern consumer, leading to a positive perception of unconventional bakery products, the latter were prepared using composite flours of durum wheat re-milled semolina added with 40/100 g of black chickpea wholemeal flour.

3.3. *Nutritional and Qualitative Properties of the Bakery Products*

Table 5 reports the nutritional features of the experimental bakery products (analytically determined data, not calculated). Bread, focaccia, and pizza crust enriched with black chickpea flour showed significantly ($p < 0.05$) higher contents of proteins, lipids, ash, and fibre than conventional products, whereas carbohydrates were significantly lower (Table 5). The slight increase in lipids was coupled to higher fibre content; therefore, the energy value of the enriched products was similar to that of the conventional products.

Table 5. Nutritional features (values per 100 g, expressed on fresh weight basis) of conventional durum wheat bread, focaccia, and pizza crust and of their black chickpea-enriched versions. DW = product prepared by using durum wheat re-milled semolina; BC = product prepared by using a composite meal containing 60/100 g of durum wheat re-milled semolina and 40/100 g of Apulian black chickpea wholemeal flour.

Parameter	Bread		Focaccia		Pizza Crust	
	DW	BC	DW	BC	DW	BC
Carbohydrates (g)	50.9 ± 2.3 [a]	44.1 ± 3.1 [b]	49.1 ± 1.8 [a]	41.7 ± 2.2 [b]	52.6 ± 2.6 [a]	46.2 ± 2.1 [b]
Proteins (g)	8.5 ± 0.1 [b]	9.9 ± 0.1 [a]	7.7 ± 0.1 [b]	10.3 ± 0.1 [a]	8.4 ± 0.1 [b]	10.5 ± 0.1 [a]
Lipids (g)	0.5 ± 0.1 [b]	1.3 ± 0.1 [a]	5.1 ± 0.1 [b]	5.8 ± 0.2 [a]	1.9 ± 0.1 [b]	2.2 ± 0.1 [a]
Fibre (g)	1.3 ± 0.1 [b]	5.2 ± 0.3 [a]	0.9 ± 0.1 [b]	3.6 ± 0.2 [a]	1.2 ± 0.1 [b]	4.6 ± 0.2 [a]
Ash (g)	1.8 ± 0.1 [b]	2.1 ± 0.1 [a]	2.1 ± 0.1 [b]	2.7 ± 0.1 [a]	1.8 ± 0.1 [b]	2.2 ± 0.1 [a]
Energy value (kJ)	1039 ± 15 [a]	1008 ± 15 [a]	1161 ± 19 [a]	1126 ± 16 [a]	1117 ± 21 [a]	1081 ± 16 [a]

Different letters in a row, for the same bakery product, indicate significant differences ($p < 0.05$).

The addition of black chickpea flour determined a significant increase ($p < 0.05$) of the content of anthocyanins and phenolic compounds in all three baked goods considered (Table 6). However, a marked reduction was observed with respect to the starting flour (reported in Table 2). The decrease of bioactives during thermal processing of *Apulian black* chickpeas was also reported during the preparation of canned sterilized puré [9]. The antioxidant activity, particularly when measured with the ABTS method, was significantly higher in chickpea-enriched bakery products than in durum wheat ones, reflecting the contents of bioactive compounds, specifically anthocyanins and carotenoids.

The variations in the dimensional parameters of bakery products induced by cooking, and their weight loss, are shown in Table 7. Usually, baking involves an increase in volume determined by the thermal expansion of the gases (air trapped during mixing and kneading, water vapor evaporated by the dough, and carbon dioxide originated by leavening). An increase in thickness, indeed, was observed, but without significant differences between conventional durum wheat and chickpea-enriched products. On the other hand, the diameter of circular products, i.e. focaccia and pizza crust, as well as length and width of bread, which had a rectangular shape, decreased with baking. The decrease was significantly more marked ($p < 0.05$) in chickpea-enriched products than in conventional durum wheat products. In bread, the conventional product had an opposite behavior showing an increase of width and maintaining its length almost constant.

Table 6. The bioactive compounds and antioxidant activity of conventional durum wheat bread, focaccia, and pizza crust and of their black, chickpea-enriched versions. DW = product prepared by using durum wheat re-milled semolina; BC = product prepared by using a composite meal containing 60/100 g of durum wheat re-milled semolina and 40/100 g of *Apulian black* chickpea wholemeal flour.

Parameter	Bread		Focaccia		Pizza Crust	
	DW	BC	DW	BC	DW	BC
Total anthocyanins [1]	n.d. [5]	7.51 ± 1.47	n.d.	3.49 ± 0.10	n.d.	4.37 ± 0.03
Total carotenoids [2]	2.69 ± 0.02 ^{b}	4.57 ± 0.45 ^{a}	2.77 ± 0.01 ^{b}	5.21 ± 0.15 ^{a}	2.68 ± 0.02 ^{b}	5.23 ± 0.16 ^{a}
Total phenolic compounds [3]	0.06 ± 0.01 ^{b}	0.09 ± 0.01 ^{a}	0.04 ± 0.01 ^{b}	0.12 ± 0.01 ^{a}	0.07 ± 0.01 ^{b}	0.12 ± 0.01 ^{a}
Antioxidant activity-ABTS [4]	1.70 ± 0.05 ^{b}	2.13 ± 0.02 ^{a}	1.44 ± 0.07 ^{b}	1.76 ± 0.03 ^{a}	1.25 ± 0.01 ^{b}	2.01 ± 0.01 ^{a}
Antioxidant activity-DPPH [4]	0.04 ± 0.03 ^{b}	0.08 ± 0.01 ^{a}	0.07 ± 0.01 ^{b}	0.10 ± 0.01 ^{a}	0.08 ± 0.01 ^{a}	0.08 ± 0.01 ^{a}

[1] Expressed as mg/kg cyanidin 3-*O*-glucoside (d.m.); [2] expressed as mg/kg β-carotene (d.m.); [3] expressed as mg/g ferulic acid (d.m.); [4] expressed as μmol/g Trolox equivalent (d.m.); [5] n.d. = not detected. Different letters in a row, for the same bakery product, indicate significant differences ($p < 0.05$).

Overall, these results agreed with the predictive analyses on dough rheology which, in turn, were due to the interference by fibre and the dilution of gluten. The high alveograph P/L ratio observed in the chickpea-added dough, in particolar, explains the limited expansion during both leavening and the first phases of baking (oven-spring). Thickness was less affected by this limitation because baking essentially induces an upward push [40].

Weight loss though, was not significantly influenced.

Table 7. Baking-induced variations of the dimensional parameters of conventional durum wheat bread, focaccia, and pizza crust and of their black, chickpea-enriched versions. DW = product prepared by using durum wheat re-milled semolina; BC = product prepared by using a composite meal containing 60/100 g of durum wheat re-milled semolina and 40/100 g of *Apulian black* chickpea wholemeal flour.

Parameter	Bread		Focaccia		Pizza Crust	
	DW	BC	DW	BC	DW	BC
Diameter variation (%)	-	-	−0.7 ± 0.1 ^{a}	−3.8 ± 0.1 ^{b}	−1.8 ± 0.0 ^{a}	−9.8 ± 0.0 ^{b}
Length variation (%)	−0.4 ± 0.7 ^{a}	−15.1 ± 1.4 ^{b}	-	-	-	-
Width variation (%)	7.7 ± 2.1 ^{a}	−6.6 ± 3.2 ^{b}	-	-	-	-
Thickness variation (%)	84.3 ± 1.6 ^{a}	83.1 ± 0.9 ^{a}	54.2 ± 1.4 ^{b}	59.3 ± 4.1 ^{a}	54.2 ± 7.2 ^{a}	66.7 ± 5.8 ^{a}
Weight loss (%)	10.3 ± 0.3 ^{a}	10.2 ± 0.4 ^{a}	9.3 ± 0.9 ^{a}	9.3 ± 0.6 ^{a}	8.9 ± 0.4 ^{a}	9.2 ± 0.4 ^{a}

Different letters in a row, for the same bakery product, indicate significant differences ($p < 0.05$).

Table 8 shows the colorimetric indices of the external and internal surface of the various bakery products prepared. Pizza was inspected only externally, due to its limited thickness (0.7–1.0 cm). The addition of black chickpea flour caused a significant ($p < 0.05$) decrease of b^* and an increase of brown index ($100 - L^*$) of all the products, which appeared grayish.

Statistically significant differences of a^* were observed only in the internal part of bread and focaccia, being higher in the chickpea-added products. All products from pure re-milled semolina, instead, were bright yellow, with values of b^* almost double those of chickpea-added products. Therefore, together with the reduced expansion degree of products during baking, color alteration is another negative effect of adding black chickpea wholemeal flour, which would require an adequate explanation to the final consumer to highlight the nutritional reasons behind it, in order not to appear too unconventional or even unpleasant (Figure 1).

Figure 1. The internal structure of bread prepared by using a composite meal containing 60/100 g of durum wheat re-milled semolina and 40/100 g of *Apulian black* chickpea wholemeal flour (**left**) and bread prepared by using only durum wheat re-milled semolina (**right**).

Table 8. Color parameters of conventional durum wheat bread, focaccia, and pizza crust and of their black, chickpea-enriched versions. DW = product prepared by using durum wheat re-milled semolina; BC = product prepared by using a composite meal containing 60/100 g of durum wheat re-milled semolina and 40/100 g of *Apulian black* chickpea wholemeal flour.

Product Type	Color Parameter		
	b^*	a^*	$100 - L^*$
Bread			
DW [1]	36.7 ± 3.2 [a]	6.7 ± 2.2 [a]	31.7 ± 6.6 [b]
BC [1]	21.4 ± 3.7 [b]	10.3 ± 4.7 [a]	46.9 ± 3.3 [a]
DW [2]	23.9 ± 1.2 [a]	0.4 ± 0.2 [b]	19.5 ± 1.9 [b]
BC [2]	12.9 ± 0.5 [b]	1.9 ± 0.1 [a]	52.7 ± 2.4 [a]
Focaccia			
DW [1]	38.3 ± 3.1 [a]	5.7 ± 2.9 [a]	35.4 ± 7.3 [b]
BC [1]	22.5 ± 3.1 [b]	11.3 ± 6.5 [a]	51.3 ± 4.9 [a]
DW [2]	25.7 ± 0.9 [a]	0.4 ± 0.2 [b]	19.5 ± 0.7 [b]
BC [2]	16.3 ± 0.7 [b]	1.7 ± 0.2 [a]	48.2 ± 1.9 [a]
Pizza crust			
DW [1]	22.4 ± 2.8 [a]	2.6 ± 1.3 [a]	25.1 ± 2.3 [b]
BC [1]	12.3 ± 2.3 [b]	1.2 ± 0.9 [a]	41.1 ± 8.9 [a]

[1] External color. [2] Internal color. Different letters in a column for the same bakery product and portion inspected, indicate significant differences ($p < 0.05$).

Table 9 shows the textural parameters of experimental bread and focaccia. Again, pizza crust was not analyzed due to its very limited thickness (0.7–1.0 cm). The chickpea-added bread and focaccia were significantly harder ($p < 0.05$) and more chewy than their counterparts made only of re-milled semolina. The springiness was very similar in all products, whereas the cohesiveness of chickpea-added bread was lower than in conventional durum wheat bread. These results, in agreement with the reduced expansion in volume, can be explained by a worse gluten formation, as already indicated by the alveograph and farinograph parameters of the starting meals, due to the richness in fibre and absence of gluten in chickpea flour.

As for the sensory profile (Table 10), all the chickpea-enriched products were significantly more consistent ($p < 0.05$) than the corresponding conventional products. The addition of chickpea flour determined the significant emergence of an odorous note of chickpeas, a darker color (both external and internal, in agreement with colorimetric data), and an increase in moisture. Saltiness and sweetness, on the other hand, were similar in all the products, as well as crumb porosity.

Table 9. Textural parameters of conventional durum wheat bread and focaccia and of their black, chickpea-enriched versions. DW = product prepared by using durum wheat re-milled semolina; BC = product prepared by using a composite meal containing 60/100 g of durum wheat re-milled semolina and 40/100 g of *Apulian black* chickpea wholemeal flour.

Parameter	Bread		Focaccia	
	DW	BC	DW	BC
Hardness (N)	7.90 ± 1.54 [b]	20.80 ± 0.80 [a]	5.50 ± 0.45 [b]	14.77 ± 2.01 [a]
Springiness	0.94 ± 0.01 [a]	0.92 ± 0.01 [a]	0.95 ± 0.03 [a]	0.93 ± 0.02 [a]
Chewiness (N)	5.59 ± 0.92 [b]	12.16 ± 1.57 [a]	3.81 ± 0.35 [b]	9.90 ± 1.75 [a]
Cohesiveness	0.74 ± 0.03 [a]	0.63 ± 0.05 [b]	0.75 ± 0.05 [a]	0.72 ± 0.07 [a]

Different letters in a row indicate significant differences ($p < 0.05$).

Greasiness, typical of focaccia but unpleasant if eccessive, was significantly more evident in the conventional focaccia than in the chickpea-added one. The inclusion of oil in the formulation, more abundant in focaccia than in pizza crust and bread, probably influenced the consistency of the focaccia as well, which was slightly softer than bread, in agreement with the results of textural analysis. Other studies reported the positive effect of oil on the consistencies and volumes of bakery products [41].

As for the pizza crust, the presence of surface bubbles was significantly greater in the conventional product than in the chickpea-added one, whose dough was less extensible and less able to retain gas, as shown by rheological analyses.

Table 10. The sensory profile of conventional durum wheat bread, focaccia, and pizza crust, and of their black chickpea-enriched versions. DW = product prepared by using durum wheat re-milled semolina; BC = product prepared by using a composite meal containing 60/100 g of durum wheat re-milled semolina and 40/100 g of *Apulian black* chickpea wholemeal flour.

Parameter	Bread		Focaccia		Pizza Crust	
	DW	BC	DW	BC	DW	BC
External color	4.4 ± 0.2 [b]	7.7 ± 0.2 [a]	4.5 ± 0.2 [b]	7.9 ± 0.1 [a]	3.8 ± 0.2 [b]	7.2 ± 0.3 [a]
Inner color	2.9 ± 0.1 [b]	7.5 ± 0.3 [a]	3.1 ± 0.5 [b]	7.1 ± 0.6 [a]	-	-
Presence of surface bubbles	-	-	-	-	2.1 ± 0.5 [a]	0.3 ± 0.1 [b]
Chickpea odor	0.1 ± 0.1 [b]	4.7 ± 0.6 [a]	0.2 ± 0.1 [b]	5.6 ± 0.8 [a]	0.1 ± 0.1 [b]	4.8 ± 0.1 [a]
Greasiness	-	-	5.1 ± 0.2 [a]	4.1 ± 0.1 [b]	-	-
Crumb elasticity [1]	5.7 ± 0.1 [a]	5.2 ± 0.1 [b]	6.7 ± 0.4 [a]	6.3 ± 0.1 [a]	6.3 ± 1.2 [a]	5.7 ± 0.9 [a]
Crumb consistency [1]	2.9 ± 0.3 [b]	5.2 ± 0.1 [a]	1.9 ± 0.2 [b]	4.6 ± 0.1 [a]	2.4 ± 0.2 [b]	4.5 ± 0.3 [a]
Crumb porosity	4.3 ± 0.3 [a]	3.9 ± 0.3 [a]	4.2 ± 0.4 [a]	4.0 ± 0.1 [a]	-	-
Crumb moisture [1]	4.6 ± 0.2 [b]	5.7 ± 0.2 [a]	4.3 ± 0.1 [b]	6.0 ± 0.4 [a]	4.5 ± 1.1 [a]	5.7 ± 0.8 [a]
Saltiness	3.0 ± 0.1 [a]	2.9 ± 0.4 [a]	2.9 ± 0.3 [a]	2.7 ± 0.2 [a]	2.3 ± 0.2 [a]	2.1 ± 0.1 [a]
Sweetness	1.2 ± 0.2 [a]	1.7 ± 0.3 [a]	1.3 ± 0.2 [a]	1.6 ± 0.3 [a]	1.2 ± 0.3 [a]	1.6 ± 0.4 [a]

[1] Evaluated on the whole product. Different letters in a row, for the same bakery product, indicate significant differences ($p < 0.05$).

4. Conclusions

On the basis of the results obtained during the characterization of the composite meals, it emerged that the addition of the wholemeal flour of *Apulian black* chickpeas to durum wheat re-milled semolina caused a decrease in the bread-making attitude, which, however, was countered by a nutritional improvement in terms of higher contents of fibre and proteins. The enriched end-products showed also higher contents of bioactive compounds and an improved antioxidant activity. The positive features should be adequately communicated to the consumer to compensate the significant negative effects of addition of chickpea flour, such as alterations of consistency and color with respect to analogous baked goods made of pure re-milled semolina.

Among the three products evaluated, the best product for consumer would be bread because of its lower content of lipids, and consequently, lower energy value. However, with enrichment levels of chickpea flour as high as 40/100 g, all three products evaluated were able to help fulfil the recent dietary guidelines, which suggest one to consume at least three legume servings per week. Adding chickpea flour to baked goods, therefore, represents a nutritionally effective strategy and a significant step forward to increase the consumption of legumes.

Author Contributions: Conceptualization, A.P. and C.S.; formal analysis, A.P., D.D.A., F.C., and C.S.; funding acquisition, A.P. and C.S.; methodology, D.D.A., G.S., and G.D.; project administration, A.P., F.C., and C.S.; software, D.D.A., G.S. and G.D.; writing—original draft, A.P. and C.S.; writing—review and editing, A.P., D.D.A., G.S., G.D., F.C., and C.S.

Funding: This work was funded by: (1) Agropolis Fondation, Fondazione Cariplo, and Daniel and the Nina Carasso Foundation through the "Investissements d'avenir" programme with reference number ANR-10-LABX-0001-01, under the "Thought for Food" Initiative (project "LEGERETE"); (2) the Italian Ministry of Education, University and Research (MIUR), program PRIN 2017 (grant number 2017SFTX3Y) "The Neapolitan pizza: processing, distribution, innovation and environmental aspects."

Acknowledgments: The Authors acknowledge Industria Molitoria Mininni (Altamura, Italy), particularly Giuseppe Galetta, Vincenzo Tricarico, and Isabella Mininni, for preparing accurately and providing flours, composite meals, and bakery products.

Conflicts of Interest: The authors declare no conflict of interest.

References

1. Sultani, M.; Gill, M.; Anwar, M.; Athar, M. Evaluation of soil physical properties as influenced by various green manuring legumes and phosphorus fertilization under rain fed conditions. *Int. J. Environ. Sci. Technol.* **2007**, *4*, 109–118. [CrossRef]
2. Mohammadi, K. Nutritional composition of Iranian *desi* and *kabuli* chickpea (*Cicer arietinum* L.) cultivars in autumn sowing. *Int. J. Biol. Biomol. Agric. Food Biotech. Eng.* **2015**, *9*, 514–517.
3. Pavan, S.; Lotti, C.; Marcotrigiano, A.R.; Mazzeo, R.; Bardaro, N.; Bracuto, V.; Ricciardi, F.; Taranto, F.; D'Agostino, N.; Schiavulli, A.; et al. A distinct genetic cluster in cultivated chickpea as revealed by genome-wide marker discovery and genotyping. *Plant Genome* **2017**, *10*. [CrossRef] [PubMed]
4. Summo, C.; de Angelis, D.; Ricciardi, L.; Caponio, F.; Lotti, C.; Pavan, S.; Pasqualone, A. Nutritional, physico-chemical and functional characterization of a global chickpea collection. *J. Food Compos. Anal.* **2019**, *84*, 103306. [CrossRef]
5. Summo, C.; de Angelis, D.; Ricciardi, L.; Caponio, F.; Lotti, C.; Pavan, S.; Pasqualone, A. Data on the chemical composition, bioactive compounds, fatty acid composition, physico-chemical and functional properties of a global chickpea collection. *Data Brief* **2019**, in press. [CrossRef]
6. Summo, C.; de Angelis, D.; Rochette, I.; Mouquet-Rivier, C.; Pasqualone, A. Influence of the preparation process on the chemical composition and nutritional value of canned purée of *kabuli* and *Apulian black* chickpeas. *Heliyon* **2019**, *5*, e01361. [CrossRef]
7. Sozer, N.; Holopainen-Mantila, U.; Poutanen, K. Traditional and new food uses of pulses. *Cereal Chem.* **2017**, *94*, 66–73. [CrossRef]
8. Wang, S.; Chelikani, V.; Serventi, L. Evaluation of chickpea as alternative to soy in plant-based beverages, fresh and fermented. *LWT Food Sci. Technol.* **2018**, *97*, 570–572. [CrossRef]
9. Meng, X.; Threinen, D.; Hansen, M.; Driedger, D. Effects of extrusion conditions on system parameters and physical properties of a chickpea flour-based snack. *Food Res. Int.* **2010**, *43*, 650–658. [CrossRef]
10. Sabanis, D.; Makri, E.; Doxastakis, G. Effect of durum flour enrichment with chickpea flour on the characteristics of dough and lasagna. *J. Sci. Food Agric.* **2006**, *86*, 1938–1944. [CrossRef]
11. Miñarro, B.; Albanell, E.; Aguilar, N.; Guamis, B.; Capellas, M. Effect of legume flours on baking characteristics of gluten-free bread. *J. Cereal Sci.* **2012**, *56*, 476–481. [CrossRef]
12. Pasqualone, A.; Caponio, F.; Summo, C.; Arapi, V. Characterisation of traditional Albanian breads derived from different cereals. *Eur. Food Res. Technol.* **2004**, *219*, 48–51.
13. Fernandez, L.M.; Berry, J.W. Rheological properties of flour and sensory characteristics of bread made from germinated chickpea. *Int. J. Food Sci. Technol.* **1989**, *24*, 103–110. [CrossRef]
14. Mohammed, I.; Adberahman, R.; Ahmed, B.S. Effects of chickpea flour on wheat pasting properties and bread making quality. *J. Food Sci. Technol.* **2014**, *51*, 1902–1910. [CrossRef] [PubMed]

15. Mohammed, I.; Ahmeda, A.R.; Sengea, B. Dough rheology and bread quality of wheat–chickpea flour blends. *Ind. Crops Prod.* **2012**, *36*, 196–202. [CrossRef]
16. Rizzello, C.G.; Calasso, M.; Campanella, D.; de Angelis, M.; Gobbetti, M. Use of sourdough fermentation and mixture of wheat, chickpea, lentil and bean flours for enhancing the nutritional, texture and sensory characteristics of white bread. *Int. J. Food Microbiol.* **2014**, *180*, 78–87. [CrossRef] [PubMed]
17. Ertop, M.H.; Coşkun, Y. Shelf-life, physicochemical, and nutritional properties of wheat bread with optimized amount of dried chickpea sourdough and yeast by response surface methodology. *J. Food Process. Preserv.* **2018**, *42*, e13650. [CrossRef]
18. Shrivastava, C.; Chakraborty, S. Bread from wheat flour partially replaced by fermented chickpea flour: Optimizing the formulation and fuzzy analysis of sensory data. *LWT Food Sci. Technol.* **2018**, *90*, 215–223. [CrossRef]
19. Zafar, T.A.; Al-Hassawi, F.; Al-Khulaifi, F.; Al-Rayyes, G.; Waslien, C.; Huffman, F.G. Organoleptic and glycemic properties of chickpea-wheat composite breads. *J. Food Sci. Technol. Mysore* **2015**, *52*, 2256–2263. [CrossRef]
20. Ficco, D.B.M.; Muccilli, S.; Padalino, L.; Giannone, V.; Lecce, L.; Giovanniello, V.; del Nobile, M.A.; de Vita, P.; Spina, A. Durum wheat breads 'high in fibre'and with reduced in vitro glycaemic response obtained by partial semolina replacement with minor cereals and pulses. *J. Food Sci. Technol.* **2018**, *55*, 4458–4467. [CrossRef]
21. Pasqualone, A. Italian durum wheat breads. In *Bread Consumption and Health*; Pedrosa Silva Clerici, M.T., Ed.; Nova Science Publisher Inc.: Hauppauge, NY, USA, 2012; pp. 57–59.
22. Giannone, V.; Giarnetti, M.; Spina, A.; Todaro, A.; Pecorino, B.; Summo, C.; Caponio, F.; Paradiso, V.M.; Pasqualone, A. Physico-chemical properties and sensory profile of durum wheat Dittaino PDO (Protected Designation of Origin) bread and quality of re-milled semolina used for its production. *Food Chem.* **2018**, *241*, 242–249. [CrossRef] [PubMed]
23. Holcomb, R. Food industry overview: Frozen pizza. *Food Technol. Fact Sheet* **2000**, *103*, 1–4.
24. Field, C. *The Italian Baker, Revised: The Classic Tastes of the Italian Countryside—Its Breads, Pizza, Focaccia, Cakes, Pastries, and Cookies*; Ten Speed Press: New York, NY, USA, 2011.
25. Pasqualone, A.; Delcuratolo, D.; Gomes, T. Focaccia Italian flat fatty bread. In *Flour and Breads and Their Fortification in Health and Disease Prevention*; Preedy, V.R., Watson, R.R., Patel, V.B., Eds.; Elsevier: Amsterdam, The Netherlands, 2011; pp. 47–58.
26. American Association of Cereal Chemists (AACC) International. *Approved Methods of Analysis*, 10th ed.; American Association of Cereal Chemists: St Paul, MN, USA, 2009.
27. Association of Official Agricultural Chemists (AOAC) International. *Official Method 991.43. Total, Soluble, and Insoluble Dietary Fiber in Foods, Enzymatic-Gravimetric Method, MES-TRIS Buffer. Official Methods of Analysis*, 16th ed.; AOAC International: Arlington, WA, USA, 1995.
28. European Parliament and Council. Regulation (EU) No 1169/2011 of the European Parliament and of the Council of 25 October 2011 on the provision of food information to consumers, amending Regulations (EC) No 1924/2006 and (EC) No 1925/2006 of the European Parliament and of the Council, and repealing Commission Directive 87/250/EEC, Council Directive 90/496/EEC, Commission Directive 1999/10/EC, Directive 2000/13/EC of the European Parliament and of the Council, Commission Directives 2002/67/EC and 2008/5/EC and Commission Regulation (EC) No 608/2004. *Off. J. Eur. Union* **2011**, *L304*, 18–63.
29. Pasqualone, A.; Bianco, A.M.; Paradiso, V.M.; Summo, C.; Gambacorta, G.; Caponio, F.; Blanco, A. Production and characterization of functional biscuits obtained from purple wheat. *Food Chem.* **2015**, *180*, 64–70. [CrossRef]
30. Pasqualone, A.; Gambacorta, G.; Summo, C.; Caponio, F.; Di Miceli, G.; Flagella, Z.; Marrese, P.P.; Piro, G.; Perrotta, C.; de Bellis, L.; et al. Functional, textural and sensory properties of dry pasta supplemented with lyophilized tomato matrix or with durum wheat bran extracts produced by supercritical carbon dioxide or ultrasound. *Food Chem.* **2016**, *213*, 545–553. [CrossRef]
31. International Organization for Standardization (ISO). *ISO 3093:2009. Wheat, Rye and Their Flours, Durum Wheat and Durum Wheat Semolina. Determination of the Falling Number According to Hagberg-Perten*; ISO: Geneva, Switzerland, 2009.

32. Pasqualone, A.; Summo, C.; Bilancia, M.T.; Caponio, F. Variation of the sensory profile of durum wheat Altamura PDO (Protected Designation of Origin) bread during staling. *J. Food Sci.* **2007**, *72*, S191–S196. [CrossRef]
33. Pasqualone, A.; Caponio, F.; Simeone, R. Quality evaluation of re-milled durum wheat semolinas used for bread-making in Southern Italy. *Eur. Food Res. Technol.* **2004**, *219*, 630–634. [CrossRef]
34. Pasqualone, A.; Laddomada, B.; Spina, A.; Todaro, A.; Guzmàn, C.; Summo, C.; Mita, G.; Giannone, V. Almond by-products: Extraction and characterization of phenolic compounds and evaluation of their potential use in composite dough with wheat flour. *LWT Food Sci. Technol.* **2018**, *89*, 299–306. [CrossRef]
35. Moussou, N.; Corzo-Martínez, M.; Sanz, M.L.; Zaidi, F.; Montilla, A.; Villamiel, M. Assessment of Maillard reaction evolution, prebiotic carbohydrates, antioxidant activity and α-amylase inhibition in pulse flours. *J. Food Sci. Technol.* **2017**, *54*, 890–900. [CrossRef]
36. Singh, U.; Kherdekar, M.S.; Jambunathan, R. Studies on *desi* and *kabuli* chickpea (*Cicer arietinum* L.) cultivars. The levels of amylase inhibitors, levels of oligosaccharides and in vitro starch digestibility. *J. Food Sci.* **1982**, *47*, 510–512. [CrossRef]
37. Pasqualone, A.; Caponio, F.; Pagani, M.A.; Summo, C.; Paradiso, V.M. Effect of salt reduction on quality and acceptability of durum wheat bread. *Food Chem.* **2019**, *289*, 575–581. [CrossRef] [PubMed]
38. Baixauli, R.; Salvador, A.; Hough, G.; Fiszman, S.M. How information about fibre (traditional and resistant starch) influences consumer acceptance of muffins. *Food Qual. Prefer.* **2008**, *19*, 628–635. [CrossRef]
39. European Parliament and European Council. Regulation (EC) No 1924/2006 of the European Parliament and of the Council of 20 December 2006 on nutrition and health claims made on foods. *Off. J. Eur. Union* **2006**, *L404*, 9–25.
40. Hossain, A.K.M.; Brennan, M.A.; Mason, S.L.; Guo, X.; Zeng, X.A.; Brennan, C.S. The effect of astaxanthin-rich microalgae "*Haematococcus pluvialis*" and wholemeal flours incorporation in improving the physical and functional properties of cookies. *Foods* **2017**, *6*, 57. [CrossRef] [PubMed]
41. Erazo-Castrejón, S.V.; Doehlert, D.C.; D'Appolonia, B.L. Application of oat oil in breadbaking. *Cereal Chem.* **2001**, *78*, 243–248. [CrossRef]

 *foods*

Article

Use of Fermented Hemp, Chickpea and Milling By-Products to Improve the Nutritional Value of Semolina Pasta

Rosa Schettino, Erica Pontonio * and Carlo Giuseppe Rizzello

Department of Soil, Plant and Food Science, University of Bari Aldo Moro, 70126 Bari, Italy; rosa.schettino@libero.it (R.S.); carlogiuseppe.rizzello@uniba.it (C.G.R.)

* Correspondence: erica.pontonio@uniba.it

Received: 8 October 2019; Accepted: 20 November 2019; Published: 22 November 2019

Abstract: A biotechnological approach including enzymatic treatment (protease and xylanase) and lactic acid bacteria fermentation has been evaluated to enhance the nutritional value of semolina pasta enriched with hemp, chickpea and milling by-products. The intense (up to circa, (ca.) 70%) decrease in the peptide profile area and (up to two-fold) increase in total free amino acids, compared to the untreated raw materials, highlighted the potential of lactic acid bacteria to positively affect their in vitro protein digestibility. Fermented and unfermented ingredients have been characterized and used to fortify pasta made under pilot-plant scale. Due to the high contents of protein (ca. 13%) and fiber (ca. 6%) and according to the Regulation of the European Community (EC) No. 1924/2006 fortified pasta can be labelled as a "source of fiber" and a "source of protein". The use of non-wheat flours increased the content of anti-nutritional factors as compared to the control pasta. Nevertheless, fermentation with lactic acid bacteria led to significant decreases in condensed tannins (ca. 50%), phytic acid and raffinose (ca. ten-fold) contents as compared to the unfermented pasta. Moreover, total free amino acids and in vitro protein digestibility values were 60% and 70%, respectively, higher than pasta made only with semolina. Sensory analysis highlighted a strong effect of the fortification on the sensory profile of pasta.

Keywords: hemp; chickpea; milling by-products; fortified pasta; lactic acid bacteria; nutritional value

---

1. Introduction

The trend in world population growth (up to 9.6 billion people by 2050) and the necessity to provide a nutritionally balanced diet and to reduce greenhouse gas emissions require relevant production increases in vegetables, as well as a transition to a diet higher in plant- rather than animal-derived proteins [1,2]. Aiming at addressing environmental concerns and meeting nutritional deficiencies and recommendations, the fortification of staple foods (e.g., bread and pasta) has been identified as an effective, sustainable and promising intervention [3,4]. To date, several studies investigated the nutritional value of additional ingredients to be used as wheat alternatives in cereal-based products [4–8]. Due to its popularity [9], pasta has been proposed as a suitable carrier of nutrients, mainly dietary fiber and proteins [4,7,10–13].

Legumes are excellent sources of proteins with high biological value and dietary fibers and they supply high levels of vitamins, minerals, oligosaccharides and phenolic compounds [14]. Moreover, thanks to their functionality (e.g., solubility and water-binding capacity), legume flours have successfully been proposed to enhance gluten-free food formulation and processing [5].

Grain germ and bran (milling by-products) comprise important sources of dietary fiber and α-tocopherol, vitamins of group B, polyunsaturated fats, minerals and different bioactive compounds with health-promoting effects [15]. Pearling by-products, thanks to the high content of dietary fiber and β-glucan, have been suggested as suitable ingredients to produce functional pasta [6]. Hemp has recently raised much interest as a sustainable food ingredient due to the high content (ca. 30%) and biological value of protein, dietary fiber content (ca. 50%) as well as the considerable content of functional compounds (e.g., phenols) with antioxidant and anti-hypertensive properties [16].

Nevertheless, the poor stability to oxidation [17], the high content of fibers and the absence of gluten may impair the wheat alternative's high nutritional value, worsening the technological and sensory profiles of the products [18]. Moreover, the presence of anti-nutritional factors (ANFs), i.e., phytic acid, condensed tannins, raffinose and trypsin inhibitors, further limit the use of such ingredients by the food industry [19]. Although different biotechnological options were suggested to overcome the drawbacks related to the use of non-wheat flours in cereal-derived foods, fermentation with lactic acid bacteria (LAB) seemed to be the best option to both decrease the ANF and improve their nutritional, technological and sensory profile [19,20].

Based on the above consideration, here, hemp and chickpea flours, and wheat milling by-products were proposed as additional ingredients to improve the nutritional quality of semolina pasta. Enzymatic pre-treatments and fermentation with LAB were evaluated as bioprocessing options to enhance protein digestibility and decrease ANF. Bioprocessed ingredients were used to manufacture fortified pasta, and the effects on the nutritional, technological, and sensory properties were investigated in comparison to samples produced with the untreated native ingredients.

2. Materials and Methods

2.1. Raw Materials, Bacterial Strains and Enzymes

Commercial hemp (*Cannabis sativa L.*) flour (Sottolestelle, San Giovanni Rotondo, Italy), chickpea grains (*Cicer arietinum L.* var. Pascià, Caporal Grani S.a.s.,Gravina di Puglia, Italy), wheat germ (Molino Rieper, Vandoies di Sotto, Italy) and bran (Molino Careccia, Stigliano, Italy) were used in this study. Chickpea, germ and bran flours were milled using a laboratory mill Ika-Werke M20 (GMBH, and Co. KG, Staufen, Germany) before use. After milling, all the flours were sieved (mesh size 500 μm) to remove the coarse fraction.

Lactobacillus plantarum LB1 and *Lactobacillus rossiae* LB5 [21] were used in this study. Strains were routinely cultivated on modified de Man Rogosa and Sharp (Oxoid, Basingstoke, Hampshire, UK) (mMRS) as reported by Rizzello et al. [21]. A commercial xylanase (880 *xylanase* u/g; Depol 761P, Biocatalysts Limited, Chicago, USA) from *Bacillus subtilis* and proteases of *Aspergillus oryzae* (500,000 hemoglobin units on the tyrosine basis/g; enzyme 1 [E1]) and *Aspergillus niger* (3000 spectrophotometric acid protease units/g; enzyme 2 [E2]), routinely used for bakery applications, supplied by BIO-CAT Inc. (Troy, VA, USA), were also used.

2.2. Proximate Chemical Composition of Raw Materials

Moisture, protein, lipids, total dietary fiber and ash of raw material were determined according to Approved Methods 44-15A, 46-11A, 30-10.01, 32-05.01, and 08-01.01 of the American Association of Cereal Chemists (AACC) [22]. Total nitrogen was corrected by 6.25, 4.99, 6.31 and 5.30 for chickpea, wheat germ, wheat bran and hemp, respectively [23,24]. Available carbohydrates were calculated as the difference [100 − (proteins + lipids + ash + total dietary fiber)]. Proteins, lipids, carbohydrates, total dietary fiber and ash were expressed as % of dry matter (d.m.).

2.3. Bioprocessing

Wheat germ and bran were mixed (1:4, *wt/wt*) prior to dough preparation. Doughs (50 g) were prepared by mixing hemp flour (62.5% *wt/wt*), or chickpea flour (62.5% *wt/wt*) or milling by-products (40.0% *w/w*) with tap water. Dough yield (DY, dough weight × 100/flour weight) was 160 for hemp (H) and chickpea (C), and 250 for milling by-products (WGB). To be used as a mixed starter for sourdough fermentation, LAB cells were harvested by centrifugation (10,000× *g*, 10 min, 4 °C), washed twice in 50 mM phosphate buffer, pH 7.0, and re-suspended in tap water used for the dough making (final cell density in the dough was ca. 7.0 log10 cfu/g). Fermentation was carried out at 30 °C for 24 h (H_F, C_F, and WGB_F).

For the enzymatic treatments, doughs with the same DY were prepared. Before mixing, xylanase was added at 50 ppm based on dough weight (H_X, C_X, and WGB_X) and proteases E1 and E2 (H_P, C_P, and WGB_P) were used at 50 ppm (25 ppm for each enzyme) on dough weight. Doughs were incubated at 30 °C for 8 h. All the doughs were prepared and incubated in triplicate.

2.4. Determination of the Protein Degradation

Aiming at selecting the optimal bioprocessing option, total free amino acid (TFAA) concentration and peptide profiles were considered as indexes of the proteolytic degradation and screening parameters. Doughs prior bioprocessing were used as the controls (H, C, and WGB).

Water/salt-soluble extracts (WSEs) from doughs were prepared as reported by Weiss et al. [25] and used for TFAA and peptides analyses. TFAA were analyzed by a Biochrom 30 series Amino Acid Analyzer (Biochrom Ltd., Cambridge Science Park, United Kingdom) with a Na-cation-exchange column (20 by 0.46 cm internal diameter), as described by Rizzello et al. [17]. For the analysis of peptides, WSE were treated with trifluoroacetic acid (TFA, 0.05% *wt/v*) and subjected to dialysis (cut-off 500 Da) to remove proteins and FAA, respectively. Then, the peptide concentration was determined by the o-phtaldialdehyde (OPA) method [26]. Peptide profiles were obtained by Reversed-Phase Fast Performance Liquid Chromatography (RP-FPLC) [26,27], using an ÄKTA FPLC equipped with a Resource RPC column and a UV detector (214 nm) (GE Healthcare Bio-Sciences AB, Uppsala, Sweden). Peptide profiles and total peak area were elaborated with Unicorn 4.0 software (GE Life Sciences).

2.5. Microbiological and Biochemical Characterization of Fermented Doughs

pH values of H, C and WGB and H_F, C_F and WGB_F were determined online by a pH meter (Model 507, Crison, Milan, Italy) with a food penetration probe. The AACC method 02-31.01 was used for the determination of total titratable acidity (TTA) of samples. Presumptive LAB was enumerated using de Man, Rogosa and Sharpe (MRS, Oxoid) agar medium, supplemented with cycloheximide (0.1 g/L). Plates were incubated, under anaerobiosis (AnaeroGen and AnaeroJar, Oxoid), at 30 °C for 48 h. WSEs from unfermented and fermented doughs were used for the determination of organic acids, peptides and TFAA concentrations. High-Performance Liquid Chromatography (HPLC) ÄKTA Purifier system (GE Healthcare, Buckinghamshire, United Kingdom) equipped with an Aminex HPX-87H column (ion exclusion, Bio-Rad, Richmond, CA, United States) and a UV detector operating at 210 nm [17] was used to quantify organic acids. The fermentation quotient (FQ) was determined as the molar ratio between lactic and acetic acids. Peptides and TFAA concentrations were determined as described above.

2.6. Pasta Making

Pasta was manufactured using a pilot plant La Parmigiana SG30 (Manufacture, Fidenza, Italy). Table 1 summarizes the ingredients and protocol used for pasta making. All the doughs had a final DY of 130, corresponding to 23% (*wt/wt*) water and 77% (*wt/wt*) dry matter (semolina, hemp, chickpea flours and milling by-products).

Unfermented and fermented doughs were obtained as described before and used as ingredients for pasta making. Due to the difference in terms of the DY (160 vs. 250) of the doughs, H, H_F, C and C_F were used at 11% (*wt/wt*), while the level of fortification of WGB and WGB_F was 17% (*wt/wt*). Unfermented and fermented doughs were mixed with durum wheat semolina and water to obtain pasta samples (*p*H, *p*H_F, *p*C and *p*C_F and *p*WGB and *p*WGB_F, respectively). A control pasta was made without fortification (*p*CT). The process is composed of four stages: i) three-steps mixing (1 min mixing/6 min hydration); ii) extrusion of final dough at 45–50 °C through a n. 76 (150 mm diameter) bronze die; iii) cutting the extruded to obtain grooved "macaroni"; and iv) drying using low temperature (55 °C) cycle (Supplementary Table S1). The proximate composition of wheat semolina was moisture, 10.2%; protein, 12.1% of d.m.; fat, 1.8% of d.m.; ash, 0.6% of. d.m.; and total carbohydrates, 75.5% of d.m.

Table 1. Formulas for pasta fortified with hemp, chickpea, and wheat germ/bran unfermented and fermented doughs: *p*H, pasta containing 11% of unfermented hemp dough (*wt*/*wt*); *p*H$_F$, pasta containing 11% of fermented hemp dough (*wt*/*wt*); *p*C, pasta containing 11% of unfermented chickpea dough; *p*C$_F$ (*wt*/*wt*), pasta containing 11% of fermented chickpea dough (*wt*/*wt*); *p*WGB, pasta containing 17% of unfermented wheat germ and bran (1:4) dough (*wt*/*wt*); *p*WGB$_F$, pasta containing 17% of fermented wheat germ and bran (1:4) dough (*wt*/*wt*); *p*CT, pasta made with durum wheat semolina.

	*p*H	*p*C	*p*WGB	*p*H$_F$	*p*C$_F$	*p*WGB$_F$	*p*CT
Semolina (%)	70.2	70.2	70.2	70.2	70.2	70.2	77
Water (%)	19.8	19.8	12.8	19.8	19.8	12.8	23
Fermented dough (%) [1]	-	-	-	11	11	17	-
Unfermented dough (%) [1]	11	11	17	-	-	-	-

[1] Having DY of 160, 11% of hemp (H and H$_F$) and chickpea (C and C$_F$) doughs contained ca. 6.8% of solids and 4.2% of water. While, 17% of wheat germ/bran doughs (WGB and WGB$_F$) having dough yield (DY) of 250 contained ca. 6.8% dry matter and 10.2% of water. Consequently, all the fortified pasta samples contained ca. 7% of non-wheat flours/milling by-products.

2.7. *In vitro Protein Digestibility (IVPD)*

The IVPD of pasta samples was determined according to Akeson and Stahmann [28,29]. In order to mimic the in vivo digestion in the gastrointestinal tract, pasta samples were subjected to a sequential enzymatic treatment. The IVPD is the percentage of the total protein solubilized after enzymatic hydrolysis. The protein quantification was made according to the Bradford method [30].

2.8. *Pasta Characterization*

2.8.1. Hydration Test, Cooking Time, Cooking Loss and Water Absorption

The method of Marti et al. [31] (ratio pasta:water of 1:20, 180 min of incubation) was used to determine the hydration at 25 °C, while the method of Schoenlecher et al. [32] was used to determine the cooking time. The optimal cooking time (OCT) corresponded to the disappearance of the white core. Cooking loss (expressed as grams of matter loss/100 g of pasta) was evaluated by determining the number of solids lost into the cooking water as proposed by D'egidio, et al. [33]. The increase in pasta weight during cooking (water absorption) was evaluated by weighing pasta before and after cooking. The results were expressed as [(W1 − W0/W0] × 100, where W1 is the weight of cooked pasta and W0 is the weight of the uncooked samples.

2.8.2. Chemical and Nutritional Profile

Chemical characteristics (determined on pasta dough prior to extrusion) and the proximal composition of pasta were determined as reported above.

The protein solubility of pasta (grinded) was evaluated under native and denaturing conditions as reported by Iametti et al. [34]. The concentration of protein and peptides was determined as reported above [7,30].

Raffinose and phytic acid concentrations were determined by using the Megazyme kit Raffinose/D-Galactose Assay Kit K-RAFGA and K-PHYT 05/07 (Megazyme International Ireland Limited, Bray, Ireland), respectively, following the manufacturer's instructions. Condensed tannins were determined using the acid butanol assay, as described by Hagerman [35].

IVDP and starch hydrolysis were determined on pasta samples at the OCT. IVPD was determined as described before. The evaluation of the starch hydrolysis rate was performed using a procedure mimicking the in vivo digestion of starch [36]. Wheat flour bread (WB) was used as the control to estimate the hydrolysis index (HI = 100). The predicted GI of all pasta samples was calculated using the following equation: $pGI = 0.549 \times HI + 39.71$ [37].

2.8.3. Texture and Color Analysis

Instrumental Texture Profile Analysis (TPA) was carried out with a TVT-300XP Texture Analyzer (TexVol Instruments, Viken, Sweden), equipped with a cylinder probe (diameter 95 mm). For the analysis, pasta samples were cooked until the OCT, left to cool at room temperature and placed in a beaker (diameter, 100 mm; height 90 mm), filled to about half volume. The selected settings were the following: test speed 1 mm/s, 30% deformation of the sample and two compression cycles. The chromaticity coordinates of the samples (obtained by a Minolta CR-10 camera) were reported in the form of a color difference, $\Delta E \times ab$ [7].

2.8.4. Sensory Analysis

A trained sensory panel ($n = 13$, aged 21–45 years) assessed the sensory profile of pasta samples. The lexicon consisted of twelve attributes as reported in Supplementary Table S2. A line scale from "not at all" (0) to "very" (10) for each attribute was used for the evaluation. Each pasta sample was cooked according to its own OCT and presented randomized in duplicate. Tap water was used to rinse the mouth between the samples. The study protocol followed the ethical guidelines of the sensory laboratory. A written informed consent was obtained from each participant.

2.9. Statistical Analysis

All analysis as well as the fermentation and enzymatic treatments were carried out in triplicate. The one-way ANOVA, using Tukey's procedure at $p < 0.05$, was performed for the data elaboration (Statistica 12.5, StatSoft Inc., Tulsa, USA). Principal component analysis (PCA) with varimax rotation was performed to visualize the sensory characteristics of the samples with Unscrambler X10.3 (Camo SA, Trondheim, Norway).

3. Results

3.1. Proximate Composition of the Raw Materials

The proximate composition of the flours is reported in Table 2. Hemp flour was characterized by the highest concentration of protein (ca. 37% of d.m.) and total dietary fiber (ca. 39.7% of d.m.), while chickpea flour was characterized by the lowest concentration of fat (ca. 4% of d.m.) (Table 2).

Table 2. Proximate composition and microbiological characterization of hemp and chickpea flours and wheat germ/bran mixture.

	Hemp	Chickpea	Wheat germ:bran (1:4)
Protein (%)	36.9 ± 0.1^{a}	25.2 ± 0.4^{b}	17.8 ± 0.5^{c}
Fat (%)	11.9 ± 0.2^{a}	3.9 ± 0.1^{c}	8.6 ± 0.7^{b}
Available carbohydrates (%)	4.2 ± 0.1	48.9 ± 1.9	34.2 ± 2.1
Total dietary fiber (%)	39.7 ± 0.1^{a}	24.3 ± 0.3^{b}	37.8 ± 1.1^{a}
Ash (%)	7.6 ± 0.2^{a}	2.9 ± 0.9^{b}	2.0 ± 0.4^{b}

Data are expressed % of dry matter. [a–c] Values in the same row with different superscript letters differ significantly ($p < 0.05$).

3.2. Proteolysis and Set-Up of the Bioprocessing

TFAA and peptide profiles were used as screening criteria for bioprocessing parameters, since they correspond to the organic nitrogen compounds released during the process from native proteins. According to the peptide profiles (Figure 1), H and WGB were characterized by a total peak area significantly lower than C (3479 ± 34 and 5194 ± 25 mAU × mL vs. 15011 ± 53 mAU × mL). As the consequence of the bioprocessing, changes were observed mainly in the range 20% to 40% of the acetonitrile gradient, while the hydrophilic zone had undergone minimal alterations (Figure 1). Enzymatic treatments led to slight increases in the total peak area of peptides. Values from 6.8% (C_P) to 9.4% (WGB_P) higher were found when proteases were used. Similarly, increases up to ca. 10% were found in C_X and WGB_X. H_X and H were characterized by a similar peptide area. On the contrary, LAB fermentation caused a significant decrease in the peptide profile area, with H_F and C_F characterized by a relevant lower area (21 and 71%, respectively) than H and C.

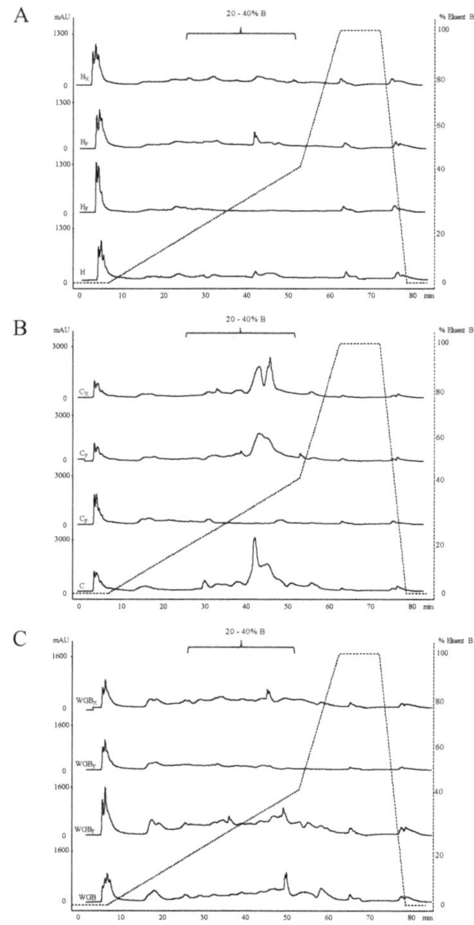

Figure 1. Reversed-Phase Fast Performance Liquid Chromatography (RP-FPLC) peptide profiles of hemp, chickpea and wheat germ/bran doughs. (**A**): H, untreated hemp dough; H_F, fermented hemp dough; H_P, hemp dough treated with proteases (E1/E2); H_X, hemp dough treated with xylanase (Depol 761P). (**B**): C, untreated chickpea dough; C_F, fermented chickpea dough; C_P, chickpea dough treated with proteases (E1/E2); C_X, chickpea dough treated with xylanase (Depol 761P). (**C**): WGB, untreated wheat germ/bran (1:4) dough; WGB_F, fermented wheat germ/bran (1:4) dough; WGB_P, wheat germ/bran (1:4) dough treated with proteases (E1/E2); WGB_X, wheat germ/bran (1:4) dough treated with xylanase (Depol 761P).

The concentration of TFAA in H, C and WGB prior to bioprocessing ranged from 802 ± 5 to 1972 ± 12 mg/Kg (Table 3). LAB and enzymes both increased TFAA concentration. Among the enzymatic treatments, proteases led to a concentration from ca. 3% (C_P) to 70% (H_P) higher than the corresponding untreated controls, while increases in the range 2 (C_X) to 40% (WGB_X) were found when xylanase was used. LAB led to the highest increases in the TFAA concentration (up to two-fold higher) (Table 3). In detail, the highest TFAA concentration was found in C_F (2457 ± 16 mg/Kg).

Table 3. Chemical characterization of unfermented/fermented hemp, chickpea and wheat germ/bran doughs. H, dough made with hemp flour; H_F, fermented dough made with hemp flour; C, dough made with chickpea flour; C_F, fermented dough made with chickpea flour; WGB, dough made with a mixture of wheat germ and bran (1:4); WGB_F, fermented dough made with a mixture of wheat germ and bran (1:4).

	H	C	WGB	H_F	C_F	WGB_F
pH	6.4 ± 0.2 ^{a}	6.1 ± 0.1 ^{a}	6.3 ± 0.5 ^{a}	4.5 ± 0.3 ^{b,c}	4.4 ± 0.2 ^{c}	4.9 ± 0.2 ^{b}
Total titratable acidity (TTA) (mL NaOH 0.1M)	4.2 ± 0.3 ^{c}	4.1 ± 0.2 ^{c}	4.0 ± 0.1 ^{c}	35.8 ± 0.6 ^{a}	28.6 ± 0.8 ^{b}	29.1 ± 0.7 ^{b}
Lactic acid (mmol/Kg)	3.3 ± 0.2 ^{d}	4.1 ± 0.2 ^{c}	4.1 ± 0.5 ^{c}	214.9 ± 5.8 ^{b}	180.7 ± 4.3 ^{c}	246.4 ± 6.2 ^{a}
Acetic acid (mmol/Kg)	1.8 ± 0.9 ^{d}	1.1 ± 0.3 ^{d}	2.6 ± 0.4 ^{c}	43.9 ± 0.9 ^{a}	42.7 ± 0.7 ^{a}	31.6 ± 0.3 ^{b}
Fermentation quotient (FQ)	1.8	3.7	1.6	4.7	4.3	7.8
Total free amino acid (TFAA) (mg/Kg)	990 ± 8 ^{d}	1972 ± 12 ^{c}	802 ± 5 ^{e}	1799 ± 11 ^{c}	2457 ± 16 ^{a}	1929 ± 13 ^{b}
Peptides (g/Kg)	0.62 ± 0.05 ^{c}	2.60 ± 0.60 ^{a}	0.81 ± 0.09 ^{b}	0.49 ± 0.59 ^{c}	0.75 ± 0.20 ^{b,c}	0.77 ± 0.10 ^{b,c}

Hemp and chickpea doughs had a DY of 160; wheat germ/bran dough had a DY of 250. The data are the means of three independent experiments ± standard deviations (n = 3). $^{a–e}$ Values in the same row with different superscript letters differ significantly ($p < 0.05$).

Although both enzymatic treatments caused a moderate increase in the peptides in treated samples, the most extensive protein degradation that occurred during fermentation suggested a more intense potential effect of the LAB on protein digestibility. According to these considerations, fermented samples were subjected to further analysis.

3.3. Chemical Characterization of Doughs

A decrease in pH of ca. 2 units was achieved in all fermented doughs (Table 3). Significant higher values of TTA were found in H_F, C_F and WGB_F compared to H, C and WGB, respectively. These changes are in accordance to the increases in lactic and acetic acids in fermented doughs. The highest concentration of lactic acid was found in WGB_F, while H_F contained the highest amount of acetic acid (Table 3). Decreases in the peptide concentrations (up to ca. 80%) were found after fermentation. The highest decrease was found in C_F, while, according to the TFAA concentration, the lowest value was found in H_F (Table 3).

3.4. Chemical, Technological and Structural Properties of Pasta

The inclusion of both unfermented and fermented ingredients affected the chemical characteristics of pasta. However, the pH and TTA values differ from *p*CT only when fermented ingredients were used (Table 4). TFAA concentration was higher in all fortified pasta, as compared to the pCT with higher extent when fermented ingredients were used. *p*WGB_F contained the highest amount (ca. ten-fold higher than *p*CT) (Table 4).

The experimental OCT of *p*CT was ca. 10 min. Decreases (from 25 to 66%) in OCT were found for fortified pasta, especially when fermented ingredients were used (Table 4). A similar trend was found in terms of water absorption; lower values (13 to 32%) were found in *p*H_F, *p*C_F and *p*WGB_F as compared to *p*CT. On the contrary, the cooking loss increased when pasta was fortified, being higher (up to 66%) when fermented ingredients were used. Hydration was also affected by the fortification; indeed, significantly higher values were found in fortified pasta especially when fermented doughs were used as ingredients. The highest value was reached in *p*WGB_F.

Table 4. Chemical, technological and structural properties of pasta fortified with unfermented/fermented hemp, chickpea and wheat germ/bran doughs: *p*H, pasta containing 11% (*wt*/*wt*; d.m.) unfermented hemp dough; pH_F, pasta containing 11% (*wt*/*wt*; d.m.) fermented hemp dough; *p*C, pasta containing 11% (*wt*/*wt*; d.m.) unfermented chickpea dough; pC_F, pasta containing 11% (*wt*/*wt*; d.m.) fermented chickpea dough; *p*WGB, pasta containing 17% (*wt*/*wt*; d.m.) unfermented wheat germ and bran (1:4) dough; $pWGB_F$, pasta containing 17% (*wt*/*wt*; d.m.) fermented wheat germ and bran (1:4) dough; *p*CT, pasta made with durum wheat semolina.

	*p*H	*p*C	*p*WGB	pH_F	pC_F	$pWGB_F$	*p*CT
Chemical Properties							
pH	6.41 ± 0.10^{a}	6.37 ± 0.21^{a}	6.42 ± 0.31^{a}	5.12 ± 0.11^{b}	5.16 ± 0.05^{b}	5.06 ± 0.08^{b}	6.47 ± 0.24^{a}
TTA (mL NaOH 0.1M)	3.01 ± 0.13^{c}	2.05 ± 0.41^{d}	1.82 ± 0.09^{d}	6.61 ± 1.01^{a}	4.01 ± 0.61^{b}	4.45 ± 0.91^{b}	2.01 ± 0.37^{d}
TFAA (mg/Kg)	238 ± 9^{e}	400 ± 8^{d}	203 ± 4^{f}	1018 ± 11^{b}	743 ± 7^{c}	1529 ± 12^{a}	102 ± 10^{g}
Technological Properties							
Optimal Cooking Time (OCT) (min)	7.0 ± 0.2^{c}	6.3 ± 0.1^{d}	8.0 ± 0.2^{b}	6.3 ± 0.1^{d}	6.0 ± 0.1^{e}	7.0 ± 0.1^{c}	10.0 ± 0.1^{a}
Water absorption (%)	94.4 ± 3.7^{c}	105.4 ± 3.2^{b}	110.6 ± 2.9^{b}	105.3 ± 4.5^{a}	99.6 ± 3.6^{b}	98.4 ± 3.2^{b}	125.1 ± 4.1^{a}
Cooking loss (% of d.m.)	4.7 ± 0.1^{d}	4.6 ± 0.1^{d}	6.3 ± 0.1^{b}	5.1 ± 0.1^{c}	5 ± 0.2^{c}	6.8 ± 0.1^{a}	4.1 ± 0.1^{e}
Structural Properties							
Hardness (N)	7.99 ± 0.51^{b}	$7.34 \pm 0.29^{b,c}$	11.6 ± 0.31^{a}	7.13 ± 0.42^{b}	4.60 ± 0.17^{d}	6.98 ± 0.18^{c}	4.24 ± 0.15^{e}
Chewiness (N)	2.00 ± 0.10^{b}	1.46 ± 0.05^{c}	2.07 ± 0.08^{b}	1.14 ± 0.08^{d}	2.03 ± 0.15^{b}	2.22 ± 0.13^{b}	2.86 ± 0.07^{a}
Cohesiveness	0.33 ± 0.03^{b}	$0.27 \pm 0.04^{b,c}$	0.26 ± 0.02^{c}	$0.29 \pm 0.02^{b,c}$	0.43 ± 0.06^{a}	0.45 ± 0.04^{a}	0.44 ± 0.04^{a}
Color Analysis							
L	52.1 ± 2.4^{c}	$65.5 \pm 4^{a,b}$	62.8 ± 3.1^{b}	51.2 ± 1.9^{c}	65.6 ± 4.3^{a}	63.2 ± 2.1^{b}	70.8 ± 1.2^{a}
a	0.6 ± 0.1^{c}	0.8 ± 0.2^{c}	$1.4 \pm 0.4^{a,b}$	$0.6 \pm 0.4^{b,c}$	0.3 ± 0.2^{c}	2.1 ± 0.6^{a}	-0.6 ± 0.2^{d}
b	9.3 ± 1.7^{c}	18.5 ± 1.6^{a}	19.9 ± 2.7^{a}	9.6 ± 0.9^{c}	22.9 ± 3.2^{a}	14.7 ± 1.5^{b}	17.7 ± 2.4^{a}
ΔE	42.9 ± 1.2^{a}	31.7 ± 4.2^{b}	35.7 ± 1.9^{b}	41.9 ± 1.1^{a}	34.7 ± 0.6^{b}	33.4 ± 2.4^{b}	26.7 ± 2.3^{c}

Hemp and chickpea doughs had a DY of 160; wheat germ/bran dough had a DY of 250. Consequently, all the fortified pasta samples contained 7% of non-wheat flours/milling by-products. The data are the means of three independent experiments ± standard deviations ($n = 3$). $^{a-g}$ Values in the same row with different superscript letters differ significantly ($p < 0.05$).

Protein solubility in phosphate buffer was very low for all samples (< 2.91 ± 0.4 mg/g), while the addition of denaturing urea corresponded to a higher protein extraction (up to three-fold). The protein solubility in a buffer containing the disulfide reducing DTT was ca. two-fold lower compared to the phosphate buffer (data not shown), nevertheless, when both urea and DTT were used, the highest protein extraction was achieved. Overall, the protein solubility of fortified pasta was higher than the pasta containing only semolina. Moreover, the values of protein solubility were higher in fermented than the corresponding unfermented samples, probably due to the more intense proteolysis.

The hardness of fortified pasta samples was higher than *p*CT. However, when fermented doughs were used, lower values were found as compared to *p*H, *p*C and *p*WGB. Chewiness decreased with the fortification. However, when fermented doughs were used, it was higher in $p\text{C}_\text{F}$ and $p\text{WGB}_\text{F}$ as compared to *p*C and *p*WGB, respectively. The inclusion of wheat substitutes led to a decrease in the cohesiveness only when unfermented doughs were used. With the only exception of $p\text{H}_\text{F}$, which showed lower value of chewiness, similar values were found between pasta containing fermented doughs and *p*CT.

Compared to *p*CT, lower lightness (L) and higher dE × ab values were found in all fortified pasta samples (Table 4). The highest "a" value, index for greenness (−)/redness (+) was observed for *p*WGB and $p\text{WGB}_\text{F}$ (Table 4).

3.5. Nutritional Properties of Pasta

As expected, the fortification with H, C and WGB improved the content of protein and total dietary fibers (Table 5). A fiber concentration higher than 6% was obtained with the fortification. Nevertheless, fortification also increased the content of the ANF, although the use of fermented doughs corresponded to lower concentration than corresponding unfermented controls. Overall, ca. ten-fold decreases in phytic acid and raffinose were observed in pasta with fermented doughs (Table 5). $p\text{C}_\text{F}$ and $p\text{H}_\text{F}$ contained the lowest amount of phytic acid and raffinose, respectively. Condensed tannins were from 23% to 59% lower in $p\text{H}_\text{F}$, $p\text{C}_\text{F}$ and $p\text{WGB}_\text{F}$ as compared to the corresponding pasta with unfermented doughs.

Pasta samples containing fermented doughs were characterized by lower values of HI (up to 79%) as compared to the corresponding unfermented ones, except for $p\text{H}_\text{F}$. The use of unfermented and fermented milling by-products led to the lowest values of HI (60.62 and 42.4, respectively). The *p*GI of unfermented and fermented pasta ranged from 72.99 to 81.27 and from 62.98 to 79.74, respectively. The lowest value was found in $p\text{WGB}_\text{F}$. A similar trend was found in terms of the IVPD. Increases from 22% to 45% were found in fortified pasta as compared to *p*CT. Values 43–64% higher than *p*CT were found in $p\text{H}_\text{F}$, $p\text{C}_\text{F}$ and $p\text{WGB}_\text{F}$ (Table 5). The fermentation led to pasta having an IVPD from 10% to 22% higher than the corresponding pH, *p*C and *p*WGB.

Table 5. Nutritional properties of pasta fortified with unfermented/fermented hemp, chickpea and wheat germ/bran doughs: *p*H, pasta containing 11% (*wt*/*wt*; d.m.) unfermented hemp dough; pH_F, pasta containing 11% (*wt*/*wt*; d.m.) fermented hemp dough; *p*C, pasta containing 11% (*wt*/*wt*; d.m.) unfermented chickpea dough; pC_F, pasta containing 11% (*wt*/*wt*; d.m.) fermented chickpea dough; *p*WGB, pasta containing 17% (*wt*/*wt*; d.m.) unfermented wheat germ and bran (1:4) dough; $pWGB_F$, pasta containing 17% (*wt*/*wt*; d.m.) fermented wheat germ and bran (1:4) dough; *p*CT, pasta made with durum wheat semolina.

	*p*H	*p*C	*p*WGB	pH_F	pC_F	$pWGB_F$	*p*CT
Protein (%)	14.77 ± 0.31 [a]	13.50 ± 0.27 [b]	13.32 ± 0.19 [b]	14.54 ± 0.29 [a]	12.95 ± 0.31	13.54 ± 0.24 [b]	12.32 ± 0.28 [c]
Fat (%)	2.61 ± 0.06 [b]	3.94 ± 0.02 [a]	2.01 ± 0.02 [c]	2.53 ± 0.02 [b]	4.02 ± 0.06 [a]	1.99 ± 0.04	1.58 ± 0.01 [d]
Available carbohydrates (%)	74.16 ± 1.02	75.48 ± 1.01	76.99 ± 0.97	73.98 ± 1.07	75.66 ± 0.98	77.11 ± 0.99	81.94 ± 0.67
Total dietary fibers (%)	6.71 ± 0.25 [a]	5.89 ± 0.31 [b]	6.10 ± 0.24 [b]	6.84 ± 0.33 [a]	6.10 ± 0.36 [b]	5.96 ± 0.32 [b]	3.05 ± 0.18 [c]
Ash (%)	1.75 ± 0.10 [a]	1.27 ± 0.13 [b]	1.58 ± 0.12 [a,b]	1.81 ± 0.13 [a]	1.34 ± 0.17 [b]	1.67 ± 0.19 [a]	1.11 ± 0.10 [b]
Phytic acid (g/100g)	1.01 ± 0.05 [a]	0.35 ± 0.01 [c]	0.12 ± 0.02 [d]	0.62 ± 0.06 [b]	0.03 ± 0.01 [e]	0.12 ± 0.02 [d]	n.d.
Raffinose (g/Kg)	0.11 ± 0.02 [c]	0.33 ± 0.02 [a]	0.17 ± 0.02 [b]	0.05 ± 0.01 [d]	0.11 ± 0.01 [c]	0.17 ± 0.02 [b]	n.d.
Condensed tannins (g/Kg)	3.43 ± 0.10 [a]	4.08 ± 0.91 [a]	2.21 ± 0.03 [c]	2.61 ± 0.09 [b]	1.96 ± 0.02 [d]	0.90 ± 0.01 [e]	n.d.
HI (%)	75.71 ±6.52 [a]	70.73 ± 4.91 [a]	60.62 ± 4.18 [b]	72.92 ± 5.17 [a]	55.61 ± 3.05 [b]	42.4 ± 3.96 [c]	75.9 ± 2.18 [a]
IVPD (%)	58.5 ± 0.6 [d]	65.61 ± 0.7 [b]	55.01 ± 0.4 [e]	64.63 ± 0.8 [c]	73.80 ± 0.7 [a]	66.83 ± 0.6 [b]	45.1 ± 0.4 [f]

Hemp and chickpea doughs had a DY of 160; wheat germ/bran dough had a DY of 250. Consequently, all the fortified pasta samples contained 7% of non-wheat flours/milling by-products. The data are the means of three independent experiments ± standard deviations ($n = 3$). [a–f] Values in the same row with different superscript letters differ significantly ($p < 0.05$). n.d.: not detected.

3.6. Sensory Analysis

Pasta was subjected to sensory analysis and the results are summarized in Figure 2. The PCA, representing 79.49% of the total variance of the data, showed that pasta samples are scattered in different parts of the plane according to the raw materials used for the production. All fortified samples were in a different part of the plane as compared to the control *p*CT, thus confirming the strong influence of the fortification on the sensory profile of pasta. Moreover, among fortified samples, the fermentation seemed to strongly affect the sensory profile of pasta only when milling by-products were used. Indeed, *p*WBG$_F$ and *p*WBG were scattered in different part of the plane. The former was characterized by a greater intensity of pungent odor and flavor and note of whole grains as compared to the corresponding *p*WBG. Only slight differences were found between *p*C and *p*C$_F$ and *p*H and *p*H$_F$, respectively. *p*C$_F$ differentiated from the former due to the most intense legume note.

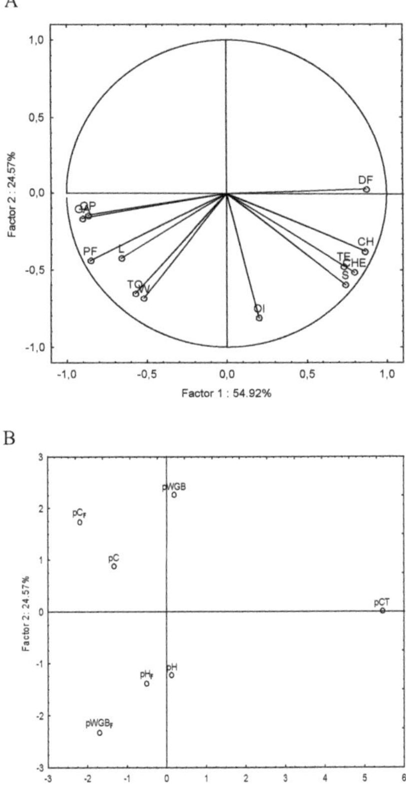

Figure 2. Loading plot (**A**) and score plot (**B**) of first and second principal components after principal component analysis (PCA) based on based on sensory analysis of pasta: *p*H, pasta containing 11% (*wt*/*wt*; d.m.) unfermented hemp dough; *p*H$_F$, pasta containing 11% (*wt*/*wt*; d.m.) fermented hemp dough; *p*C, pasta containing 11% (*wt*/*wt*; d.m.) unfermented chickpea dough; *p*C$_F$, pasta containing 11% (*wt*/*wt*; d.m.) fermented chickpea dough; *p*WGB, pasta containing 17% (*wt*/*wt*; d.m.) unfermented wheat germ and bran (1:4) dough; *p*WGB$_F$, pasta containing 17% (*wt*/*wt*; d.m.) fermented wheat germ and bran (1:4) dough; *p*CT, pasta made only with semolina and water. The data are the means of three independent experiments ± standard deviations (n = 3). The attributes used in sensory analysis were: GA, general acceptability; OI, Odor Intensity; OP, Odor pungent; CH, Color Heterogeneity; S, Sapidity; PF, Pungent Flavor; DF, Delicate Flavor; TE, Texture; CHE, Chewability; ST, Stacking; L, Legume; W, Whole; TO, Toasted.

4. Discussion

The need for a diversified, balanced and healthy diet and the continued emphasis on the importance of dietary proteins and fibers are pressuring food companies and researchers to develop new products. Pasta is an important staple food; compared to other wheat-based foods, it is characterized by a lower glycemic index (GI), and it has been identified as a suitable carrier of bioactive compounds in daily diet [4,5,20]. Novel pasta recipes including the replacement of wheat flour with alternative flours, as well as the inclusion of pre-fermented ingredients have been recently proposed [20].

Here, hemp, chickpea and milling by-products were used to fortify semolina pasta. Aiming at improving the protein bio-accessibility and digestibility of the non-wheat flours and milling by-products before pasta making, treatments with food–grade enzymes and fermentation were investigated. Proteases from *A. niger*, commercial xylanase (Depol 761P) and LAB (*L. plantarum* LB1 and *L. rossiae* LB5) have been used as pre-treatment of the raw materials. The first selection of the more suitable bioprocess option was carried out by the evaluation of the peptide profiles and the TFAA concentration.

When hemp, chickpea and milling by-products were subjected to enzymatic treatments, a moderate increase in the peptides was observed as the consequence of the proteolysis of the native proteins (proteases) [38] and the release of soluble compounds from the fibrous cellular compartments (xylanase) [39]. On the contrary, fermentation led to a decrease in the peptides (up to ca. 70% lower) and a relevant increase in the TFAA (up to ca. 80% higher), suggesting an intense proteolysis operated by both endogenous and bacterial proteases and peptidases on proteins and their derivatives. It has largely been reported that the biological acidification operated by LAB lead to the activation of endogenous proteases which start the primary proteolysis where medium-sized polypeptides are released and subjected to LAB peptidase activities [40]. Based on these results, fermentation was chosen as the optimal bioprocessing option and its effects on hemp, chickpea, and milling by-products further investigated.

Aiming at investigating the suitability of hemp, chickpea and milling by-products as food ingredients, unfermented and fermented doughs were included in pasta formulation. In order to limit the weakening of the gluten network, the level of fortification was kept below 30% [7,41]. Nevertheless, the cooking performances and textural properties of fortified products were affected by the inclusion of the additional ingredients. A decrease in OCT and the increase in the cooking loss observed in fortified pasta might be due to the lower quality of the gluten network [20]. Overall, fortified pasta was characterized by values of hardness higher than control. However, the magnitude was lower when the fermentation was used as pre-treatment, regardless of the raw material. Data from panel test highlighted that the fortification affected the flavor of pasta. Indeed, *p*CT was characterized by the more intense delicate flavor, while legume, toasted and whole flavors were identified in fortified pasta, according to the raw materials used. The high intensity of pungent flavor and odor, due to the fermentation, also contributed to the differentiation among fortified pasta and *p*CT.

The fortification led to a pasta rich in fiber and protein, regardless of the fermentation process. Indeed, more than 13% of protein as well as ca. 6% (d.m) of fiber were achieved. According to EC Regulation [42] on nutrition and health claims on food products, experimental fortified pasta can be labelled as a "source of fiber" and a "source of protein". Nevertheless, increases in the ANF, as compared to control pasta were found as a result of the fortification. The fermentation with selected LAB led to significant degradation (to traces) of the phytic acid, raffinose and condensed tannins as compared to the corresponding unfermented samples. Phytic acid (Myo-inositol 1,2,3,4,5,6 hexakis [dihydrogen phosphate]) is considered ANF due to the binding capacity towards essential dietary minerals, proteins and starch, thus reducing their bioavailability. The degradation of phytic acid during fermentation is achieved mainly through plant phytases [19] activated by LAB acidification. Moreover, a specific role of the organic acids on phytase activity has recently been proposed [43]. Cation chelation from organic acids may inhibit the aggregation of minerals and other molecules by phytic acid, thereby increasing their digestibility [43]. When present at high concentration, i.e., in legumes, raffinose is considered an ANF. However, LAB contribute to its enzymatic hydrolysis during fermentation [44], thus increasing product digestibility and reducing digestive discomfort [45]. The degradation of condensed tannins through LAB has already been proposed. It involves several enzymatic activities such as tannase, polyphenol oxidase and decarboxylase [46].

Beside the improvements in terms of ANF, an ca. 20% higher IVPD was found in fermented pasta samples as compared to the corresponding unfermented ones. Control pasta was characterized by an IVPD value 45% and 64% lower than the fortified samples (unfermented and fermented, respectively). Pasta containing non-wheat flours and milling by-products had a lower value of *p*GI compared to control, probably due to the higher concentration of dietary fibers and resistant starch, and a further decrease was found when the fermented flours were used. This effect could be attributed to biological acidification, which is among the main factors that decreases the starch hydrolysis rate and HI [36].

5. Conclusions

The study highlights the suitability of the fortification as a tool to improve the nutritional quality of pasta. Nevertheless, the pre-treatment of the non-wheat flours seems to be necessary to overcome the nutritional, structural and sensory drawback related to the use of such ingredients. Lactic acid bacteria fermentation has successfully been used to include hemp and chickpea flours and milling by-products in pasta making. LAB contributed to the increase in free amino acid content and decrease in phytic acid, raffinose and condensed tannins as compared to the corresponding unfermented doughs containing pasta. Moreover, fermentation improved protein digestibility and decreased the starch hydrolysis rate. Structural properties, cooking quality and sensory profiles were strongly affected by the fortification. Aiming at limiting the loss of rheological properties and cooking quality caused by the incorporation of non-wheat ingredients, further optimization of the technological processes may be needed.

Supplementary Materials: The following are available online at http://www.mdpi.com/2304-8158/8/12/604/s1. Table S1: Drying cycle used for making pasta; Table S2: List and definition of the attributes used for the sensory analysis made on pasta samples.

Author Contributions: R.S. carried out the laboratory work; E.P. contributed to the draft and critically revised the final manuscript; C.G.R. was the scientific advisor, responsible for the research funding and designed the experimental work. All authors read and approved the final manuscript.

Funding: This research received no external funding.

Acknowledgments: Biocatalysts (Biocatalysts Limited, Chicago, USA) is gratefully thanked for having provided the Depol 761P.

Conflicts of Interest: The authors declare no conflict of interest.

References

1. Kc, K.B.; Dias, G.M.; Veeramani, A.; Swanton, C.J.; Fraser, D.; Steinke, D.; Lee, E.; Wittman, H.; Farber, J.M.; Dunfield, K.; et al. When too much isn't enough: Does current food production meet global nutritional needs? *PLoS ONE* **2018**, *13*, e0205683. [CrossRef] [PubMed]
2. Banovic, M.; Arvola, A.; Pennanen, K.; Duta, D.E.; Brückner-Gühmann, M.; Lähteenmäki, L.; Grunert, K.G. Foods with increased protein content: A qualitative study on European consumer preferences and perceptions. *Appetite* **2018**, *125*, 233–243. [CrossRef] [PubMed]
3. Mannar, M.V.; Hurrell, R.F. *Food Fortification in a Globalized World*, 1st ed.; Academic Press: New York, NY, USA, 2018.
4. Spinelli, S.; Padalino, L.; Costa, C.; Del Nobile, M.A.; Conte, A. Food by-products to fortified pasta: A new approach for optimization. *J. Clean. Prod.* **2019**, *215*, 985–991. [CrossRef]
5. Foschia, M.; Horstmann, S.W.; Arendt, E.K.; Zannini, E. Legumes as Functional Ingredients in Gluten-Free Bakery and Pasta Products. *Annu. Rev. Food Sci. Technol.* **2017**, *8*, 75–96. [CrossRef] [PubMed]
6. Marconi, E.; Graziano, M.; Cubadda, R. Composition and Utilization of Barley Pearling By-Products for Making Functional Pastas Rich in Dietary Fiber and β-Glucans. *Cereal Chem. J.* **2000**, *77*, 133–139. [CrossRef]

7. Rizzello, C.G.; Verni, M.; Koivula, H.; Montemurro, M.; Seppa, L.; Kemell, M.; Katina, K.; Coda, R.; Gobbetti, M. Influence of fermented faba bean flour on the nutritional, technological and sensory quality of fortified pasta. *Food Funct.* **2017**, *8*, 860–871. [CrossRef]
8. Cankurtaran, T.; Bilgiçli, N. Influence of wheat milling by-products on some physical and chemical properties of filled and unfilled fresh pasta. *J. Food Sci. Technol.* **2019**, *56*, 2845–2854. [CrossRef]
9. International Pasta Organization. *The World Pasta Industry Status Report 2014*; International Pasta Organization: Rome, Italy, 2014.
10. Foschia, M.; Peressini, D.; Sensidoni, A.; Brennan, M.A.; Brennan, C.S. How combinations of dietary fibres can affect physicochemical characteristics of pasta. *LWT* **2015**, *61*, 41–46. [CrossRef]
11. Lu, X.; Brennan, M.A.; Serventi, L.; Liu, J.; Guan, W.; Brennan, C.S. Addition of mushroom powder to pasta enhances the antioxidant content and modulates the predictive glycaemic response of pasta. *Food Chem.* **2018**, *264*, 199–209. [CrossRef]
12. Martínez, M.L.; Marín, M.A.; Gili, R.D.; Penci, M.C.; Ribotta, P.D. Effect of defatted almond flour on cooking, chemical and sensorial properties of gluten-free fresh pasta. *Int. J. Food Sci. Technol.* **2017**, *52*, 2148–2155. [CrossRef]
13. Parvathy, U.; Bindu, J.; Joshy, C.G. Development and optimization of fish-fortified instant noodles using response surface methodology. *Int. J. Food Sci. Technol.* **2017**, *52*, 608–616. [CrossRef]
14. Roy, A.; Kucukural, A.; Zhang, Y. I-TASSER: A unified platform for automated protein structure and function prediction. *Nat. Protoc.* **2010**, *5*, 725–738. [CrossRef] [PubMed]
15. Krishnaiah, D.; Sarbatly, R.; Nithyanandam, R. A review of the antioxidant potential of medicinal plant species. *Food Bioprod. Process.* **2011**, *89*, 217–233. [CrossRef]
16. Teh, S.-S.; Bekhit, A.E.-D.A.; Carne, A.; Birch, J. Antioxidant and ACE-inhibitory activities of hemp (Cannabis sativa L.) protein hydrolysates produced by the proteases AFP, HT, Pro-G, actinidin and zingibain. *Food Chem.* **2016**, *203*, 199–206. [CrossRef]
17. Rizzello, C.G.; Nionelli, L.; Coda, R.; Di Cagno, R.; Gobbetti, M. Use of sourdough fermented wheat germ for enhancing the nutritional, texture and sensory characteristics of the white bread. *Eur. Food Res. Technol.* **2009**, *230*, 645–654. [CrossRef]
18. Coda, R.; Di Cagno, R.; Gobbetti, M.; Rizzello, C.G. Sourdough lactic acid bacteria: Exploration of non-wheat cereal-based fermentation. *Food Microbiol.* **2014**, *37*, 51–58. [CrossRef]
19. Gobbetti, M.; De Angelis, M.; Di Cagno, R.; Calasso, M.; Archetti, G.; Rizzello, C.G. Novel insights on the functional/nutritional features of the sourdough fermentation. *Int. J. Food Microbiol.* **2019**, *302*, 103–113. [CrossRef]
20. Montemurro, M.; Coda, R.; Rizzello, C.G. Recent Advances in the Use of Sourdough Biotechnology in Pasta Making. *Foods* **2019**, *8*, 129. [CrossRef]
21. Rizzello, C.G.; Nionelli, L.; Coda, R.; De Angelis, M.; Gobbetti, M. Effect of sourdough fermentation on stabilisation, and chemical and nutritional characteristics of wheat germ. *Food Chem.* **2010**, *119*, 1079–1089. [CrossRef]
22. AACC. Approved Methods of Analysis. 2010. Available online: http://methods.aaccnet.org/ (accessed on 18 December 2015).
23. Mariotti, F.; Tomé, D.; Mirand, P.P. Converting Nitrogen into Protein—Beyond 6.25 and Jones' Factors. *Crit. Rev. Food Sci. Nutr.* **2008**, *48*, 177–184. [CrossRef]
24. FAO Food and Nutrition paper 77. Food Energy—Methods of Analysis and Conversion Factors. In Proceedings of the Food and Agriculture Organization of the United Nations Technical Workshop Report, Rome, Italy, 3–6 December 2002.
25. Weiss, W.; Vogelmeier, C.; Görg, A. Electrophoretic characterization of wheat grain allergens from different cultivars involved in bakers' asthma. *Electrophoresis* **1993**, *14*, 805–816. [CrossRef] [PubMed]
26. Church, F.C.; Swaisgood, H.E.; Porter, D.H.; Catignani, G.L. Spectrophotometric Assay Using o-Phthaldialdehyde for Determination of Proteolysis in Milk and Isolated Milk Proteins. *J. Dairy Sci.* **1983**, *66*, 1219–1227. [CrossRef]
27. Rizzello, C.G.; Cassone, A.; Coda, R.; Gobbetti, M. Antifungal activity of sourdough fermented wheat germ used as an ingredient for bread making. *Food Chem.* **2011**, *127*, 952–959. [CrossRef] [PubMed]
28. Akeson, W.R.; Stahmann, M.A. A Pepsin Pancreatin Digest Index of Protein Quality Evaluation. *J. Nutr.* **1964**, *83*, 257–261. [CrossRef]

29. Rizzello, C.G.; Calasso, M.; Campanella, D.; De Angelis, M.; Gobbetti, M. Use of sourdough fermentation and mixture of wheat, chickpea, lentil and bean flours for enhancing the nutritional, texture and sensory characteristics of white bread. *Int. J. Food Microbiol.* **2014**, *180*, 78–87. [CrossRef]
30. Bradford, M.M. A rapid and sensitive method for the quantitation of microgram quantities of protein utilizing the principle of protein-dye binding. *Anal. Biochem.* **1976**, *72*, 248–254. [CrossRef]
31. Marti, A.; Fongaro, L.; Rossi, M.; Lucisano, M.; Pagani, A.M. Quality characteristics of dried pasta enriched with buckwheat flour. *Int. J. Food Sci. Technol.* **2011**, *46*, 2393–2400. [CrossRef]
32. Schoenlechner, R.; Drausinger, J.; Ottenschlaeger, V.; Jurackova, K.; Berghofer, E. Functional Properties of Gluten-Free Pasta Produced from Amaranth, Quinoa and Buckwheat. *Plant Foods Hum. Nutr.* **2010**, *65*, 339–349. [CrossRef]
33. D'egidio, M.G.; Mariani, B.M.; Nardi, S.; Novaro, P.; Cubadda, R. Chemical and technological variables and their relationships: A predictive equation for pasta cooking quality. *Cereal Chem.* **1990**, *67*, 275–281.
34. Iametti, S.; Bonomi, F.; Pagani, M.A.; Zardi, M.; Cecchini, C.; D'Egidio, M.G. Properties of the Protein and Carbohydrate Fractions in Immature Wheat Kernels. *J. Agric. Food Chem.* **2006**, *54*, 10239–10244. [CrossRef]
35. Hagerman, A.E. Acid Buthanol Assay for Proanthocyanidins. In *The Tannin Handbook 2002*; Hagerman, A.E., Ed.; Miami University: Oxford, UK, 2002.
36. De Angelis, M.; Damiano, N.; Rizzello, C.G.; Cassone, A.; Di Cagno, R.; Gobbetti, M. Sourdough fermentation as a tool for the manufacture of low-glycemic index white wheat bread enriched in dietary fibre. *Eur. Food Res. Technol.* **2009**, *229*, 593–601. [CrossRef]
37. Capriles, V.D.; Arêas, J.A.G. Effects of prebiotic inulin-type fructans on structure, quality, sensory acceptance and glycemic response of gluten-free breads. *Food Funct.* **2013**, *4*, 104–110. [CrossRef] [PubMed]
38. Rizzello, C.G.; De Angelis, M.; Di Cagno, R.; Camarca, A.; Silano, M.; Losito, I.; De Vincenzi, M.; De Bari, M.D.; Palmisano, F.; Maurano, F.; et al. Highly Efficient Gluten Degradation by Lactobacilli and Fungal Proteases during Food Processing: New Perspectives for Celiac Disease. *Appl. Environ. Microbiol.* **2007**, *73*, 4499–4507. [CrossRef] [PubMed]
39. Arte, E.; Rizzello, C.G.; Verni, M.; Nordlund, E.; Katina, K.; Coda, R. Impact of Enzymatic and Microbial Bioprocessing on Protein Modification and Nutritional Properties of Wheat Bran. *J. Agric. Food Chem.* **2015**, *63*, 8685–8693. [CrossRef] [PubMed]
40. Gänzle, M.G.; Loponen, J.; Gobbetti, M. Proteolysis in sourdough fermentations: Mechanisms and potential for improved bread quality. *Trends Food Sci. Technol.* **2008**, *19*, 513–521. [CrossRef]
41. Torres, A.; Frias, J.; Granito, M.; Vidal-Valverde, C. Fermented Pigeon Pea (Cajanus cajan) Ingredients in Pasta Products. *J. Agric. Food Chem.* **2006**, *54*, 6685–6691. [CrossRef] [PubMed]
42. EFSA Panel on Dietetic Products, Nutrition and Allergies (NDA). Scientific Opinion on the substantiation of a health claim related to cocoa flavanols and maintenance of normal endothelium-dependent vasodilation pursuant to Article 13(5) of Regulation (EC) No 1924/2006. *EFSA J.* **2012**, *10*, 2809. [CrossRef]
43. Vieira, B.; Caramori, J.; Oliveira, C.; Correa, G. Combination of phytase and organic acid for broilers: Role in mineral digestibility and phytic acid degradation. *World's Poult. Sci. J.* **2018**, *74*, 711–726. [CrossRef]
44. Coda, R.; Kianjam, M.; Pontonio, E.; Verni, M.; Di Cagno, R.; Katina, K.; Rizzello, C.G.; Gobbetti, M. Sourdough-type propagation of faba bean flour: Dynamics of microbial consortia and biochemical implications. *Int. J. Food Microbiol.* **2017**, *248*, 10–21. [CrossRef]
45. Waters, D.M.; Mauch, A.; Coffey, A.; Arendt, E.K.; Zannini, E. Lactic acid bacteria as a cell factory for the delivery of functional biomolecules and ingredients in cereal-based beverages: A review. *Crit. Rev. Food Sci. Nutr.* **2015**, *55*, 503–520. [CrossRef]
46. Filannino, P.; Di Cagno, R.; Gobbetti, M. Metabolic and functional paths of lactic acid bacteria in plant foods: Get out of the labyrinth. *Curr. Opin. Biotechnol.* **2018**, *49*, 64–72. [CrossRef] [PubMed]

 *foods*

Article

Development and Optimization of Djulis Sourdough Bread Using Taguchi Grey Relational Analysis

Pei-Ling Chung, Ean-Tun Liaw, Mohsen Gavahian and Ho-Hsien Chen *

Department of Food Science, National Pingtung University of Science and Technology, Neipu 91201, Pingtung, Taiwan; plchung@tajen.edu.tw (P.-L.C.); alexliaw@mail.npust.edu.tw (E.-T.L.); mohsengavahian@yahoo.com (M.G.)

* Correspondence: hhchen@mail.npust.edu.tw; Tel.: +886-8-7740402

Received: 5 July 2020; Accepted: 12 August 2020; Published: 20 August 2020

Abstract: Bakery products made from naturally fermented sourdough show a diversified flavor and nutritional profile. Djulis (*Chenopodium formosanum*), known as red quinoa or Taiwan djulis, originally cultivated by Taiwanese indigenous people in mountain areas in eastern and southern Taiwan, has a high nutritional value and characteristic properties. In the present study, a new bakery product (djulis sourdough bread) was developed and a combination of the Taguchi method coupled with grey theory was utilized to optimize the baking parameters (product formulation). Five main factors, i.e., djulis sourdough (A), hulled djulis (B), oil type (C), a mixture of bread flour (wet gluten content of 29.0%) and a high-gluten flour (wet gluten content of 35.5%) (D), and honey (E), (each at four levels) were chosen for the Taguchi experiment design ($L_{16}(4)^5$). Dependent parameters were the data from texture profile analysis (brittleness, springiness, cohesiveness, gumminess, and chewiness), color analysis (L^*, a^*, and b^*), and sensory evaluation (appearance, aroma, bitterness, sourness, chewiness, and overall acceptance) of the final product. Taguchi grey relational analysis successfully determined the optimal conditions based on combined parameters (5 factors), which highlighted the advantages of this innovative optimization technique. The result shows that the optimal formula for producing a djulis sourdough bread with the best texture, color, and sensory qualities was A3B1C1D2E2, i.e., 20% djulis sourdough, 0% addition of hulled djulis, 8% unsalted butter, 80% wheat flour + 20% high-gluten flour, and 10% honey, respectively. Such a novel application could be a reference for improving the quality of bakery products in the industry. Moreover, it seems that the new bakery product developed in this study has good potential to be commercially produced after further nutritional and economic analysis.

Keywords: bakery products; bread; djulis; food quality; optimization; product development; Taguchi grey relational analysis; texture profile analysis; sensory attributes; sourdough

1. Introduction

Djulis (*Chenopodium formosanum* Koidz.) belongs to the Amaranthaceae family and *Chenopodium* genus. According to Encyclopedia Britannica, the Amaranthaceae family includes about 175 genera and more than 2500 species. Many species, including beets and quinoa, are considered staple food crops, and some are cultivated as garden ornamental plants [1]. Among them, amaranth species are located mainly in tropical and subtropical areas. Grain amaranth yields tiny seeds that can be used as a grain to make flour, porridge, and other foods. For instance, amaranth grain can be processed to be added into several products including baby food, cakes, and cookies. In addition, amaranth grain has a high concentration of lysine, that is, an essential amino acid for the biosynthesis of proteins, which is vital for human tissue development and healing. Furthermore, this grain is rich in calcium, phosphorus, iron, potassium, zinc, vitamin E, and vitamin B-complex [2]. Previous studies have shown bioactive effects for species of the Amaranthaceae family [3]. For example, Sánchez-Urdaneta et al. fed rats with breads made with amaranth (*Amaranthus dubius* Mart. ex Thell) flour and observed that consumption of amaranth-enriched bread enhanced lipid profiles of rats and prevented metabolic and cardiovascular diseases due to its hypoglycemic and hypolipidemic effects [4]. Moreover, an in vivo study revealed that phenylpropanoid extract from *Halosarcia indica* (Willd.) has analgesic and anti-inflammatory effects on Wistar albino rats [5].

In Taiwan, djulis is also called Taiwan djulis or Taiwan red quinoa and has different strains with diverse colors. This plant has been cultivated mainly in eastern and southern areas of Taiwan (Taitung and Pingtung) as a cereal crop and was previously used for worship purposes and decorations in seasonal festivals by Taiwanese indigenous people (Taiwan aborigines). In recent years, physicochemical and bioactive characteristics as well as preventive healthcare applications of this crop get attention [6]. For instance, Hong et al. demonstrated that djulis extract could protect skin from UV-induced damage [7]. Additionally, Lee et al. found that the early stages of chemically induced colon carcinogenesis were suppressed in mice after feeding them djulis for 10 weeks [8]. However, there is limited information about the possibility of using this crop for developing a bakery product with optimal characteristics.

The Taguchi method is a systematic approach for experimental design and analysis. Recently, this approach has gained popularity to be used in various sectors of the industry for new product development and quality improvement in an economical way. Previous work demonstrated that the Taguchi method was used for the optimization of the conditions for submerged culture at a laboratory-scale study and resulted in the development of an upscaled fermentation process that could yield a high concentration of monacolin K [9]. Although the application of other optimization approaches (e.g., response surface methodology) has been widely explored for food processing [10,11], there are only limited studies in the literature that explored the applicability of the Taguchi method for developing new bakery products. Moreover, it seems that the limitations of other optimization approaches (e.g., response surface methodology; RSM) can be addressed by the application of the Taguchi method. For example, A study conducted by Chen et al. showed the applicability of the Taguchi technique for the quality improvement of egg–shortening cakes [12]. Basically, when applying the Taguchi method in process optimization, the optimum combination is determined based on one quality characteristic at a time. However, in practice, in production lines usually the process involves more than one characteristic (multiple objectives). Consequently, the effects of non-linear interactions between control factors exist and cannot be ignored. Therefore, the grey system theory is an approach that can be employed to optimize multi-characteristic processes.

The theory of the grey system was proposed by Professor Deng Julong in the 1980s [13]. Grey relational analysis (GRA), procured from grey system theory, is a measurement technique to determine the relationship between sequences through the analysis of a limited number of data [14,15]. The relational grade is defined as measuring the relevance between or two sequences or two systems and can be used to describe the trend relationship between a reference sequence (objective or ideal sequence) and a comparative sequence in a specific system. In such studies, a relational grade approaching 1 suggests that the reference sequence and the comparative sequence tend toward concordance. On the other hand, a relational grade approaching 0 indicates that the reference sequence and the comparative sequence do not tend toward concordance completely. This technique requires small quantities of data, and the data are not restricted to specific statistical distributions, which makes GRA superior to classical statistical methods. In previous studies, Chen et al. used GRA successfully to investigate the adulterated cases of commercial soybean sauces [16]; Chen et al. employed GRA effectively to identify and classify undried roselle samples frozen at −20 °C, and roselle samples dried correctly at 20, 50, 75, and 85 °C, respectively [17]. Moreover, associating the Taguchi method with GRA has also shown a powerful tool to optimize the multiple performance characteristics in the food manufacturing process. Chen et al. applied the Taguchi grey relational analysis method to optimize the fish drying process based on performance characteristics such as color measurement value (L^*, a^*, b^*), thiobarbituric acid value (TBA), and shear stress value [18]. Chung et al. used a grey-based Taguchi approach to improve beneficial monacolin K, *Monascus* pigment synthesis, and to decrease the adverse metabolite, citrinin, in the fermentation of *Monascus purpureus* [19].

Despite the progress in the application of grey system theory in several fields, this application of this innovative approach is an ongoing topic in the food industry. For example, the grey system theory has not been well-explored for optimization of products such as sourdough breads, which are believed to have improved shelf life and sensory properties such as flavor, aroma, and texture (mainly due to fermentation by yeasts, *Aspergillus*, and lactic acid bacteria) [20,21]. Specifically, developing a new djulis sourdough-based bread need a tremendous optimization that has not been explored in the literature. Therefore, this study aims to develop a new naturally fermented bread product (djulis sourdough bread) with acceptable palatability by utilizing the Taguchi–GRA method as an innovative optimization approach that can serve as a potential practice for industrial baking. In this regard, such a novel approach was utilized through an overall evaluation process for the optimization of djulis sourdough bread manufacturing based on 14 characteristics (objectives) including attributes obtained by texture profile analysis (TPA), measurement of L^*, a^*, and b^* values by a colorimeter, and sensory evaluation.

2. Materials and Methods

2.1. Questionnaire and Experimental Strategy

Thirty experienced bakers were invited to reexamine the standard formula used for making general commercial round-top white bread set up by the China Grain Products Research and Development Institute (New Taipei City, Taiwan) [22]. This standard formula is as follows (ratio of materials): 100% high-gluten flour (HGF), 10% fine granulated sugar, 8% butter, 4% fresh yeast, 54% water, 12% egg, 2% salt, and 4% milk powder. Based on the questionnaire responses, feedback, and discussions collected from these 30 experienced bakers, five potential influential factors that would affect bakery product quality were selected and identified for the experimental design and were then used to develop the djulis sourdough bread. These five factors were the addition/non-addition of djulis sourdough, honey, wheat flour (WF) or HGF, addition/non-addition of hulled djulis, and butter/oil. The abovementioned evaluations were performed to design an experiment with a specified number of tests as well as specified ranges for each test.

2.2. Experimental Preparation

Before running the experiment, a bunch of organic djulis was added and mixed with leftover baguette dough for four hours to form a djulis sourdough. Then the djulis sourdough was cultured in a refrigerator at 4–7 °C. Re-culturing was performed every seven days. In this regard, an appropriate amount of djulis sourdough was added into flour and water, kept at room temperature for three hours, and then refrigerated for the continuation of the low-temperature fermentation process.

The preparation of djulis sourdough bread was based on formulas derived from the factor/level assignments. Three fermentation processes were conducted for djulis sourdough preparation, i.e., the first fermentation was run at 28 °C, 75% of humidity for 60 min; after cutting and rounding, the second fermentation was run again at 28 °C, 75% of humidity for 15 min. Then after the appearance shaping, 38 °C, 75% of humidity for 50 min were applied in the third fermentation. After fermentation, the sourdough was baked for 35 min at 180–200 °C (top and bottom heat). The flow chart for making the djulis sourdough bread is shown in Figure 1.

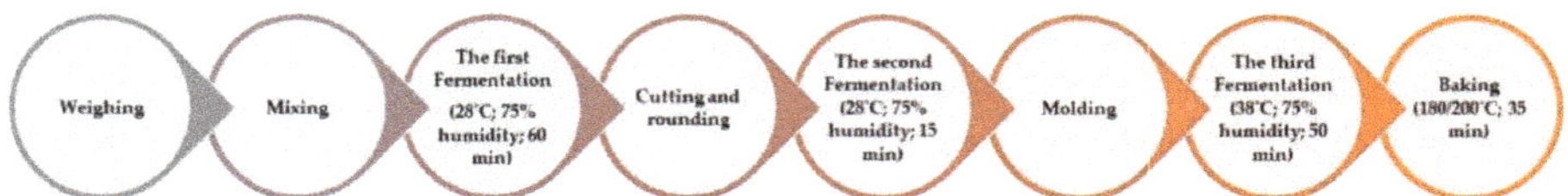

Figure 1. A flow chart representing djulis sourdough bread preparation in the present study.

2.3. Materials

Djulis was collected from Machia Township (Pingtung County, Taiwan). A WF sample with a wet gluten content of 29.0%, protein content of 10.0%, and ash content of 0.6% and an HGF with a wet gluten content of 35.5%, protein content of 12.5%, and ash content of 0.4% were obtained from Yuan Shan Food Co., Ltd. (Pingtung, Taiwan); Anchor unsalted butter, originally from New Zealand, obtained from Tehmag Foods Co. (New Taipei City, Taiwan); camellia oil (Yuan Shan Food Co., Ltd., Pingtung, Taiwan); Italian olive oil obtained from Tehmag Foods Co. (New Taipei City, Taiwan); lard (I-MEI Foods Co., Ltd., Pingtung, Taiwan); honey (Longan honey, The Chen's Honey, Pingtung, Taiwan); eggs (PX Mart, Pingtung, Taiwan); Anchor full cream milk powder, originally from New Zealand, obtained from Yu Hsuan Inc. (Pingtung, Taiwan); salt (Yu Hsuan Inc., Pingtung, Taiwan); charcoal-filtered water (PX Mart, Pingtung, Taiwan); and yeast (Yu Hsuan Inc., Pingtung, Taiwan).

2.4. Sensory Evaluation and Instrumental Measurement

2.4.1. Sensory Evaluation Analysis

The seven-point hedonic scale was adopted to assess the overall round-top djulis sourdough bread acceptability. Such a test was performed by 60 participants (between 20 and 35 years old) who were chosen randomly from students of the National Pingtung University of Science and Technology (NPUST). Appropriate guidance and training for all panelists were provided before conducting the sensory evaluation. Four categories of sensory attributes were stated on the score sheet as follows: appearance, smell/taste (aroma, bitterness, and sourness), texture (chewiness), and overall acceptability. Each item was scored between 1 and 7 (1: dislike extremely, 2: dislike moderately, 3: dislike slightly, 4: neither like nor dislike, 5: like slightly, 6: like moderately, 7: like extremely). Items could not be scored more than once. There were a total number of 16 slices of bread that needed to be practiced by a panelist per day. After the evaluation had been completed for one slice of bread, a 15 s interval was required before the next practice. During each evaluation, the external appearance of the sliced bread was first observed and scored. Afterward, the scoring was performed one by one for aroma, bitterness, sourness, chewiness, and overall acceptability. The sensory evaluation tests were performed in triplicate ($n = 3$) on three days, meaning that each member of the sensory evaluation panel (60 members) need to repeat the tests three times on different days. The final score of each item was calculated and obtained by averaging the three-replicate data.

2.4.2. Texture Profile Analysis

Oven-fresh round-top djulis loaves were subjected to TPA using a texture analyzer (TA-XT-Plus, Stable Micro Systems, Ltd., Godalming, UK) based on a standard method according to approved methods of the American Association of Cereal Chemists (AACC), as described by Amigo et al. [23]. First, each sample was sliced into 1.25-cm-thick slices, the slices at both ends were discarded, and the slices from the middle portion of each sample were used for analysis. During the testing process, two bread slices were stacked and analyzed using a 36-mm-diameter probe, with a compression ratio of 25% and a probe compression speed of 5 m/s. Each type of sample was compressed four times using eight slices from the middle portion, and data related to brittleness, springiness, cohesiveness, gumminess, and chewiness were recorded and calculated using the developed software (TA-XT-Plus, Stable Micro Systems, Ltd., Godalming, UK) [24,25].

2.4.3. Colorimeter Analysis

After cooling for one hour, round-top djulis loaves were sliced into 1.25-cm-thick slices, and slices from the middle portion of each sample were subjected to color and lightness measurements using a colorimeter (Minolta CR 310, Konica Minolta Sensing Singapore Pte. Ltd., Jurong East, Singapore). Color and lightness values were expressed as L^*, a^*, and b^*, with L^* representing lightness (L^* for brightest white = 100) or darkness (L^* for darkest black = 0); a^* representing the red/green component ($+a^*$: red, $-a^*$: green); and b^* representing the yellow/blue component ($+b^*$: yellow, $-b^*$: blue). The average value of six measurements was used for each parameter [26].

2.5. Data Analysis Models

Analyses of multiple quality characteristics of the baked djulis sourdough bread samples were performed using the novel combination approach, the Taguchi–GRA method, as shown in the following.

2.5.1. Taguchi Method

With the Taguchi method [27], an orthogonal array is first constructed by assigning known or assumed control factors and noise factors. Accordingly, the optimal parameter levels are determined with the minimum number of experiments. The orthogonal array is denoted by $L_n(X^m)$, where n is the number of columns of the array (i.e., the number of parameter and level combinations in the experiment), X is the number of levels, and m is the number of rows of the array (i.e., the number of factors). The orthogonal array used in the present study is denoted by $L_{16}(4^5)$, meaning that five control factors with four levels were used in 16 bakery product experiments. The five control factors used in this study were djulis sourdough (A), hulled djulis (B), butter/oil (C), Taiwan flour (D), and honey (E). Four levels (the ratio of formula) were set for each control factor (Table 1) in which the common ingredients were 3.5% fresh yeast, 54% distilled water, 12% egg, 2% salt, and 4% milk powder. The selected orthogonal array $L_{16}(4^5)$ and factor/level assignments are shown in Table 2 as the mean of three replicates.

Table 1. Parameter design factors and levels *.

Factors/Level	1	2	3	4
A. Djulis sourdough	0%	10%	20%	30%
B. Hulled djulis	0%	0.5%	1%	1.5%
C. Butter/oil (8%)	Unsalted butter	Camellia oil	Lard	Olive oil
D. WF + HGF **	WF 100%	WF 80% + HGF 20%	WF 60% + HGF 40%	WF 40% + HGF 60%
E. Honey	8%	10%	12%	14%

* The common ingredients were 3.5% fresh yeast, 54% distilled water, 12% egg, 2% salt, and 4% milk powder.
** WF: wheat flour; HGF: high-gluten flour.

2.5.2. Calculation of S/N (Signal-to-Noise Ratio) Values

Experimental data of those multiple quality characteristics in the orthogonal table were used to calculate the signal-to-noise ratio (S/N ratio, η). The S/N ratio did create a transformation function of the repetition data to another value and was used as a measure of the variation present in the experiment. S/N is a function indicator that measures performance, with higher S/N values indicating smaller quality losses. There are three types of quality characteristics for S/N values: nominal-the-best, smaller-the-best, and larger-the-best. In this study, we aimed to find the optimal operational parameters for the manufacturing of djulis sourdough bread retaining taste, nutrients, and supple flavors. Accordingly, the larger-the-best loss function was, therefore, used to calculate the S/N ratio as described in Equation (1).

$$\eta = -10 \log\left(\frac{1}{n}\sum_{i=0}^{n} 1/y_i^2\right) \quad (1)$$

where y_i is the ith value of the quality attribute, and n is the number of trials.

Table 2. L_{16} (4^5) orthogonal array and factor/level assignments *.

Factor/Level L_{16} (4^5)	A	B	C	D	E
1	1	1	1	1	1
2	1	2	2	2	2
3	1	3	3	3	3
4	1	4	4	4	4
5	2	1	2	3	4
6	2	2	1	4	3
7	2	3	4	1	2
8	2	4	3	2	1
9	3	1	3	4	2
10	3	2	4	3	1
11	3	3	1	2	4
12	3	4	2	1	3
13	4	1	4	2	3
14	4	2	3	1	4
15	4	3	2	4	1
16	4	4	1	3	2

* A: djulis sourdough; B: hulled djulis; C: butter/oil (8%); D: wheat flour + high-gluten flour; E: honey.

2.5.3. Algorithm of GRA

GRA was used to develop multiple quality characteristics of the djulis sourdough bread, to assess the optimal combination of parameters that best satisfies all the quality characteristics and to proceed with overall evaluation. These quality characteristics (dependent parameters) include color values, sensory attributes, and textural property. Data pretreatment was performed before employing GRA for data normalization, i.e., normalizing the raw data or their S/N ratios in the range of 0–1. All these characteristics and their S/N ratios were in the nature of the larger-the-better characteristics.

According to the literature [12,13,16,17], normalized functions could firstly be represented as Equation (2).

$$X_i^*(k) = \frac{X_i(k) - \min[X_i(k)]}{\max[X_i(k)] - \min[X_i(k)]} \quad (2)$$

where $X_i^*(k)$ is normalized raw data, $X_i^*(k)$ is a comparative sequence with kth entities, $i = 1, \ldots, m$; $k = 1, \ldots, n$; and $\max[X_i(k)]$ and $\max[X_i(k)]$ are the maximum and minimum ones in the comparative sequence.

The grey relational grade (GRG) could depict the degree of relationship between a reference sequences (objective sequence or ideal sequence) and a comparative sequence in which it is comprised of 14 characteristics including attributes obtained by TPA; measurement of L^*, a^*, and b^* values; and sensory evaluation. Equations (3)–(5) were used for the calculations related to GRA.

Let X_0 (k) be the reference sequence with kth entities, that is,

$$X_0\,(k) = \{x_0\,(1), x_0\,(2), \ldots, x_0\,(n)\}, \quad (3)$$

where $k = 1, 2, 3, \ldots, n$.

Let $X^*j\ (k)$ be the compared sequence; each X^*_j possess the same number of entities as X_0, that is,

$$X^*_j(k) = \{x^*_j\ (1), x^*_j\ (2), \ldots, x^*_j\ (n)\}, \tag{4}$$

where $k = 1, 2, 3, \ldots, n$.

The grey relational coefficient between the reference sequence of X_0 and the compared sequence X^*_j and at the kth entity are described in Equation (5):

$$\gamma\left(X_0(k), X^*{}_j(k)\right) = \frac{\Delta\min + \xi\Delta\max}{\Delta_{0j}(k) + \xi\Delta\max} \tag{5}$$

where

i. $\Delta_{0j}(k)$ is the absolute difference value between X_0 and X^*_j at the kth entity, that is, $\Delta_{0j}\ (k) = \left|x_0(k) - x^*{}_j(k)\right|$,
ii. $\Delta\max = \overset{\max}{\forall j}\ \overset{\max}{\forall k}\, \Delta_{0j}\ (k)$,
iii. $\Delta\min = \overset{\min}{\forall j}\ \overset{\min}{\forall k}\, \Delta_{0j}\ (k)$,
iv. $\xi \in [0, 1]$ is the distinguishing parameter in controlling the resolution between $\Delta\max$ and $\Delta\min$. For this case, the value of 0.5 was selected.

The optimal settings of process parameters combine multiple quality characteristics into one integrated numerical value, that is, GRG. This parameter for the sequence of X^*_j is represented in Equation (6).

$$\Gamma_{0j} = \Gamma\left(X_0, X_j\right) = \sum_{k=1}^{n} w_k \gamma\left(X_0(k), X_j(k)\right) \tag{6}$$

where W_k is the kth weighting of γ_{0j}.

The value of the GRG (Γ_{0j} in Equation (6)) represents the level of similarity between the comparative sequence X^*_j (the jth of the experimental trials) and the referential sequence X_0. The GRG of each experimental trial can be treated as a response (Γ_j) for each row of the orthogonal array of Table 2. The response graph can be set up by grouping the response values of the corresponding same factor levels of the column in the array, taking the sum, and dividing by the number of responses, as follows:

$$L_j = \frac{\sum_{j=1}^{n} \Gamma_j}{n} \tag{7}$$

where Γ_j is the response value of corresponding same factor levels of the column in the array, L_j is the mean response of the corresponding factor level.

2.6. Statistical Analysis

All the experiments were analyzed in triplicate. The analysis of variance (one-way ANOVA) was applied to the date to determine the significance of influences of control factors used for making djulis sourdough bread and was performed using an SPSS Statistics V.22.0 for Windows Statistical package (IBM Corporation, Armonk, NY, USA). The differences were significant statistically when $p < 0.05$ using the Duncan multiple range tests.

3. Results and Discussion

3.1. Appearance and Bread Volume

Sixteen round-top djulis sliced bread samples were prepared using different formulas based on the $L_{16}(4^5)$ orthogonal array and factor/level assignments (Tables 1 and 2). These samples were numbered as samples No. 1–16 as can be seen in Figure 2. The first four types of bread in Table 2 (samples No.

1–4 in Figure 2) were prepared without the addition of djulis sourdough. The average length, width, and height of these bread were 30.3, 10.3, and 13.1 cm, respectively. The other 12 types of samples that were prepared with the addition of djulis sourdough (samples No. 5–16 in Figure 2) had smaller sizes, that is, the average length, width, and height of these sourdough breads were 29.9, 10.0, and 11.88 cm, respectively. These observations suggest that the incorporation of djulis in the sourdough bread may reduce the loaf rising. Similarly, previous studies showed that change in the formulation may affect the loaf volume [28,29]. Such changes could be related to several parameters including the effect of the formulation on the gluten network as well as on the fermentation process.

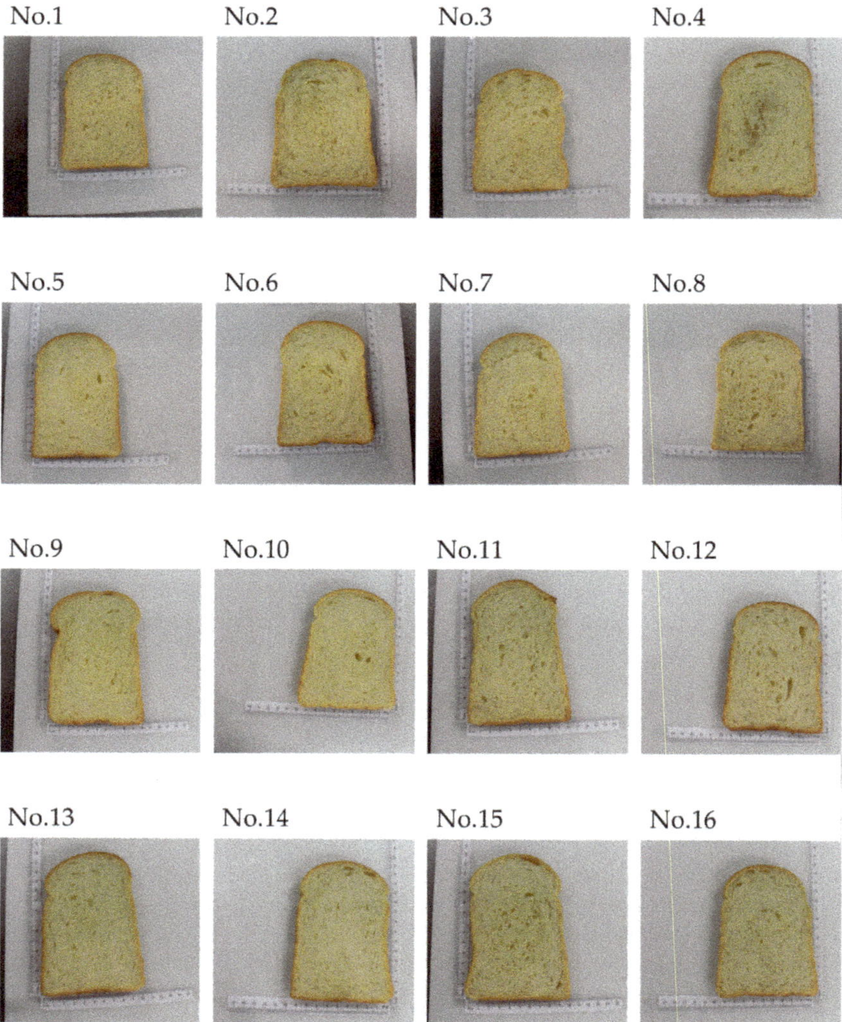

Figure 2. Sixteen numbered round-top djulis bread sliced samples (No. 1 to No. 16) made of different formulas as described in Table 2. No.: number.

3.2. Sensory Evaluation, Texture, and Color Analysis

The raw sensory data from questionnaires (Supplementary Materials; Figure S1) were used to calculate the results of the sensory evaluation of the 16 sliced bread samples (Table 3). According to the results, the trial No. 9 (20% djulis sourdough, 0% addition of hulled djulis, 8% lard, 40% WF + 60% HGF, and 10% honey) consistently gained higher scores on sensory attributes, including appearance, aroma, bitterness, sourness chewiness, and overall acceptance (Table 3). As djulis would release a bitter taste, the sensory evaluation of trial No. 9 showed the highest score on bitterness (less bitter, in the level of like slightly). In terms of sourness, sample No. 9 got the highest score, which showed that incorporation of djulis in the formulation can affect the sourness of the bread. This could be related to both the direct effect of djulis on the final taste of the product as well as its effect on the fermentation process and fermentation products. Similarly, for the appearance the bread, the sensory evaluation team preferred the appearance of sample No. 9. As can be seen in Figure 2, this sample has a unique distribution of air bubbles and color, which is related to the different formulation compared to other samples. Additionally, the panelists found sample No. 4 more chewy, which is in line with the observation about the air bubble distribution (Figure 2) as well as the reduced volume of the sample. This observation depicted that the sensory panelists in this study did not like bitterness in djulis sourdough sliced bread. Nevertheless, it could not be concluded that trial No. 9 was the best product among all bread samples only based on these sensory observations. Objective criterion data from instrumental analyses such as TPA and colorimeter analysis are also equally crucial parameters that should be included in the calculation (Table 3). In other words, an overall evaluation, based on both sensory and instrumental data, is necessary to determine the best set among others. In such cases, different results might be generated depending on the objectives of interest. According to the instrumental data, the addition of djulis affected the color values and textural properties of the final product. It was observed that the addition of djulis can alter the bread color values. Moreover, the instrumental texture analysis data were in line with those of the sensory evaluation. For example, similar to the panelists, TPA also confirmed that sample No. 9 is among the chewiest samples. The results of the present study were in line with those reported in the literature. For example, researchers observed a notable impact of formulation on the textural attributes of bread [28,29].

3.3. Calculation of S/N Values of Quality Characteristics

Table 3 shows data on the 14 quality characteristics obtained from sensory evaluation, texture analysis, and colorimetric analysis. Using the Taguchi method, the values of quality characteristics were transformed into S/N values (Table 4), which were then used to determine a formulation with the best quality and lowest variance. When the Taguchi method is used for process optimization, in some cases, a single quality characteristic is set as the target. As a result, experimental results can be shown as a simple linear relationship through the calculation of S/N values, and the best experimental combination can be directly determined from the response graph of the Taguchi orthogonal array. However, in practical production lines of the bakery industry, the investigation of a single quality objective is extremely rare. It means that, for food processing and product development, a number of parameters affect the overall quality of the product. Therefore, process optimization based on a single quality parameter usually cannot provide practical information for the industry. In the present study, 16 experimental trials with multiple quality objectives had to be investigated at a time. As the different quality characteristics had different units and attributes, data incomparability existed in the sequences. Therefore, data preprocessing was required to convert the data of the sequence with distinct scale and dimension into ones having a consistent unified scale and no dimension. The optimization practice in this study has successfully taken into account various processing and quality parameters. Therefore, the grey relational analysis could be employed with available comparable sequences [17].

Table 3. Raw data of 14 attributes of 16 trials based on an orthogonal array $L_{16}(4^5)$ (n = 3).

	Experiment Results													
	Sensory Evaluation data						Texture Analysis data						Color Values		
TN *	Appearance	Aroma	Bitterness	Sourness	Chewiness	Overall Acceptance	Gumminess (kgf)	Chewiness (kgf × mm)	Brittleness (kgf)	Springiness (mm)	Cohesiveness (Ratio)	L^*	a^*	b^*
1	5.37 ± 1.09 **	5.52 ± 0.91	3.97 ± 0.97	3.08 ± 1.16	3.92 ± 1.83	5.85 ± 0.72	0.10 ± 0.01	4.62 ± 0.16	2.13 ± 0.23	5.28 ± 0.11	0.73 ± 0.00	67.42 ± 0.52	−0.03 ± 0.03	14.88 ± 0.15
2	5.83 ± 1.43	2.07 ± 0.76	3.17 ± 1.18	2.83 ± 1.11	3.40 ± 1.62	4.85 ± 0.98	0.16 ± 0.03	4.26 ± 0.15	1.92 ± 0.19	5.48 ± 0.08	0.61 ± 0.01	54.78 ± 0.13	0.80 ± 0.09	14.14 ± 0.27
3	5.77 ± 1.11	4.35 ± 1.10	3.10 ± 0.96	3.10 ± 1.34	3.67 ± 1.79	5.28 ± 1.20	0.09 ± 0.04	3.44 ± 0.43	1.98 ± 0.12	5.19 ± 0.09	0.72 ± 0.01	64.72 ± 0.36	0.13 ± 0.16	14.61 ± 0.52
4	5.40 ± 1.36	4.43 ± 1.04	3.17 ± 0.86	3.15 ± 1.40	3.37 ± 1.76	4.37 ± 1.36	0.06 ± 0.03	3.30 ± 0.19	1.96 ± 0.14	4.92 ± 0.06	0.74 ± 0.06	59.62 ± 0.46	−0.44 ± 0.03	12.15 ± 0.27
5	5.68 ± 1.23	2.17 ± 0.67	3.32 ± 1.16	3.45 ± 0.89	4.25 ± 1.29	3.17 ± 1.40	0.10 ± 0.03	3.94 ± 0.67	1.55 ± 0.26	5.54 ± 0.44	0.82 ± 0.12	52.59 ± 0.34	0.15 ± 0.01	13.22 ± 0.35
6	5.78 ± 1.03	5.08 ± 1.15	3.77 ± 1.12	3.70 ± 1.09	4.45 ± 1.23	3.98 ± 1.48	0.10 ± 0.02	3.99 ± 0.82	1.56 ± 0.28	5.56 ± 0.36	0.81 ± 0.10	50.77 ± 0.22	0.89 ± 0.02	14.52 ± 0.37
7	5.20 ± 1.35	4.58 ± 1.28	3.90 ± 1.20	3.95 ± 1.04	4.48 ± 1.34	4.62 ± 1.21	0.07 ± 0.02	3.77 ± 0.35	1.53 ± 0.27	5.58 ± 0.33	0.78 ± 0.06	61.22 ± 0.18	−0.70 ± 0.04	10.68 ± 0.06
8	5.47 ± 1.37	3.97 ± 1.29	3.85 ± 1.01	3.80 ± 1.07	4.60 ± 1.13	3.88 ± 1.13	0.07 ± 0.03	3.75 ± 0.18	1.54 ± 0.26	4.94 ± 0.18	0.79 ± 0.14	64.34 ± 0.13	−0.42 ± 0.06	11.51 ± 0.22
9	6.97 ± 0.16	6.17 ± 0.77	5.47 ± 1.23	5.58 ± 1.38	6.97 ± 0.14	6.28 ± 0.78	0.19 ± 0.05	4.94 ± 0.30	1.67 ± 0.08	5.64 ± 0.57	0.97 ± 0.01	61.31 ± 0.78	−0.55 ± 0.03	12.10 ± 0.19
10	5.57 ± 1.31	4.55 ± 1.30	4.30 ± 1.05	4.25 ± 0.92	4.85 ± 1.16	5.13 ± 1.30	0.08 ± 0.03	3.16 ± 0.54	1.67 ± 0.26	5.51 ± 0.35	0.79 ± 0.04	67.00 ± 0.17	−0.62 ± 0.08	13.20 ± 0.19
11	5.85 ± 1.17	5.45 ± 0.90	5.00 ± 0.91	5.23 ± 0.75	6.83 ± 0.39	5.95 ± 0.87	0.19 ± 0.03	3.42 ± 0.42	1.67 ± 0.33	5.26 ± 0.12	0.79 ± 0.11	67.92 ± 0.40	−0.32 ± 0.04	12.61 ± 0.27
12	5.98 ± 0.98	2.75 ± 1.13	3.95 ± 1.48	4.15 ± 0.99	4.93 ± 1.15	5.08 ± 1.64	0.08 ± 0.01	3.02 ± 0.50	1.66 ± 0.23	5.30 ± 0.57	0.79 ± 0.04	62.02 ± 0.15	0.07 ± 0.02	10.45 ± 0.06
13	5.03 ± 3.64	4.68 ± 1.14	5.33 ± 1.05	5.05 ± 1.07	6.10 ± 0.67	5.92 ± 0.95	0.06 ± 0.03	3.88 ± 0.32	1.78 ± 0.36	5.50 ± 0.55	0.79 ± 0.04	72.10 ± 0.25	−0.70 ± 0.06	12.75 ± 0.04
14	5.52 ± 1.26	4.62 ± 1.03	3.90 ± 1.18	4.28 ± 1.02	5.08 ± 1.06	4.95 ± 1.22	0.08 ± 0.03	3.34 ± 0.65	1.79 ± 0.08	5.63 ± 0.65	0.79 ± 0.04	65.92 ± 0.34	−0.49 ± 0.05	13.01 ± 0.61
15	5.17 ± 1.60	1.95 ± 0.78	3.27 ± 1.06	3.68 ± 1.19	3.92 ± 1.21	2.52 ± 0.94	0.09 ± 0.04	3.56 ± 0.90	1.80 ± 0.15	5.38 ± 0.11	0.89 ± 0.09	68.55 ± 0.24	−0.91 ± 0.03	9.87 ± 0.14
16	5.37 ± 1.42	4.88 ± 0.94	3.57 ± 1.02	3.83 ± 1.04	4.90 ± 1.03	4.20 ± 1.47	0.16 ± 0.04	4.99 ± 0.98	1.83 ± 0.23	5.53 ± 0.02	0.79 ± 0.11	66.55 ± 0.13	−0.54 ± 0.02	12.76 ± 0.34

* TN: trial number; L^*: lightness; a^*: green–red coordinate; b^*: blue–yellow coordinate; kgf: kilogram-force. ** The results represent the average value of three replications (n = 3) followed by standard deviation (SD), i.e., mean ± SD. For sensory evaluation data, n = 3 means that 60 panelists repeated the sensory tests on 3 different days.

Table 4. Signal-to-noise ratio response table.

S/N Ration														
	Sensory Evaluation						Texture Analysis					Color Values		
Trial No.	Appearance	Aroma	Bitterness	Sourness	Chewiness	OA	Gumminess (kgf *)	Chewiness (kgf × mm)	Brittleness (kgf)	Springiness (mm)	Cohesiveness (Ratio)	L^*	a^*	b^*
1	14.60 **	14.84	11.98	9.77	11.87	15.34	−20.00	13.29	6.57	14.45	−2.73	36.58	−30.46	23.45
2	15.31	6.32	10.02	9.04	10.63	13.71	−15.92	12.59	5.67	14.78	−4.29	34.77	−1.94	23.01
3	15.22	12.77	9.83	9.83	11.29	14.45	−20.92	10.73	5.93	14.30	−2.85	36.22	−17.72	23.29
4	14.65	12.93	10.02	9.97	10.55	12.81	−24.44	10.37	5.85	13.84	−2.62	35.51	−7.13	21.69
5	15.09	6.73	10.42	10.76	12.57	10.02	−20.00	11.91	3.81	14.87	−1.72	34.42	−16.48	22.42
6	15.24	14.12	11.53	11.36	12.97	12.00	−20.00	12.02	3.86	14.90	−1.83	34.11	−1.01	23.24
7	14.32	13.22	11.82	11.93	13.03	13.29	−23.10	11.53	3.69	14.93	−2.16	35.74	−3.10	20.57
8	14.76	11.98	11.71	11.60	13.26	11.78	−23.10	11.48	3.75	13.87	−2.05	36.17	−7.54	21.22
9	16.86	15.81	14.76	14.93	16.86	15.96	−14.42	13.87	4.45	15.03	−0.26	35.75	−5.19	21.66
10	14.92	13.16	12.67	12.57	13.71	14.20	−21.94	9.99	4.45	14.82	−2.05	36.52	−4.15	22.41
11	15.34	14.73	13.98	14.37	16.69	15.49	−14.42	10.68	4.45	14.42	−2.05	36.64	−9.90	22.01
12	15.53	8.79	11.93	12.36	13.86	14.12	−21.94	9.60	4.40	14.49	−2.05	35.85	−23.10	20.38
13	14.03	13.40	14.53	14.07	15.71	15.45	−24.44	11.78	5.01	14.81	−2.05	37.16	−3.10	22.11
14	14.84	13.29	11.82	12.63	14.12	13.89	−21.94	10.47	5.06	15.01	−2.05	36.38	−6.20	22.29
15	14.27	5.80	10.29	11.32	11.87	8.03	−20.92	11.03	5.11	14.62	−1.01	36.72	−0.82	19.89
16	14.60	13.77	11.05	11.66	13.80	12.46	−15.92	13.96	5.25	14.85	−2.05	36.46	−5.35	22.12

* kgf: kilogram-force; OA: overall acceptance; L^*: lightness; a^*: green–red coordinate; b^*: blue–yellow coordinate. ** The results represent the average value of three replications ($n = 3$).

3.4. Grey Relational Analysis

Before using GRA, the S/N values of the various quality characteristics were preprocessed and converted into normalized values ranging from 0 to 1 through Equation (2), as shown in Table 5. The normalized data possessed good consistency and satisfied the three basic conditions for sequence comparison mentioned in the previous section. As the larger-the-best S/N values were calculated for the multiple quality characteristics of the present study, the value of the reference sequence for GRA was set to 1. Therefore, the normalized GRG closer to 1 in the sequence indicated greater closeness to the target value.

With the calculation of GRG, the distinguishing coefficient ζ is set within the range of zero to one ($0 < \zeta \leq 1$). This ensures that the maximum sequence difference Δ_{max} does not become excessively large and causes a loss of the influencing power of the minimum sequence difference Δ_{min}. Although excessively high or low values of ζ will lead to linear biases in data [13], the main function of ζ is to adjust the degree of contrast between the background value and the object being tested. Therefore, the value can be adjusted based on actual needs, as changes in the value of ζ only lead to changes in relative values without affecting the order of the GRG [13]. In the present investigation, a value of 0.5 was used for ζ.

Weightings were assigned to the quality characteristics. Fourteen quality characteristics, including appearance, smell/taste (aroma, bitterness, and sourness), texture (chewiness), overall acceptance, TPA attributes (gumminess, chewiness, brittleness, springiness, and cohesiveness), and colorimetric values (L, a^*, and b^*), were classified into four categories based on the characteristics preferred by consumers who purchase bakery products, and a weighting of 1/4 was assigned to each category (Table 6). The reference sequence (ideal values) was chosen as $X_0\ (k) = (1, 1, 1, 1, 1, 1, 1, 1, 1, 1, 1, 1, 1, 1)$, in which all 14 characteristics employed the concept of the larger-the-better for this work. All 16 sequences were treated as comparative sequences, and each sequence was composed of 14 characteristics (entities). The GRGs of 16 comparative sequence, which were calculated according to Equations (2) and (6), are presented in the last column of Table 6. Then, the grey relational analysis user interface was developed accordingly, to perform computer computing instead of manual calculation (Figure 3).

Table 5. Data preprocessing for each performance characteristic ($n = 3$).

Data Pretreatment														
	Sensory Evaluation						Texture Analysis					Color Values		
Trial No.	Appearance	Aroma	Bitterness	Sourness	Chewiness	OA	Gumminess (kgf)	Chewiness (kgf × mm)	Hardness (kgf)	Springiness (mm)	Cohesiveness (ratio)	L^*	a^*	b^*
1	0.20	0.90	0.44	0.12	0.21	0.92	0.44	0.85	1.00	0.52	0.39	0.81	0.00	1.00
2	0.45	0.05	0.04	0.00	0.01	0.72	0.85	0.69	0.69	0.79	0.00	0.22	0.96	0.88
3	0.42	0.70	0.00	0.13	0.12	0.81	0.35	0.26	0.78	0.39	0.36	0.69	0.43	0.96
4	0.22	0.71	0.04	0.16	0.00	0.60	0.00	0.18	0.75	0.00	0.42	0.46	0.79	0.51
5	0.37	0.09	0.12	0.29	0.32	0.25	0.44	0.53	0.04	0.87	0.64	0.10	0.47	0.71
6	0.43	0.83	0.34	0.39	0.38	0.50	0.44	0.55	0.06	0.90	0.61	0.00	0.99	0.94
7	0.10	0.74	0.40	0.49	0.39	0.66	0.13	0.44	0.00	0.92	0.53	0.53	0.92	0.19
8	0.26	0.62	0.38	0.43	0.43	0.47	0.13	0.43	0.02	0.03	0.56	0.68	0.77	0.37
9	1.00	1.00	1.00	1.00	1.00	1.00	1.00	0.98	0.26	1.00	1.00	0.54	0.85	0.50
10	0.31	0.74	0.58	0.60	0.50	0.78	0.25	0.09	0.26	0.83	0.56	0.79	0.89	0.71
11	0.46	0.89	0.84	0.90	0.97	0.94	1.00	0.25	0.26	0.49	0.56	0.83	0.69	0.60
12	0.53	0.30	0.43	0.56	0.52	0.77	0.25	0.00	0.25	0.54	0.56	0.57	0.25	0.14
13	0.00	0.76	0.95	0.85	0.82	0.94	0.00	0.50	0.46	0.82	0.56	1.00	0.92	0.62
14	0.28	0.75	0.40	0.61	0.56	0.74	0.25	0.20	0.47	0.99	0.56	0.74	0.82	0.67
15	0.08	0.00	0.09	0.39	0.21	0.00	0.35	0.33	0.49	0.65	0.81	0.86	1.00	0.00
16	0.20	0.80	0.25	0.45	0.52	0.56	0.85	1.00	0.54	0.86	0.56	0.77	0.85	0.63

kgf: kilogram-force; OA: overall acceptance; L^*: lightness; a^*: green–red coordinate; b^*: blue–yellow coordinate.

Table 6. Grey relational analysis (GRA) calculation model ($\zeta = 0.5$).

Trial No. *	Appearance	Aroma	Bitterness	Sourness	Chewiness	OA	Gumminess	Chewiness	Hardness	Springiness	Cohesiveness	L^*	a^*	b^*	Grey Relational Grade
Weighting	1/16	1/16	1/16	1/16	1/8	1/8	1/20	1/20	1/20	1/20	1/20	1/12	1/12	1/12	
Reference Sequence	1	1	1	1	1	1	1	1	1	1	1	1	1	1	
1	0.38	0.84	0.47	0.36	0.39	0.87	0.47	0.77	1.00	0.51	0.45	0.72	0.33	1.00	0.6163
2	0.48	0.35	0.34	0.33	0.34	0.64	0.77	0.61	0.61	0.70	0.33	0.39	0.93	0.80	0.5439
3	0.46	0.62	0.33	0.37	0.36	0.72	0.44	0.40	0.69	0.45	0.44	0.62	0.47	0.92	0.5354
4	0.39	0.63	0.34	0.37	0.33	0.56	0.33	0.38	0.67	0.33	0.46	0.48	0.70	0.50	0.4690
5	0.44	0.36	0.36	0.41	0.42	0.40	0.47	0.52	0.34	0.79	0.58	0.36	0.49	0.63	0.4597
6	0.47	0.75	0.43	0.45	0.45	0.50	0.47	0.53	0.35	0.83	0.56	0.33	0.99	0.89	0.5710
7	0.36	0.66	0.46	0.50	0.45	0.60	0.37	0.47	0.33	0.86	0.52	0.52	0.87	0.38	0.5290
8	0.40	0.57	0.45	0.47	0.47	0.49	0.37	0.47	0.34	0.34	0.53	0.61	0.69	0.44	0.4839
9	1.00	1.00	1.00	1.00	1.00	1.00	1.00	0.96	0.40	1.00	1.00	0.52	0.77	0.50	0.8675
10	0.42	0.65	0.54	0.55	0.50	0.69	0.40	0.35	0.40	0.75	0.53	0.71	0.82	0.63	0.5861
11	0.48	0.82	0.76	0.84	0.95	0.89	1.00	0.40	0.40	0.49	0.53	0.75	0.62	0.55	0.7131
12	0.52	0.42	0.47	0.53	0.51	0.68	0.40	0.33	0.40	0.52	0.53	0.54	0.40	0.37	0.4881
13	0.33	0.68	0.92	0.77	0.73	0.89	0.33	0.50	0.48	0.73	0.53	1.00	0.87	0.57	0.7026
14	0.41	0.67	0.46	0.56	0.53	0.66	0.40	0.38	0.49	0.97	0.53	0.66	0.73	0.60	0.5855
15	0.35	0.33	0.36	0.45	0.39	0.33	0.44	0.43	0.50	0.59	0.73	0.78	1.00	0.33	0.4930
16	0.38	0.71	0.40	0.47	0.51	0.53	0.77	1.00	0.52	0.78	0.53	0.69	0.77	0.57	0.6016

* OA: overall acceptance; L^*: lightness; a^*: green–red coordinate; b^*: blue–yellow coordinate.

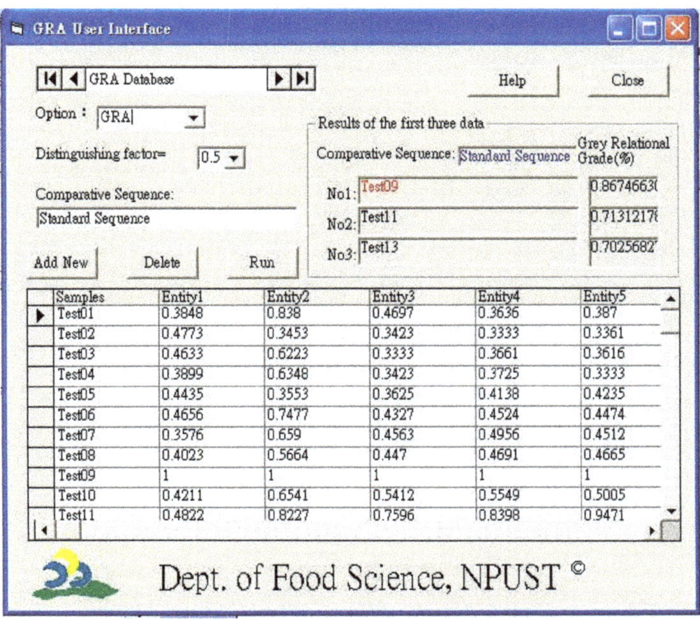

Samples	Entity1	Entity2	Entity3	Entity4	Entity5
Test01	0.3848	0.838	0.4697	0.3636	0.387
Test02	0.4773	0.3453	0.3423	0.3333	0.3361
Test03	0.4633	0.6223	0.3333	0.3661	0.3616
Test04	0.3899	0.6348	0.3423	0.3725	0.3333
Test05	0.4435	0.3553	0.3625	0.4138	0.4235
Test06	0.4656	0.7477	0.4327	0.4524	0.4474
Test07	0.3576	0.659	0.4563	0.4956	0.4512
Test08	0.4023	0.5664	0.447	0.4691	0.4665
Test09	1	1	1	1	1
Test10	0.4211	0.6541	0.5412	0.5549	0.5005
Test11	0.4822	0.8227	0.7596	0.8398	0.9471

(**A**)

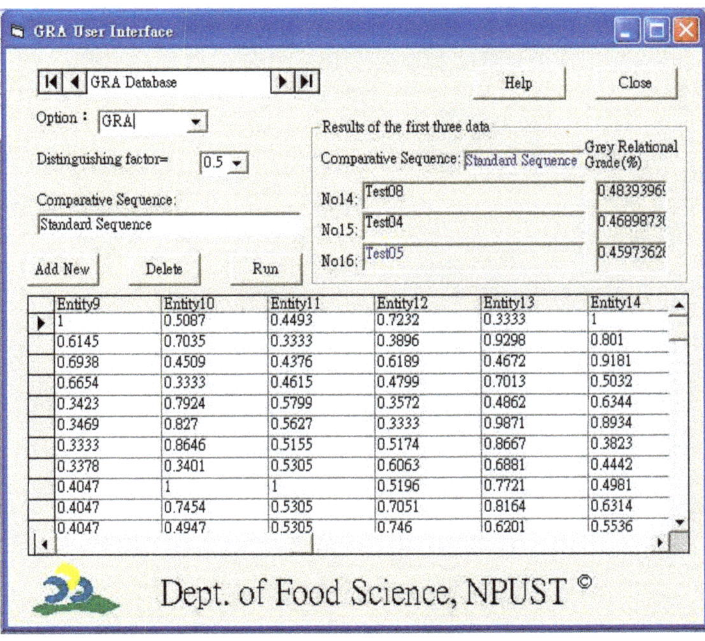

Entity9	Entity10	Entity11	Entity12	Entity13	Entity14
1	0.5087	0.4493	0.7232	0.3333	1
0.6145	0.7035	0.3333	0.3896	0.9298	0.801
0.6938	0.4509	0.4376	0.6189	0.4672	0.9181
0.6654	0.3333	0.4615	0.4799	0.7013	0.5032
0.3423	0.7924	0.5799	0.3572	0.4862	0.6344
0.3469	0.827	0.5627	0.3333	0.9871	0.8934
0.3333	0.8646	0.5155	0.5174	0.8667	0.3823
0.3378	0.3401	0.5305	0.6063	0.6881	0.4442
0.4047	1	1	0.5196	0.7721	0.4981
0.4047	0.7454	0.5305	0.7051	0.8164	0.6314
0.4047	0.4947	0.5305	0.746	0.6201	0.5536

(**B**)

Figure 3. (**A**) First three results obtained directly from the GRA user interface; (**B**) last three results obtained directly from the GRA user interface.

3.5. Parameter Optimizaion

3.5.1. Optimal Factors and Levels

The larger the GRG, the closer the product quality to the objective value. For instance, in Table 6, experimental trial No. 9 seems to be acceptable and closer to the reference sequence (ideal sequence), in which the highest GRG, 0.8675, was obtained. Accordingly, the order of trials (No. 1–No. 16) was rearranged as No. 9 (0.8675) > No. 11 (0.7131) > No. 13 (0.7026) > No. 1 (0.6163) > No. 16 (0.6016) > No. 10 (0.5861) > No. 14 (0.5855) > No. 6 (0.5710) > No. 2 (0.5439) > No. 3 (0.5354) > No. 11 (0.5290) > No. 15 (0.4930) > No. 12 (0.4881) > No. 8 (0.4839) > No. 4 (0.4690) > No. 5 (0.4597). However, non-linear interrelations were observed between the processing parameters in bakery manufacturing. A previous investigation showed that the mean GRG of each experimental trial could be regarded as the response and processed to determine optimal combinations of process parameter levels when a system with multiple performance characteristics is evaluated [18].

Based on the $L_{16}(4^5)$ orthogonal array (Table 2), the GRG for all 16 experimental trials (Table 6), and Equation (7), mean responses of the control level were calculated by taking the sum of the same levels in the column divided by the number of levels. For instance, values of A1, A2, A3, and A4 were 0.541 ((0.6163 + 0.5439 + 0.5354 + 0.4690)/4), 0.511 ((0.4597 + 0.5710 + 0.5290 + 0.4839)/4), 0.664 ((0.8675 + 0.5861 + 0.7131 + 0.4881)/4), and 0.596 ((0.7026 + 0.5855 + 0.4930 + 0.6016)/4), respectively. Likewise, mean responses were calculated for all factor levels of B, C, D, and E to generate a response graph (Figure 4). In this response graph, the highest value of the level for each factor represents the strongest effect. Therefore, optimal parameters were selected based on the highest response values in Figure 4, which were A3 (20% djulis sourdough), B1 (0% addition of hulled djulis), C1 (8% unsalted butter), D2 (80% WF + 20% HGF), and E2 (10% honey). The suggested condition (A3B1C1D2E2) is a generated optimal factor-level combination, which is not among the 16 experimental trials that are listed in Table 2. Although the present study mainly focused on the technological, physical, and sensory properties of the newly developed djulis sourdough bread, it should be noted that the new product developed in the present study can possess unique nutritional values considering previous reports about the bioactive effects of djulis [7]. Further investigation about the nutritional profile, bioactive effects, and potential healthcare applications of this product can be investigated in future studies. Similarly, studies regarding the properties of sourdough bread incorporated with other strains of djulis can be suggested as researchers showed that bread ingredients (e.g., wheat flour) originated from different locations possess various bioactive compounds that can affect the nutritional characteristics of the final product, which might be the case for djulis sourdough bread [30].

3.5.2. ANOVA Analysis

The results of ANOVAs indicate the degrees of influence over multiple quality characteristics (Table 7). As GRA provides a comprehensive analysis of the various characteristics, in the way of balanced consideration of the point of view of consumers, the influences of conflicting factors had already been weakened during the analysis process. Table 7 shows that all five factors (addition/non-addition of djulis sourdough, addition/non-addition of hulled djulis, butter/oil type, WF + HGF, and honey) significantly influenced the quality characteristics of the naturally leavened sourdough bread developed in the present study.

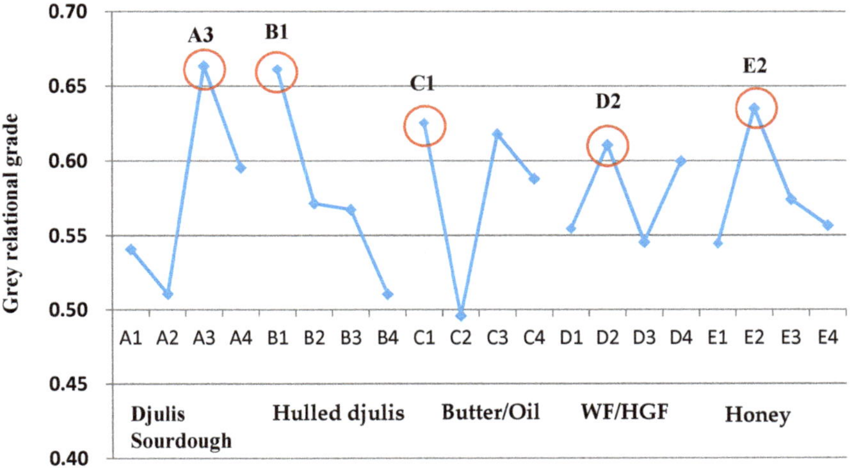

Figure 4. Response graph of grey relational grades of djulis sourdough bread. WF: wheat flour; HGF: high-gluten flour.

Table 7. The summary of ANOVA results.

	SS *	DF **	Variance	F-Ratio	Confidence (%)	Significant
A	0.1622	3	0.0541	116.98	100.00%	***
B	0.1399	3	0.0466	100.91	100.00%	***
C	0.1280	3	0.0427	92.32	100.00%	***
D	0.0379	3	0.0126	27.33	100.00%	***
E	0.0584	3	0.0195	42.14	100.00%	***
Error	0.0148	32	0.0005			
Total	0.5411	47				

* SS: sum of squares. ** DF: degrees of freedom. *** $p < 0.001$.

4. Conclusions

The present study demonstrated that a combination of Taguchi and grey relational analysis, i.e., a Taguchi–GRA approach, could be employed to investigate the effects of processing parameters on the quality of djulis sourdough bread and to identify the optimal settings for manufacturing new bakery products when multiple characteristics are involved. Such multiple characteristics of bread (e.g., aroma, color, and texture) are important for consumers. At this moment, it seems that no systematic approach has been implanted in the bakery industry that can consider multiple objectives at a time for developing new products. Therefore, the Taguchi–GRA approach, which was explored in this study, could be a prospective optimization technique that can be implemented in the bakery and other sectors of the food industry. The novel Taguchi–GRA approach introduced in this study could provide a reference of the basis for the enhancement of consumer-oriented products in the bakery industry. Furthermore, it was depicted that, sometimes, sensory evaluation could not be the only decisive approach to determine the optimal bakery products. Therefore, a combination of instrumental and sensory analysis can provide realistic data to develop a product with optimal quality parameters. Further studies in the nutritional aspects of such an innovative product can be suggested for future studies.

Supplementary Materials: The following are available online at http://www.mdpi.com/2304-8158/9/9/1149/s1, Figure S1: The raw sensory data from questionnaires.

Author Contributions: Conceptualization, H.-H.C. and P.-L.C.; methodology, H.-H.C. and E.-T.L.; software, H.-H.C. and P.-L.C.; validation, H.-H.C., E.-T.L. and M.G.; formal analysis, H.-H.C.; investigation, P.-L.C.; resources, H.-H.C.; data curation, H.-H.C. and P.-L.C.; writing—original draft preparation, H.-H.C. and P.-L.C.; writing—review and editing, H.-H.C. and M.G.; visualization, H.-H.C. and M.G.; supervision, H.-H.C.; project administration, H.-H.C.; funding acquisition, H.-H.C.. All authors have read and agreed to the published version of the manuscript.

Funding: This research received no external funding.

Conflicts of Interest: The authors declare no conflict of interest.

References

1. Augustyn, A.; Zeidan, A.; Zelazko, A.; Eldridge, A.; McKenna, A.; Tikkanen, A.; Schreiber, B.A.; Duignan, B.; Mahajan, D.; Promeet, D. Amaranthaceae Plant Family. Available online: https://www.britannica.com/plant/Amaranthaceae (accessed on 20 July 2020).
2. Berkelaar, D.; Alemu, J. Grain Amaranth. *ECHO Dev. Notes* **2006**, *91*, 1–8.
3. Li, P.H.; Chan, Y.J.; Lu, W.C.; Huang, D.W.; Chang, T.C.; Chang, W.H.; Nie, X.B.; Jiang, C.X.; Zhang, X.L. Bioresource Utilization of Djulis (*Chenopodium formosanum*) Biomass as Natural Antioxidants. *Sustainability* **2020**, *12*, 5926. [CrossRef]
4. Sánchez-Urdaneta, A.B.; Montero-Quintero, K.C.; González-Redondo, P.; Molina, E.; Bracho-Bravo, B.; Moreno-Rojas, R. Hypolipidemic and Hypoglycaemic Effect of Wholemeal Bread with Amaranth (Amaranthus dubius Mart. ex Thell.) on Sprague Dawley Rats. *Foods* **2020**, *9*, 707. [CrossRef]
5. Bhanuvalli, R.S.; Lotha, R.; Sivasubramanian, A. Phenyl propanoid rich extract of edible plant Halosarcia indica exert diuretic, analgesic, and anti-inflammatory activity on Wistar albino rats. *Nat. Prod. Res.* **2018**, *34*, 1616–1620. [CrossRef]
6. Lu, W.-C.; Chan, Y.-J.; Tseng, F.-Y.; Chiang, P.-Y.; Li, P.-H. Production and Physicochemical Properties of Starch Isolated from Djulis (*Chenopodium formosanum*). *Foods* **2019**, *8*, 551. [CrossRef]
7. Hong, Y.-H.; Huang, Y.-L.; Liu, Y.-C.; Tsai, P.-J. Djulis (Chenopodium formosanum Koidz.) Water Extract and Its Bioactive Components Ameliorate Dermal Damage in UVB-Irradiated Skin Models. *BioMed Res. Int.* **2016**, *2016*, 1–8. [CrossRef]
8. Lee, C.-W.; Chen, H.-J.; Xie, G.-R.; Shih, C.-K. Djulis (Chenopodium Formosanum) Prevents Colon Carcinogenesis via Regulating Antioxidative and Apoptotic Pathways in Rats. *Nutrients* **2019**, *11*, 2168. [CrossRef]
9. Chung, C.C.; Chen, H.H.; Hsieh, P.C. Application of the taguchi method to optimize monascus spp. culture. *J. Food Process Eng.* **2007**, *30*, 241–254. [CrossRef]
10. Al-Hilphy, A.R.; Ali, H.I.; Al-Iessa, S.A.; Lorenzo, J.M.; Barba, F.J.; Gavahian, M. Optimization of process variables on physicochemical properties of milk during an innovative refractance window concentration. *J. Food Process. Preserv.* **2020**, e14782. [CrossRef]
11. Al-Hilphy, A.R.; Al-Shatty, S.M.; Al-Mtury, A.A.A.; Gavahian, M. Infrared-assisted oil extraction for valorization of carp viscera: Effects of process parameters, mathematical modeling, and process optimization. *LWT* **2020**, *129*, 109541. [CrossRef]
12. Chen, H.H.; Huang, T.C.; Hsieh, C.F.; Wang, C.Y. An Application of Taguchi Technique for Egg Shortening Cakes Processing Improvements. *Taiwan. J. Agric. Chem. Food Sci.* **2000**, *38*, 214–222. (In Chinese)
13. Deng, J. *The Primary Methods of Grey System Theory*; Huazhong University of Science and Technology Press: Wuhan, China, 2005.
14. Chen, H.-H.; Chung, C.-C.; Hsu, C.-H.; Huang, T.-C. Application of Grey Prediction in a Solar Energy-Assisted Photocatalytic Low-Pressure Drying Process. *Dry. Technol.* **2010**, *28*, 1097–1106. [CrossRef]
15. Chung, C.C.; Chen, H.H.; Ting, C.H. Grey prediction fuzzy control for pH processes in the food industry. *J. Food Eng.* **2010**, *96*, 575–582.
16. Chen, H.H.; Huang, T.C.; Su, Y.M. Applying Grey Relational Analysis to Assist PCA for Testing the Quality of Soybean Sauces. *J. Grey Syst.* **2000**, *3*, 25–35. (In Chinese)
17. Chen, H.-H.; Tsai, P.-J.; Chen, S.-H.; Su, Y.-M.; Chung, C.-C.; Huang, T.-C. Grey relational analysis of dried roselle (*Hibiscus Sabdariffa* L.). *J. Food Process. Preserv.* **2005**, *29*, 228–245. [CrossRef]
18. Chen, H.H.; Wang, C.Y.; Huang, T.C. The Application of Grey-Taguchi Method in Food Processing: A Model Study of Fly Fish Drying Process. *Taiwan. J. Agric. Chem. Food Sci.* **2004**, *42*, 207–214.

19. Chung, C.C.; Chen, H.H.; Hsieh, P.C. Optimization of the Monascus purpureus Fermentation Process Based on Multiple Performance Characteristics. *J. Grey Syst.* **2008**, *11*, 85–96.
20. Corsetti, A.; Settanni, L. Lactobacilli in sourdough fermentation. *Food Res. Int.* **2007**, *40*, 539–558. [CrossRef]
21. Bourré, L.; McMillin, K.; Borsuk, Y.; Boyd, L.; Lagassé, S.; Sopiwnyk, E.; Jones, S.; Dyck, A.; Malcolmson, L. Effect of adding fermented split yellow pea flour as a partial replacement of wheat florin bread. *Legume Sci.* **2019**, *1*, 1–10.
22. Hsu, H.C.; Huang, D.H.; Gu, D.T. *Cake and Pastry Technology*, 3rd ed.; China Grain Products Research and Development Institute: Taipei, Taiwan, 1999.
23. Amigo, J.M.; Alvarez, A.D.O.; Engelsen, M.M.; Lundkvist, H.; Engelsen, S.B. Staling of white wheat bread crumb and effect of maltogenic α-amylases. Part 1: Spatial distribution and kinetic modeling of hardness and resilience. *Food Chem.* **2016**, *208*, 318–325. [CrossRef]
24. Sangnark, A.; Noomhorm, A. Effect of dietary fiber from sugarcane bagasse and sucrose ester on dough and bread properties. *LWT* **2004**, *37*, 697–704. [CrossRef]
25. Gavahian, M.; Chu, Y.-H.; Farahnaky, A. Effect of ohmic and microwave cooking on textural softening and physical properties of rice. *J. Food Eng.* **2019**, *243*, 114–124. [CrossRef]
26. Al-Hooti, S.N.; Sidhu, J.S.; Al-Saqer, J.M. Utility of cie tristimulus system in measuring the objective crumb color of high-fiber toast bread formulations. *J. Food Qual.* **2000**, *23*, 103–116. [CrossRef]
27. Stufken, J.; Peace, G.S. Taguchi Methods: A Hands-On Approach. *Technometrics* **1994**, *36*, 121. [CrossRef]
28. Ren, Y.; Linter, B.R.; Foster, T.J. Starch replacement in gluten free bread by cellulose and fibrillated cellulose. *Food Hydrocoll.* **2020**, *107*, 105957. [CrossRef]
29. Olojede, A.; Sanni, A.; Banwo, K. Rheological, textural and nutritional properties of gluten-free sourdough made with functionally important lactic acid bacteria and yeast from Nigerian sorghum. *LWT* **2020**, *120*, 108875. [CrossRef]
30. Hernandez-Espinosa, N.; Laddomada, B.; Payne, T.; Huerta-Espino, J.; Govindan, V.; Ammar, K.; Ibba, M.I.; Pasqualone, A.; Guzmán, C. Nutritional quality characterization of a set of durum wheat landraces from Iran and Mexico. *LWT* **2020**, *124*, 109198. [CrossRef]

Article

Effects of the Addition of Flaxseed and Amaranth on the Physicochemical and Functional Properties of Instant-Extruded Products

Jazmin L. Tobias-Espinoza [1,2], Carlos A. Amaya-Guerra [2], Armando Quintero-Ramos [1,*], Esther Pérez-Carrillo [3], María A. Núñez-González [2], Fernando Martínez-Bustos [4], Carmen O. Meléndez-Pizarro [1], Juan G. Báez-González [2] and Juan A. Ortega-Gutiérrez [5]

1 Facultad de Ciencias Químicas, Universidad Autónoma de Chihuahua, Nuevo Campus Universitario, Circuito Universitario, C.P. 31125 Chihuahua, Chih., Mexico; jazletobias@gmail.com (J.L.T.-E.); cmelende@uach.mx (C.O.M.-P)

2 Departamento de Investigación y Posgrado, Facultad de Ciencias Biológicas, Universidad Autónoma de Nuevo León, Ciudad Universitaria, C.P. 66450 San Nicolás de los Garza, N.L., Mexico; numisamaya@hotmail.com (C.A.A.-G.); maria.nunezgn@uanl.edu.mx (M.A.N.-G.); juan.baezgn@uanl.edu.mx (J.G.B.-G.)

3 Centro de Biotecnología-FEMSA, Escuela de Ingeniería y Ciencias, Tecnológico de Monterrey, Av. Eugenio Garza Sada 2501 Sur, C.P. 64849 Monterrey, N.L., Mexico; perez.carrillo@tec.mx

4 Centro de Investigación y de Estudios Avanzados del Instituto Politécnico Nacional Unidad-Querétaro, Libramiento Norponiente 2000, Fracc. Real de Juriquilla, C.P. 76230 Santiago de Querétaro, Qro., Mexico; fmartinez@cinvestav.mx

5 Facultad de Zootecnia y Ecología, Universidad Autónoma de Chihuahua, Periférico R. Almada Km 1, C.P. 33820 Chihuahua, Chih., Mexico; jortega@uach.mx

* Correspondence: aquinter@uach.mx or aqura60@gmail.com; Tel.: +52-614-236-6000

Received: 2 May 2019; Accepted: 27 May 2019; Published: 30 May 2019

Abstract: The addition of flaxseed and amaranth on the physicochemical, functional, and microstructural changes of instant-extruded products was evaluated. Six mixtures with different proportions of amaranth (18.7–33.1%), flaxseed (6.6–9.3%), maize grits (55.6–67.3%) and minor ingredients (4.7%) were extruded in a twin-screw extruder. Insoluble and soluble fiber contents in extrudates increased as the proportions of amaranth and flaxseed increased. However, the highest flaxseed proportion had the highest soluble fiber content (1.9%). Extruded products with the highest proportion of flaxseed and amaranth resulted in the highest dietary fiber content and hardness values (5.2 N), which was correlated with the microstructural analysis where the crystallinity increased, resulting in larger, and more compact laminar structure. The extruded products with the highest maize grits proportion had the highest viscosity, expansion, and water absorption indexes, and the lowest water solubility index values. The mixtures with amaranth (18.7–22.9%), flaxseed (8.6–9.3%), and maize grits (63.8–67.3%) resulted in extruded products with acceptable physicochemical and functional properties.

Keywords: extruded products; flaxseed; amaranth; dietary fiber; extrusion-cooking

1. Introduction

Currently, an increasing trend in the demand for processed foods that include pro-health compounds such as soluble fiber is occurring due to evidence of potential health benefits to consumers. Reduction in various types of chronic diseases such as cancer, cardiovascular disease, type II diabetes, and various gastrointestinal disorders are among them [1]. The development of products and processes that incorporate high-fiber ingredients without altering the physical, functional, and sensory properties of the processed foods are of interest and desirable to meet current consumer trends. A technological alternative for the incorporation of ingredients in processed products is the extrusion-cooking process, a very versatile technique widely used for the development of breakfast cereals and instant foods [2]. Some extruded products marketed as breakfast cereals have significant caloric value. One of the strategies that has allowed the food industry to reduce the energetic density and produce more healthy products has been the incorporation of dietary fiber. However, the addition of dietary fiber during extrusion, especially if it is insoluble, results in products with less expansion and crispness and a higher bulk density and hardness, which are properties less preferred by consumers [3]. These characteristics can be explained by the interactions of fiber with starch that impact the mechanisms of starch gelatinization. These, in turn, are related to water absorption of the extrudates and other physicochemical transformations that occur during extrusion, such as viscoelastic properties associated with the stabilizing membranes of the bubbles formed during bubble growth in the final product [4]. Adding insoluble fiber to extruded products has been shown to decrease the proportion of starch in the food matrix, thereby reducing the water absorption capacity and, in turn, the viscosity caused by the gelatinization of the starch, which results in a reduction in the expansion of the final product [3]. A sectional reduction in the expansion of extruded products had been reported by Brennan et al. [5], due to an increase in insoluble fiber results in structures with a high number of small cells and a high cell density. Several authors reported that the bulk density is increased by adding insoluble fiber to extruded products [5,6]. Also, it has been reported that extrusion process causes significant effects on the dietary fiber content, as breakage of structural polysaccharides or complex carbohydrates formation [3]. These effects include the formation of resistant starch that may occur during extrusion and the formation of covalent interactions between macronutrients and insoluble components (such as insoluble fiber) that make the extrudates indigestible by amylase or protease activity [7].

The development of ready-to-eat products with high fiber content (6 g of fiber/100 g of product) [8], is based on the use of ingredients that meet this requirement and at the same time, provide numerous health benefits. Flaxseed and amaranth have been used specifically in extruded products due to their health benefits for consumers, but, usually, they are used individually [9–13]. Amaranth contains a good balance of amino acids, including the essential amino acid lysine, which is present in limited amounts in most cereals [14,15]. Flaxseed is low in carbohydrates (sugars and starches), high in fiber and protein, and rich in polyunsaturated fatty acids, particularly alpha-linolenic acid (ALA or ALN) and linoleic acid (AL), known as omega-3 and omega-6 essential fatty acids, respectively [16,17]. An important component of these two grains is soluble fiber, considered a functional ingredient because it generates high-viscosity products by causing the gelation of chyme. Chyme acts as a network to capture glucose and cholesterol molecules in their passage through the gut, hindering their absorption and thereby decreasing blood glucose and cholesterol levels, resulting in beneficial health effects [18].

Despite the health benefits of amaranth and flaxseed, little information has been reported on the combined effect of both ingredients on the physicochemical and functional properties of extruded products. Therefore, the aim of this study was to develop an extruded product with high-fiber ingredients and to evaluate the effects of the addition of flaxseed and amaranth on the physicochemical and functional properties of instant-extruded products.

2. Materials and Methods

2.1. Materials

Grains of amaranth (*Amaranthus hypochondriacus* L.), flaxseed (*Linum usitatissimum* L.) and minor ingredients such as sucralose, cocoa, and cinnamon were obtained from a local distribution store (Chihuahua, Chihuahua, Mexico). Also, maize grits number 4 (GPC, Muscatine, IA, USA) was used for the extrusion food matrix. The amaranth and flaxseed grains were milled in a roller mill (Zhengzhou Chengli Grain & Oil Machinery Co., Ltd., model 6F-2240, Zhengzhou, China), separately and sieved in a mesh number 35 (model RX-24, Tyler industrial products, Mentor, OH, USA) to obtain a particle size of 0.5 mm. All materials were stored in plastic bags at room temperature until their use.

2.2. Chemicals

Hydrochloric acid 37.2%, sulfuric acid 97.9%, hexane 99.8%, ethanol 99.9%, and boric acid 99.5% were all analytical grade and obtained from J.T. Baker (Mexico City, Mexico). Analytical grade sodium hydroxide (97.0%) was obtained from Sigma-Aldrich (St. Louis, MO, USA). The selenium reaction mixture was obtained from Merck (Darmstadt, Germany). The kit for soluble and insoluble dietary fiber was obtained from Sigma-Aldrich (St. Louis, MO, USA) [19].

2.3. Methods

2.3.1. Mixtures Preparation

Different proportions of amaranth and flaxseed flours were mixed with maize grits and fixed minor ingredients (sucralose, cocoa, and cinnamon). The ingredients were mixed in an industrial mixer (Bathammex, Mexico City, Mexico) for five minutes obtaining six different mixtures as shown in Table 1. The proportion of each ingredient in each mixture was determined considering fat and crude fiber contents around 5.6% and 2.5%, respectively; this according to the literature [2,20] to obtain extruded products with acceptable physicochemical and sensory characteristics.

Table 1. Proportion of ingredients of the different mixtures *.

Treatment	Flaxseed (%)	Amaranth (%)	Maize Grits (%)	Minor Ingredients (%)
Mixture 1	8.2	24.7	62.4	4.7
Mixture 2	7.3	29.3	58.7	4.7
Mixture 3	6.6	33.1	55.6	4.7
Mixture 4	9.3	18.7	67.3	4.7
Mixture 5	8.6	22.9	63.8	4.7
Mixture 6	7.9	26.4	61	4.7

* Minor ingredients: sucralose (1.6%), cocoa (2.5%), cinnamon (0.6%).

2.3.2. Extrusion Process

For the extrusion process, we used a twin-screw corotating extruder (BCTM-30, Bühler, AG, Uzwil, Switzerland) with a 600 mm length, a length to diameter ratio (L/D) of 20:1, a die opening of 4 mm, the screw configuration was selected specifically to create high levels of shear. The mixtures were fed to the extruder at a rate of 7.5 kg/h and were processed at a speed of 272 rpm at moisture content of 0.22 kg water/kg dry matter, which was adjusted within the extruder, at a temperature of 150 °C. It was controlled at the final stage of the extruding chamber by using a TT-137N water heater (Tool-temp, Sulgen, Switzerland). All extrudates were dried at 120 °C for 15 min in an air convection oven (Electrolux 10 GN/1, Stockholm, Sweden) at air cross-flow velocity of 1.5 $m{\cdot}s^{-1}$ until the extrudates reached a range moisture level of 0.017–0.031 kg $H_2O{\cdot}kg\ ss^{-1}$. The extrudates were packed and stored at room temperature (25 °C) until evaluation.

2.4. Analytical Methods

2.4.1. Proximate Analyses

The starting materials and extruded products were analyzed for moisture, protein, fat, crude fiber, and ash content according to methods 950.02, 960.52, 920.39, 962.09, and 923.03 of AOAC [19], respectively. Carbohydrates mass was calculated by difference. The analyses were carried out in triplicate for each treatment, and the results were expressed in g/100 g.

2.4.2. Insoluble and Soluble Dietary Fiber

The insoluble and soluble fiber in the ingredients, the mixtures and the extruded products were determined with the total dietary fiber assay kit (Sigma-Aldrich, St. Louis, MO, USA) according to method 991.43 of AOAC [19]. The analyses were carried out in triplicate for each treatment, and the results were expressed in g/100 g.

2.5. Functional Properties of the Extruded Products

2.5.1. Water Absorption and Water Solubility Indexes

The water absorption index (WAI) and water solubility index (WSI) were determined in triplicate following the procedures described by Anderson et al. [21]. The methods measure the quantity of water incorporated in the flour and the soluble solids that dissolve in water at 30 °C. Samples were weighed (2.5 g) into plastic tubes and mixed with 30 mL of distilled water. The samples were manually shaken, the slurries were centrifuged for 10 min at 3200× *g* (Thermo IEC model CL3-R, Thermo Scientific, Waltham, MA, USA), and the supernatant was decanted into pre-weighed porcelain capsules. Capsules were dried for 24 h at 105 °C and weighed. The gel remaining in the tubes after decanting the supernatant was weighed. The ratio between gel-forming solids and soluble solids was measured as grams of water per gram of flour. The WAI was calculated as a percentage of remaining gel weight compared to the pre-dried weight from the extruded products. The WSI was calculated as a percentage of the dried supernatant weight compared to the pre-dried weight from the extruded products.

2.5.2. Bulk Density

The bulk density (BD) was determined according to Jin et al. [22] in which the ground extrudate (40/60 mesh) was poured into a cylindrical container. Excess extrudate was scraped off, and the net weight of the powder was divided by the volume of the container. Bulk density was expressed in kilograms per liter (kg L^{-1}). The analysis was performed in triplicate, and mean values were reported.

2.5.3. Expansion Index

The expansion index (EI) was reported as the ratio of extruded product diameter and the diameter of the die hole [23]. Values were reported as means of 60 measurements.

2.6. Physical Properties of the Extruded Products

2.6.1. Textural Measurement: Hardness and Crispness

The evaluation of the hardness and crispness of the extrudates was performed according to the method described by Ding et al. [24], and carried out using a Texture Analyzer TA.XT (Texture Technologies Corporation, Scarsdale, New York/Stable Micro Systems, Haslemere, Surrey, UK) configured with a 2 mm punch at a crosshead speed of 5 mm/s and a travel distance of 15 mm. Twenty-four extruded unit samples were taken randomly from each treatment and analyzed. A force time curve was recorded and analyzed by the Texture Exponent 32 (Surrey, UK) program to calculate the maximum force (N) to determine the hardness and the area under the curve (N/mm) to determine the crispness.

2.6.2. Pasting Properties of the Extruded Products

The amylographic viscosity profile was determined according to Sánchez-Madrigal et al. [25], with some modifications, using a Rapid Visco Analyzer (RVA SUPER 4 (Newport Scientific, Sydney, Australia). Flour sample suspensions were prepared by weighing 4 g of milled and dried (50 ± 2 °C, 12 h) extrudates with a 7.5 to 8.5% moisture content and a small particle size (0.25 mm) into an RVA canister and individually adjusting each sample to a total weight of 28 g using distilled water. The rotating paddles were held at 50 °C for 1 min to stabilize the temperature and ensure uniform dispersion and heated to 92 °C at a rate of 5.6 °C/min, which was held constant for 5 min. The dispersion was cooled to 50 °C at the same rate and was held at 50 °C for 1 min. The maximum viscosity (MaxV) at 92 °C, the minimum viscosity (MinV or lowest viscosity at the end of heating constant period at 92 °C) and the final viscosity (FinV attained during cooling to or holding at 50 °C) were recorded. The total setback viscosity or viscosity of retrogradation (final viscosity minus minimum viscosity) was calculated from these parameter values. The viscosity with RVA was obtained in RVU units (1 RVU = 10 centipoises). Each treatment was performed twice.

2.7. Scanning Electron Microscopy

This analysis was performed according to the method described by Sánchez-Madrigal et al. [25]. Flours of each extruded cereal with a particle size <0.15 mm and a moisture content of 1% were stuck to stubs and coated with a gold layer in a high vacuum using a Denton vacuum evaporator (Desk II), set to a pressure of 7.031×10^{-2} kg cm^{-2}. The samples were examined using a scanning electron microscope (JSM-5800LV, JEOL, Akishima, Japan) equipped with a secondary electron detector at an acceleration rate of 10 kV.

2.8. Statistical Methods

A univariate analysis of variance was performed adjusting a model that included the main effects and their interaction (Minitab 16). When the effect of the interaction factor or the main effects was significant (α 0.05); means comparison was performed by Tukey's test [26].

3. Results and Discussion

3.1. Raw Materials Characterization

Proximate analysis showed a significant difference ($p < 0.05$) between the raw materials for each of the components (Table 2) and indicated that they were of high nutritional value. Is important to highlight that flaxseed had the highest protein, fat, and fiber content, with the lowest carbohydrate content. Whereas the amaranth had protein content twice as maize grits. These values (%) are consistent with those reported in the literature [16,27], where amaranth contains a good balance of amino acids, including lysine, an essential amino acid that is not found in most cereals [14]. Flaxseed has been reported to be low in carbohydrates (sugars and starches), high in quality protein, fiber and rich in polyunsaturated fatty acids [16]. Maize grits were the main source of carbohydrates, as shown in Table 2. Meanwhile amaranth and flaxseed showed the highest dietary fiber contents (Table 3), these results agree with Morris [16] and Cervantes [27]. The soluble fiber content in the ingredients was in the following relevance order: flaxseed (9%), amaranth (1.3%) and maize grits (0.71%).

Table 2. Proximate composition of the raw materials and the extruded products *.

Component (%)	Amaranth	Flaxseed	Maize grits	Mixture 1	Mixture 2	Mixture 3	Mixture 4	Mixture 5	Mixture 6
Moisture	1.4 ± 0.01 c	5.3 ± 0.01 b	11.3 ± 0.01 a	3.0 ± 0.13 a	2.1 ± 0.13 bc	2.4 ± 0.13 ab	3.1 ± 0.13 a	1.7± 0.13 c	2.8 ± 0.13 ab
Crude Fat	7.1 ± 0.18 b	37.2 ± 0.18 a	0.99 ± 0.18 c	2.6 ± 0.06 b	2.7 ± 0.06 b	2.6 ± 0.06 b	2.7 ± 0.06 b	2.4 ± 0.06 b	3.0 ± 0.06 a
Crude Fiber	3.2 ± 0.05 b	16.3 ± 0.05 a	0.64 ± 0.05 c	1.5 ± 0.05 b	1.7 ± 0.05 ab	1.9 ± 0.05 a	1.7 ± 0.05 ab	1.7 ± 0.05 ab	1.9 ± 0.05 a
Ash	3.0 ± 0.03 b	3.3 ± 0.03 a	0.55 ± 0.03 c	1.6 ± 0.02 c	1.7 ± 0.02 ab	1.8 ± 0.02 a	1.6 ± 0.02 c	1.6 ± 0.02 c	1.7 ± 0.02 b
Crude Protein	17.4 ± 0.2 b	22.4 ± 0.2 a	8.7 ± 0.2 c	12.2 ± 0.08 ab	12.4 ± 0.08 a	12.0 ± 0.08 bc	12.3 ± 0.08 ab	11.7 ± 0.08 c	12.3 ± 0.08 ab
Carbohydrates	67.9	15.5	77.8	79.1	79.4	79.3	78.6	80.9	78.3

* Means ± standard error (SE). Means by files for raw materials and extruded products, with different letters show significant difference, contrast test ($p < 0.05$). Carbohydrates were calculated by difference.

Table 3. Dietary fiber content of the raw materials and the mixtures without extruding *.

Component (%)	Amaranth	Flaxseed	Maize grits	Mixture 1	Mixture 2	Mixture 3	Mixture 4	Mixture 5	Mixture 6
SDF	1.3 ± 0.23 b	9.0 ± 0.23 a	0.71 ± 0.23 b	0.6 ± 0.1 b	1.3 ± 0.1 a	1.3 ± 0.1 a	1.5 ± 0.1 a	1.6 ± 0.1 a	1.2 ± 0.1 a
IDF	11.9 ± 0.15 b	50.6 ± 0.15 a	7.3 ± 0.15 c	9.3 ± 0.05 cd	9.7 ± 0.05 c	8.9 ± 0.05 d	11.6 ± 0.05 b	12.0 ± 0.05 b	12.6 ± 0.05 a
TDF	13.2 ± 0.32 b	59.6 ± 0.32 a	8.0 ± 0.32 c	9.9 ± 0.11 d	11.0 ± 0.11 c	10.2 ± 0.11 d	13.1 ± 0.11 b	13.6 ± 0.11 ab	13.8 ± 0.11 a

* Means ± standard error (SE). Means by files for raw materials and mixtures without extruding, with different letters show significant difference, contrast test ($p < 0.05$). SDF, soluble dietary fiber; IDF, insoluble dietary fiber; TDF, total dietary fiber.

3.2. Extrudate Characterization

Proximate analysis of different extruded products is shown in Table 2. Chemical characteristics were significantly affected ($p < 0.05$) for amaranth and flaxseed additions. The extruded products had high percentage of protein compared to other commercial extruded cereals, which typically have protein content between 5 and 8% [28]. This is due to the contributions of protein of amaranth (17.4%) and flaxseed (22.4%) (Table 2). Similar protein content for extruded amaranth with maize grits were reported [10]. The addition of amaranth could lead to a good balance of amino acids because it contains lysine, an essential amino acid, which is not found in most cereals [14]. Whereas flaxseed protein is rich in arginine, aspartic acid and glutamic acid and deficient in lysine [16].

The extrudates had a desirable crude fiber content (<2%; Table 2), with acceptable physicochemical and sensory characteristics for the consumer without a negative effect on the caloric and nutrient content, which could especially benefit young children [2]. Additionally, the fat content in all extruded products was below the minimum acceptable value of 5.6% [20] to reach desirable characteristic in extruded products. From the nutritional point of view (<5%) a low-fat product was obtained.

3.3. Dietary, Insoluble and Soluble Fiber

Table 3 shows the soluble and insoluble fiber content of the different mixtures before being subjected to the extrusion process. The addition of flaxseed significantly affected the soluble and total dietary fiber content. After the extrusion process of the mixtures, the results for the dietary fiber content showed a significant difference between treatments ($p < 0.05$), indicating that variation in the proportions of ingredients (amaranth and flaxseed) and their interaction, significantly affected the content of insoluble and soluble fiber in extruded products (Figure 1). Additionally, it was observed that the extrusion process caused an increase in the soluble fiber content and a decrease in insoluble fiber compared to the non-extruded mixtures (Table 3, Figure 1). The extruded products from the mixtures 3, 4, and 5 had the highest percentage of soluble fiber but they did not present a significant difference. This can be explained by the fact that extrudates 4 and 5 had the highest content of flaxseed (9.3% and 8.6% respectively), which is a significant source of soluble fiber, reaching up to 9% (Table 3). In turn, extrudate 3 had the highest percentage of amaranth and flaxseed (41.7%), which resulted in high soluble fiber content. The rest of the extrudates (mixtures 1, 2 and 6) had a low percentage of soluble fiber without showing significant differences among them (Figure 1) because they had low percentages of flaxseed. Various studies had shown that dietary fiber, especially soluble fiber in extrudates, increases when they are subjected to the extrusion process [29]. Additionally, other biomolecules such as starch undergo structure changes, leading to the formation of resistant starch, another possible mechanism causing the increase in fiber during the extrusion process [7]. On the other hand, total dietary fiber content could decrease due to the fact that during the extrusion process, shear stress caused by high screw speed, combined with high process temperatures causes chemical bond breakage from complex carbohydrates, releasing molecules as xylose, glucose, arabinose, oligosaccharides, and, preferentially, slightly branched arabinoxylans which are solubilized [29,30]. Similarly, this decrease in total dietary fiber content could be observed in some of our treatments.

A food product is considered high in dietary fiber when it contains >6% [8]. Therefore, the extruded products presented high fiber content (Figure 1), due to the addition of amaranth and flaxseed, which contain a high percentage of dietary fiber; 13.1 and 59.6% respectively (Table 3).

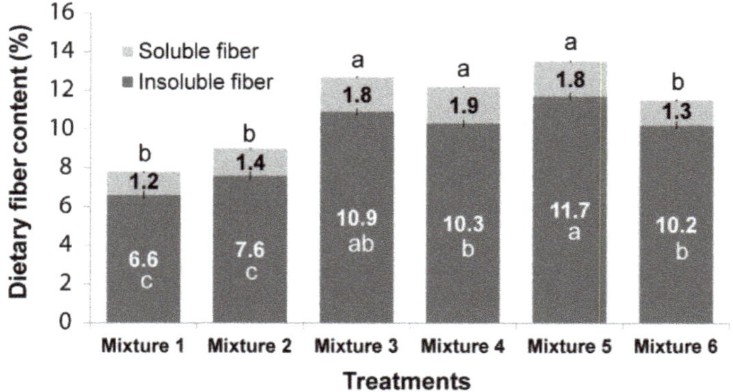

Figure 1. Insoluble and soluble dietary fiber content of extruded products. Means ± standard error (SE). SE insoluble fiber, 0.2; SE soluble fiber, 0.05. Means by columns and colors with different letters show significant differences based on contrast tests ($p < 0.05$).

3.4. Water Absorption and Water Solubility Indexes

The WAI in the extruded products was significantly affected ($p < 0.05$) by the addition of flaxseed and amaranth and their interaction, in the mixtures (Table 4). The extruded cereal from mixture 4 (9.3% flaxseed, 18.7% amaranth, and 67.3% maize grits) resulting in high WAI values due to its high maize grits (cornstarch) content which, during the extrusion process, undergoes pronounced changes in gelatinization properties, favoring a higher water absorption. This is consistent with the amylographic viscosity profile, where the extrudates with the highest values of viscosity have the highest values of WAI (Figure 2). On the other hand, low WAI values resulted for extrudate from mixture 3 (6.6% flaxseed, 33.1% amaranth, and 55.6% maize grits), which had the lowest maize grits (cornstarch) content but the highest proportion of high-fiber ingredients (39.7% amaranth and flaxseed). Similar results were obtained in extrudates from mixtures 2 and 6 (Table 4). Similar effects of fiber addition on extruded products were reported by Altan et al. [31] for the extrusion of barley mixtures with tomato pomace.

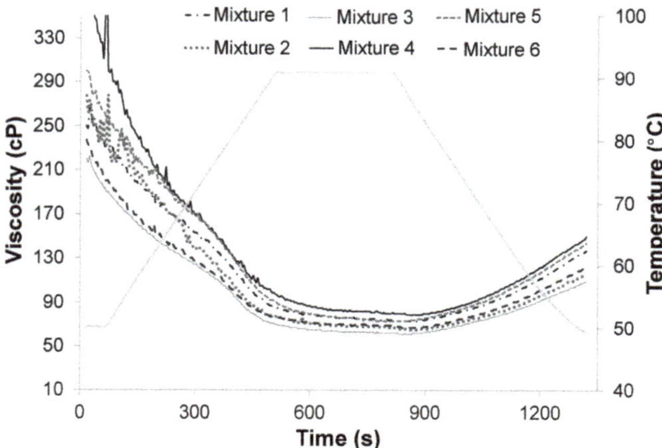

Figure 2. Amylographic viscosity profile of extruded products obtained with Rapid Visco Analyzer (RVA).

WSI is an indicator of the degradation of molecular components: an example is the amount of soluble polysaccharide of starch released after extrusion, which is a measurement of the degree of conversion of starch during extrusion [24]. Table 4 shows significant effects on the WSI of the extruded products by the addition of amaranth and flaxseed, in the mixtures. Extrudates from mixtures 2, 5, and 6, with a high fiber content (Table 4), had the highest WSI values (0.5). Whereas the extrudate from mixture 1, with a low fiber content, and the extrudate from mixture 4, with a high starch content, had the lowest WSI values (0.46 and 0.45, respectively). The increase in the fiber content caused an increase in the WSI values, which can be attributed to the rupture of the structural polysaccharides by the extrusion process [29,30]. A similar finding was reported by Ganorkar et al. [13] and Altan et al. [31].

3.5. Bulk Density

The bulk density in extruded cereals shows significant changes ($p < 0.05$) due to the addition of the ingredients in the mixtures (Table 4). The extrudate from mixture 6, with a high fat content, presented the lowest density value: this can be attributed to the fat's low-density and the oil contained in cereals, which are emulsified during extrusion due to the high pressure reached during the process. The fine drops of fat are coated by starches and proteins, leaving the fat encapsulated and causing a decrease in the density [32]. The extruded products from the mixtures 2 and 4 were high in protein and presented the highest density values. The rigid tertiary structure, high cohesiveness, high molecular weights, and structural functions in cereal proteins such as corn, can increase the density of food products [33]. Ryu et al. [23], reported that the density of an extruded product is strongly affected by water, fiber, fat, and starch content. The extruded products from the mixtures 6 and 3, with similar amaranth contents (26.4 and 33.1 g/100 g respectively), resulted in the lowest density values (4.6 and 4.7 kg L^{-1}, respectively; Table 4). These results were consistent with those found by Ilo et al. [9], who evaluated the effect of extrusion-cooking process on the properties of extruded rice flour and amaranth blends. They observed that amaranth had an important influence on the product density, resulting in a minimum density value at amaranth content of 30 g/100 g. Another report had shown increases in the bulk density values during the extrusion of rice flour and corn fortified with flaxseed [12].

Table 4. Functional and physical properties of the extruded products *.

Treatments	BD (kg L^{-1})	EI	WSI	WAI	Hardness (N)	Crispness (N/mm)	MaxV (cp)	MinV (cp)	FinV (cp)	Setback Viscosity (cp)
Mixture 1	0.5 ^{a}	3.11 ^{c}	0.46 ^{bc}	2.9 ^{b}	5.0 ^{ab}	26.9 ^{a}	86.2 ^{b}	69.9 ^{abc}	151.0 ^{a}	81.1 ^{a}
Mixture 2	0.49 ^{abc}	3.10 ^{c}	0.5 ^{a}	2.5 ^{c}	4.9 ^{bc}	25.4 ^{a}	75.3 ^{bc}	65.7 ^{bc}	124.5 ^{bc}	58.9 ^{c}
Mixture 3	0.47 ^{cd}	3.06 ^{d}	0.49 ^{ab}	2.5 ^{c}	4.7 ^{c}	24.9 ^{a}	66.7 ^{c}	63.2 ^{c}	111.1 ^{c}	48.0 ^{c}
Mixture 4	0.49 ^{ab}	3.33 ^{a}	0.45 ^{c}	3.4 ^{a}	4.7 ^{c}	25.7 ^{a}	99.3 ^{a}	79.8 ^{a}	162.5 ^{a}	82.7 ^{a}
Mixture 5	0.47 ^{bcd}	3.17 ^{b}	0.5 ^{a}	2.9 ^{b}	5.2 ^{a}	26.7 ^{a}	85.6 ^{b}	75.2 ^{ab}	154.8 ^{a}	75.1 ^{ab}
Mixture 6	0.46 ^{d}	3.09 ^{c}	0.5 ^{a}	2.5 ^{c}	4.8 ^{bc}	26.1 ^{a}	76.1 ^{bc}	70.2 ^{abc}	130.5 ^{b}	60.5 ^{bc}

* Means ± standard error (SE). SE expansion index, 0.007; SE bulk density, 0.006; SE WSI, 0.007; SE WAI, 0.05; SE Hardness, 0.16; SE Crispness, 1.3; SE MaxV, 2.4; SE MinV, 2.3; SE FinV, 2.8; SE setback viscosity, 2.9. Means by columns with different letters show significant difference, contrast test ($p < 0.05$). BD, bulk density; EI, expansion index; WSI, water solubility index; WAI, water absorption index; MaxV, maximum viscosity; MinV, minimum viscosity; FinV, final viscosity.

3.6. Expansion Index

The addition of the different ingredients in their different proportions, and their interaction, significantly affected the expansion of extruded products ($p < 0.05$) (Table 4). The EI of instant-extruded products is very important since it is directly related to consumer acceptability; related typically to an inflated, lightweight and crunchy structure [34], mainly attributed to the presence of starch in the final extruded products [35]. The extrudate of the mixture 4 resulted in the highest ($p < 0.05$) EI (3.33), followed by the extruded mixture 5 (3.17) due to a higher percentage of maize grits (starch), and finally the extruded products containing less maize grits (mixtures 1, 2 and 6) did not have significant difference between them and they presented the lowest EI values (Table 4). It is important to note that the mixtures 4 and 5 (with higher expansion values), contained a higher starch amount and a lower percentage of the ingredients with high dietary fiber content: amaranth and flaxseed. Mixtures 1, 2, and 6 with higher amaranth and flaxseed content presented the lowest EI, due to their high dietary fiber content, which affects the expansion of extruded products. The effect of the fiber on the expansion of the extruded products depends mainly on its interactions with the starch and, therefore, on the type and amount of fiber. Insoluble fiber significantly reduces expansion volumes and increases the density of extruded products. Conversely, soluble fiber leads to better expansion volumes, unaffecting the bulk density of the extruded products compared to the insoluble fiber components [3]. The difference in expansion behavior between soluble and insoluble fiber can be explained by their interactions with starch, differences in water absorption and plasticizing behavior, but also by the physicochemical transformations they undergo during extrusion [3]. This is consistent with the results reported by Altan et al. [31], who made extruded barley using tomato pomace as fiber source, noting that additions of tomato pomace, provoked a decrease of the EI on the final products. Similar results were reported by other authors [35–37].

3.7. Textural Measurement (Hardness and Crispness)

An important quality parameter of ready-to-eat extrudates is texture. Table 4 shows the values of hardness and crispness of extruded products made from the different mixtures of ingredients. The crispness of the extruded products was not significantly different ($p > 0.05$) among the various mixtures. Similar findings have been reported for corn extruded with amaranth, where the amaranth content from 20 to 35% had no substantial effect on crispiness [11]; this amaranth percentage was close to the one used in this study (18.7–33.1%).

However, the hardness of the cereals was significantly affected ($p < 0.05$) by the added ingredients (flaxseed and amaranth) and their interaction. Extruded products from mixture 1and 5 had the highest hardness ($p < 0.05$) attributed to their high dietary fiber content. As was described above, the addition of dietary fiber leads to reduced expansion volumes and increases in density of the extruded products, inducing harder textures and less crispiness [3]. This result is consistent with the results reported by Ganorkar and Jain [12], who showed that an increase in added flaxseed caused an increase in the hardness of the extruded products. A similar finding was reported by Brennan et al. [5] where they showed that increases in the wheat bran content up to 15% in extruded breakfast cereals causes breaking force increases. In contrast, it has also been reported that soluble fibers, such as inulin, deliver a more favorable texture compared to insoluble fibers, such as bran fiber [6]. This was corroborated by Brennan et al. [5], who observed a slight hardness change when adding either inulin or guar gum to extruded corn flour. This can be corroborated by our results, where the extruded products from mixture 3 and 4 (Figure 1) contained the highest percentage of soluble fiber and the lowest hardness values (4.7 N; Table 4).

3.8. Pasting Properties of the Extruded Products

The addition of amaranth and flaxseed to the mixtures significantly affected ($p < 0.05$) the amylographic viscosity profile (RVA) of the extruded products (Table 4). MaxV is the peak viscosity where the highest degree of starch gelatinization occurs. The MaxV values obtained for each of the extruded products were very low, due the damage in the starch granules during the extrusion process [38]; this led to a notorious decrease of viscosity values in the extruded products, as shown in Figure 2. A similar trend on other viscosity parameters (MinV, FinV, and setback viscosity) values is shown in Table 4. The mixture 4, with the highest starch content (67.3% maize grits), had the highest MaxV value ($p < 0.05$); whereas mixture 3, with the lowest percentage of starch (55.6% maize grits), had the lowest value ($p < 0.05$). On the other hand, it is possible that the formation of complex structures during extrusion-cooking through interactions between starch-lipid complexes and/or starch-protein walls prevents adequate gelatinization of the starch [39,40]. Another factor influencing the pasting properties of extruded products is the presence of dietary fiber, which leads to a decrease in the fraction of water-swelling starch, due to its replacement by the fiber [37]. All these factors limited a complete starch gelatinization, causing a decrease in viscosity. Similar results were observed in a study where increases in amaranth in rice-amaranth blends generally decreases the viscosity of extruded products [9]. Similar findings were reported for extruded cornstarch blends with whey protein concentrate and Agave tequilana fiber [37].

Additionally, low setback values were found for all extruded products, indicating low rates of starch retrogradation and syneresis. During cooling, reassociation of starch molecules, especially amylose, result in viscosity increase favoring the final viscosity. This phase is commonly described as the setback region during which retrogradation and reordering of starch molecules occurs [41].

3.9. Scanning Electron Microscopy

The scanning electron micrographs revealed the impact of the different ingredients (Figure 3). After extrusion, it can be observed that combination of shear force and temperature inside the barrel caused microstructural changes in the extrudates of the different treatments [42]. The microstructural analysis shows that the addition of amaranth and flaxseed, increased the fiber content in the mixtures, resulting in compact agglomerates, increased crystallinity, and larger, more compact laminar structures (Figure 3c–f). This can be attributed to high protein (12%) and fiber (9–13%) content in the extrudates. Fiber tends to rupture cell walls and promotes breakage of air cells during extrusion, which prevents matrices from expanding [43], resulting in harder textures, higher densities and more compact structures as shown in the micrographs (Figure 3c–f). Similar results were found by Zhang et al. [29] and Cueto et al. [42].

Figure 3. Micrographs of the extruded products from: (**a**) Mixture 1, (**b**) Mixture 2, (**c**) Mixture 3, (**d**) Mixture 4, (**e**) Mixture 5, (**f**) Mixture 6.

4. Conclusions

This research shows that different levels of amaranth and flaxseed in the development of extruded products had a significant impact on their functional and physicochemical properties. The extruded products obtained had high protein content (>12%), which is higher than in the commercial breakfast cereals. Besides these characteristics, the obtained extruded products presented a healthy fat content (<5%) and a high content of soluble and insoluble dietary fiber.

Another important ingredient was the maize grits, which was the source of starch, and it served as a basis to produce some expansion level in the extrudates. Extruded products with low levels of starch (maize grits) and high levels of fiber in the mixtures resulted in extrudates with low EI and high hardness values. These results suggest that the extrusion-cooking process of high-fiber flours such as flaxseed (8.6–9.3%) and amaranth (18.7–22.9%) in mixture with maize grits (63.8–67.3%) and minor ingredients results in an extruded product of good nutritional quality, with suitable functional and physicochemical characteristics.

Author Contributions: Conceptualization, J.L.T.-E. and A.Q.-R.; investigation, J.L.T.-E.; writing—original draft, J.L.T.-E. and A.Q.-R.; methodology, J.G.B.-G., C.A.A.-G. and F.M.-B.; writing—review and editing, E.P.-C., M.A.N.-G., C.O.M.-P.; performed the analysis, data curation, J.A.O.-G.

Funding: This research received no external funding.

Acknowledgments: The authors acknowledge the Universidad Autónoma de Nuevo León and the Universidad Autónoma de Chihuahua for supporting this investigation. This paper is based on the postgraduate studies of Jazmin Leticia Tobias Espinoza, who was supported by a PhD scholarship from the National Council of Science and Technology of Mexico (CONACYT).

Conflicts of Interest: The authors declare no conflict of interest.

References

1. Kaur, K.; Jha, A.; Sabikhi, L.; Singh, A. Significance of coarse cereals in health and nutrition: A review. *J. Food Sci. Tech.* **2012**, *51*, 1429–1441. [CrossRef]
2. Harper, J. *Extrusion of Foods*; CRC Press Inc: Boca Raton, FL, USA, 1981; Vol. II.
3. Robin, F.; Schuchmannnb, H.; Palzer, S. Dietary fiber in extruded cereals: Limitations and opportunities. *Trends Food Sci. Tech.* **2012**, *28*, 23–32. [CrossRef]
4. Robin, F.; Théoduloz, C.; Gianfrancesco, A.; Pineau, N.; Schuchmann, H.; Palzer, S. Starch transformation in bran-enriched extruded wheat flour. *Carbohyd. Polym.* **2011**, *85*, 65–74. [CrossRef]
5. Brennan, M.; Monro, J.; Brennan, C. Effect of inclusion of soluble and insoluble fibers into extruded breakfast cereal products made with reverse screw configuration. *Int. J. Food Sci. Technol.* **2008**, *43*, 2278–2288. [CrossRef]
6. Blake, O. Effect of molecular and supramolecular characteristics of selected dietary fibers on extrusion expansión. Ph.D. Thesis, Purdue University, West Lafayette, Indiana, December 2006.
7. Esposito, F.; Arlotti, G.; Bonifati, A.; Napolitano, A.; Vitale, D.; Fogliano, V. Antioxidant activity and dietary fiber in durum wheat bran by-products. *Food Res. Int.* **2005**, *38*, 1167–1173. [CrossRef]
8. Codex Alimentarius. Guidelines for Use of Nutrition and Health Claims. Available online: http://www.fao.org/ag/humannutrition/32444-09f5545b8abe9a0c3baf01a4502ac36e4.pdf (accessed on 28 March 2019).
9. Ilo, S.; Liu, Y.; Berghofer, E. Extrusion cooking of rice flour and amaranth blends. *Lebensm. Wiss. Technol.* **1999**, *32*, 79–88. [CrossRef]
10. Guerra-Matías, A.; Arêas, J. Glycemic and insulinemic responses in women consuming extruded amaranth (*Amaranthus cruentus* L.). *Nutr. Res.* **2005**, *25*, 815–822.
11. Ramos-Diaz, J.; Suuronen, J.; Deegan, K.; Serimaa, R.; Tuorila, H.; Jouppila, K. Physical and sensory characteristics of corn-based extruded snacks containing amaranth, quinoa and kañiwa flour. *LWT-Food Sci. Technol.* **2015**, *64*, 1047–1056. [CrossRef]
12. Ganorkar, P.; Jain, R. Development of flaxseed fortified rice-corn flour blend based extruded product by response surface methodology. *J. Food Sci. Tech.* **2015**, *52*, 5075–5083. [CrossRef]
13. Ganorkar, P.; Patel, J.; Shan, V.; Rangrej, V. Defatted flaxseed meal incorporated corn-rice flour blend based extruded product by response surface methodology. *J. Food Sci. Tech.* **2016**, *53*, 1867–1877. [CrossRef]
14. Pedersen, B.; Kalinowski, L.; Eggum, B. The nutritive value of amaranth grain (*Amaranthus caudatus*): 1. Protein and minerals of raw and processed grain. *Plant Food Hum. Nutr.* **1987**, *36*, 309–324. [CrossRef]
15. Bressani, R. Composition and nutritional properties of amaranth. In *Amaranth Biology, Chemistry, and Technology*; Paredes-Lopez, O., Ed.; CRC Press: Boca Raton, FL, USA, 2018; pp. 185–205.
16. Morris, D. *Flax-A Health and Nutrition Primer*; Flax Council of Canada: Winnipeg, MB, Canada, 2007; pp. 9–21.
17. Liu, J.; Shim, Y.Y.; Tse, T.J.; Wang, Y.; Reaney, M.J.T. Flaxseed gum a versatile natural hydrocolloid for food and non-food applications. *Trends Food Sci. Tech.* **2018**, *75*, 146–157. [CrossRef]
18. Wood, P. Cereals β-glucans in diet and health. *J. Cereal Sci.* **2007**, *46*, 230–238. [CrossRef]
19. Association of Official Analytical Chemists. *Official Methods of Analysis of AOAC International*; AOAC: Washington, DC, USA, 1998.
20. Lusas, E.; Riaz, M. Introduction to extrusion and extrusion principles. In *Feeds and Pet Food Extrusion Manual*; Riaz, M., Ed.; Texas A & M University: College Station, TX, USA, 1999.
21. Anderson, R.; Conway, H.; Pfeifer, V.; Griffin, E. Gelatinization of corn grits by roll and extrusion-cooking. *Cereal Sci. Today* **1969**, *14*, 4–12.

22. Jin, Z.; Hsieh, F.; Huff, H. Extrusion cooking of corn meal with soy fiber, salt and sugar. *Cereal Chem.* **1994**, *71*, 227–234.
23. Ryu, G.; Neumann, P.; Walker, C. Effects of some baking ingredients on physical and structural properties of wheat flour extrudates. *J. Cereal Chem.* **1993**, *70*, 291–297.
24. Ding, Q.; Ainsworth, P.; Tucker, G.; Marson, H. The effect of extrusion conditions on the physicochemical properties and sensory characteristics of rice-based expanded snacks. *J. Food Eng.* **2005**, *66*, 283–289. [CrossRef]
25. Sánchez-Madrigal, M.; Melendez-Pizarro, C.; Martínez-Bustos, F.; Ruiz-Gutierrez, M.; Quintero-Ramos, A.; Márquez-Melendez, R.; Lardizábal-Gutiérrez, D.; Campos-Venegas, K. Structural, functional, thermal and rheological properties of nixtamalised and extruded blue maize (*Zea mays* L.) flour with different calcium sources. *Int. J. Food Sci. Technol.* **2013**, *49*, 578–586.
26. Montgomery, D.C. *Design and Analysis of Experiments*; Wiley: New York, NY, USA, 2004.
27. Cervantes, R. *Catálogo de propiedades nutrimentales, nutracéuticas y medicinales de amaranto. Secretaría de Desarrollo Rural del Estado de Puebla México, Coordinación General de cadenas productivas, 2007*; Secretaría de Desarrollo Rural del Estado de Puebla México: Puebla, México.
28. Pérez-Lizaur, A.; Palacios-González, B.; Castro-Becerra, A.; Flores-Galicia, I. *Sistema Mexicano de Alimentos Equivalentes (Mexican equivalent food system)*; Fomento de Nutrición y Salud, AC: México DF, México, 2014; pp. 31–36.
29. Zhang, M.; Bai, X.; Zhang, Z. Extrusion process improves the functionality of soluble dietary fiber in oat bran. *J. Cereal Sci.* **2011**, *54*, 98–103. [CrossRef]
30. Gualberto, D.; Bergman, C.; Kazemzadeh, M.; Weber, C. Effect of extrusion processing on the soluble and insoluble fiber, phytic acid contents of cereal brans. *Plant Food Hum. Nutr.* **1997**, *51*, 187–198. [CrossRef]
31. Altan, A.; McCarthy, K.; Maskan, M. Evaluation of snack foods from barley-tomato pomace blends by extrusion processing. *J. Food Eng.* **2008**, *84*, 231–242. [CrossRef]
32. Valls Porta, A. El Proceso de Extrusión en Cereales y Habas de Soja I. Efecto de la Extrusión Sobre la Utilización de Nutrientes. Available online: http://www.ucv.ve/fileadmin/user_upload/facultad_agronomia/Extrusi%C3%B3n_y_su_efecto.pdf (accessed on 11 March 2019).
33. Gálvez, A.; Flores, I.; González, A. Proteínas vegetales (Vegetables proteins). In *Química de los Alimentos (Food Chemistry)*, 4th ed.; Badui, S., Ed.; Pearson Educación: Estado de México, México, 2006; pp. 222–227.
34. Ainsworth, P.; Ibanoglu, S.; Plunkett, A.; Ibanoglu, E.; Stojceska, V. Effect of brewers spent grain addition and screw speed on the select physical and nutritional properties of an extruded snack. *J. Food Eng.* **2007**, *81*, 702–709. [CrossRef]
35. Chinnaswamy, R.; Hanna, M. Physicochemical and macreomolecular properties of starch-cellulose fiber extrudates. *Food Struct.* **1991**, *10*, 229–239.
36. Huth, M.; Dongowski, G.; Gebhardt, E.; Flamme, W. Functional properties of dietary fibre enriched extrudates from barley. *J. Cereal Sci.* **2000**, *32*, 115–128. [CrossRef]
37. Santillán-Moreno, A.; Martínez.Bustos, F.; Castaño-Tostado, E.; Amaya-Llano, S. Physicochemical characterization of extruded blends of corn starch-whey protein concentrate-agave tequilana fiber. *Food Bioprocess Technol.* **2011**, *4*, 797–808.
38. Duvarci, O.C.; Yazar, G.; Dogan, H.; Kokini, J.L. Linear and non-linear rheological properties of foods. In *Handbook of Food Engineering*; Heldman, D.R., Lund, D.B., Sabliov, C., Eds.; CRC press: Boca Raton, FL, USA 2018.
39. Wang, S.; Kim, Y. The effect of protein and fiber on the kinetics of starch gelatinization and melting in waxy corn flour. *Starch/Stärke.* **1998**, *50*, 419–423. [CrossRef]
40. Thachil, M.T.; Chovksey, M.K.; Gudipati, V. Amylose-lipid complex formation during extrusion cooking: effect of added lipid type and amylose level on corn-based puffed snacks. *Int. J. Food Sci. Technol.* **2014**, *49*, 309–316. [CrossRef]
41. Ragaee, S.; Abdel-Aal, E. Pasting Properties of starch and protein in selected cereals and quality of their food products. *Food Chem.* **2006**, *95*, 9–18. [CrossRef]

42. Cueto, M.; Porras-Saavedra, J.; Farroni, A.; Alamilla-Beltrán, L.; Schöenlechner, R.; Schleining, G.; Buera, P. Physical and mechanical properties of maize extrudates as affected by the addition of chia and quinoa seeds and antioxidants. *J. Food Eng.* **2015**, *167*, 139–146. [CrossRef]
43. Vadukapuram, N.; Hall III, C.; Tulbek, M.; Niehaus, M. Physicochemical properties of flaxseed fortified extruded bean snack. *Int. J. Food Sci.* **2014**, *2014*, 478018. [CrossRef]

Article

Brosimum alicastrum Sw. (Ramón): An Alternative to Improve the Nutritional Properties and Functional Potential of the Wheat Flour Tortilla

Rodrigo Subiria-Cueto [1], Alfonso Larqué-Saavedra [2], María L. Reyes-Vega [3], Laura A. de la Rosa [1], Laura E. Santana-Contreras [1], Marcela Gaytán-Martínez [3], Alma A. Vázquez-Flores [1], Joaquín Rodrigo-García [1], Alba Y. Corral-Avitia [1], José A. Núñez-Gastélum [1] and Nina R. Martínez-Ruiz [1,*]

[1] Instituto de Ciencias Biomédicas, Universidad Autónoma de Ciudad Juárez, Anillo Envolvente del Pronaf y Estocolmo s/n, C.P. Ciudad Juárez, Chihuahua 32310, Mexico; al183351@alumnos.uacj.mx (R.S.-C.); ldelaros@uacj.mx (L.A.d.l.R.); lsantana@uacj.mx (L.E.S.-C.); alma.vazquez@uacj.mx (A.A.V.-F.); jogarcia@uacj.mx (J.R.-G.); acorral@uacj.mx (A.Y.C.-A.); jose.nunez@uacj.mx (J.A.N.-G.)

[2] Unidad de Recursos Naturales, Centro de Investigación Científica de Yucatán, A.C. (CICY), Calle 43, No. 130 x 32 y 34, Chuburná de Hidalgo, C.P. Mérida, Yucatán 97205, Mexico; larque@cicy.mx

[3] Programa de Posgrado en Alimentos, Universidad Autónoma de Querétaro, Cerro de las Campanas s/n, Col. Las Campanas, Santiago de Querétaro, Querétaro 76100, Mexico; luzrega@icloud.com (M.L.R.-V.); marcelagaytanm@yahoo.com.mx (M.G.-M.)

* Correspondence: nmartine@uacj.mx; Tel.: +52-656-688-1800 (ext. 1979)

Received: 23 October 2019; Accepted: 21 November 2019; Published: 24 November 2019

Abstract: The wheat flour tortilla (WFT) is a Mexican food product widely consumed in the world, despite lacking fiber and micronutrients. Ramón seed flour (RSF) is an underutilized natural resource rich in fiber, minerals and bioactive compounds that can be used to improve properties of starchy foods, such as WFT. The study evaluated the impact of partial replacement of wheat flour with RSF on the physicochemical, sensory, rheological and nutritional properties and antioxidant capacity (AC) of RSF-containing flour tortilla (RFT). Results indicated that RFT (25% RSF) had higher dietary fiber (4.5 times) and mineral (8.8%; potassium 42.8%, copper 33%) content than WFT. Two sensory attributes were significantly different between RTF and WFT, color intensity and rollability. RFT was soft and it was accepted by the consumer. Phenolic compounds (PC) and AC were higher in RFT (11.7 times, 33%–50%, respectively) than WFT. PC identification by ultra-performance liquid chromatography quadrupole time of flight mass spectrometry (UPLC-QTOF-MS) showed that phenolic acids esterified with quinic acid, such as chlorogenic and other caffeoyl and coumaroyl derivatives were the major PC identified in RSF, resveratrol was also detected. These results show that RSF can be used as an ingredient to improve nutritional and antioxidant properties of traditional foods, such as the WFT.

Keywords: nutritional value; antioxidant capacity; phenolic compounds; sensory properties; functional foods

---

1. Introduction

Tortilla is an iconic food in the Mexican and Central American diet. Mexico is the main tortilla consumer around the world (75 kg/person) [1], with 11.5 million tons consumption per year [2]. The growing demand of tortilla has caused its globalization, representing nowadays a food of importance in countries such as USA or China. Tortilla is usually made from wheat or corn [3]. On the one hand, the corn tortilla provides proteins, calcium, potassium and carbohydrates, which makes this food an important vehicle to provide nutrients to people with malnutrition. On the other hand, the wheat flour tortilla (WFT) provides more calories (47%), proteins, lipids and carbohydrates and less dietary fiber and micronutrients than corn tortilla [4], which makes this type of tortilla a hypercaloric and a poorly nutritious food. However, WFT has a wide demand in northern Mexico and USA and it is highly consumed by children, youth and adults from ethnic and low-income populations contributing to exacerbate states of undernourishment [5] or obesity [6]. Malnutrition is one of the main public health problems affecting many countries [7] and is associated, in many cases, with the high intake of hypercaloric foods with low dietary fiber, minerals and vitamins [6]. Therefore, it is of interest to make culturally rooted foods, such as tortilla, with greater nutritional value in order to contribute to a healthier diet of the population; particularly for groups in poverty and/or malnutrition.

Brosimum alicastrum Sw. (ramón) is a tree of the Mexican tropics whose fruit and seed have high nutritional value. This tree was appreciated by the Mayan culture from 300 to 900 A.D. [8]. The flour obtained from the seed (RSF) is characterized by high protein, dietary fiber and micronutrient content. *Brosimum alicastrum* Sw. is considered as an underexploited natural resource in Mexico with potential nutritional and functional properties [9], which can be incorporated into foods with little nutritional value. In this context, the aim of this study was to evaluate the impact of partial replacement of wheat flour with RSF on the physicochemical, sensory, rheological and nutritional properties and antioxidant capacity of RSF-containing flour tortilla (RFT). The hypothesis proposed was that the partial replacement of WF with RSF in the tortilla improves its nutritional contribution in dietary fiber and micronutrient content, increases its antioxidant capacity and the RFT is sensory accepted by the consumer compared as well as a tortilla made with 100% WF.

2. Materials and Methods

2.1. Materials

The ramón seed flour (RSF) (*B. alicastrum* Sw.) was provided by CICY (Herbarium Roger Orellana, Centro de Investigación Científica de Yucatán A.C.). The seeds were collected from growing ramón trees at rancho Xoccheila (20°33′ N; 89°34′ W), municipality of Sacalum, Yucatán. The seeds were dried in the sun, the testa was removed and the seed was ground to a fine flour. The wheat flour (WF), dry (salt, baking powder) and moist (shortening) ingredients were purchased in the local markets of Ciudad Juárez, Chihuahua, México.

2.2. Tortilla Development

Different formulations were proposed partially replacing the content of WF for RSF (20%, 25%, 28%, 30% and 40%). The ramón flour tortilla (RFT) and the wheat flour tortilla (WFT) were developed following the previously reported method with some modifications [10]. Extra water was added to the RSF doughs until a more elastic texture was obtained. Microbiological quality in total coliforms bacteria (CC), aerobics mesophilic bacteria (AC), yeast and mold (YM) was determined in tortillas samples according to the plate count method (3M™, Petrifilm™, Minneapolis, MN, USA). Briefly, tortilla samples were diluted 1:10 in saline solution (0.9%), homogenized and one mL plated onto Petrifilm. Plates were incubated at 35 ± 1 °C for 24 h for CC, 48 h for AC and 25 ± 1 °C for YM. Official Mexican regulation was observed for the limits of CC, AC and YM in tortillas [11].

2.3. Proximate Composition, pH, Activity Water (Aw) and Titratable Acidity

All tortilla samples were homogenized using a commercial blender (Nutribullet®). The samples were analyzed in triplicate following the AOAC methods [12]: ash was determined in muffle furnace (Felisa®, Model FE-340, Jalisco, México) at 550 °C for 5 h; crude protein by Kjeldahl method (Labconco®, Model RapidStill II, Kansas city, MO, USA); fat by Soxhlet method (Soxtec™, Model 2043, Foss™, Hilleroed, Denmark); total carbohydrates by difference method; crude fiber by gravimetric method, dietary fiber by enzymatic-gravimetric assay, water activity in AQUA LAB® (Model Serie 3, Meter Food, Washington, D.C., USA) equipment; pH and titratable acidity by potentiometric method (Accumet®,

Model AB15 Plus, Westford, MA, USA). Moisture analysis was performed by the AOAC method [12] with the following modifications: it was determined in an oven (VWR®, Model 1324, Irving, TX, USA) at 105 °C for 8 h.

2.4. Extraction and Quantification of Minerals

The mineral content (Cu, Zn, K, Fe and Na) was obtained from the ashes of the samples following the method by the AOAC with minor modifications. Flours and tortilla samples (5 g) were calcined in a muffle furnace (Felisa®, Model FE-340, Guadalajara, Jalisco, México) for 8 h at 550 °C. Subsequently, 3 mL of HNO_3 (SCP®, Quebec, Canada) (0.2%, *v/v*) was added and taken to dryness on a hot plate (100 °C). The samples were calcined again for 2 h at 550 °C until white ashes were obtained, 5 mL of 6 M HCl (JT Baker®, Fisher Scientific, West Palm Beach, Florida, USA) were added and dried. Finally, 20 mL of HNO_3 (0.2%, *v/v*) (Merck®, Toluca, Estado de México, Mexico) were added and samples were transferred into plastic conical tubes (Corning®, Merck, Toluca, Estado de México, México). The samples were analyzed using atomic absorption spectroscopy (Perkin Elmer®, Model AAnalyst 200, Madrid, Spain) with acetylene flame adjusting the wavelength for each mineral [12].

2.5. Extraction and Quantification of Carotenoids

The content of carotenoids was determined following the method by Moreno-Escamilla et al. [13] with some modifications. Flours and tortillas were dried at 45 °C in a vacuum oven (Shel Lab®, Model VWR A-143, Tualatin, OR, USA) at 20 mm Hg for 36 h. Next, they were ground using a commercial blender (Nutribullet®) and kept at −18 °C in darkness for no more than 48 h. Subsequently, 0.3 g of dry ground sample were mixed with 10 mL of acetone (JT Baker®, Fisher Scientific, West Palm Beach, FL, USA), sonicated (Branson®, Model 5800, Fisher Scientific, West Palm Beach, Florida, USA) for 20 min and centrifuged (Eppendorf®, Model 5810 R, Alto da lapa, Sâo Paulo, Brazil) at 3500 rpm for 10 min. The supernatant was recovered, and the residue was extracted two more times under the same conditions. The absorbance of the combined supernatants was read in an ultraviolet–visible (UV–Vis) microplate reader (BioRad®, Model XMark, Ciudad de México, México) at a wavelength of 474 nm. Results were reported as milligrams of β-carotenoids per 100 g of sample.

2.6. Extraction and Quantification of Ascorbic Acid

Ascorbic acid determination was realized according to the technique described by Alvarez-Parrilla et al. [14], with some modifications. Extracts were obtained by mixing 0.2 g of flour or tortilla samples (DW) with 5 mL of 5% metaphosphoric acid (Merck®, Toluca, Estado de México, México), the mixture was sonicated (Branson®, Model 5800 Fisher Scientific, West Palm Beach, FL, USA) for 20 min in dark conditions, and centrifuged (Eppendorf®, Model 5810 R, Alto da lapa, Sâo Paulo, Brazil) at 3500 rpm for 10 min. For quantification, 300 μL of each supernatant was taken and mixed with 200 μL of 6.65% tricloriacetic acid (Merck®, Toluca, Estado de México, México) and 75 μL of the DNPH (2,4-dinitrophenylhydrazine) reagent (Merck®, Toluca, Estado de México, México) in 100 mL of 5 M H_2SO_4 (JT Baker®, West Palm Beach, Fisher Scientific, FL, USA), then incubated for 3 h at 37 °C and 0.5 mL of H_2SO_4 (JT Baker®, Fisher Scientific, West Palm Beach, FL, USA) (65%, *v/v*). Absorbance was measured in the UV-Vis microplate reader (BioRad®, Model xMark, Ciudad de México, México) at 520 nm, using ascorbic acid as standard. Results were reported as milligrams of ascorbic acid per 100 g of sample.

2.7. Sensory Characterization

Tortillas (RFT and WFT) were sensory characterized with a descriptive analysis by a trained panel of 8 judges. The attributes in the olfactory phase were odor intensity and descriptors determined by focus group technique. In the oral phase, mouthfeel characteristics were evaluated: such as cohesiveness, hardness, moistness, adhesiveness and astringency; taste: such as sour, salty and bitter. Color and texture attributes, such as elasticity, firmness and rollability were also evaluated. All tests were conducted in individual booths and the judges used a 150 mm linear scale, labeled at the end as "Not all ... " and "Extremely ... " for each attribute or descriptor. Each judge was provided with slices of tortilla (10 g), previously heated in a microwave for 30 s, and they were placed in 2 oz plastic cups, identified with three-digit random numbers. The samples were served to the judges in a balanced and randomized form, together with evaluation sheets. Judges rinsed their mouths with purified water (Alaska®, Chihuahua, Mexico) at the beginning and between samples for the oral phase and they used eye covers in all tests, except in the visual phase. Pantone® scale was used in color test. Two attributes or descriptors were evaluated per session of 60 min, standards for each attribute or descriptor were used at the beginning of the test and each test was performed by duplicate [15,16].

2.8. Consumer Acceptance Test

An acceptance test was carried out to evaluate the consumer degree of liking for RFT and WFT. The test was performed in 120 consumers using a 9-point hedonic scale, ranging from "Like extremely" to "Dislike extremely". Participants were given 5 g of RFT or WFT (freshly made) and kept warm in thermal containers (35 ± 2 °C) for a limited time of 15 min. Two samples were evaluated by session and each sample was presented monadically in disposable dishes (12 cm diameter), labeled with three-digit random numbers. Participants rinsed their mouths with purified water (Alaska®, Chihuahua, Mexico) at the beginning of the session and between samples. They tasted each sample and indicated on the hedonic scale the degree of liking for the sample [16].

2.9. Rheological Measurements

For the rheological characterization, texture profile analysis (TPA) and cohesiveness tests were performed in the RFT and WFT dough samples using the methods described by Reyes-Vega et al. [17] and Flores-Farías et al. [18] with modifications. Spherical dough fractions (2.0 cm diameter) were tested to obtain gumminess (N), hardness (N), adhesiveness (J), elasticity (mm) and chewiness (Nm). Cohesiveness was performed in dough samples (20 g) placed in a cylindrical container and a penetration speed of 2.0 $cm{\cdot}s^{-1}$ was applied. In the tortilla samples (2 × 6 cm), cutting, elongation and extensibility tests were made to obtain the force (N) required to separate it, the elongation (mm), the distance that can be stretched before breaking (mm), work necessary for extension (J) and cohesiveness (N) following the methods by Kelekei et al. [19] and Suhendro et al. [20]. All measurements were carried out in a texture analyzer (Lloyd Instruments™-Model TAPlus AMETEK, Elancourt, France), adjusting different probes for each test.

2.10. Content of Total Phenolic Compounds and Flavonoids

Extraction of phenolic compounds was performed following the methodology established by Alvarez-Parrilla et al. [14] with modifications. Flour and tortilla samples were dried, ground and stored as described for the extraction and quantification of carotenoids. Next, 5 g of the ground samples were sonicated for 30 min with 10 mL of methanol (JT Baker®, Fisher Scientific, West Palm Beach, FL, USA) (80%, *v*/*v*), centrifuged at 3500 rpm for 15 min and the supernatant was collected by filtration, adjusted to a volume of 25 mL, and kept in refrigeration at 4 °C for less than 12 h, until analysis. Total phenolic content (TPC) was determined by the Folin–Ciocalteu method and total flavonoids (TF) were determined by the $AlCl_3$ method, according to the methodology described by de la Rosa et al. [21]. Results were expressed as milligrams of gallic acid equivalents (GAE) and milligrams of catechin equivalents (CE) per 100 g of sample (DW), respectively.

2.11. Antioxidant Capacity

The FRAP (ferric ion reducing antioxidant power), DPPH (2,2-diphenyl-1-picryl-hydrazyl-hydrate free radical) and ABTS (2,2′-azinobis-(3-ethylbenzothiazoline-6-sulfonate) assays were used to quantify the antioxidant capacity of the methanolic extracts of flour and tortillas samples, according to the methodology described by Alvarez-Parrilla et al. [14], de la Rosa et al. [21] and Brand-Williams et al. [22].

The extracts were obtained as described for the content of total phenolic compounds and flavonoids. For the FRAP assay, the FRAP reagent was prepared by mixing 0.3 M acetate buffer (Hycel®, Zapopan, Jalisco, México) with 10 mM TPTZ (2,4,6-tripyridyl-s-triazine; Acros Organics®, Morris Plains, NJ, USA) dissolved in 40 mM HCl (Hycel®, Zapopan, Jalisco, México), and 20 mM $FeCl_3$ (Hycel®, Zapopan, Jalisco, México); in a ratio 10:1:1, v / v / v. The FRAP reagent was heated at 37 °C for 30 min and the assay was performed by mixing 180 μL of the FRAP reagent with 24 μL of sample in microplate wells. Absorption was measured at 595 nm every 60 s for 30 min in the UV–Vis microplate reader. The results were reported in millimole Trolox equivalent/g dry weight sample.

The DPPH assay was performed by mixing 50 μL of sample with 200 μL of the DPPH radical (190 μM in methanol; Merck®, Toluca, Estado de México, México) in microplate wells and absorbance was read at 515 nm for 10 min in the UV–Vis microplate reader.

For the ABTS assay, ABTS radical cation was prepared by diluting ABTS salt (7 mM; Merck®, Toluca, Estado de México, México) and $K_2S_2O_8$ (2.45 mM; Merck®, Toluca, Estado de México, México) in phosphate buffered saline (PBS, pH 7.4, 0.15 M KCl; Merck®, Toluca, Estado de México, México), and the solution was incubated in refrigeration for 16 h. Then 12 μL of the sample was mixed with 285 μL of the ABTS radical cation in microplate wells and the absorbance was read at 734 nm for 30 min in the UV–Vis microplate reader. Results of DPPH and ABTS assays were reported as inhibition percentage.

2.12. *Identification of Individual Phenolic Compounds in Ramón Seed Flour (RSF) by Ultra-Performance Liquid Chromatography Quadrupole Time of Flight Mass Spectrometry (UPLC-QTOF-MS)*

Phenolic extracts of RSF, were cleaned up by solid phase extraction (SPE) using a C_{18} column (SupelCo, Merck®, Darmstadt, Germany) to reduce the presence of sugars and small organic acids. Briefly, 6 mL (1 volume) of extract was washed with 2 volumes of water (fractions discarded) and phenolic compounds were eluted with 2 volumes of pure methanol. Solvent was eliminated in a rotary evaporator and the dried extract re-dissolved in methanol high-performance liquid chromatography (HPLC) grade (2 mg/mL), passed through a nylon filter (0.45 μm) and applied into the chromatography system. Separation and identification of phenolic compounds was carried out by using an Infinity II LC System equipped with a photodiode array detector with a binary solvent pump and autosampler (Agilent Technologies, California, USA). Separation of individual phenolic compounds was carried out using a rapid-resolution high-definition (RRHD) reverse-phase C_{18} column (2.1 × 50 mm; 1.8 particle; ZORBAX Eclipse Plus®, Agilent, California, USA) at 25 °C, with a pre-column cartridge. The samples (1 μL) were injected and elution of phenolic compounds was completed in 12 min with a linear gradient and constant flow rate of 0.4 mL/min, as described before by Torres-Aguirre et al. [23]. The mobile phase consisted of solvent A (formic acid, 0.1% *v/v*, from Tedia®, Fairfield, Ohio, USA) and solvent B (acetonitrile from Tedia®, Fairfield, Ohio, USA). The linear gradients were as follows: 0–4 min, 90% A, 4–6 min, 70% A, 6–8 min, 62% A, 8–8.5 min, 40% A, 8.5–9.5 min, 90% A. Elution of phenolic compounds was detected at 255, 275 and 320 nm.

The LC equipment was coupled to a quadrupole time of flight (Q-TOF) mass spectrometer with electrospray ionization (ESI) source. The mass spectrometer was operated in negative mode and specific conditions as follows: capillary voltage of 4500 V, gas nebulizer pression 30 psi, dry gas (nitrogen) temperature of 340 °C and flow at 13 L/min. Mass range was monitored from 100 to 3000 m/z. Phenolic compounds were identified by comparing the accurate mass and isotopic distribution of their molecular ions $[M - H]^-$, and in some cases their retention times with those of commercial standards, and compounds listed in a specialized database, using a find by database algorithm in the MassHunter Workstation qualitative analysis, version B. 07.00 (Agilent Technologies, Santa Clara, CA, USA).

2.13. *Statistical Analysis*

Data from microbiological analysis were compared with the permissible limits established by the Mexican regulation [9]. Physicochemical parameters, minerals, vitamins and phytochemical data were analyzed by Student *t*-test. Data from sensory descriptive analysis were analyzed using a repeated measure analysis of variance (ANOVA) and Fisher's multiple comparisons (LSD) and data from acceptance test were analyzed Chi square test. All the analyses were carried out with XLSTAT program, version 2016.05 (Addinsoft®, Paris, France). The results are presented in mean values ± standard deviation (SD). The criterion for statistical significance was $p < 0.05$.

3. Results and Discussion

3.1. Tortilla Preparation and Food Safety

The preliminary sensory acceptance tests were considered for the selection of the RFT formulation, establishing the addition of 25% RSF for the tortilla. The lack of gluten in RSF influenced the properties of elasticity and firmness imparted by prolamines (gliadin) and glutelins (glutenin) [24]. Therefore, slight changes were made in the steps to prepare the tortilla (RSF), in order to obtain better physical characteristics. First, the kneading time was increased to 20 min and more water was added until a flexible and firm dough was obtained. In this way, higher hydration was achieved, breaking the endosperm protein bodies and promoting covalent and non-covalent interactions between larger polypeptides [25]. Second, water incorporation increased the elasticity and adhesiveness in the dough, so it was necessary to change the use of the rolling pin for a manual tortilla press, which is a simple, traditional and low-cost procedure. Third, salt content was increased by 25% in order to increase the hydration property of the proteins in RSF. Also, sodium facilitates denaturalization of proteins exposing their hydrophilic groups and increasing their degree of aggregation. This effect is generated by the increase of ionic force producing a reduction in protein solubility [26]. Finally, the cooking time was adjusted for RFT (26 s per side) according to the other changes made in the process, in order to avoid an impact on the consistency and rheology of the tortilla.

Microbiological analysis showed that both tortillas (RFT and WFT) were within the permissible limits, according to the Mexican legislation, in colony forming units of aerobic mesophilic bacteria (AC) (8×10^1 and 1×10^1 CFU/g, respectively. Permissible limit: 10,000 CFU/g), total coliforms bacteria (CC) ($<1 \times 10^1$ CFU/ g in each sample. Permissible limit: <30 CFU/g) and yeasts and molds (YM) ($<10 \times 10^1$ CFU /g. Permissible limit: 300 CFU/g) [11]. These microorganisms are indicators of management conditions or efficiency of the food preparation. They timely warn of inadequate handling or contamination that increases the risk for the presence of pathogenic microorganisms in the product [27].

3.2. Physicochemical and Micronutrient Characterization

Chemical and micronutrient composition of flours and tortillas is presented in Table 1. In flours, the contents of protein, ash, crude, and dietary fiber were higher in RSF than in commercial WF. In minerals, RSF was 2.5 times higher in copper ($p = 0.01$), 8 times higher in potassium ($p < 0.01$) and 2.3 times higher in sodium ($p < 0.01$) than WF. RSF was equal to WF in iron content ($p = 0.13$) and 6 times lower in zinc content ($p < 0.01$). RSF showed higher acidity and Vitamin C content (3.8 times) than WF, but carotenoid content was equal in both flours.

In tortilla, RFT retained more moisture (2.9%) than WFT, reflecting the extra water added in the wetting process, but the water activity was the same in both samples. Protein content was equal in RFT and WFT, showing RFT protein content was not affected by the partial substitution with RSF. RFT had an increase in the mineral content, showing an increase in copper (1.5 times higher) ($p = 0.02$) and potassium (1.8 times higher) in RFT ($p < 0.01$), while the content of iron ($p = 0.84$) and zinc ($p = 0.81$) remained similar. Dietary fiber was 4.5 times higher in RFT (14% DRV/100 g product) than in WFT (3% DRV/100 g product).

Table 1. Physicochemical characteristics of flour and tortillas samples.

Parameter	RSF	WF	RFT	WFT
Calories (Kcal)	336	349	350	357
Water (%)	13.3 ± 0.14 [a]	12.9 ± 0.02 [b]	25.0 ± 0.07 [a]	22.1 ± 0.10 [b]
Protein (%)	11.5 ± 0.39 [a]	9.6 ± 0.16 [b]	7.3 ± 0.11 [a]	7.6 ± 0.10 [a]
Fat (%)	0.6 ± 0.00 [a]	0.6 ± 0.01 [a]	9.6 ± 0.08 [a]	8.9 ± 0.02 [b]
Ashes (%)	3.4 ± 0.11 [a]	0.6 ± 0.02 [b]	3.4 ± 0.06 [a]	3.1 ± 0.02 [b]
Total Carbohydrates (%)	71.2 ± 0.56 [b]	76.3 ± 0.17 [a]	54.6 ± 0.16 [b]	58.3 ± 0.15 [a]
Crude Fiber (%)	3.4 ± 0.13 [a]	0.4 ± 0.00 [b]	0.9 ± 0.15 [a]	0.2 ± 0.01 [b]
Dietary Fiber (%)	13.0 ± 0.21 [a]	1.6 ± 0.00 [b]	3.6 ± 0.20 [a]	0.8 ± 0.01 [b]
pH	5.5 ± 0.01 [a]	5.9 ± 0.01 [b]	6.3 ± 0.01 [b]	6.6 ± 0.02 [a]
Activity water (Aw)	0.3 ± 0.02 [a]	0.2 ± 0.01 [b]	0.9 ± 0.02 [a]	0.9 ± 0.01 [a]
Titratable acidity (% CAE) *	0.004 ± 0.00 [a]	0.001 ± 0.00 [b]	0.002 ± 0.00 [a]	0.001 ± 0.00 [b]
Cupper (mg/100 g)	0.5 ± 0.10 [a]	0.2 ± 0.00 [b]	0.3 ± 0.00 [a]	0.2 ± 0.00 [b]
Potassium (mg/100 g)	1256.0 ± 12.00 [a]	159.0 ± 5.00 [b]	367.6 ± 13.00 [a]	210.0 ± 10.00 [b]
Iron (mg/100 g)	4.0 ± 0.70 [a]	5.0 ± 0.20 [a]	3.9 ± 0.20 [a]	4.1 ± 1.20 [a]
Zinc (mg/100 g)	1.0 ± 0.10 [b]	6.0 ± 0.10 [a]	4.5 ± 1.60 [a]	4.9 ± 0.00 [a]
Sodium (mg/100 g)	47.0 ± 0.10 [a]	20.0 ± 0.10 [b]	369.2 ± 0.40 [b]	378.0 ± 0.60 [a]
Vitamin C mg (Ascorbic acid/100g)	2.3 ± 0.01 [a]	0.6 ± 0.06 [b]	0.9 ± 0.09 [a]	0.7 ± 0.03 [a]
Carotenoids (mg β-carotenoids/100 g)	1.2 ± 0.10 [a]	0.9 ± 0.10 [a]	1.0 ± 0.00 [a]	0.9 ± 0.00 [a]

Mean ± SD. RSF—ramón seed flour, WF—wheat flour, RFT—ramón flour tortilla, WFT—wheat flour tortilla. * CAE—citric acid equivalent (0.064). Comparison between flours and between tortillas. Different letters indicate significant difference ($p < 0.05$).

Little information has been published on the nutritional composition of the *Brosimum alicastrum* Sw. seed. A study carried out by Carter [8] showed that the dietary fiber content in ramón's seeds from different countries (México, Honduras and Guatemala) was from 4.91 to 21.71 g/100 g (dry weight). In this study, RSF had a dietary fiber content of 13.0 g/100 g (fresh weight) or 14.9 g /100 g (dry weight), which is within the range reported by other authors [9,28]. Considering this fiber content, and according to the FDA criteria, RSF should be considered as food ingredient rich in dietary fiber (52.1% DRV/100 g) [29]. In comparing the mineral content of the flours, it is important to consider that commercial WF is enriched with folic acid, iron, and zinc, in accordance with Mexican regulations [11], which indicates that RSF is naturally rich in iron and would not need to be enriched to meet the legal iron requirements. This is particularly important for groups in poverty, for example iron deficiency is the main cause of anemia in Mexico, and is associated with a low intake of food from animal origin and a high intake of corn, with a high content of phytates that inhibit the iron absorption [30]. On the other hand, it is the first time that vitamin C or total carotenoid content of RSF has been reported. Both values were lower than those reported in lettuce using the same analytical technique [13], so we considered RSF was not a good source of these compounds.

Substitution of 25% WF by RSF had a positive impact in the mineral and dietary fiber content of RFT, so the new formulation showed a better nutrient profile and could have a positive impact on the consumer's health. Minerals have important biological roles in the organism; for example, copper is an essential cofactor involved in the maturation of connective tissue, the synthesis of neurotransmitters and the prevention of cardiovascular diseases [31] while potassium participates in insulin secretion, creatine phosphorylation, carbohydrate metabolism, protein synthesis, nerve transmission and muscle contraction [32,33]. The potassium content in RFT was similar to that of bananas (358 mg/100 g) and higher than tomatoes (237 mg/100 g), which are considered as foods rich in this mineral [34]. The dietary fiber content in RFT (14% DRV/100 g) was higher than in corn tortilla (3.1% DRV/ 100 g product) [34], so RST can be considered as a good source of dietary fiber [29]. Guevara-Arauza et al. [35] reported a high fiber content in a tortilla added with nopal fiber (16.7%); nevertheless, this tortilla would provide 45% less protein than RFT. The importance of dietary fiber in food is well known; its consumption has shown benefits throughout the digestive process. Different physiological and prebiotic effects at the colon level make this nutrient a key component of a healthy diet [36]. Finally, the partial replacement of 25% WF with RSF did not provide a significant increase of Vitamin C or carotenoids in RFT.

3.3. Sensory Attributes and Consumer Acceptance

The sensory characterization of both tortillas (RFT and WFT) in different attributes is shown in Figure 1. For a better interpretation of the intensity linear scale (150 mm), five intensity levels were considered: low (L, 0 to 37 mm), medium low (ML, 38 to 74 mm), medium (M, 75 mm), medium-high (MH, 76 to 112 mm) and high (H, 113 to 150 mm).

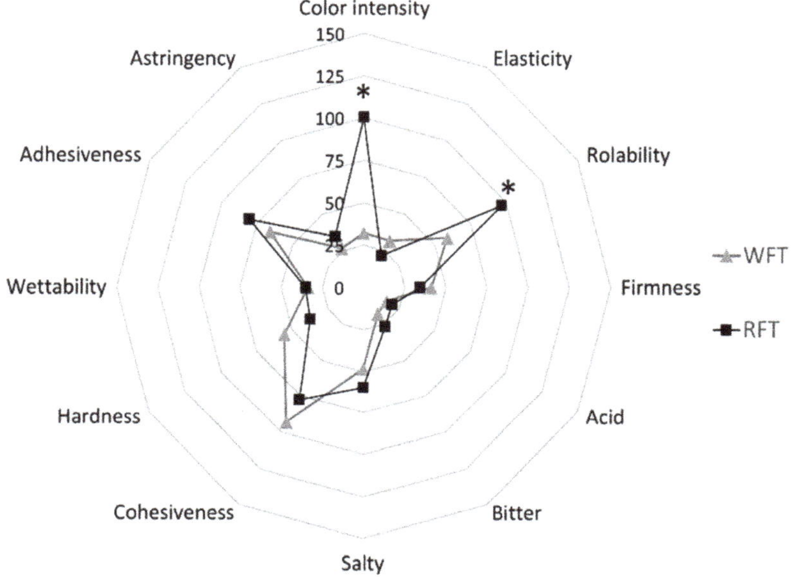

Figure 1. Sensory profile of tortillas. RFT—ramón flour tortilla, WFT—wheat flour tortilla. Intensity linear scale (150 mm). * Significant difference at $p < 0.05$.

Of all the attributes evaluated in both tortillas, only two were perceived as significantly different between them: color and rollability. In color, RFT showed a MH intensity level (100.4 ± 9.4 mm), significantly different from WFT that was situated in L intensity level (32.2 ± 10.1 mm) ($p < 0.01$). Pantone® scale was used to identify the color tones in both tortillas. RFT was located between Pantone codes # C 728C and 729C, indicating a light brown color, while WFT was classified with Pantone code # C 7401C, which corresponds to a light cream color. In tactile tests, RFT presented higher rollability (MH, 96.8 ± 23.1 mm) compared to WFT, which was in ML intensity (58.6 ± 24.8 mm) ($p < 0.01$). In the case of hardness, a trend was observed and WFT was perceived to be slightly harder (ML, 55.4 ± 24.0 mm) than RFT (L, 37.5 ± 14.7 mm) ($p = 0.09$). RSF color tones may be caused by the presence of tannins which are present in the maturation stages of certain seeds that change from green to brown due to morphological changes in the vascular tissue [37]; while WFT color tones are caused by Maillard reactions [38]. The increase of salt in RFT, added to promote hydration, was not detected in the salty taste. Cohesiveness and adhesiveness, which are directly related to the viscoelastic properties that gluten provides to create a support matrix capturing water and air molecules [39], were not affected by the water increase in RFT.

In the olfactory phase, no significant differences were identified between tortillas, although RFT was perceived with a higher odor to whole wheat (MH, 90.0 ± 38.0 mm) ($p = 0.06$) and a less intense flour odor (ML, 52.8 ± 26.9 mm) ($p = 0.07$) in comparison to WFT (ML, 54.8 ± 25.9 mm and MH, 79.1 ±

27.0 mm, respectively). The smell of dough (RFT, 72.8 ± 34.5 mm and WFT 61.1 ± 36.4 mm) ($p = 0.52$) and toast (RFT, 62.8 ± 28.4 mm and WFT, 44.4 ± 27.5 mm) ($p = 0.20$) was found in a ML intensity in both tortillas (Figure 2).

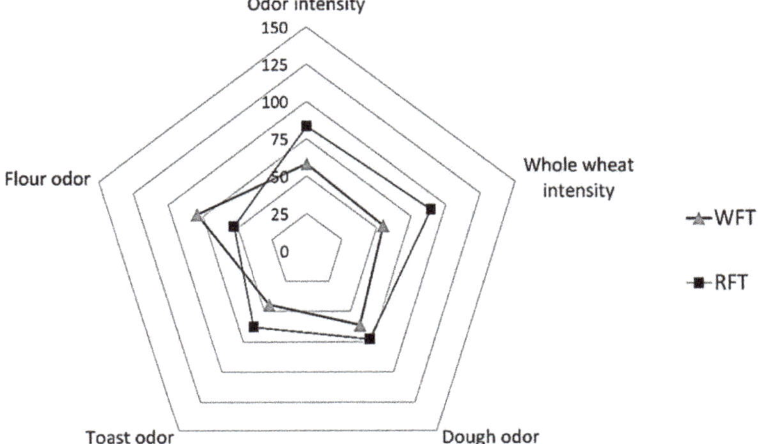

Figure 2. Sensory odor profile of tortillas. RFT—ramón flour tortilla, WFT—wheat flour tortilla. Intensity linear scale (150 mm).

The final consumer acceptance was tested on 120 participants. The 9-point hedonic scale was divided into three areas (like, neutral and dislike). RFT and WFT were both liked by the consumer ($p > 0.05$) (Figure 3). However, 36% of consumers showed dislike for RFT and 10% for WFT ($p < 0.01$). RFT and WTF were placed in the following categories on the 9-point hedonic scale: "Like extremely" 2.5 and 4%, respectively ($p = 0.72$), "Like very much" 10.8 and 15%, ($p = 0.44$); "Like moderately" 22.5 and 40.8%, ($p < 0.01$); "Like slightly" 20.8 and 15.8%, ($p = 0.40$); "Neither like nor dislike" 7.5 and 14.2%, ($p = 0.14$) and "Dislike slightly" 20% and 5%, ($p < 0.01$). In general, 64% of consumers accepted RFT. Some consumer comments indicated that RST was milder, with balanced salt content, and whole wheat odor. These comments confirm the attributes evaluated by the judges, describing RST as softer ($p = 0.06$) and having a whole wheat odor too ($p = 0.09$).

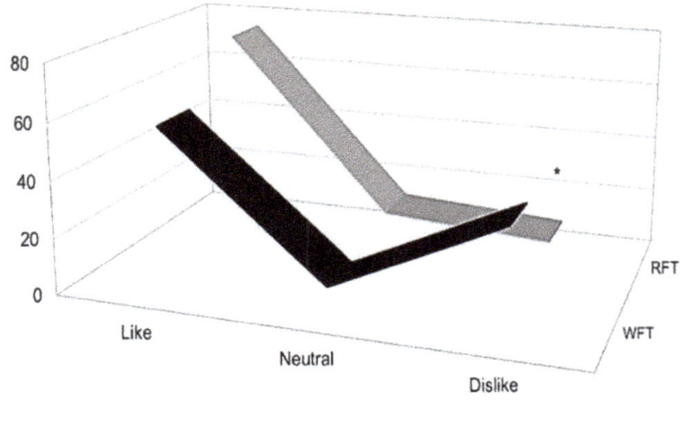

Figure 3. Consumer acceptance tortillas. RFT—ramón flour tortilla, WFT—wheat flour tortilla. * Significant difference at $p < 0.05$.

RFT had a sensory acceptance similar to corn tortilla added with soy and amaranth (60%) [40] or corn tortilla added with *Brosimum alicastrum* [41]. Innovation in products with high cultural value could be an important factor for greater consumer acceptance. Studies in this regard are necessary in the development of functional foods appealing to different populations or ethnic groups.

3.4. Rheological Characterization

Data from the TPA test showed that there was a significant difference in gumminess ($p < 0.01$), adhesiveness ($p = 0.01$), elasticity ($p = 0.01$) and chewiness ($p < 0.01$) in the tortilla doughs (Table 2). RSF addition generated a less gummy, elastic and softer dough; requiring less force during mastication compared to the WF dough or other foods, such as white bread (10.8 Nm), lettuce (9.8 Nm) and carrot (12.7 Nm) [42]. It was also less adhesive than corn tortilla dough substituted with bean and amaranth flour (−0.36 N) and nopal flour and algae (−0.32 N) [43]. These results might be related to the extra water addition that participates in the dough structural changes and in the absorption capacity of flour starch granules, which affect the adhesiveness and hardness mass viscoelastic and tension properties [44]. In the same way, it has been reported that the starch granules of RSF are similar to the granules of potato, which possess superior gelling properties and better capacity of water retention than wheat starch [45].

Table 2. Rheological characteristics of tortillas dough samples.

Characteristic	RSF Dough	WF Dough
Gumminess (N)	10.2 ± 3.2 [a]	39.1 ± 1.0 [b]
Hardness (N)	34.1 ± 10.0 [a]	60.5 ± 14.0 [a]
Adhesiveness (J)	0.03 ± 0.0 [a]	0.003 ± 0.0 [b]
Elasticity (mm)	1.8 ± 0.0 [a]	1.6 ± 0.1 [b]
Chewiness (Nm)	1.9 ± 0.6 [a]	6.3 ± 1.5 [b]
Cohesiveness (Nm)	3.5 ± 0.1 [a]	4.0 ± 0.1 [b]

Mean ± standard deviation (SD). N—Newton, J—joul, mm—millimeter, Nm—newton for meter, RFS—ramón seed flour, WF—wheat flour RSF—ramón seed flour, WF—wheat flour. Comparison between doughs. Different letters indicate significant difference ($p < 0.05$).

The cohesiveness test was also performed on the tortilla dough, where RFT was slightly less cohesive than WFT dough. This indicates that the particles of the dough are less strongly bound in the dough with RSF [46] and therefore less effort is required to deform or break it. Therefore, it is considered that these behaviors could be produced by the primary structure of the RSF proteins, that could be low in thiol groups. In this way, there is a possibility that the lack of thiol–thiol reactions to form disulfide bridges, would make the dough collapse [47].

Cut test for the RFT required less strength and work than for the WFT ($p < 0.01$). Thereby, RSF provides more softness to the tortilla (Table 3), which means that it does not require a great effort to cut it with the incisors unlike other functional tortillas added with bean flour with amaranth (5.22 J) or nopal with algae (6.51 J) [43]. Another factor that could modify the rheology of the tortilla is the size of the starch granules of RSF (6.5 to 15 μm), which are smaller than those provided by wheat flour (11 to 41 μm), so it can be retro-degraded quicker in the cooking process [48,49]. In the multi-directional elongation and extensibility test, the same behavior was found as in the previous tests. The WFT was 10 times more resistant than RFT ($p < 0.01$) and with a similar elongation ($p = 0.74$), whereas extensibility in RFT required less force during extension ($p < 0.01$), it was less cohesive ($p = 0.01$) and it needed half the work to extend ($p = 0.01$). All these rheological characteristics are due to the interactions between globular proteins and starch molecules. The lower cohesiveness and other related rheological characteristics of RFT in comparison with WFT could be explained by considering that RSF proteins have a higher molecular weight, which would affect their extension properties generated by the release of CO_2 during cooking and, hence, reduce the force of intermolecular interactions in the RFT [50,51]. A complete characterization of RSF proteins could help to better explain these behaviors and the rheological properties of RFT. Another factor that could be responsible for the lack of resistance in RFT may be the molecular rearrangement of polysaccharides with water during the cooking process, compromising the viscoelasticity and gelling of starch granules [52].

Table 3. Rheological characteristics in tortilla samples.

Characteristic	RFT	WFT
Cutting force (N)	−0.8 ± 0.1 [a]	2.6 ± 0.3 [b]
Cutting work (J)	0.04 ± 0.0 [a]	0.15 ± 0.0 [b]
Elongation force (N)	0.3 ± 0.1 [a]	3.1 ±0 0.0 [b]
Elongation distance (mm)	3.9 ± 0.4 [a]	4.1 ± 0.5 [a]
Extensibility max force (N)	0.8 ± 0.0 [a]	1.5 ± 0.2 [b]
Rupture distance (mm)	3.3 ± 0.3 [a]	3.2 ± 1.0 [a]
Cohesiveness (N)	3.6 ± 0.2 [a]	4.0 ± 0.0 [b]
Work to max extension (J)	0.002 ± 0.0 [a]	0.004 ± 0.0 [b]

Mean ± SD. RFT—ramón flour tortilla, WFT—wheat flour tortilla. Comparison between tortillas. Different letters indicate significant difference ($p < 0.05$).

3.5. Polyphenolic Quantification and Antioxidant Capacity

RSF and RFT reported higher phenolic content than WF and WFT, respectively ($p < 0.1$). RSF showed higher total flavonoid content than WF ($p < 0.01$) and RFT and WFT were equal between them ($p = 0.08$) (Table 4). Compared with other studies, the total phenol content of RSF was higher than that reported by Tokpunar [53] for RSF (24.6 mg GAE/g), this could be due to genetic and environmental differences between *B. alicastrum* trees, which are wild trees with an extensive geographical distribution [9,24]. The phenolic content of RSF was also higher than that of other seeds, such as walnut (15.6–16.3 mg GAE/g sample), pecan nut (12.8–20.2 mg GAE/g sample), pistachio (8.7–16.6 mg GAE/g sample) or almond (2.4–4.2 mg GAE/g sample) [54] and 73 times higher than WF.

Table 4. Phytochemical content and antioxidant capacity of flour and tortillas samples.

			Antioxidant Capacity		
Samples	TPC	TF	DPPH	ABTS	FRAP
	mg GAE/g	mg CE/g	mmol TEAC/100 g	mmol TEAC/100 g	mmol TEAC/100 g
RSF	65.8 ± 2.26 [a]	4.4 ± 0.18 [a]	0.9 ± 0.09 [a]	14.3 ± 0.10 [a]	0.41 ± 0.04 [a]
WF	0.9 ± 0.02 [b]	0.1 ± 0.01 [b]	0.0 ± 0.01 [b]	0.3 ± 0.09 [b]	0.04 ± 0.00 [b]
RFT	21.1 ± 1.50 [a]	0.7 ± 0.10 [a]	0.3 ± 0.01 [a]	0.4 ± 0.01 [a]	0.04 ± 0.00 [a]
WFT	1.8 ± 0.02 [b]	0.5 ± 0.10 [a]	0.2 ± 0.01 [b]	0.2 ± 0.00 [b]	0.04 ± 0.01 [a]

Mean ± SD. All values are presented on dry weight basis. TPC—total phenolic compounds, TF—total flavonoids, DPPH -.2,2-diphenyl-1-picryl-hydrazyl-hydrate free radical assay, ABTS - 2,2′-azinobis-(3-ethylbenzothiazoline-6-sulfonate assay, FRAP -ferric ion reducing antioxidant power assay. GAE—gallic acid equivalents, CE—catechin equivalents, TEAC—trolox equivalents antioxidant capacity, RSF—ramón seed flour, WF—wheat flour, RFT—ramón flour tortilla, WFT—wheat flour tortilla. Comparison between flours and between tortillas. Different letters indicate significant difference ($p < 0.05$).

The content of phenolic compounds in RFT was 12 times higher than WFT and also exceeded the content of the previously mentioned nuts, and was similar to blackberry (27.1 mg GAE/g) [55]. In the same way, RFT was higher in flavonoids content than WFT, or other foods, such as blackberry (0.6 mg CE/g) [56], pistachio and almond (0.14 and 0.15 mg CE/g, respectively) [57]. In this sense, partial substitution of RSF in the tortilla samples provided a high phenolic and flavonoid contribution.

In relation to antioxidant capacity, RSF showed 48 times more activity than WF in the ABTS assay. Also, it was higher than blackberry (11.4 mmol TEAC/100g) [56] and equal to walnut (13.7 mmol TEAC/100 g) [58]. In the case of the tortillas, RFT reported twice as much antioxidant activity as WFT. The ABTS assay showed the highest values of antioxidant activity in all the samples. ABTS is sensitive to the presence of hydrophilic and lipophilic compounds, and a good correlation can be usually observed between the content of total phenols and antioxidant capacity [59], so it is a suitable technique to clarify the impact of RSF on the potential functional properties of RFT.

3.6. Identification of Individual Phenolic Compounds in RSF by UPLC-QTOF-MS

In addition to the nutritional value of RSF, the presence of a high content and diversity of phenolic compounds, many of which are strong free-radical scavengers, provides this product with added potential as a functional food. Twenty phenolic compounds were tentatively (comparison of high-resolution m/z value and isotope distribution) or positively (MS data plus retention time of available standards) identified in RSF (Table 5).

Many of them are phenolic acids esterified with quinic acid. Compounds 3, 5 and 8 with Rt = 0.53, 0.94 and 1.54 min and a m/z = 353.0885, 353.0884, and 353.0882 were tentatively identified as isomers of caffeoylquinic acid, a derivate of caffeic acid esterified with quinic acid differing only in the position of esterification (C3, C4 or C5 of the aryl ring of quinic acid). Only compound **5** was positively identified as chlorogenic acid (Figure 4C). Compounds 16 and 17 were tentative identified as dicaffeoylquinic acid isomers with the same m/z = 515.1282 and 515.1206. These compounds have a structure of two moieties of caffeic acid and one of quinic acid linked through esterification at different positions into the aryl ring of quinic acid (Figure 4B). Compound 10 (Rt = 2.40 min, m/z = 367.1036) was tentatively identified as 3-O-feruloylquinic acid (Figure 4G). Compound **6** (m/z = 137.0244 and a Rt = 1.03 min) was tentatively identified as any of the possible isomers (*o-*,*m-*,*p-*) of hydroxybenzoic acid (Figure 4E). Compounds 9 and 12, wich eluted at Rt = 1.66 and 2.91 min respectively, with a m/z = 337.0936 and 337.0928, were tentatively identified as isomers of coumaroylquinic acid (Figure 4D). Compound 18 (Rt = 4.96 min, m/z = 499.1259) was tentatively identified as a coumaroyl-caffeoylquinic acid derivate (Figure 4A). Compound **1** (Rt = 0.47 and m/z = 329.0886) was tentatively identified as vanillic acid glucoside and was the only glycosylated phenolic acid (Figure 4F).

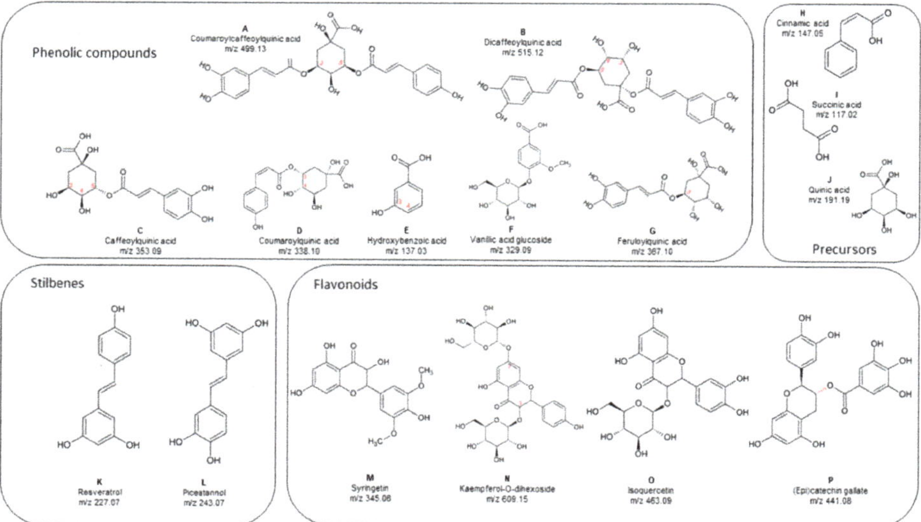

Figure 4. Structures of phenolic compounds tentatively or positively identified in RSF extract, by ultra-performance liquid chromatography quadrupole time of flight mass spectrometry (UPLC-QTOF-MS) analysis. Numbers in red indicate the possible isomers of each structure. Names of compounds (**A–P**) are provided in the figure and description of their identification by MS is given in the text.

Table 5. Retention times, and characteristic ions of phenolic compounds found in RSF extracts.

Compound	Tentative Identification	Formula	Rt (min)	m/z $[M-H]^-$	Measured Mass	Exact Mass	Δm ppm	Abundance
1	Vanillic acid glucoside	$C_{14}H_{18}O_9$	0.47	329.0886	330.0961	330.0951	3.02	5170
2	Succinic acid	$C_4H_6O_4$	0.47	117.0195	118.0265	118.0266	−0.84	2542
3	Caffeoylquinic acid	$C_{16}H_{18}O_9$	0.53	353.0885	354.0960	354.0951	2.54	86,163
4	Catechin gallate	$C_{22}H_{18}O_{10}$	0.59	441.0813	442.0886	442.0902	−3.16	1111
5	Chlorogenic acid *	$C_{16}H_{18}O_9$	0.94	353.0884	354.0958	354.0951	1.97	122,874
6	*m*−Hydroxybenzoic acid	$C_7H_6O_3$	1.03	137.0244	138.0318	138.0317	0.72	11,067
7	Quinic acid	$C_7H_{12}O_6$	1.52	191.0563	192.0632	192.0634	−1.04	13,344
8	Caffeoylquinic acid	$C_{16}H_{18}O_9$	1.54	353.0882	354.0956	354.0951	1.41	51,125
9	p−coumaroylquinic acid	$C_{16}H_{18}O_8$	1.66	337.0936	338.1006	338.1002	1.18	8509
10	3−O−Feruloyl quinic acid	$C_{17}H_{20}O_9$	2.40	367.1036	368.1109	368.1107	0.54	17,141
11	Syringetin	$C_{17}H_{14}O_8$	2.54	345.0628	346.0703	346.0689	4.04	5800
12	p−coumaroylquinic acid	$C_{16}H_{18}O_8$	2.91	337.0928	338.1001	338.1002	−0.29	8910
13	Cinnamic acid	$C_9H_8O_2$	2.97	147.0448	148.0519	148.0524	−3.37	4009
14	Kaempferol−O−dihexoside	$C_{27}H_{30}O_{16}$	3.73	609.1457	610.1544	610.1534	1.63	3113
15	Isoquercetin	$C_{21}H_{20}O_{12}$	3.86	463.0894	464.0968	464.0955	2.80	2519
16	Dicaffeoylquinic acid	$C_{25}H_{24}O_{12}$	4.16	515.1204	516.1282	516.1268	2.71	51,207
17	Dicaffeoylquinic acid	$C_{25}H_{24}O_{12}$	4.29	515.1206	516.1276	516.1268	1.55	82,208
18	*p*−Coumaroyl−caffeoylquinic acid	$C_{25}H_{24}O_{11}$	4.96	499.1259	500.1329	500.1319	1.99	3648
19	Piceatannol	$C_{14}H_{12}O_4$	5.58	243.0673	244.0748	244.0736	1.63	1112
20	Resveratrol	$C_{14}H_{12}O_3$	7.06	227.0723	228.0784	228.0786	−0.87	1669

Abbreviations: Rt. Retention time * Identification confirmed by commercial standards.

These compounds, linked by esterification with quinic acid, were the most numerous phenolic compounds in the RSF extract, representing half of the total compounds detected. Free quinic acid was also found (Compound 7, Figure 4J). This was expected since quinic acid is an organic acid abundant in plant tissues, that participates in metabolic routes as synthesis of lignin [60]. Succinic and cinnamic acids were also identified tentatively (Compounds 1 and 13, Figure 4I,H, respectively), both organic acids are precursors in the biosynthesis of flavonoids, isoflavones, and stilbenes [61].

Only a few flavonoids, stilbenes and their glycosylated derivates were identified in the RSF extract. Two stilbenes were tentatively identified, compound **20** (Rt = 7.06 and m/z = 227.0723) as cis or trans resveratrol (Figure 4K), and compound **19** (Rt = 5.58 and m/z = 243.0673) as piceatannol (Figure 4L), a more hydroxylated derivative of resveratrol. Compound 11 (Rt = 2.54 and m/z = 345.0628) was tentatively identified at as syringetin, a dimethylated flavonoid (Figure 4M). Compounds 14 and 15 (Rt = 3.73 and 3.86 with m/z = 609.1457 and 463.0894, respectively) were tentatively identified as glycosylated flavonoids: kaempferol 3-dihexoside (Figure 4N), whose structure contains two sugar moieties and isoquercetin, that is formed with quercetin and glucose (Figure 4O). Finally, one of the few free phenolics tentatively identified in RSF with Rt = 0.59 and m/z = 441.0813, was catechin gallate (Figure 4P). As described, the most abundant portion of phenolic compounds in the RSF extract consisted of phenolic acids esterified with quinic acid, while a minor portion was more heterogeneous containing flavonoids, stilbenes, and flavan-3-ols, mostly esterified with sugar moieties.

Only one study has been published identifying phenolic compounds in RSF, where hydroxycinnamic, gallic, vanillic, caffeic, and coumaric acids were identified in acid-hydrolyzed methanol extract and one flavonoid, epicatechin, was released by a continuous alkaline extraction [62]. In comparison with this previous study, a greater number of phenolic compounds were identified in RSF methanol extract in the present work. This was made possible by using the UPLC-QTOF-MS equipment whose main advantage is to provide a tentative identification of compounds for which no commercial standards are available, by using information about the molecular weight of individual phenolic compounds. To our knowledge, this is the first study to report a more detailed profile of phenolic compounds in ramón seeds using this technique, allowing for the identification of their native structures without alterations by hydrolytic reactions. This is important because acid and alkaline hydrolysis can degrade the glycosyl or ester linkages, but at the same time can fragment the structure of the phenolic compounds and alter their subsequent identification [63]. It is worth mentioning that no hydrolysable or condensed tannins were identified in RSF, which should be considered beneficial since tannins, despite their high antioxidant capacity are also known to interfere in nutrient absorption, which is not desirable in a product aimed at groups in poverty and/or malnutrition.

Derivatives of phenolic acids esterified with quinic acid, such as chlorogenic acid and the caffeoyl, coumaryl and feruloyl derivatives, found to be abundant in ramón seed, are typically found in coffee beans, which are recognized as a good source of healthy antioxidant compounds with health benefits like cardiovascular risk prevention. Special mention deserves those derivatives in which one quinic acid is esterified with two phenolic acid moieties; for example, dicaffeoil quinic acid, which has been found in medicinal plants and has a greater antioxidant activity than free phenolic acids [64]. The presence of stilbenes, such as resveratrol is also worth mentioning, although their abundance was low, and their identity should be confirmed. Resveratrol is a phenolic compound that demonstrates a wide range of health benefits as antioxidant, anti-inflammatory, and anti-proliferative in cancer cells. It has been identified in foods like peanuts, grapes and their products like roasted peanut butter and wine [65]. So, the fact that RSF can be considered a novel source of resveratrol for foods that regularly do not have it, such as tortilla or other food formulations, endows this seed with meaningful functional potential. The small number of flavonoids like syringetin, kaempferol and isoquercetin can further increase the healthy properties of RSF.

4. Conclusions

According to the results obtained in this study, *Brosimum alicastrum* Sw. seed flour (RSF) was a good source of protein, dietary fiber and minerals, such as copper and potassium, and a natural iron resource comparable to iron-fortified wheat flour (WF). In addition, RSF has a high antioxidant capacity (AC) and is rich in phenolic compounds, mainly chlorogenic acid and other phenolic acids esterified with quinic acid, although stilbenes and flavonoids were also present. The partial substitution of RSF (25%) in a wheat flour tortilla (RFT) increased dietary fiber, copper and potassium content. The sensory characteristics of RFT were like those of traditional flour tortilla (WFT), except for the light brown color and higher rollability. RFT was soft and less cohesive and it was accepted by 64% of consumers. Also, RFT increased 12 times its content of total polyphenolic compounds and twice its AC compared to WFT. Therefore, ramón seed flour improves the nutritional value of wheat flour tortilla and may provide potential functional properties that contribute to a healthier diet.

5. Patents

Patent request: MX/a/2018/011397. Dough of wheat flour and *Brosimum alicastrum* Sw. (ramón) flour for elaboration of food products, preferably tortilla.

Author Contributions: Conceptualization, N.R.M.-R.; Data curation, A.L.-S.; Formal analysis, R.S.-C., M.L.R.-V., L.A.d.l.R., A.A.V.-F. and N.R.M.-R.; Investigation, R.S.-C., A.L.-S., L.A.d.l.R. and N.R.M.-R.; Methodology, M.L.R.-V., L.E.S.-C., M.G.-M., A.A.V.-F., J.R.-G., J.A.N.-G. and N.R.M.-R.; Project administration, N.R.M.-R.; Resources, A.L.-S.; Supervision, L.E.S.-C., M.G.-M., J.R.-G. and A.Y.C.-A.; Writing—Original draft, R.S.-C.; Writing—Review and editing, L.A.d.l.R., A.Y.C.-A. and N.R.M.-R.

Funding: Part of this work was supported by the CONACyT (project CB-2016-286449).

Acknowledgments: The authors thank Consejo Nacional de Ciencia y Tecnología (CONACyT) for funding Rodrigo Subiria Cueto's master scholarship in Universidad Autónoma de Ciudad Juárez, Chihuahua, México.

Conflicts of Interest: The authors declare no conflict of interest.

References

1. Solano, R. GRUMA: Inicio de Cobertura. Available online: https://www.monex.com.mx/portal/download/reportes/Inicio%20de%20Cobertura%20de%20Gruma%20(Septiembre%202018).pdf (accessed on 12 December 2018).
2. CEDRSSA Consumo. *Distribución y Producción de Alimentos: El Caso Del Complejo Maíz-Tortilla*; Centro de Estudios para el Desarrollo Rural Sustentable y la Soberanía Alimentaria: Ciudad de México, Mexico, 2014.
3. Calleja, M.; Basilia, M. La tortilla como identidad culinaria y producto de consumo global. *Región Soc.* **2016**, *28*, 161–194. [CrossRef]
4. USDA. *Full Report (All Nutrients) 12005, Seeds, Breadnut Tree Seeds, Dried*; U.S. Department of Agriculture: Washington, DC, USA, 2016.
5. Rivera, J.Á. Deficiencias de micronutrimentos en México: Un problema invisible de salud pública. *Salud Pública Mex.* **2012**, *54*, 101–102.
6. Levy, T.; Amaya, M.A.; Cuevas, L. Desnutrición y obesidad: Doble carga en México. *Rev. Digit. Univ.* **2015**, *16*, 1–17.
7. Peláez, R.B. Desnutrición y enfermedad. *Nutr. Hosp.* **2013**, *6*, 10–23.
8. Carter, C.T. Chemical and Functional Properties of *Brosimum alicastrum* Seed Powder (Maya Nut, Ramón Nut). Master's Thesis, Clemson University, Clemson, SC, USA, 2015.
9. Larqué, A. El sector forestal en apoyo a la "Cruzada contra el Hambre" La inclusión de *Brosimum alicastrum* (Ramón) como estudio de caso. *For. XXI* **2014**, *11*, 11–12.
10. Waniska, R.D. Processing of flour tortillas. In *Tortillas Wheat Flour and Corn Products*; Rooney, L.W., Serna-Saldivar, S.O., Eds.; AACC International: Washington, DC, USA, 2016; pp. 125–145.
11. *Secretaria de Salud Norma Oficial Mexicana NOM-247-SSA1-2008, Productos y Servicios. Cereales y Sus Productos. Cereales, Harinas de Cereales, Sémolas o Semolinas. Alimentos a Base de: Cereales, Semillas Comestibles, de Harinas, Sémolas o Semolinas o Sus Mezclas*; Gobierno Federal: Ciudad de México, Mexico, 2008.

12. AOAC. *Official Methods of Analysis of AOAC International*, 17th ed.; Horwitz, W., Ed.; Association of Official Analytical Chemists: Washington, DC, USA, 2000.
13. Moreno-Escamilla, J.O.; Alvarez-Parrilla, E.; De La Rosa, L.A.; Núñez-Gastélum, J.A.; González-Aguilar, G.A.; Rodrigo-García, J. Effect of Different Elicitors and Preharvest Day Application on the Content of Phytochemicals and Antioxidant Activity of Butterhead Lettuce (*Lactuca sativa* var. capitata) Produced under Hydroponic Conditions. *J. Agric. Food Chem.* **2017**, *65*, 5244–5254. [CrossRef]
14. Alvarez-Parrilla, E.; de la Rosa, L.A.; Amarowicz, R.; Shahidi, F. Antioxidant Activity of Fresh and Processed Jalapeño and Serrano Peppers. *J. Agric. Food Chem.* **2011**, *59*, 163–173. [CrossRef]
15. Meilgaard, M.; Civille, G.V.; Carr, T.B. *Sensory Evaluation Techniques*, 5th ed.; CRC Press: New York, NY, USA, 2015.
16. Lawless, H.; Heymann, H. *Sensory Evaluation of Food Principles and Practices*, 2nd ed.; Springer: New York, NY, USA, 2010.
17. Reyes-Vega, M.L.; Peralta-Rodríguez, R.D.; Anzaldúa-Morales, A.; Figueroa-Cárdenas, J.D.; Martínez-Bustos, F. Relating Sensory Textural Attributes of Corn Tortilla to Some Instrumental Measurements. *J. Texture Stud.* **1998**, *29*, 361–373. [CrossRef]
18. Flores-Farías, R.; Martínez-Bustos, F.; Salinas-Moreno, Y.; Chang, Y.K.; Hernández, J.G.; Ríos, E. Physicochemical and rheological characteristics of commercial nixtamalised Mexican maize flours for tortillas. *J. Sci. Food Agric.* **2000**, *80*, 657–664. [CrossRef]
19. Kelekei, N.N.; Pascut, S.; Waniska, R.D. The effects of storage temperature on the staling of wheat flour tortillas. *J. Cereal Sci.* **2003**, *37*, 377–380. [CrossRef]
20. Suhendro, E.L. Tortilla bending technique: An objective method for corn tortilla texture measurement. *Cereal Chem.* **1998**, *75*, 854–858. [CrossRef]
21. de la Rosa, L.A.; Alvarez-Parrilla, E.; Shahidi, F. Phenolic compounds and antioxidant activity of kernels and shells of Mexican pecan (*Carya illinoinensis*). *J. Agric. Food Chem.* **2011**, *59*, 152–162. [CrossRef] [PubMed]
22. Brand-Williams, W.; Cuvelier, M.E.; Berset, C. Use of a free radical method to evaluate antioxidant activity. *LWT-Food Sci. Technol.* **1995**, *28*, 25–30. [CrossRef]
23. Torres-Aguirre, G.A.; Muñoz-Bernal, Ó.A.; Álvarez-Parrilla, E.; Núñez-Gastélum, J.A.; Wall-Medrano, A.; Sáyago-Ayerdi, S.G.; de la Rosa, L.A. Optimización de la extracción e identificación de compuestos polifenólicos en anís (*Pimpinella anisum*), clavo (*Syzygium aromaticum*) y cilantro (*Coriandrum sativum*) mediante HPLC acoplado a espectrometría de masas. *TIP Rev. Espec. Cienc. Químico-Biol.* **2018**, *21*, 103–115.
24. Ramírez-Sánchez, S.; Ibáñez-Vázquez, D.; Gutiérrez-Peña, M.; Ortega-Fuentes, M.S.; García-Ponce, L.L.; Larqué-Saavedra, A. El ramón (*Brosimum alicastrum* Swartz) una alternativa para la seguridad alimentaria en México. *AGRO Product.* **2017**, *10*, 80–83.
25. Raza, H.; Pasha, I.; Shoaib, M.; Zaaboul, F.; Niazi, S.; Aboshora, W. Review on functional and rheological attributes of kafirin for utilization in gluten free baking industry. *J. Food Sci. Nutr. Res.* **2017**, *4*, 150–157.
26. Remondetto, G.; Añón, M.C.; González, R.J. Hydration Properties of Soybean Protein Isolates. *Braz. Arch. Biol. Technol.* **2001**, *44*, 425–431. [CrossRef]
27. Bullermann, L.B.; Bianchini, A. The microbiology of Cereal and Cereal Products. *Food Qual.* **2011**, *18*, 26–29.
28. Cravioto, R.; Massieu, G.; Guzman, G.G.; Calvo, D.J. Valor nutritivo de plantas alimenticias de Yucatán. *Boletín. Sanit. Panam. Nutr.* **1952**, *32*, 328–339.
29. FDA Title 21 Food and Drugs-Food Labeling. Code of Federal Regulations. Available online: https://www.accessdata.fda.gov/scripts/cdrh/cfdocs/cfcfr/CFRSearch.cfm?fr=101.54 (accessed on 16 November 2019).
30. Cruz-Góngora, V.; De Villalpando, S.; Mundo-Rosas, V.; Shamah-Levy, T. Prevalencia de anemia en niños y adolescentes mexicanos: Comparativo de tres encuestas nacionales. *Salud Publica Mex.* **2013**, *55*, 180–189.
31. Frei, B.; Higdon, J.V. Antioxidant activity of tea polyphenols in vivo: Evidence from animal studies. *J. Nutr.* **2003**, *133*, 3275–3284. [CrossRef] [PubMed]
32. Weaver, C.M. Potassium and Health. *Am. Soc. Nutr.* **2013**, *4*, 368–377. [CrossRef] [PubMed]
33. Ringer, J.; Bartlett, Y. The significance of potassium. *Pharm. J.* **2007**, *278*, 497–500.
34. USDA. *Food Data Central*; U.S. Department of Agriculture: Washington, DC, USA, 2018.

35. Guevara-Arauza, J.C.; Órnelas Paz, J.J.; Rosales Mendoza, S.; Soria Guerra, R.E.; Paz Maldonado, L.M.T.; Pimentel González, D.J. Biofunctional activity of tortillas and bars enhanced with nopal. Preliminary assessment of functional effect after intake on the oxidative status in healthy volunteers. *Chem. Cent. J.* **2011**, *5*, 1–10. [CrossRef]
36. Escudero, E.; González, P. La fibra dietética. *Nutr. Hosp.* **2006**, *21*, 61–72.
37. García-Estévez, I.; Alcalde-Eon, C.; Puente, V.; Escribano-Bailón, M.T. Enological tannin effect on red wine color and pigment composition and relevance of the yeast fermentation products. *Molecules* **2017**, *22*, 2046. [CrossRef]
38. Miranda, G.; Ventura, J.; Suárez, S.; Fuertes, C. Actividad citotóxica y antioxidante de los productos de la reacción de Maillard de los sistemas modelo D-glucosa-glicina y D-glucosa-L-lisina. *Rev. Soc. Quím. Perú.* **2007**, *4*, 215–225.
39. Ribeiro, F.A. Wheat Flour Tortilla: Quality Prediction and Study of Physical and Textural Changes during Storage. Master's Thesis, Texas A&M University, Calgary, TX, USA, 2009.
40. Martínez-Vázquez, J.I.; Pérez-Carrera, S.N.; Quiroz-Ramírez, M.A.; Espitia-Orozco, F.J. Mejoramiento de la calidad proteica de toretillas hechas a base de maíz adicionadas con soya y amaranto. *Investig. Desarro. Cienc. Tecnol. Aliment.* **2017**, *2*, 312–316.
41. Domínguez-Zarate, P.A.; García-Martínez, I.; Güemes-Vera, N.; Totosaus, A. Textura, color y aceptación sensorial de tortillas y pan producidos con harina de ramón (*Brosimum alicastrum*) para incrementar la fibra dietética total. *Cienc. Tecnol. Agropecu.* **2019**, *20*, 699–719. [CrossRef]
42. Bourne, M. *Food Texture and Viscosity: Concept and Measurement*, 2nd ed.; Taylor, S.L., Ed.; Academic Press: New York, NY, USA, 2002.
43. Vázquez, J.A. Desarrollo de Tortillas de Maíz Fortificadas Con Fuentes de Proteína y Fibra y Su Efecto Biológico en Un Modelo Animal. Ph.D. Thesis, Universidad Autónoma de Nuevo León, Autónoma de Nuevo León, Mexico, 2013.
44. Rodríguez, E.; Fernández, A.; Ayala, A. Reología y textura de masas: Aplicaciones en trigo y maíz. *Rev. Ing. Investig.* **2005**, *57*, 72–75.
45. Pérez-Pacheco, E.; Moo-Huchin, V.M.; Estrada-León, R.J.; Ortiz-Fernández, A.; May-Hernández, L.H.; Ríos-Soberanis, C.R.; Betancur-Ancona, D. Isolation and characterization of starch obtained from *Brosimum alicastrum* Swarts Seeds. *Carbohydr. Polym.* **2014**, *101*, 920–927. [CrossRef] [PubMed]
46. Hleap, J.I.; Velasco, V.A. Análisis de las propieades de textura durante el almacenamiento de salchcihas elaboradas a partir de tilapia roja (*Oreochromis* sp.). *Biotecnol. Sect. Agropecu. Agroind.* **2010**, *8*, 46–56.
47. de la Vega, G. Proteínas de la harina de trigo: Clasificación y propiedades funcionales. *Temas Cienc. Tecnol.* **2009**, *13*, 27–32.
48. Edwards, W.P. *The Science of Bakery Products*; The Royal Society of Chemistry: Cambridge, UK, 2007.
49. Mesas, J.M.; Alegre, M.T. El pan y su proceso de elaboración. *Cienc. Tecnol. Aliment.* **2002**, *3*, 307–313. [CrossRef]
50. Wannerberger, L.; Eliasson, A.; Sindberg, A. Interfacial behaviour of secalin and rye flour-milling streams in comparison with gliadin. *J. Cereal Sci.* **1997**, *25*, 243–252. [CrossRef]
51. Badui, S. *Química de los Alimentos*, 5th ed.; López Ballesteros, G., Ed.; Pearson Educación: Naucalpan de Juárez, Estado de México, Mexico, 2012.
52. Moo-Huchin, V.M.; Cabrera-Sierra, M.J.; Estrada-León, R.J.; Ríos-Soberanis, C.R.; Betancur-Ancona, D.; Chel-Guerrero, L.; Ortiz-Fernández, A.; Estrada-Mota, I.A.; Pérez-Pacheco, E. Determination of some physicochemical and rheological characteristics of starch obtained from *Brosimum alicastrum* swartz seeds. *Food Hydrocoll.* **2015**, *45*, 48–54. [CrossRef]
53. Tokpunar, H.K. Chemical Composition and Antioxidant Properties of Maya Nut (*Brosimum alicastrum*). Master's Thesis, Clemson University, Clemson, SC, USA, 2010.
54. Chen, O.; Blumberg, J. Phytochemical composition of nuts. *Asia Pac. J. Clin. Nutr.* **2008**, *17*, 329–332.
55. Marinova, D.; Ribarova, F.; Atanassova, M. Total phenolics and flavonoids in Bulgarian fruits and vegetables. *J. Univ. Chem. Technol. Metall.* **2005**, *40*, 255–260.
56. Sariburun, E.; Sahin, S.; Demir, C.; Türkben, C.; Uylaser, V. Phenolic content and antioxidant activity of raspberry and blackberry cultivars. *J. Food Sci.* **2010**, *75*, 328–335. [CrossRef]
57. Alasalvar, C.; Shahidi, F. Natural antioxidants in tree nuts. *Eur. J. Lipid Sci. Technol.* **2009**, *111*, 1056–1062. [CrossRef]

58. Pellegrini, N.; Serafini, M.; Salvatore, S.; Del Rio, D.; Bianchi, M.; Brighenti, F. Total antioxidant capacity of spices, dried fruits, nuts, pulses, cereals and sweets consumed in Italy assessed by three different in vitro assays. *Mol. Nutr. Food Res.* **2006**, *50*, 1030–1038. [CrossRef] [PubMed]
59. Kuskoski, E.M.; Asuero, A.G.; Troncoso, A.M.; Mancini-Filho, J.; Fett, R. Aplicación de diversos métodos químicos para determinar actividad antioxidante en pulpa de frutos. *Ciência Tecnol. Aliment.* **2005**, *25*, 726–732. [CrossRef]
60. Clifford, M.N.; Jaganath, I.B.; Ludwig, I.A.; Crozier, A. Chlorogenic acids and the acyl-quinic acids: Discovery, biosynthesis, bioavailability and bioactivity. *Nat. Prod. Rep.* **2017**, *34*, 1391–1421. [CrossRef] [PubMed]
61. Dias, M.I.; Sousa, M.J.; Alves, R.C.; Ferreira, I.C.F.R. Exploring plant tissue culture to improve the production of phenolic compounds: A review. *Ind. Crops Prod.* **2016**, *82*, 9–22. [CrossRef]
62. Ozer, H.K. Phenolic compositions and antioxidant activities of Maya nut (*Brosimum alicastrum*): Comparison with commercial nuts. *Int. J. Food Prop.* **2017**, *20*, 2772–2781. [CrossRef]
63. Vazquez-Flores, A.A.; Wong-Paz, J.E.; Lerma-Herrera, M.A.; Martinez-Gonzalez, A.I.; Olivas-Aguirre, F.J.; Aguilar, C.N.; Wall-Medrano, A.; Gonzalez-Aguilar, G.A.; Alvarez-Parrilla, E.; de la Rosa, L.A. Proanthocyanidins from the kernel and shell of pecan (*Carya illinoinensis*): Average degree of polymerization and effects on carbohydrate, lipid, and peptide hydrolysis in a simulated human digestive system. *J. Funct. Foods* **2017**, *28*, 227–234. [CrossRef]
64. Choi, J.Y.; Lee, J.W.; Jang, H.; Kim, J.G.; Lee, M.K.; Jong, J.T.; Lee, M.S.; Hwang, B.Y. Quinic acid esters from Erycibe obtusifolia with antioxidant and tyrosinase inhibitory activities. *Nat. Prod. Res.* **2019**. [CrossRef]
65. Dybkowska, E.; Sadowska, A.; Świderski, F.; Rakowska, R.; Wysocka, K. The occurrence of resveratrol in foodstuffs and its potential for supporting cancer prevention and treatment. A review. *Rocz. Panstw. Zakl. Hig.* **2018**, *69*, 5–14.

Article

Sorghum–Insect Composites for Healthier Cookies: Nutritional, Functional, and Technological Evaluation

Temitope D. Awobusuyi, Muthulisi Siwela * and Kirthee Pillay

Department of Dietetics and Human Nutrition, University of KwaZulu-Natal, Pietermaritzburg 3209, South Africa; temmybusuyi@gmail.com (T.D.A.); pillayk@ukzn.ac.za (K.P.)
* Correspondence: siwelam@ukzn.ac.za; Tel.: +27-33-260-5459

Received: 9 September 2020; Accepted: 25 September 2020; Published: 9 October 2020

Abstract: Protein-energy malnutrition (PEM) is a major health concern in sub-Saharan Africa (SSA). Relying on unexploited and regionally available rich sources of proteins such as insects and sorghum might contribute towards addressing PEM among at-risk populations. Insects are high in nutrients, especially protein, and are abundant in SSA. Sorghum is adapted to the tropical areas of SSA and as such it is an appropriate source of energy compared with temperate cereals like wheat. It is necessary to assess whether cookies fortified with sorghum and termite would be suitable for use in addressing PEM in SSA. Whole grain sorghum meal and termite meal were mixed at a 3:1 ratio (*w*/*w* sorghum:termite) to form a sorghum–termite meal blend. Composite cookies were prepared where the sorghum–termite blend partially substituted wheat flour at 20%, 40%, and 60% (sorghum–termite blend:wheat flour (*w*/*w*). The functional and nutritional qualities of the cookies were assessed. Compared with the control (100% wheat flour), the cookies fortified with sorghum and termite had about double the quantity of protein, minerals, and amino acids. However, with increased substitution level of the sorghum–termite blend, the spread factor of the cookies decreased. There is a potential to incorporate sorghum and termite in cookies for increased intake of several nutrients by communities that are vulnerable to nutrient deficiencies, especially PEM.

Keywords: protein energy malnutrition; insect; sorghum; wheat

---

1. Introduction

The World Health Organization (WHO) estimates that about 60% of all deaths occurring among children under five years of age in developing countries could be attributed to malnutrition [1]. Protein-energy malnutrition (PEM) results from deficiency in dietary protein and/or energy in varying proportions [2]. Sub-Saharan Africa (SSA) continues to lead in bearing the brunt of PEM and globally, the prevalence has continued to rise up to 47.3% with the worst increase occurring in regions of Africa [3,4]. Therefore, an improvement in nutrition is needed to decrease the high mortality and morbidity rates associated with PEM [5].

Sorghum, a drought-tolerant staple, contributes to the diet of over half a billion people in the regions, where maize struggles to grow if there are very limited agronomic intervention technologies [6,7]. Baked products, such as bread and cookies are part of the leading foods world-wide, including in the sub-Saharan African region. Therefore, they are the most appropriate vehicles to deliver vital nutrients, for example protein to vulnerable populations [8]. In SSA, the limitation of producing wheat due to the less conducive climatic conditions and the exorbitant demand on forex for its importation creates a necessity to try to substitute wheat with the well adapted cereals like sorghum in baked products [9]. However, sorghum grain is low in protein content, while the essential amino acids lysine, threonine, and tryptophan are limited [10,11]. Hence, if sorghum were to be used to partially substitute wheat in baked goods, it would be necessary to find an accessible, yet high quality source of nutrients, including protein, to complement it.

Insects are a traditional source of food in several parts of the world. They are especially rich in protein, calcium, iron, and zinc [12,13]. The energy content of insects is comparable to that of meat, except for pork, because of its particularly high fat content [14]. Furthermore, given that food insecurity is prevalent in SSA, the use of insects in this region, where they are already being consumed, although not at a nutritionally significant level, should be promoted to serve as an alternative protein source, in particular. Thus, the incorporation of insect in popular, staple foods to complement staple cereal grains should be considered.

Cookies are an energy dense and shelf-stable, popular baked ready-to-eat snack, consumed by both children and adults globally [15,16]. The main ingredients in cookie baking include wheat flour, fat, sugar, butter, and water. Other added ingredients may be optional or added to improve organoleptic attributes [17]. With the afore-mentioned nutritional advantages of insects and the popularity of cookies, it might be advantageous to complement sorghum with insects in partial replacement of wheat flour in cookies to contribute to addressing PEM in developing regions, especially SSA.

Sorghum–legume cookies in which sorghum was combined with sunflower and peanut flours, to increase the protein content of the cookies have been reported [18]. In addition, Mridula et al. [19] reported that acceptable cookies could be developed with wheat–sorghum composite flours with up to 50% sorghum substitution level. Several studies have demonstrated the potential for supplementing wheat flour with sorghum in bread, and cookies, and other snacks [9,20,21]. The influence of finger millet flour [22], fibers from different cereals [23], maltodextrin, and guar gum [24] on the rheological properties of dough and quality of cookies has also been reported. However, it appears that the compositing of wheat, sorghum, and termite to make cookies has not been reported. Therefore, this study aimed to determine the effect of partially substituting wheat flour with a sorghum–termite blend on the nutritional composition and functional properties of cookies.

2. Materials and Methods

2.1. Preparation of Sorghum and Termite Meal Blend

Winged termites (*Macrotermes belliscosus*), harvested during the harmattan season were purchased from Oja-oba main market in Ondo State, Nigeria, and used in this study. The termites were de-winged and cleaned three times to remove soil and dirt. They were oven-dried at 40 °C for 8 h [25]. Dried insects were milled into a meal with a blender to a particle size of ≤1 mm, vacuum packed, labelled and stored at −4 °C until analysis. The termites were washed, dried in the oven, and milled into a meal. Sorghum grain was purchased and cleaned to make sure they were free of dirt. A mill fitted with a 0.4 mm screen was used to grind whole grain sorghum meal into a meal [9]. Both the sorghum and termite meal were used to substitute wheat flour at varying proportions (Table 1).

Table 1. Ratios of ingredients (wheat:sorghum and termite) for cookie formulation.

Ingredient	Relative Concentration (% *w/w*)			
Wheat flour	100	80	60	40
Sorghum meal	0	15	30	45
Termite meal	0	5	10	15
Identity of cookie sample	C0 (control)	C20	C40	C60

Sorghum:termite meal (ratio 3:1) replaced wheat flour at 0, 20, 40, and 60% (*w/w*) levels.

2.2. Methods

2.2.1. Preparation of Cookie Samples

Cookies fortified with sorghum and termite as well as the control were prepared according to the method described by de Jager [26], with minimal modification. The sorghum meal and insect meal were mixed at a 3:1 (*w*/*w*) ratio to form a sorghum–termite blend. The ratio 3:1 was chosen after preliminary trials in the laboratory revealed that other substitution levels of sorghum and termite resulted in cookies that were too brittle. Experimental cookies were prepared where wheat flour was partially substituted with different proportions of the sorghum–termite blend, 20%, 40%, and 60% (*w*/*w*), separately. Cookies (100% wheat) in which no sorghum or insect was added served as the control (Table 1). About 200 g of sugar, 5 mL of vanilla essence, 5 mL of salt, 50 g of powdered milk, and 10 mL baking powder were sieved and mixed together with 480 g wheat flour for three to five minutes. About 250 g of margarine was added to the mixture and kneaded for two minutes to form a firm dough. The dough was rolled out, cut into desired shapes, and transferred into the oven. Cookies were baked at 150° C in a preheated oven for 20 min. Cookies were crushed to a particle size of $\leq$ 1 mm for chemical analyses and then stored.

2.2.2. Nutritional Composition

The nutritional composition of cookies was determined by standard methods stated below.

Ash

Ash was determined using the Association of Official Analytical Chemists (AOAC) official method 923.03 [27].

Protein

Crude protein was measured using the AOAC official method 968.06 [28].

Glycemic (Available) Carbohydrate Content

Glycemic carbohydrates were calculated by difference.

Fat

Fat content was determined according to the AOAC Official Method 920.39C [29].

Fiber

Fiber was done based on the method reported by Saha et al. [22].

Gross Energy

Gross energy was determined according to the AOAC Official Method 935.42 [30].

Selected Minerals

Mineral content was determined by the AOAC Official Method 6.1.2 [31].

Amino Acids

The amino acid profile of the cookie samples was analysed by the Waters API Quattro Micro Method, which consists of a column C18, 1.7 µm, 2.1 × 100 mm and a binary solvent manager. Samples (400 mg) were subjected to AccQ-Tag Ultra Derivatization kit (Waters^{TM}, Johannesburg, South Africa); 10 µL of the undiluted sample was added to the Waters AccQ-Tag kit constituents and placed in a heating block at a temperature of 55 °C for 10 min. Injection volume was 1 µL and gradient separation was performed using Solvents A and B from the Waters Accutag kit.

In Vitro Protein Digestibility

In vitro protein digestibility was determined using the method described by Hamaker et al. [32].

2.2.3. Physical Characteristics

Physical quality parameters of the cookie samples, such as diameter, thickness, spread ratio, and spread factor were determined using the procedure of the American Association of Cereal Chemists [33].

Texture

Texture analysis of cookies fortified with sorghum and termite (Figure 1) was done using a TA-XT plus 100C model texture analyzer (Stable Micro Systems, Godalming, UK). The cookies were measured for hardness and fracturability using a three-point bend rig attachment at a 3.0 mm/s cross head speed for a 5 mm distance and a 5 kg load cell.

Figure 1. Cookies fortified with sorghum and termite. C0: control (100% wheat cookies); C20: 15% sorghum and 5% termite substitution level; C40: 30% sorghum and 10% termite substitution level; C60: 45% sorghum and 15% termite substitution level.

2.2.4. Functional PropertiesWater and Oil Absorption Capacity

Water and Oil Absorption Capacity

The water and oil absorption capacities were determined by the method of Sosulski et al. [34].

Bulk Density

The process reported by Okaka and Potter [35] was used to determine the bulk density of the cookie flours.

2.2.5. Statistical Analysis

The resulting data was analysed using the Statistical Package for Social Science (SPSS version 20.0 SPSS Inc., Chicago, IL, USA). One-way analysis of variance (ANOVA) was done; and separation of means was by Fisher Least Significance Difference (LSD) test. A p-value of ≤0.05 was considered significant.

3. Results

3.1. Proximate Composition

Protein, fat, and carbohydrates constitute the major nutrients in cookie samples (Figure 1). The proximate composition of the cookies fortified with the sorghum–termite blend is presented in Table 2. As expected, the highest protein content was observed in cookies containing 60% sorghum and termite substitution level, which contained the most insect. The results showed that fat content increased with increasing concentrations of the sorghum–termite blend. The fiber and ash content of cookie samples fortified with the sorghum–termite blend was also significantly higher than the control. Carbohydrate content was higher in the control cookies than in the cookies fortified with the sorghum–termite blend. The gross energy of the cookies fortified with the sorghum–termite blend was higher than the control.

Table 2. The sorghum–termite blend on the proximate composition (g/100 g) and gross energy (kJ) of cookies [1].

Sample	Ash	Protein	Fat	CHO	Dietary Fiber	Energy
C0	1.7 ^{d} ± 0.5	10.5 ^{d} ± 0.4	14.3 ^{d} ± 0.4	54.4 ^{a} ± 0.4	8.3 ^{d} ± 0.5	1180.2 ^{d} ± 0.5
C20	3.5 ^{c} ± 0.6	36.4 ^{c} ± 0.4	22.3 ^{c} ± 0.5	23.4 ^{b} ± 0.4	13.2 ^{a} ± 0.5	2032.8 ^{c} ± 0.5
C40	4.0 ^{b} ± 0.5	38.3 ^{b} ± 0.5	25.2 ^{b} ± 0.4	19.4 ^{c} ± 0.5	10.3 ^{c} ± 0.4	2108.4 ^{b} ± 0.3
C60	4.2 ^{a} ± 0.4	41.0 ^{a} ± 0.4	28.2 ^{a} ± 0.4	17.2 ^{d} ± 0.5	13.0 ^{b} ± 0.5	2217.6 ^{a} ± 0.2

Mean (±*SD*) of three determinations; CHO: glycemic carbohydrates; Means with different superscripts in a column vary significantly ($p < 0.05$); [1] Values are on dry matter basis. kJ; refers to Kilojoules; C0: control (100% wheat cookies); C20: 15% sorghum and 5% termite substitution level; C40: 30% sorghum and 10% termite substitution level; C60: 45% sorghum and 15% termite substitution level.

3.2. Mineral Composition

The minerals abundant in the cookie samples were iron, phosphorus, and magnesium (Table 3). The incorporation of the sorghum–termite blend substantially increased the mineral concentration of the cookies. Zinc and iron were abundant in the cookie samples fortified with the sorghum–termite blend. Cookies containing 60% of the sorghum–termite blend had the highest concentration of these minerals compared with the cookie samples with lower concentrations of the sorghum–termite blend.

Table 3. Selected mineral elements in cookies fortified with sorghum–termite blend (mg/100 g) [1].

Minerals	C0	C20	C40	C60	Recommended Daily Intake [2]	
					Children	Adults
Ca	2.5 ^{d} ± 0.4	5.5 ^{c} ± 0.5	8.6 ^{b} ± 0.5	10.8 ^{a} ± 0.5		
Fe	2.4 ^{d} ± 0.6	28.5 ^{c} ± 0.4	34.2 ^{b} ± 0.5	37.4 ^{a} ± 0.4	4.2	19.6
Mn	1.2 ^{d} ± 0.5	13.2 ^{c} ± 0.4	17.3 ^{b} ± 0.4	24.5 ^{a} ± 0.4		
Cu	0.8 ^{d} ± 0.6	1.4 ^{c} ± 0.5	2.4 ^{b} ± 0.5	3.8 ^{a} ± 0.6	0.9	1.3
K	1.8 ^{d} ± 0.6	12.5 ^{c} ± 0.4	18.6 ^{b} ± 0.5	22.9 ^{a} ± 0.4		
Na	0.8 ^{d} ± 0.6	8.1 ^{c} ± 0.5	13.5 ^{b} ± 0.6	16.9 ^{a} ± 0.5		
Zn	2.5 ^{d} ± 0.5	8.4 ^{c} ± 0.6	10.4 ^{b} ± 0.4	14.8 ^{a} ± 0.4	2.4	2.5
P	0.8 ^{d} ± 0.5	22.5 ^{c} ± 0.5	31.2 ^{b} ± 0.4	37.6 ^{a} ± 0.5		
Mg	1.8 ^{d} ± 0.4	24.3 ^{c} ± 0.5	29.6 ^{b} ± 0.6	33.5 ^{a} ± 0.5		

Mean (±*SD*) of three determinations. Means with different superscripts in a column vary significantly ($p < 0.05$). [1] Values are on dry matter basis. [2] Food and Agriculture Organization (FAO)/World Health Organization (WHO) 2006 Recommended daily intake for children and adults. C0: control (100% wheat cookies); C20: 15% sorghum and 5% termite substitution level; C40: 30% sorghum and 10% termite substitution level; C60: 45% sorghum and 15% termite substitution level.

3.3. Amino Acid Content

Table 4 shows that cookies containing the sorghum–termite blend had substantial amounts of amino acids and lysine and leucine were the major amino acids present. Cookies fortified with the sorghum–termite blend showed the highest increase across all amino acids when compared with the control.

Table 4. Essential amino acid composition of cookies fortified with sorghum–termite blend (mg/100 g) [1].

Amino Acids	C0	C20	C40	C60	FAO Reference Pattern [2]	
					Children	Adults
Histidine	15	21	32	43		15
Lysine	10	22	37	49	75	45
Tyrosine	18	32	36	42		
Cysteine	18	22	29	33		6
Tryptophan	10	18	24	32	4.6	6
Methionine	18	20	25	29	34	16
Isoleucine	28	35	40	46	37	30
Phenylalanine	25	33	39	44	34	30
Threonine	21	30	34	46	44	23
Leucine	27	48	58	63	56	59
Valine	27	42	45	47	41	39

[1] Values are on dry matter basis; [2] Food and Agriculture Organization/World Health Organization (2007). C0: control (100% wheat cookies); C20: 15% sorghum and 5% termite substitution level; C40: 30% sorghum and 10% termite substitution level; C60: 45% sorghum and 15% termite substitution level.

3.4. In Vitro Protein Digestibility

The in vitro protein digestibility (IVPD) results are presented in Table 5. The addition of the sorghum–termite blend improved the digestibility of cookies, which increased by 23.8% with the incorporation of the sorghum–termite blend from 67% to 83%.

Table 5. In vitro protein digestibility of cookies fortified with sorghum–termite blend.

Samples	IVPD g/100 g	% Increase
C0	$67^{d} \pm 0.4$	-
C20	$73^{c} \pm 0.2$	8.95
C40	$79^{b} \pm 0.2$	17.9
C60	$83^{a} \pm 0.3$	23.8

Mean (± *SD*) of three determinations. Means with different superscripts in a column vary significantly ($p < 0.05$). C0: control (100% wheat cookies); C20: 15% sorghum and 5% termite substitution level; C40: 30% sorghum and 10% termite substitution level; C60: 45% sorghum and 15% termite substitution level.

3.5. Physical Characteristics

The physical characteristics of cookies fortified with the sorghum–termite blend are shown in Table 6. The results showed that the addition of sorghum–termite blend reduced the weight of cookies, compared with the control. Cookies fortified with 60% sorghum–termite blend (24.5 g) recorded the highest weight loss when compared with the control (29.5 g). There was a progressive increase in the thickness of the cookies fortified with the sorghum–termite blend (7.5 mm, 7.7 mm, and 7.9 mm for sample C20, C40, and C60, respectively), when compared with the control (7.3 mm).

Table 6. Physical qualities of cookies fortified with sorghum–termite blend.

Samples	Weight (g)	Diameter (mm)	Thickness (mm)	Spread Factor
C0	$29.5^{a} \pm 0.5$	$45.8^{a} \pm 0.2$	$7.3^{c} \pm 0.6$	$6.2^{a} \pm 0.4$
C20	$27.4^{b} \pm 0.7$	$44.4^{b} \pm 0.4$	$7.5^{c} \pm 0.7$	$5.9^{b} \pm 0.7$
C40	$25.4^{c} \pm 0.7$	$42.3^{c} \pm 0.6$	$7.7^{b} \pm 0.4$	$5.5^{c} \pm 0.5$
C60	$24.5^{c} \pm 0.7$	$41.6^{c} \pm 0.5$	$7.9^{a} \pm 0.5$	$5.2^{c} \pm 0.2$

Mean (± *SD*) of three determinations. Means with different superscripts in a column vary significantly ($p < 0.05$). C0: control (100% wheat cookies); C20: 15% sorghum and 5% termite substitution level; C40: 30% sorghum and 10% termite substitution level; C60: 45% sorghum and 15% termite substitution level.

3.6. Texture and Colour

The effect of incorporating sorghum and termite on the texture and colour of cookies is shown in Table 7. The results showed that as the substitution level of the sorghum–termite blend increased, the cookie hardness decreased. Further, cookies became darker with increasing concentration of the sorghum–termite blend. The L* (lightness) values decreased and the a* (redness) values increased, while b* (yellowness) values were similar across all cookie samples.

Table 7. Texture and colour of cookies fortified with sorghum–termite blend.

Sample	Hardness (N)	Fracturability (mm)	Colour		
			Hunter L*	a*	b*
C0	$36.4^{a} \pm 5.5$	$0.31^{d} \pm 0.8$	$47.3^{a} \pm 0.5$	$12.7^{d} \pm 0.6$	$8.1^{d} \pm 0.6$
C20	$29.4^{d} \pm 8.6$	$0.64^{c} \pm 1.2$	$46.2^{b} \pm 1.2$	$14.5^{c} \pm 0.8$	$8.3^{c} \pm 0.7$
C40	$31.7^{c} \pm 9.9$	$0.72^{b} \pm 0.3$	$44.5^{c} \pm 0.5$	$17.4^{b} \pm 1.2$	$8.4^{b} \pm 1.5$
C60	$32.1^{b} \pm 0.4$	$1.14^{a} \pm 0.1$	$41.2^{d} \pm 0.6$	$18.8^{a} \pm 0.7$	$8.6^{a} \pm 0.7$

Mean (± *SD*) of three determinations. Means with different superscripts in a column vary significantly ($p < 0.05$). L* = Lightness, a* = Redness, b* = Yellowness. C0: control (100% wheat cookies); C20: 15% sorghum and 5% termite substitution level; C40: 30% sorghum and 10% termite substitution level; C60: 45% sorghum and 15% termite substitution level.

3.7. Water and Oil Absorption Properties of Cookie Flours

The oil absorption capacity (OAC) varied between 1.43 and 1.65 g oil/g flour (Figure 2) while the water absorption (WAC) varied from 1.55 g to 1.78 g water/g flour. The results revealed that the higher the sorghum and termite substitution level, the higher the absorption capacities of the flours.

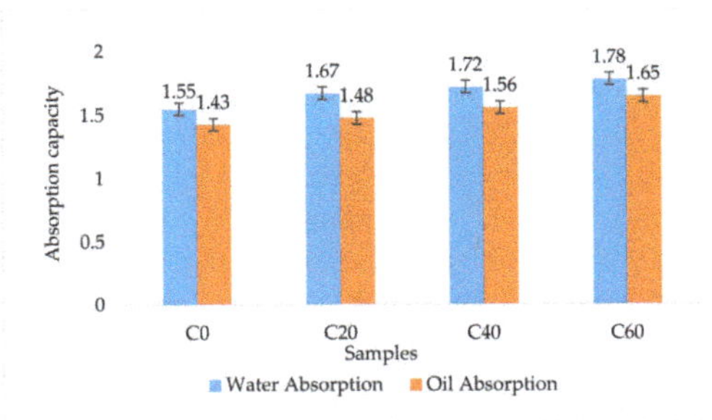

Figure 2. Water and oil absorption capacity of cookie flours. Water absorption is expressed as g water/g flour. Oil absorption is expressed as g oil/g flour. Error bar values are actual values obtained. C0: control (100% wheat cookies); C20: 15% sorghum and 5% termite substitution level; C40: 30% sorghum and 10% termite substitution level; C60: 45% sorghum and 15% termite substitution level.

3.8. Bulk Density of Cookie Flours

The packed bulk density (PBD) and loose bulk density (LBD) varied (Figure 3). The PBD was 0.69 g/mL for the control and 0.96 g/mL for the 60% composite flour containing the sorghum–termite blend, which was much higher than the control. The LBD ranged between 0.59 and 0.87 g/mL. All composite flours containing wheat, sorghum and termite had relatively higher LBD than the wheat flour (control).

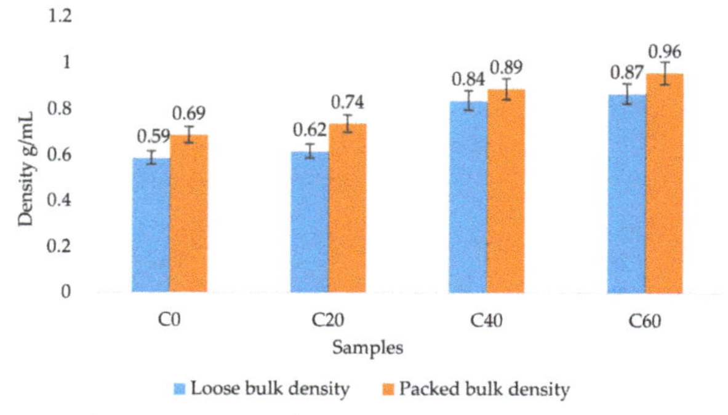

Figure 3. Bulk density of cookie flours. Error bar values are actual values obtained. C0: control (100% wheat cookies); C20: 15% sorghum and 5% termite substitution level; C40: 30% sorghum and 10% termite substitution level; C60: 45% sorghum and 15% termite substitution level.

4. Discussion

4.1. Proximate Composition

Protein content was significantly higher in all cookies fortified with the sorghum–termite blend (36.4 to 41.0 g/100 g), compared with the control (10.5 g/100 g). This result agrees with Koffi-Niaba et al. [36], who reported that supplementing sorghum with termite flour significantly improved the protein content from 9.6% in the control to 21.7%. A report by Kinyuru et al. [37] working on the nutritional quality of wheat buns enriched with edible termites also found that the buns showed a significant increase in protein content (47.5%), when compared with buns without termite substitution. It has been reported that fortifying sorghum with defatted soy flour significantly enhanced the protein quality and content of cookies [21]. Similar results have also been reported by Omoba and Omogbemile [38]. Further, as the concentration of the sorghum–termite blend increases, a substantial increase was observed in the fat content. This is expected as termites, the insect used in this study ranked among the highest in fat concentration [39,40]. Fibre also recorded a significant increase across all cookies containing sorghum–termite blend due to the addition of insects (10.2 to 13.5 g/100 g). Fiber is beneficial in the human diet to reduce the risk of heart disease, blood pressure, obesity, and lower cholesterol levels [41].

The ash content of the cookie samples fortified with the sorghum–termite blend (3.5 to 4.2 g/100 g) was significantly higher than the control (1.7 g/100 g). However, C0 (control), had the highest carbohydrate content (54%). Cookies containing 60% sorghum–termite blend had the highest energy value (2217.6 kJ/100 g) and the lowest was recorded in the control (C0) (gross energy: 1180.2 kJ/100 g). This result could be attributed to the high fat content of the experimental cookies. The energy value of edible insects has been reported to depend mainly on their fat content [38]. Similar results were reported by Koffi-Niaba et al. [36], they obtained between 1626 kJ/100 g and 1712 kJ/100 g in sorghum-based cookies fortified with termites. Overall, these are promising results and developing countries, where PEM remains a major problem, could benefit from the consumption of these cookies.

4.2. Mineral Composition

Cookie samples fortified with the sorghum–termite blend had increased mineral content compared with the control (Table 3). The calcium content was significantly higher (5.5 to 10.8 mg/100 g) when compared with the control (2.5 mg/100 g). The iron content varied significantly, with the highest level (37.4 mg/100 g) found in sample C60 (45% sorghum, 15% insect substitution level). Although insects are known to have high levels of minerals [40], the higher levels noted in this study could also be partially attributed to the high iron content in sorghum [42]. Zinc levels were also significantly high across all cookies fortified with the sorghum–termite blend (8.4 to 14.8 mg/100 g), compared with the control (2.5 mg/100 g). This supports Yhoungaree et al. [43], who stated that insects are a valuable source of iron and zinc. A previous study used caterpillar cereal to prevent anemia and stunting in infants. The authors found that the caterpillar cereal produced had appropriate macro and micronutrient contents and concluded that it could be used for complementary feeding [44]. This further supports the results in this study, which showed that consumption of insects, could prove to be a valuable measure to combat deficiencies of iron and zinc in developing countries.

In this study, phosphorous content was highest in the cookie sample containing 60% sorghum–termite blend (37.6 mg/100 g) and lowest in the control (0.8 mg/100 g). Magnesium ranged from 24.3 to 33.5 mg/100 g in cookies fortified with the sorghum–termite blend, which was significantly higher than the control (1.8 mg/100 g). Copper ranged from 0.8 mg/100 g in the control to 3.8 mg/100 g in sample C60 (45% sorghum and 15% termite substitution level). Manganese ranged between 13.2 and 24.5 mg/100 g in cookies fortified with the sorghum–termite blend, which was higher than the control (1.2 mg/100 g). Overall, the addition of sorghum and termite substantially increased the mineral profile of the cookies. This study demonstrates the potential of edible insects to increase the intake of minerals, which are well reported to be deficient and causing severe public health problems in poor populations [45]. Given the worldwide deficiencies of these minerals among human population groups [46], insect-fortified cookies would supply the amount of iron and zinc required for basic body functions. Further, the bioavailability of minerals from insects is likely to be higher than that from plant foods because their nutrients are easily assimilated by the human body and there are no antinutritional factors such as phytic acid and oxalic acid in insects [47,48]. Previous studies found an appreciable concentration of minerals: calcium, iron, magnesium, copper, potassium, sodium, and zinc, and a particularly high iron content present in termites (*syntermes* soldiers) [49]. Chakravorty et al. [50] also reported that insects; *Oecophylla smaragdina* and *Odontotermes* sp. can serve as a source of micronutrients such as Fe, Zn, Cu, and Mn. Therefore, the consumption of insects should be encouraged, especially among rural communities with low animal protein intake, to contribute to meeting their nutritional requirements.

4.3. Amino Acid Profile

Lysine was significantly higher in cookies fortified with the sorghum–termite blend. The increase in lysine could also be attributed to the addition of the insect (termite) meal. It has been reported that the most concentrated essential amino acid found in termites was lysine [51]. As stated earlier, cereal grains are important staples in diets globally but are generally lysine deficient [39]. Therefore, supplementing cereal-based foods with insect is recommended as it would improve the lysine content of the foods. The lysine content reported in this study accounted for 80% of the recommended intake for children and 100% requirements for adults. The fact that insects are a traditional food in most developing regions, including SSA [52] is an advantage. Tryptophan and threonine known to be deficient in cereal proteins were also significantly higher in cookies fortified with the sorghum and termite meal. Tryptophan content was lowest in the control (10 mg/100 g) and ranged between 18 mg/100 g and 32 mg/100 g in cookies fortified with the sorghum–termite blend. Threonine content in cookies fortified with the sorghum–termite blend was between 30 mg/100 g and 46 mg/100 g, compared with 21 mg/100 g in the control. High concentrations of methionine and cysteine were also found (20 to 29 mg/100 g and 22 to 33 mg/100 g, respectively), in comparison to the control (18 mg/100 g and 18 mg/100 g methionine and cysteine, respectively). Histidine content ranged from 21 to 43 mg/100 g. Histidine is a precursor of histamine, which is present in small quantities in cells. Histamine communicates messages to the brain, triggers the release of stomach acid to aid digestion, and is released after an injury or allergic reaction as part of the body's immune response [53]. Further, children grow poorly if there is an absence of histidine in their diet [53]. Therefore, cookies fortified with sorghum–termite blend developed in this study could be a good source of histidine required by children. Isoleucine (35 to 46 mg/100 g), leucine (48 to 63 mg/100 g), phenylalanine (33 to 44 mg/100 g), tyrosine (32 to 42 mg/100 g), and valine

(42 to 47 mg/100 g) were also present in abundance. Overall, the concentrations of amino acids obtained in this study are higher compared with the concentrations found in meat sources such as beef, pork, and chicken meat [54]. Although it has been reported that the nutritional composition of insects may vary due to their feeding habits and harvesting season [51], values reported in this study could be largely dependent on the environmental factors from where the termites were purchased. As compared with the 2007 Food and Agriculture Organization (FAO)/World Health Organization (WHO) standard, the concentrations of the essential amino acids in all cookies fortified with the sorghum–termite blend were generally higher than the pattern of amino acid requirements for both children and adults [55] (Table 4). Hence, the developed cookies would be able to contribute to the essential amino acids in the human diet. Although most of the amino acids reported exceeded the requirements, they can be viewed as beneficial, especially for population groups whose staple diet consists of maize and wheat, which may lack some of the important amino acids [56].

4.4. *In Vitro Protein Digestibility*

In vitro protein digestibility (IVPD) ranged from 67% to 83% (Table 5). The reason for the higher IVPD in the cookies fortified with the sorghum–termite blend compared with the control is likely due to the higher digestibility of insect protein. Insect protein is highly digestible, and a range of 77% to 98% digestibility has been reported [39,57]. The results of this study are similar to the results reported by Ajayi [58], who reported high digestibility for winged termites (83.41%) and soldier termites (81.10%).

4.5. *Physical Characteristics*

The results show that as the sorghum–termite blend substitution level increased, cookie diameter decreased. Cookies containing 60% of the sorghum–termite blend had the lowest spread factor. The experimental cookies had a lower spread factor relative to the control most likely due to dilution of gluten, which is essential in the expansion of baked products [59–61].

4.6. *Texture and Colour*

The decrease in hardness and high fracturability of cookies fortified with the sorghum–termite blend may be attributed to dilution of gluten in the experimental cookies, because sorghum and termite do not contain gluten proteins. Gluten, which is formed during the dough mixing process and coagulated into a fiber-like foam, is responsible for the mechanical structure of baked products [61]. Furthermore, coarse particles may have been introduced by the increased fiber content of cookies, interfering with the homogeneity of the dough and cookie structure, thereby resulting in lower hardness values [62]. Additionally, the high crumbliness and fragility of the experimental cookies could also be due to high level of bran and the absence of gluten [63]. Table 7 shows that the 100% wheat cookies (control) were the hardest and least fracturable of all the cookies. The C0 (control) recorded the lowest fracturability value, while the maximum fracturability value was recorded for C60 (45% sorghum and 15% termite substitution level). Higher fracturability for cookies enriched with fiber has also been reported [64]. A previous study by Awobusuyi et al. [65], reported that fortifying wheat with a sorghum–insect meal did not compromise the product quality or acceptability, as the texture of the cookie samples containing the sorghum–termite meal was liked and rated the same as that of the control (100% wheat cookies).

The cookies darkened (decreasing Hunter L* values) with increased concentration of the sorghum–termite blend. This is likely due to the darker colour of the sorghum insect blend compared with wheat flour. Awobusuyi et al. [65], reported that the acceptability of sorghum–insect cookies and the colour acceptability of cookie samples supplemented with 5% sorghum–termite meal was higher when compared with the control (100% wheat biscuits) and cookies with higher concentrations of termites. It has also been reported that cookies made with whole meal sorghum resulted in cookies with a darker colour [63]. In addition, the increased protein content of the experimental cookies would result in production of higher levels of Maillard reaction products, the majority of which are brown pigments [66].

4.7. Water and Oil Absorption Properties of Cookie Flours

The results revealed that the higher the sorghum–termite blend substitution level, the higher the water absorption capacity (WAC). Water absorption capacity is a product's ability to interact with water under restricted conditions [67]. Previous reports have suggested that flours with high water absorption capacity as the composite flours of this study would be beneficial in bakery products, as this could prevent staling by reducing moisture loss [68]. Similarly, oil absorption capacity (OAC) increased with increasing substitution level of sorghum–termite blend. Oil absorption capacity (OAC) refers to the capability of flour to absorb oil [69]. This is vital because oil acts as a flavour retainer and improves the mouth feel of cookies [70]. The observed trend of an increase in OAC with an increase in the concentration of the termite meal in biscuits may be attributed to the high protein content of the termite meal. The main chemical component affecting OAC in foods is protein, which is composed of both hydrophilic and hydrophobic parts. Hydrophobic proteins possess superior binding of lipids—non-polar amino acid side chains predominant in hydrophobic proteins can form a hydrophobic interaction with hydrocarbon chains of lipids, and thereby increase OAC [71,72]. The blends in this study are potentially valuable in the structural interaction in food, which is also important in developing new food products and the extension of shelf life, particularly in baked foods or other food products where fat absorption is desired [73,74].

4.8. Bulk Density of Cookie Flours

The results show that all flours containing the sorghum–termite blend had relatively higher packed bulk density (PBD) and loose bulk density (LBD) than the control. This indicated that the sorghum–termite blend had a higher bulk density than wheat flour. As explained earlier, bulk density provides information on the porosity of a product and can affect the choice and design of the packaging materials [42].

5. Conclusions

Cookies fortified with a sorghum–termite blend have the potential to serve as a protein, energy, and nutrient-rich supplementary food to address PEM. The results of the present study suggest that a sorghum–termite blend can be successfully incorporated into cookies up to a level of 60% (15% insect) and yield cookies of high nutritional value. The contribution of the experimental cookies to dietary iron, zinc, and lysine would be of particular significance as deficiencies of these nutrients remain a problem in Africa, especially in countries in SSA.

Author Contributions: Conceptualization, T.D.A., M.S. and K.P; methodology, T.D.A., M.S. and K.P.; investigation; T.D.A., M.S. and K.P.; resources, M.S. and K.P.; data curation, T.D.A.; writing—original draft preparation, T.D.A.; and writing—review and editing, M.S. and K.P. All authors have read and agreed to the published version of the manuscript.

Funding: This research received no external funding.

Conflicts of Interest: The authors declare no conflict of interest.

References

1. Ubesie, A.C.; Ibeziako, N.S.; Ndiokwelu, C.I.; Uzoka, C.M.; Nwafor, C.A. Under-five protein energy malnutrition admitted at the university of Nigeria teaching hospital, Enugu: A 10-year retrospective review. *Nutr. J.* **2012**, *11*, 1–7. [CrossRef] [PubMed]
2. Franco, V.; Hotta, J.K.; Jorge, S.M.; Dos Santos, J.E. Plasma fatty acids in children with grade III protein-energy malnutrition in its different clinical forms: Marasmus, marasmic kwashiorkor and kwashiorkor. *J. Trop. Paediatr.* **1999**, *45*, 71–75. [CrossRef] [PubMed]

3. World Bank. The Global Burden of Disease: Main Findings for Sub-Saharan Africa. 2013. Available online: https://www.worldbank.org/en/region/afr/publication/global-burden-of-diseasefindings-for-sub-saharan-africa (accessed on 24 August 2019).
4. Chukwu, E.O.; Fiase, T.M.; Haruna, H.; Dathini, H.; Kever, T.R.; Lornengen, E.M. Predisposing factors to protein energy malnutrition among children (0–5years) in umuogele umuariam community in Obowo local government area of Imo state, Nigeria. *J. Nutr. Health Food Eng.* **2017**, *6*, 108–115.
5. Anwar, F.; Khomsan, A.; Sukandar, D.; Riyadi, H.; Mudjajanto, E.S. High participation in posyandu nutrition program improved children nutritional status. *Nutr. Res. Pract.* **2010**, *4*, 208–214. [CrossRef] [PubMed]
6. International Crops Research Institute for the Semi-Arid Tropics (ICRISAT). Crops: Sorghum. 2009. Available online: www.icrisat.org/sorghum/sorghum.html (accessed on 13 June 2019).
7. Doggett, H. *Sorghum*, 2nd ed.; Wiley: London, UK, 1988; pp. 260–282.
8. Bulusu, S.; Laviolette, L.; Mannar, V.; Reddy, V. Cereal fortification programs in developing countries. In *Issues in Complementary Feeding*, 1st ed.; Agostoni, C., Brunser, O., Eds.; Karger AG: Basel, Switzerland, 2007; pp. 91–105.
9. El-Khalifa, O.E.; El-Tinay, A. Effect of cystine on bakery products from wheat-sorghum blends. *Food Chem.* **2002**, *77*, 133–137. [CrossRef]
10. Shewry, P.R. Wheat. *J. Exp. Bot.* **2009**, *60*, 1537–1553. [CrossRef] [PubMed]
11. Taylor, J.R.N.; Schüssler, L. The protein compositions of the different anatomical parts of sorghum grain. *J. Cereal Sci.* **1986**, *4*, 361–369. [CrossRef]
12. Ghosh, S.; Lee, S.M.; Jung, C.; Meyer-Rochow, V.B. Nutritional composition of five commercial edible insects in South Korea. *J. Asia-Pac. Entomol.* **2017**, *20*, 686–694. [CrossRef]
13. Ghosh, S.; Jung, C.; Meyer-Rochow, V.B. Nutritional value and chemical composition of larvae, pupae, and adults of worker honey bee, *Apis mellifera* ligustica as a sustainable food source. *J. Asia-Pac. Entomol.* **2016**, *19*, 487–495. [CrossRef]
14. Sirimungkararat, S.; Saksirirat, W.; Nopparat, T.; Natongkham, A. Edible Products From Eri Silkworm (*Samia ricini* D.) and Mulberry Silkworm (*Bombyx mori* L.) in Thailand. In *Forest Insects as Food: Humans Bite Back*, 2nd ed.; Durst, P.B., Johnson, D.V., Leslie, R.N., Shono, K., Eds.; RAP: Chiang Mai, Thailand, 2010; pp. 189–200.
15. Florence, S.P.; Asna, U.; Asha, M.R.; Jyotona, R. Sensory, physical and nutritional qualities of cookies prepared from pearl millet (Opennisetum typhoideum). *Food Process. Technol.* **2014**, *5*, 1–6.
16. Okaka, J.C. *Handling, Storage and Processing of Plant Foods*, 2nd ed.; OCJANCO Academic Publishers: Enugu, Nigeria, 2009; p. 132.
17. Chinma, C.E.; Gernah, D.I. Physicochemical and sensory properties of cookies produced from cassava/soya bean/mango composite flours. *J. Food Technol.* **2007**, *5*, 256–260.
18. Hikeezi, D. Cookie Making Using Dehulled High Tannin Sorghum Fortified with Peanut and/or Sunflower Flours. Master's Thesis, Kansas State University, Manhattan, KS, USA, 1994.
19. Mridula, D.; Gupta, R.K.; Manikantan, M.R. Effect of incorporation of sorghum flour to wheat flour on quality of biscuits fortified with defatted soy flour. *Am. J. Food Technol.* **2007**, *2*, 428–434.
20. Yousif, A.; Nhepera, D.; Johnson, S. Influence of sorghum flour addition on flat bread in vitro starch digestibility, antioxidant capacity and consumer acceptability. *Food Chem.* **2012**, *134*, 880–887. [CrossRef] [PubMed]
21. Serrem, C.A.; Kock, H.L.D.; Taylor, J.R.N. Nutritional quality, sensory quality and consumer acceptability of sorghum and bread wheat biscuits fortified with defatted soy flour. *Int. J. Food Sci. Technol.* **2011**, *46*, 74–83. [CrossRef]
22. Saha, S.; Gupta, A.; Singh, S.R.K.; Bharti, N.; Singh, K.P.; Mahajan, V.; Gupta, H.S. Compositional and varietal influence of finger millet flour on rheological properties of dough and quality of biscuit. *LWT Food Sci. Technol.* **2011**, *44*, 616–621. [CrossRef]
23. Sudha, M.L.; Vetrimani, R.; Leelavathi, K. Influence of fibre from different cereals on the rheological characteristics of wheat flour dough and on biscuit quality. *Food Chem.* **2007**, *100*, 1365–1370. [CrossRef]
24. Sudha, M.; Srivastava, A.K.; Vetrimani, R.; Leelavathi, K. Fat replacement in soft dough biscuits: Its implications on dough rheology and biscuit quality. *J. Food Eng.* **2007**, *80*, 922–930. [CrossRef]

25. Akullo, J.; Agea, J.G.; Obaa, B.B.; Acai, J.O.; Nakimbugwe, D. Process development, sensory and nutritional evaluation of honey spread enriched with edible insect flour. *Afr. J. Food Sci.* **2017**, *11*, 30–39.
26. De Jager, H.C. *Foods and Cookery*, 2nd ed.; Government Printers: Pretoria, South Africa, 1991; p. 215.
27. Association of Official Analytical Chemists (AOAC) International. *Official Methods of Analysis*, 18th ed.; Association of Official Analytical Chemists: Arlington, VA, USA, 2005.
28. Association of Official Analytical Chemists (AOAC) International. *Official Methods of Analysis*, 17th ed.; Association of Official Analytical Chemists: Arlington, VA, USA, 2002.
29. Pearson, D.; Egan, H.; Sawyer, R.; Kirk, R.S. *Chemical Analysis of Foods, Pearson's Chemical Analysis of Foods*, 8th ed.; Churchill Livingstone Publishers: Edinburg, UK, 1981; p. 24.
30. Association of Official Analytical Chemists (AOAC) International. *Official Methods of Analysis*, 17th ed.; Association of Official Analytical Chemists: Gaithersburg, MD, USA, 1998.
31. Association of Official Analytical Chemists (AOAC) International. *Official Methods of Analysis*, 14th ed.; Association of Official Analytical Chemists: Arlington, VA, USA, 1984.
32. Hamaker, B.R.; Kirleis, A.W.; Butler, L.G.; Axtell, J.D.; Mertz, E.T. Improving the invitro protein digestibility of sorghum with reducing agents. *Proc. Natl. Acad. Sci. USA* **1987**, *84*, 626–628. [CrossRef]
33. American Association of Cereal Chemists (AACC). *Approved methods of the American Association of Cereal Chemists*, 10th ed.; AACC International Press: St. Paul, MN, USA, 2000.
34. Sosulski, F.W.; Ganat, M.D.; Slinkard, A.F. Functional properties of ten legumes flours. *Can. Inst. Food Sci. Technol.* **1976**, *9*, 66–74. [CrossRef]
35. Okaka, J.C.; Potter, N.N. Functional and storage properties of cowpea-wheat flour blends in bread making. *J. Food Sci.* **1977**, *42*, 828–833. [CrossRef]
36. Koffi-Niaba, P.V.; Kouakoul, B.; Gildas, G.K.; Thierry, A.; N'golo, K.; Dago, G. Quality characteristics of biscuits made from sorghum and defatted *Macrotermes subhyalinus*. *Int. J. Biosci.* **2013**, *3*, 58–69.
37. Kinyuru, J.N.; Kenji, G.M.; Njoroge, M.S. Process development, nutrition and sensory qualities of wheat buns enriched with edible termites (*Macrotermes subhylanus*) from lake victoria region, Kenya. *Afr. J. Food Agric. Nutr. Dev.* **2009**, *9*, 1739–1750. [CrossRef]
38. Omoba, O.S.; Omogbemile, A. Physicochemical properties of sorghum biscuits enriched with defatted soy flour. *Br. J. Appl. Sci. Technol.* **2014**, *3*, 1246–1256. [CrossRef]
39. Bukkens, S.G.F. Insects in the human diet: Nutritional aspects. In *Ecological Implications of Mini Livestock*; Role of rodents, frogs, snails, and insects for sustainable development; Paoletti, M.G., Ed.; Science Publishers: Enfield, NH, USA, 2005.
40. Bukkens, S.G.F. The nutritional value of edible insects. *Ecol. Food Nutr.* **1997**, *36*, 287–319. [CrossRef]
41. Rehinan, Z.; Rashid, M.; Shah, W.H. Insoluble dietary fibre components of food legumes as affected by soaking and cooking processes. *Food Chem.* **2004**, *85*, 245–249. [CrossRef]
42. Esan, O.T.; Adesanwo, A.S.; Dauda, O.A.; Arise, A.K.; Sotunde, A.J. Chemical composition and sensory qualities of wheat-sorghum date cookies. *Croat. J. Food Technol. Biotechnol. Nutr.* **2017**, *12*, 71–76.
43. Yhoungaree, J.; Puwastien, P.; Attig, G.A. Edible insects in Thailand: An unconventional protein source? *Ecol. Food Nutr.* **1997**, *36*, 133–149. [CrossRef]
44. Bauserman, M.; Lokangaka, A.; Gado, J. A cluster-randomized trial determining the efficacy of caterpillar cereal as a locally available and sustainable complementary food to prevent stunting and anaemia. *Public Health Nutr.* **2015**, *18*, 1785–1792. [CrossRef]
45. Hongo, T.A. Micronutrient malnutrition in Kenya. *Afr. J. Food Agric. Nutr. Dev.* **2003**, *3*, 1–11.
46. Michaelsen, K.F.; Hoppe, C.; Roos, N.; Kaestel, P.; Stougaard, M.; Lauritzen, L.; Molgaard, C.; Girma, T.; Friis, H. Choice of foods and ingredients for moderately malnourished children 6 months to 5 years of age. *Food Nutr. Bull.* **2009**, *30*, 343–404. [CrossRef] [PubMed]
47. Ayieko, M.; Kinyuru, J.; Ndong'a, M.; Kenji, G. Nutritional value and consumption of black ants (*Carebara vidua* Smith) from the Lake Victoria region in Kenya. *Adv. J. Food Sci. Technol.* **2012**, *4*, 39–45.
48. Christensen, D.L.; Orech, F.O.; Mungai, M.N.; Larsen, T.; Friss, H.; Aagaard-Hansen, J. Entomophagy among the luo of Kenya: A potential mineral source? *Int. J. Food Sci. Nutr.* **2006**, *57*, 198–203. [CrossRef] [PubMed]
49. Paoletti, M.G.; Buscardo, E.; Vanderjagt, D.J.; Pastuszyn, A.; Pizzoferato, L.; Huang, Y.S.; Chung, L.T.; Glew, R.H.; Millson, M.; Cerda, H. Nutrient content of termites (Syntermes soldiers) consumed by Makirtare Amerindians of the Alto Orinoco of Venezuela. *Ecol. Food Nutr.* **2003**, *42*, 173–187. [CrossRef]

50. Chakravorty, J.; Ghosh, S.; Megu, K.; Jung, C. Nutritional and anti-nutritional composition of *Oecophylla smaragdina* (Hymenoptera: Formicidae) and *Odontotermes* sp. (Isoptera: Termitidae): Two preferred edible insects of Arunachal Pradesh, India. *J. Asia-Pac. Entomol.* **2016**, *19*, 711–720. [CrossRef]
51. Olaleye, A.A.; Adubiaro, H.O.; Olatoye, R.A. Amino acids profile of bee brood, soldier termite, snout beetle larva, silkworm larva and pupa: Nutritional implications. *Adv. Anal. Chem.* **2017**, *7*, 31–38.
52. Food and Agriculture Organization of the United Nations (FAO). Edible Insects: Future Prospects for Food and Feed Security. 2013. Available online: http://www.fao.org/3/i3253e/i3253e06.pdf (accessed on 11 July 2019).
53. Ogungbenle, H.N.; Olaleye, A.A.; Ayeni, K.E. Amino acid composition of three fresh water fish samples commonly found in South Western states of Nigeria. *Elixir Food Sci.* **2013**, *62*, 17411–17415.
54. Ekpo, K.E.; Onigbinde, A.O. Nutritional potentials of the larva of *Rhynchophorus phoenicis*. *Pak. J. Nutr.* **2005**, *4*, 287–290.
55. Food and Agriculture Organization/United Nations/World Health Organization (FAO/UN/WHO). Protein and Amino Acid Requirements in Human Nutrition. 2007. Available online: https://apps.who.int/iris/bitstream/handle/10665/43411/WHO_TRS_935_eng.pdf?ua=1 (accessed on 28 April 2019).
56. Pencharz, P.B.; Elango, R.; Ball, R.O. An Approach to defining the upper safe limits of amino acid intake. *J. Nutr.* **2008**, *138*, 1996S–2002S. [CrossRef]
57. Ramos-Elorduy, J.; Pino Moreno, J.M.; Prado, E.E.; Perez, M.A.; Otero, J.L. Nutritional value of edible insects from the state of Oaxaca, Mexico. *J. Food Compos. Anal.* **1997**, *10*, 142–157. [CrossRef]
58. Ajayi, O.E. Biochemical analyses and nutritional content of four castes of subterranean termites, *Macrotermes subhyalinus* (Rambur) (Isoptera: Termitidae): Differences in digestibility and anti-nutrient contents among castes. *Int. J. Biol.* **2012**, *4*, 54–59. [CrossRef]
59. Maache-Rezzoug, Z.; Bouvier, J.; Allaf, K.; Patras, C. Effect of principle ingredients and rheological behavior of biscuit dough on quality of biscuits. *J. Food Eng.* **1998**, *35*, 23–42. [CrossRef]
60. Stauffer, C.E. Principles of dough formation. In *Technology of Bread Making*; Cauvain, S.P., Young, L.S., Eds.; Springer Science & Business Media: New York, NY, USA, 2007; pp. 299–322.
61. Goeseart, H.; Brijs, K.; Veraverbeke, W.S.; Courtin, C.M.; Grubruers, K.; Delcour, J.A. Wheat flour constituents: How they impact on bread quality and how to impact their functionality. *Trends Food Sci. Technol.* **2005**, *16*, 12–30. [CrossRef]
62. Saleem, Q.; Wildman, R.D.; Huntley, J.M.; Whitworth, M.B. Material properties of semi-sweet biscuits for finite element modelling of biscuit cracking. *J. Food Eng.* **2005**, *68*, 19–32. [CrossRef]
63. Chiremba, C.; Taylor, J.R.N.; Duodu, K.G. Phenolic content, antioxidant activity and consumer acceptability of sorghum cookies. *Cereal Chem.* **2009**, *86*, 590–594. [CrossRef]
64. Brennan, C.S.; Samyue, E. Evaluation of starch degradation and textural characteristics of dietary fibre enriched biscuits. *Int. J. Food Prop.* **2004**, *7*, 647–657. [CrossRef]
65. Awobusuyi, T.D.; Pillay, K.; Siwela, M. Consumer Acceptance of Biscuits Supplemented with a Sorghum–Insect Meal. *Nutrients* **2020**, *12*, 895. [CrossRef]
66. Mohamed, A.; Rayas-Duarte, P.; Xu, J. Hard red spring wheat /C-TRIM 20 bread formulation, processing and texture analysis. *Food Chem.* **2008**, *107*, 516–524. [CrossRef]
67. Singh, U. Functional properties of grain legume flours. *J. Food Sci.* **2001**, *38*, 191–199.
68. Okpala, L.; Okoli, E.; Udensi, E. Physicochemical and sensory properties of cookies made from blends of germinated pigeon pea, fermented sorghum, and cocoyam flours. *Food Sci. Nutr.* **2013**, *1*, 8–14. [CrossRef]
69. Abu, J.O.; Muller, K.; Duodu, K.G.; Minnaar, A. Gamma irradiation of cowpea (*Vigna unguiculata*) flours and pastes: Effects on functional, thermal and molecular properties of isolated proteins. *Food Chem.* **2006**, *95*, 138–147. [CrossRef]
70. Aremu, M.O.; Olaofe, O.; Akintayo, E.T. Functional properties of some Nigerian varieties of legume seed flours and flour concentration effect on foaming and gelation properties. *J. Food Technol.* **2007**, *5*, 109–115.
71. Lawal, O.S.; Adebowale, K.O. Effect of acetylation and succinylation on solubility profile, water absorption capacity, oil absorption Capacit and emulsifying properties of muncuna bean (*Mucuna pruiens*) protein concentrate. *Nahr. Food* **2004**, *48*, 129–136. [CrossRef]
72. Jitngarmkusol, S.; Hongsuwankul, J.; Tananuwong, K. Chemical compositions, functional properties, and microstructure of defatted macadamia flours. *Food Chem.* **2008**, *110*, 23–30. [CrossRef]

73. Falade, K.O.; Olugbuyi, A.O. Effects of maturity and drying methods on the physico-chemical and reconstitution properties of plantain flour. *Int. J. Food Sci. Technol.* **2010**, *45*, 170–178. [CrossRef]
74. Ubbor, S.C.; Akobundu, E.N.T. Quality characteristics of cookies from composite flours of watermelon seed, cassava and wheat. *Pak. J. Nutr.* **2009**, *8*, 1097–1102. [CrossRef]

Article

Determination of the Sensory Characteristics of Traditional and Novel Fortified Blended Foods Used in Supplementary Feeding Programs

Sirichat Chanadang [1,2] and Edgar Chambers IV [1,*]

[1] Faculty of Agricultural Product Innovation and Technology, Srinakharinwirot University, Bangkok 10110, Thailand

[2] Department of Food, Nutrition, Dietetics and Health, College of Human Ecology, Kansas State University, Manhattan, KS 66506, USA

* Correspondence: eciv@ksu.edu

Received: 30 June 2019; Accepted: 15 July 2019; Published: 17 July 2019

Abstract: Despite the wide use of traditional non-extruded fortified blended foods (FBFs), such as corn soy blend plus (CSB+), in supplementary feeding programs, there is limited evidence of its effectiveness on improving nutritional outcomes and little information on actual sensory properties. Fifteen novel extruded FBFs were developed with variations in processing and ingredients in order to improve the quality of food aid products based on the Food Aid Quality Review (FAQR) recommendations. Descriptive sensory analysis was performed to determine the effects of the processing parameters and ingredients on the sensory properties of traditional and novel FBFs. The extrusion process affected the aroma and flavor of the tested products. Novel FBFs from the extrusion process had more pronounced toasted characteristics, probably because of the high temperature used during extrusion. The ingredient composition of the FBFs also had a significant impact on the sensory properties of the products. The addition of sugar to novel FBFs leads to a significant increase in sweetness, which could improve acceptance. The level of lipids in binary blends appeared to be mainly responsible for the bitterness of the product. In addition, legumes, which were a primary ingredient, contributed to the beany characteristics of the products. The higher amounts of legume used in the formulations led to beany characteristics that could be perceived from the products and could be a negative trait depending on consumers' prior use of legume-based products.

Keywords: fortified blended foods (FBFs); sensory; food aid; extrusion; cereal; legume; infant; child; porridge

---

1. Introduction

Food insecurity around the world is always increasing due to many causes, including growing populations, poverty, and natural disasters [1]. The State of Food Insecurity in the World, 2015 reported that approximately 795 million people in the world were undernourished in 2014–2016 [2]. Fortified blended foods (FBFs) were developed in the 1960s by the United States Agency for International Development (USAID, Washington, DC, USA) to provide a supplement for young children who suffered from moderate acute malnutrition in many developing countries around the world [3,4]. The most commonly distributed cereal based FBF by USAID is a corn-soy blend (CSB) which consists of corn (75–80%) as a source of carbohydrate and soy (20–25%) as a source of protein. Although FBFs form an important part of the food aid ration, there is limited evidence of their abilities in treating young children with malnutrition [3–5] and little information on their sensory properties.

The Food Aid Quality Review (FAQR) in 2011 by Webb et al. [6] recommended changing the formulation of existing FBFs in order to improve their nutritional quality. These recommendations included adding animal-source protein to promote linear growth of children, increasing fat content through the addition of vegetable oil, adding a flavor enhancer to formulations to improve the acceptability of FBFs, and upgrading micronutrient compositions in FBFs. In addition, the decortication of cereals and legumes used in FBFs is recommended in order to reduce the fiber content and eliminate phenolic compounds that can reduce the energy intake and lower protein digestibility and mineral absorption [5].

Another recommendation from Webb et al. [6] was to increase the solids content of FBFs to 20% to increase the nutrient content. However, porridge prepared from the current FBFs at this concentration is too viscous for consumption by infants and young children [7]. Mothers normally add more water into porridge to make it more drinkable before feeding it to their children, which results in a low nutritional value and energy density [5]. Extrusion cooking of starchy ingredients for FBFs can result in less viscous cooked porridge, making them more ideal for delivering higher density energy meals at lower viscosities for infants and young children [8]. Extruded products also require short cooking times and less fuel [5], which makes them more valuable to people with limited time and energy sources.

Webb et al. [6] also encouraged the exploration of new grains or legumes that could be used beyond the traditional FBFs, including CSBs and wheat-soy blends (WSB). Corn has been used as the main staple for current FBFs because it is a good source of starch, plant-based protein, dietary fiber, B vitamins, and is available in bulk for the food aid program [9,10]. However, the high demand of corn for many uses, especially for fuel production, makes the prices increase [11] and this directly affects food aid commodities. Heat-treated soy in full fat form or defatted flour is primarily used as a source of protein in FBFs. However, soy may contain high levels of anti-nutritional factors, such as phytate and phytoestrogen, which have unknown long-tern health effects [9].

Sorghum has been examined as a potential alternative ingredient in FBFs with a number of advantages over corn, including higher levels of protein, fat, and some micronutrients when processed properly [12,13]. Cowpea has also been considered as an alternative legume that can be used in FBFs because of the high levels of protein, energy, and other nutrients [14]. Sorghum and cowpea are cultivated and consumed as part of human foods in many parts of developing countries [14,15]. Therefore, populations in these areas should be familiar with the flavor of sorghum and cowpea, which makes these good candidates for use in FBFs. Moreover, both sorghum and cowpea are mostly non-genetically modified organism (GMO) crops, which allows them to be used in many countries that have banned the use of GMO products.

Recent work has shown that various types of extruded FBFs made with sorghum or corn and cowpea or soy are at least as preferred as CSB+ by children in Tanzania [16]. In addition, data has shown that the shelf-life of such products is generally 24 months or greater [17], far exceeding the required shelf-life for such products.

Based on the recommendations of the FAQR, fifteen newly formulated, extruded FBFs with varied processing techniques and ingredients were developed. The objective of this study was to determine the effects of processing techniques (extrusion vs. non-extrusion, milling type, decortication process, and the step of adding antioxidant to the product) and ingredients on the sensory properties of traditional and novel FBFs.

2. Materials and Methods

2.1. Samples

Fifteen novel extruded FBFs and one current-non extruded FBF were used in this study. These products were potential variations in a large feeding trial in Tanzania to test sorghum cowpea blends against other products [18].

2.1.1. Novel Extruded Fortified Blended Foods

Fifteen possible extruded FBFs varied in milling type, decortication process, the order of adding antioxidant to the blends, and ingredients are shown in Table 1.

The whole grains—sorghum varieties V1 (Fontanelle 4525), V2 (738Y), V3 (217X Burgundy) (Nu Life Market, Scott City, KS, USA), and corn (Agronomy Foundation Seed, Kansas State University, Manhattan, KS, USA) were used for pilot milling at Hall Ross Flour Mill (Kansas State University, Manhattan, KS, USA) to obtain whole and decorticated flours. Soybeans (Kansas River Valley Experiment Field, Kansas State University, Manhattan, KS, USA) and cowpea grains (LPD Enterprises LLC, Olathe, KS, USA) were milled at Hall Ross Flour Mill (Kansas State University, Manhattan, KS, USA). Commercially milled whole and decorticated sorghum flour variety V1 were obtained from Nu Life Market, Scott City, KS, USA. Commercially milled degermed corn flour and whole corn flour were purchased from Agricor, Marion, Indiana, USA. Defatted soy flour was purchased from American Natural Soy, Cherokee, IA, USA.

The cereal/legume flours were blended. For seven sorghum-cowpea blends, one of the three sorghum varieties of flour, whole or decorticated, was mixed with cowpea flour. For five sorghum-soy blends, sorghum variety V1, whole or decorticated, was mixed with low fat (1.85%), medium fat (6.94%), or full fat (16.93%) soybean flour. For three corn-soy blends, whole or degermed corn flour with medium fat and full fat soybean flour were used. All binary blends were extruded on a single screw extruder X-20 (Wenger Manufacturing Inc., Sabetha, KS, USA) at a screw speed ranging from 500–550 rpm with 18–24% process moisture. The extrudates were cut at the die exit with a face-mounted five blade rotary knife, and dried in a Wenger double pass Dryer/Cooler (Series 4800, Wenger Manufacturing Inc., Sabetha, KS, USA) at 104 °C for 10 min.

The dried extrudates were ground using a Schutte Buffalo Hammer mill (Buffalo, NY, USA). The ground binary blends were then mixed with sugar (Domino Foods, Inc., Yonkers, NY, USA), whey protein concentrate WPC80 (Davisco Foods International, Inc., Eden Prarie, MN, USA), antioxidant (BHA, butylated hydroxyanisole and BHT, butylated hydroxytoluene), vitamins and minerals (Research Products Company, Salina, KS, USA), and non-GMO soybean oil (Zeeland Farm Services, Inc., Zeeland, MI, USA). The composition of all blends is shown in Table 2.

2.1.2. Current Non-Extruded Fortified Blended Food

Corn soy blend plus (CSB+) was produced by Bunge Milling (St. Louis, MO, USA) according to the USDA commodity requirements [19] (Table 2).

2.2. *Sample Preparation*

All products were prepared into porridges, which are the most common dishes prepared from cereal-based commodities for children in developing countries [20–22], with 20% solids content according to the recommendation from [6].

A weighted FBF flour (200 g) was mixed with cold water (400 mL) to prevent the formation of lumps. The mixture was then added to boiling water (400 mL), brought back to a boil, cooked with continuous stirring with a wooden spoon for 2 min for extruded FBFs and 10 min for non-extruded FBFs. The sample was removed from the stovetop and cooled to a temperature of 45 °C, which is the typical consumption temperature by infant and young children [23].

Table 1. List of processing and ingredients used for each extruded fortified blended food (FBF).

Treatment		Product Code [1]	Cereal			Legume
			Cereal Type	Variety	Milling Type	
1	Sorghum-Cowpea blend	SCB-V1 com	Sorghum-Decorticated	White-Fontanelle 4525	Commercial	Cowpea
2		SCB-V1	Sorghum-Decorticated	White-Fontanelle 4525	Pilot	Cowpea
3		SCB-V2	Sorghum-Decorticated	White-738Y	Pilot	Cowpea
4		SCB-V3	Sorghum-Decorticated	Red-217X Burgundy	Pilot	Cowpea
5		WSCB-V1	Sorghum-Whole	White-Fontanelle 4525	Pilot	Cowpea
6		WSCB-V2	Sorghum-Whole	White-738Y	Pilot	Cowpea
7		WSCB-V3	Sorghum-Whole	Red-217X Burgundy	Pilot	Cowpea
8	Sorghum-Soy blend	SS'B-V1 com	Sorghum-Decorticated	White-Fontanelle 4525	Commercial	Soybean—High Fat
9		WSSB-V1	Sorghum-Whole	White-Fontanelle 4525	Pilot	Soybean—Low Fat
10		WSS'B-V1 com	Sorghum-Whole	White-Fontanelle 4525	Commercial	Soybean—High Fat
11		WSS'B-V1 com (pre-anti)	Sorghum-Whole	White-Fontanelle 4525	Commercial	Soybean—High Fat
12		WSS"B-V1	Sorghum-Whole	White-Fontanelle 4525	Pilot	Soybean—Full Fat
13	Corn-Soy blend	CS'B com	Corn-Degermed		Commercial	Soybean—High Fat
14		WCS'B com	Corn-Whole		Commercial	Soybean—High Fat
15		WCS"B	Corn-Whole		Pilot	Soybean—Full Fat

[1] W = Whole, first S = Sorghum flour, first C = Degermed corn flour, second S = Low-fat soy flour, S' = Medium-fat soy flour, S" = Full-fat soy flour, second C = Cowpea flour, V1 and V2 = White variety of sorghum, V3 = Red variety of sorghum, com = Commercial milling, (pre-anti) = Antioxidant had been added to the binary blend before extrusion process.

Table 2. Composition of extruded FBFs and non-extruded FBFs.

Ingredients (%)	Extruded FBFs [1]			Non-Extruded FBF
	Sorghum-Cowpea Blends (SCB)	Sorghum-Soy Blends (SSB)	Corn-Soy Blends (CSB)	Corn Soy Blend Plus (CSB+)
Sorghum flour	24.7	47.6		
Cowpea flour	38.6			
Corn flour			48.1	
Corn (White or Yellow)				78.5
Whole soybeans				20.0
Soy flour		15.7	15.2	
Sugar	15.0	15.0	15.0	
Whey Protein Concentrate (WPC80)	9.5	9.5	9.5	
Soybean oil	9.0	9.0	9.0	
Vitamin & Mineral Premix	3.1	3.1	3.1	
Antioxidant [2]	0.1	0.1	0.1	
Vitamin/Mineral				0.2
Tri-Calcium Phosphate				1.2
Potassium Chloride				0.2

[1] For extruded FBFs with full-fat soy, WPC80 was increased from 9.5 to 13.0%, and soybean oil was decreased from 9 to 5.5%. [2] Antioxidant was a mixture of 50% butylated hydroxyanisole (BHA) and 50% butylated hydroxytoluene (BHT).

2.3. Descriptive Sensory Analysis

Descriptive sensory analysis was conducted by six highly-trained panelists at the Center for Sensory Analysis and Consumer Behavior, Manhattan, Kansas USA. All of these panelists had completed 120 h of general descriptive analysis panel training, and had over 2000 h of evaluation experience with a wide array of food products, including cereal-based products.

Sixteen sensory attributes, including 6 aroma and 10 flavor, were evaluated in all samples (Table 3). Some of the same attributes were used in Chanadang et al. [23].

Fifty grams of each prepared porridge was served in a 4 oz styrofoam cup (Dart container corporation, Mason, MI, USA) and labeled with a three-digit code for each panelist. All samples were evaluated on a numerical scale of 0–15 with 0.5 increments, where 0 represents none and 15 represents extremely high. The samples were prepared and evaluated in triplicate in a randomized order.

2.4. Data Analysis

Sixteen sensory attributes were evaluated for all porridge samples, however, panelists did not detect rancid or painty characteristics in any samples. Therefore, twelve sensory attributes, besides rancid and painty characteristics, were reported and analyzed in this study.

Data for each sensory attribute was analyzed by a one-way ANOVA mixed effect model (SAS version 9.4, The SAS Institute Inc., Cary, NC, USA) using PROC GLIMMIX to determine significant differences ($p \leq 0.05$) among porridge samples. Tukey's HSD test was used at the 5% level of significance to locate significant effects of the sample on each sensory property. Principal component analysis (PCA) was performed in order to visualize the relationship among sensory attributes and samples using Unscrambler® X 10.5 (Camo, Magnolia, TX, USA).

Table 3. Aroma and flavor attributes, definitions, and references for descriptive analysis of porridge prepared from FBFs.

Attribute	Definition	Reference $
Aroma		
Overall Grain *	A general term used to describe the aromatics which includes musty, dusty, slightly brown, slightly sweet and is associated with harvested grains and dry grain stems.	Cereal Mix(dry) = 7.5. Preparation: Mix $\frac{1}{2}$ cup of each General Mills Rice Chex, Wheaties and Quaker Quick Oats. Put in a blender and "pulse" blend into small particles. Serve 2 Tablespoon in a 12 oz brandy snifter, covered with a watch glass.
Toasted *	A moderately browned/baked impression.	Crushed Post Shredded wheat = 2.5. Preparation: Crush $\frac{1}{4}$ cup of Shredded wheat and served in a 12 oz brandy snifter, covered with a watch glass. Crushed General Mills Cheerios = 7.0. Preparation: Crush $\frac{1}{4}$ cup of Cheerios and serve in a 12 oz brandy snifter, covered with a watch glass.
Beany	Aromatic characteristic of beans and bean products, includes musty/earthy, musty/dusty, sour aromatics, bitter aromatics, starchy and green/pea pod, nutty or brown.	Cooked Soy Bean = 4.0. Preparation: Soak $\frac{1}{2}$ cup of soy bean overnight and boil the bean 2.5 h. Serve 1 table spoon of cooked soy bean in a 12 oz brandy snifter, covered with a watch glass. Bush Pinto Beans (canned) = 7.0. Preparation: Drain beans and rinse with de-ionized water Place one table spoon in a 12 oz brandy snifter, covered with a watch glass.
Musty Overall *	A combination of one or more aromatic impressions characterized to some degree as being somewhat dry, dusty, damp, earthy, stale, sour, or moldy. If identifiable, attribute will be listed.	1,2,4Trimethoxybenzene 50,000 ppm = 4.0. Preparation: Dip an Orlandi Perfumer Strip #27995 2.2 cm (second marking line) into solution and place dipped end up in a Fisherbrand Disposable Borosilicate Glass Tubes with Threaded End (15 × 150 mm) cap.
Rancid	A somewhat heavy aromatic characteristic of old, oxidized, decomposing fat and oil. The aromatics may include painty, varnish, or fishy.	Microwaved Wesson vegetable oil (4 min at high) = 2.5. Preparation: Microwave 1 $\frac{1}{2}$ cups oil on high power for 4 min. Let cool and serve $\frac{1}{4}$ cup in a 12 oz brandy snifter covered with a watch glass. Microwaved Wesson vegetable oil (5 min at high) = 5.0. Preparation: Microwave 1 $\frac{1}{2}$ cups oil on high power for 5 min. Let cool and serve $\frac{1}{4}$ cup in a 12 oz brandy snifter covered with a watch glass.
Painty	The aromatic associated with rancid oil and fat, typically in the late stages of rancidity.	Microwaved Wesson vegetable oil (4 min at high) = 2.5. Preparation: Microwave 1 $\frac{1}{2}$ cups oil on high power for 4 min. Let cool and pour into 1 oz cups. Serve covered. Microwaved Wesson vegetable oil (5 min at high) = 4.5. Preparation: Microwave 1 $\frac{1}{2}$ cups oil on high power for 5 min. Let cool and pour into 1 oz cups.
Flavor		
Overall Grain *	A general term used to describe the light dusty/musty aromatics associated with grains such as corn, wheat, bran, rice, oats and soybean.	Cereal Mix (dry) = 8.0. Preparation: Mix $\frac{1}{2}$ cup of each General Mills Rice Chex, Wheaties and Quaker Quick Oats. Put in a blender and "pulse" blend into small particles. Serve in 1 oz cup.

Table 3. *Cont.*

Attribute	Definition	Reference $
Toasted *	A moderately browned/baked impression.	Post Shredded Wheat (Spoon size) = 3.5. Preparation: Serve in 3.25 oz cup.General Mills Cheerios = 7.0. Preparation: Serve in 3.25 oz cup.
Beany	Aromatic characteristic of beans and bean products, includes musty/earthy, musty/dusty, sour aromatics, bitter aromatics, starchy and green/pea pod, nutty or brown.	Cooked Soy Bean = 4.0. Preparation: Soak $\frac{1}{2}$ cup of soy bean overnight and boil the bean 2.5 h. Serve in 1 oz cup. Bush Pinto Beans (canned) = 7.5. Preparation: Drain beans and rinse with de-ionized water Serve in 1 oz cup.
Musty *	Aromatics associated with wet grain and damp earth.	Cooked American Beauty elbow macaroni = 5.0. Preparation: Bring 3 cups water to a rapid boil. Add 1 cup pasta and stir, returning to a rapid boil. Cook 6 min, stirring occasionally. Drain and place into 3.25 oz cups.
Rancid	A somewhat heavy aromatic characteristic of old, oxidized, decomposing fat and oil. The aromatics may include painty, varnish, or fishy.	Microwaved Wesson vegetable oil (4 min at high) = 3.0. Preparation: Microwave 1 $\frac{1}{2}$ cups oil on high power for 4 min. Let cool and serve in 1 oz cup. Microwaved Wesson vegetable oil (5 min at high) = 5.0. Preparation: Microwave 1 $\frac{1}{2}$ cups oil on high power for 5 min. Let cool and serve in 1 oz cup.
Painty	The aromatic associated with rancid oil and fat, typically in the late stages of rancidity.	Microwaved Wesson vegetable oil (4 min at high) = 0.0. Preparation: Microwave 1 $\frac{1}{2}$ cups oil on high power for 4 min. Let cool and serve in 1 oz cup. Microwaved Wesson vegetable oil (5 min at high) = 3.0. Preparation: Microwave 1 $\frac{1}{2}$ cups oil on high power for 5 min. Let cool and serve in 1 oz cup.
Sweet *	A fundamental taste factor of which sucrose is typical.	2% Sucrose Solution = 2.0 4% Sucrose Solution = 4.0
Salt *	Fundamental taste factor of which sodium chloride is typical.	0.15% Sodium Chloride Solution = 1.5 0.20% Sodium Chloride Solution = 2.5
Bitter *	The fundamental taste factor associated with a caffeine solution.	0.01% Caffeine Solution = 2.0 0.02% Caffeine Solution = 3.5 0.035% Caffeine Solution = 5.0 0.05% Caffeine Solution = 6.5
Astringent *	The drying, puckering sensation on the tongue and other mouth surfaces.	0.050% alum solution = 2.5 0.100% alum solution = 5.0

$ 0 to 15 point numeric scale with 0.5 increments was used to rate the intensities of the sample and references. * From Chanadang and others (2016).

3. Results and Discussion

The results showed that six out of twelve sensory attributes were significantly different among porridge samples ($p \leq 0.05$), including toasted and beany aroma and flavor, sweetness, and bitterness (Table 4).

Porridges prepared from novel extruded FBFs appeared to be higher in toasted aroma and flavor than non-extruded FBF (CSB+), although, not all novel extruded FBFs were significantly different from CSB+ in this sensory characteristic ($p > 0.05$). The high temperature used in the extrusion process might be the main reason for the increased toasted characteristic in extruded FBFs. Extrusion cooking of cereal normally involves thermally induced reactions, including the Maillard reaction, that could generate chemical compounds that correspond to a desirable aroma and flavor of the products [24,25], including such aspects as "toasted" sensory properties. Parker et al. [26] reported that extruded cereal samples with high levels of Maillard reaction products, such as pyrazines and sulfur-containing alicyclic compounds, were generally described as having a desirable toasted or roasted cereal aroma and flavor. Besides the extrusion process, other processing parameters, including types of milling, decortication process, and the step of adding antioxidant to the blends, did not show significant effects on sensory properties of FBFs in this study.

The composition of FBFs seemed to be another important factor that affected the sensory properties of the products. Porridges prepared from sorghum-cowpea blends, especially WSCB-V3, had significantly higher intensity in beany aroma and flavor ($p \leq 0.05$) than the ones prepared from sorghum-soy and corn-soy blends. Beany characteristics are often found in legume-containing products and are attributed to the action of the lipoxygenase enzyme, which catalyzes the lipid oxidation of linolenic and linoleic fatty acids [27,28]. Since all of the products in this study contained legumes (either soybeans or cowpea), the difference in intensity in beany characteristics among products was primarily due to the amount of legume used in each blend. This probably explains why sorghum-cowpea blends with higher amounts of legume (38.6% cowpea) were higher in beany aroma and flavor.

The variety of sorghum used in FBFs might be another factor that affected the beany property of the products. The blend containing whole red sorghum flour (WSCB-V3) was significantly higher in beany flavor than the rest of the FBFs, except for the one that contained decorticated red sorghum flour (SCB-V3). Vara-Ubol et al. [29] indicated that beany was considered as a combination of attributes, including musty/dusty, musty/earthy, sour aromatics, and characterizing attributes such as green/pea pod, nutty or brown. Red sorghum varieties were reported to have higher dusty flavor [30] and porridges made with red sorghum were also reported to have higher overall flavor intensity [31]. FBFs with red sorghum variety in this study might be higher in dusty flavor or overall intensity, and that resulted in an increased intensity of beany characteristics.

Porridges prepared from various FBFs were also significantly different in sweetness ($p \leq 0.05$). As expected, novel extruded FBFs with the addition of 15% sugar were significantly higher in sweetness than the traditional non-extruded FBF (CSB+) ($p \leq 0.05$). The addition of sugar into the FBFs formulation was not only to provide energy, but could also to increase the palatability and consumption of the products [6]. Iuel-Brockdorf et al. [32] also found that products with a sweeter flavor received better ratings in terms of child and caregiver acceptability.

Salt was significantly different among the FBFs porridges ($p \leq 0.05$), however, it was only a small difference (lower than 0.5 points on a 15 point scale). The higher intensity of salt in novel extruded FBFs was probably due to the higher amount of vitamin and mineral premix that had been added into the formulation. Gilbertson et al. [33] indicated that the taste system plays important roles in nutrient identification and salty taste reflects the recognition of minerals in foods. The study by Teillet et al. [34] also found that a more salty taste was found in waters with higher mineral contents.

Table 4. Mean scores [1] (standard error) of sensory attributes for porridges prepared from FBFs.

Treatment [2]	Overall Grain (a) [3]	Toasted (a)	Beany (a)	Musty Overall (a)	Overall Grain (f)	Toasted (f)	Beany (f)	Musty (f)	Sweet (f)	Salt (f)	Astringent (f)	Bitter (f)
SCB-V1 com	7.14 (0.07)	3.53 ^{ab4} (0.18)	3.28 ^{abc} (0.19)	3.36 (0.16)	7.36 (0.07)	2.97 ^{abc} (0.20)	3.58 ^{bcd} (0.15)	4.47 (0.15)	2.11 ^{a} (0.16)	1.42 ^{ab} (0.15)	2.64 (0.17)	2.89 ^{d} (0.18)
SCB-V1	7.17 (0.07)	3.89 ^{ab} (0.29)	3.28 ^{abc} (0.24)	3.11 (0.21)	7.36 (0.10)	3.28 ^{abc} (0.27)	3.64 ^{bc} (0.24)	4.36 (0.22)	2.03 ^{a} (0.15)	1.39 ^{ab} (0.15)	2.81 (0.25)	3.17 ^{bcd} (0.18)
SCB-V2	7.25 (0.08)	4.47 ^{a} (0.23)	3.19 ^{abc} (0.13)	3.17 (0.18)	7.42 (0.09)	3.22^{abc} (0.16)	3.64^{bc} (0.16)	4.33 (0.21)	2.00 ^{a} (0.19)	1.31 ^{ab} (0.14)	2.81 (0.15)	3.08 ^{cd} (0.20)
SCB-V3	7.22 (0.07)	4.53 ^{a} (0.30)	3.36 ^{ab} (0.19)	2.94 (0.19)	7.36 (0.09)	3.75 ^{a} (0.31)	4.19 ^{ab} (0.14)	4.69 (0.22)	1.97 ^{a} (0.12)	1.58 ^{a} (0.20)	2.68 (0.19)	3.31 ^{bcd} (0.13)
WSCB-V1	7.11 (0.08)	4.28 ^{ab} (0.18)	3.25 ^{abc} (0.19)	3.19 (0.18)	7.39 (0.08)	3.50 ^{ab} (0.23)	3.50 ^{bcde} (0.17)	4.58 (0.23)	2.00 ^{a} (0.10)	1.58^{a} (0.15)	2.78 (0.18)	3.28 ^{bcd} (0.20)
WSCB-V2	7.22 (0.07)	3.83 ^{ab} (0.24)	3.14 ^{abc} (0.18)	3.06 (0.18)	7.44 (0.08)	3.11 ^{abc} (0.23)	3.64 ^{bc} (0.18)	4.75 (0.22)	2.03 ^{a} (0.12)	1.50 ^{ab} (0.16)	2.72 (0.19)	2.97 ^{d} (0.19)
WSCB-V3	7.19 (0.07)	3.67 ^{ab} (0.23)	3.89 ^{a} (0.17)	3.44 (0.21)	7.33 (0.08)	3.33 ^{abc} (0.27)	4.44 ^{a} (0.19)	4.39 (0.24)	2.06 ^{a} (0.14)	1.47 ^{ab} (0.20)	2.83 (0.23)	3.36 ^{bcd} (0.20)
SS′B-V1 com	6.92 (0.10)	3.56 ^{ab} (0.21)	2.72 ^{bc} (0.21)	3.47 (0.22)	7.17 (0.07)	2.75 ^{abc} (0.11)	3.19 ^{cde} (0.10)	4.75 (0.18)	1.89 ^{a} (0.14)	1.31 ^{ab} (0.13)	2.97 (0.19)	3.31 ^{bcd} (0.17)
WSSB-V1	6.92 (0.06)	2.97 ^{b} (0.14)	2.69 ^{bc} (0.21)	3.19 (0.20)	7.19 (0.08)	2.36 ^{c} (0.13)	3.39 ^{cde} (0.21)	4.94 (0.25)	1.97^{a} (0.17)	1.58 ^{a} (0.18)	2.75 (0.13)	3.47 ^{bcd} (0.14)
WSS′B-V1 com	7.03 (0.16)	3.72 ^{ab} (0.21)	2.61 ^{bc} (0.16)	3.22 (0.18)	7.14 (0.17)	2.69 ^{bc} (0.14)	3.11 ^{cde} (0.15)	4.69 (0.21)	2.17 ^{a} (0.18)	1.44 ^{ab} (0.18)	2.67 (0.17)	3.22 ^{bcd} (0.18)
WSS′B-V1 com (pre-anti)	7.06 (0.08)	3.58 ^{ab} (0.19)	2.75 ^{bc} (0.18)	3.36 (0.13)	7.19 (0.06)	3.00 ^{abc} (0.16)	3.28 ^{cde} (0.16)	4.72 (0.18)	1.94 ^{a} (0.15)	1.44 ^{ab} (0.15)	2.86 (0.18)	3.31 ^{bcd} (0.21)
WSS″B-V1	7.00 (0.07)	3.00 ^{b} (0.17)	2.56 ^{c} (0.18)	3.75 (0.20)	7.25 (0.09)	2.50 ^{bc} (0.16)	3.17 ^{cde} (0.18)	4.94 (0.21)	1.86 ^{a} (0.08)	1.64 ^{a} (0.18)	3.06 (0.19)	3.81 ^{abc} (0.19)
CS′B com	6.94 (0.09)	3.89 ^{ab} (0.22)	2.64 ^{bc} (0.11)	3.28 (0.18)	7.19 (0.12)	2.53 ^{bc} (0.14)	3.03^{cde} (0.14)	4.42 (0.27)	2.03 ^{a} (0.17)	1.47 ^{ab} (0.17)	2.72 (0.23)	3.58 ^{bcd} (0.17)
WCS′B com	7.08 (0.14)	4.22 ^{ab} (0.24)	2.58 ^{bc} (0.16)	3.22 (0.19)	7.17 (0.11)	3.19 ^{abc} (0.19)	2.89 ^{de} (0.10)	4.50 (0.19)	2.11 ^{a} (0.19)	1.67 ^{a} (0.19)	2.67 (0.22)	3.89 ^{ab} (0.20)
WCS″B	7.03 (0.10)	4.50 ^{a} (0.24)	2.64 ^{bc} (0.18)	3.17 (0.14)	7.06 (0.07)	3.03 ^{abc} (0.14)	2.89 ^{de} (0.17)	4.75 (0.23)	1.89 ^{a} (0.21)	1.69 ^{a} (0.21)	3.08 (0.18)	4.53 ^{a} (0.20)
CSB+	7.33 (0.11)	2.97 ^{b} (0.20)	2.75 ^{bc} (0.19)	3.22 (0.15)	7.17 (0.11)	2.36 ^{c} (0.10)	2.83 ^{e} (0.15)	4.36 (0.18)	0.86 ^{b} (0.13)	1.14 ^{b} (0.13)	2.28 (0.14)	3.39 ^{cde} (0.18)

[1] Scores are based on a 0–15-point numeric scale with 0.5 increments (0 = none and 15 = extremely high). Each mean score intensity was calculated from six panelists with three replicates. [2] W = Whole, first S = Sorghum flour, first C = Degermed corn flour, second S = Low-fat soy flour, S′ = Medium-fat soy flour, S″ = Full-fat soy flour, second C = Cowpea flour, V1 and V2 = White variety of sorghum, V3 = Red variety of sorghum, com = Commercial milling, (pre-anti) = Antioxidant had been added to the binary blend before extrusion process. [3] (a) = Aroma, (f) = Flavor [4]. Average for each parameter with a different letter in the same column were significantly different ($p \leq 0.05$) between treatments.

Porridge prepared from binary blends with higher levels of lipids, e.g., whole corn with full-fat soybean blend (WCS″B), was significantly higher in bitterness than most of the FBFs porridges ($p \leq 0.05$). The high temperature used in the extrusion process could have accelerated the degradation of lipids, and the degraded lipids appeared to be associated with unpleasant flavors, such as astringent, bitter, and rancid [24,35,36]. WCS″B, which had high levels of lipid, was more likely to have a higher amount of degraded lipid after the extrusion process, and this could result in the higher bitter taste of the cooked porridge.

Principal component analysis (PCA) of twelve sensory attributes helped to visualize the differences among porridge samples (Figure 1). PC1 accounted for 39% of the variation, and seemed to differentiate among samples according to beany, toasted, grain, musty, and bitter attributes. PC2 accounted for 25% of the variation, and seemed to differentiate among samples according to flavor attributes, including astringency, sweetness, and saltiness. Current non-extruded FBF (CSB+) was separated from novel extruded FBFs due to the lower intensity in sweetness, saltiness, and astringency. Extruded corn-soy blends and extruded sorghum-soy blends were grouped together and had more pronounced bitter and musty attributes. As previously mentioned, the extruded products containing higher amount of lipids were more bitter ($p \leq 0.05$) because of the high possibility of having more degraded lipids. However, it must be noted that the lipids certainly were not degraded enough to produce marked changes in shelf-life [17]. Phenolic compounds, which can be found in sorghum, are responsible for the bitterness of many similar foods and may cause a negative effect on products' acceptability [35,37]. Therefore, the higher amount of sorghum (47.6% sorghum) used in sorghum-soy blend formulations was another reason that made those blends higher in bitter taste. This effect also was found in 20% solids FBFs made of sorghum without added sugar [38].

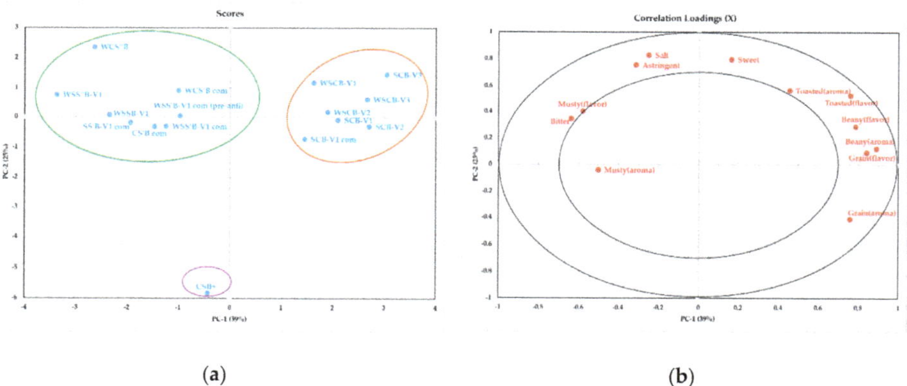

Figure 1. Principal component analysis of the porridges prepared from FBFs and sensory attributes (**a**) Score plot. (**b**) Correlation loading plot. For the FBFs, W = Whole, first S = Sorghum flour, first C = Degermed corn flour, second S = Low-fat soy flour, S′ = Medium-fat soy flour, S″ = Full-fat soy flour, second C = Cowpea flour, V1 and V2 = White variety of sorghum, V3 = Red variety of sorghum, com = Commercial milling, (pre-anti) = Antioxidant had been added to the binary blend before extrusion process. CSB+ represents the control sample (current non-extruded FBF).

All extruded sorghum-cowpea blends were grouped together. They were mainly characterized by toasted, grain, and beany attributes. The sorghum-cowpea binary blend that was used to make extruded sorghum-cowpea blends had lower levels of lipids compared to sorghum-soy and corn-soy binary blends [39]. Feng and Lee [40] reported that during extrusion, the lipid worked as a lubricant, and decreased the temperature in the extruder barrel. The lower amount of lipids in the sorghum-cowpea blend contributed to higher friction between the particles in the mix and the screw surface, and directly related to a higher temperature in the extruder barrel. The higher temperature during the extrusion process could probably generate higher levels of chemical compounds from the Maillard reaction, which were responsible for desirable attributes, such as cereal-like, toasted, or roasted aromas [24,26].

4. Conclusions

The results from this study clearly identified the effects of the extrusion process and ingredients used on the sensory properties of the products. Novel FBFs from the extrusion process had more pronounced toasted characteristics due to the higher temperature during extrusion. The type of milling, decortication process, and the step of adding antioxidant to the blends did not show effects on the sensory properties of FBFs in this study. Adding sugar and increasing the amount of vitamin-mineral premix in the novel FBFs formulation increased the sweetness and saltiness of the products, respectively, as expected, which is not surprising given that caregivers have been shown to add sugar to current unsweetended FBFs. The level of lipids in binary blends was mainly responsible for the bitterness of the product. In addition, legumes, such as soybeans and cowpeas, were the main ingredient that contributed to the beany characteristics of the products. The higher amount of legume used in the formulations, the more beany characteristics that could be perceived from the products.

Author Contributions: Conceptualization, E.C.; Data curation, E.C. and S.C.I.; Formal analysis, S.C.I. Funding acquisition, E.C.; Investigation, S.C.I.; Methodology, E.C. and S.C.I.; Project administration, E.C.; Supervision, S.C.I.; Writing-original draft, S.C.I.; Writing-review & editing, E.C. Both authors contributed to the design, conduct, and writing of the manuscript.

Funding: Agricultural Research Service: FFE-621-2012/033-00.

Acknowledgments: This work was funded, in part, by the USDA Foreign Agricultural Service under the Micronutrient Fortified Food Aid Products Pilot (MFFAPP) program, contract number #FFE-621-2012/033-00. The authors would like to thank the staff of the Center for Sensory Analysis and Consumer Behavior at Kansas State University who assisted in conducting this study. The authors wish to acknowledge Sajid Alavi and Brian Lindshield for their work in planning the larger project and for help in selecting the products that would be tested in this study.

Conflicts of Interest: The authors declare no conflict of interest.

References

1. Hill, R.V.; Smith, L.C.; Wiesmann, D.M.; Frankenberger, T.; Gulati, K.; Quabili, W.; Yohannes, Y. *The World's Most Deprived: Characteristics and Causes of Extreme Poverty and Hunger*; International Food Policy Research Institute: Washington, DC, USA, 2007; Volume 43.
2. FAO; IFAD; WFP. *The State of Food Insecurity in the World 2015. Meeting the 2015 International Hunger Targets: Taking Stock of Uneven Progress*; Food and Agriculture Organization Publications: Rome, Italy, 2015.
3. Perez-Exposito, A.B.; Klein, B.P. Impact of fortified blended food aid products on nutritional status of infants and young children in developing countries. *Nutr. Rev.* **2009**, *67*, 706–718. [CrossRef] [PubMed]
4. De Pee, S.; Bloem, M.W. Current and potential role of specially formulated foods and food supplements for preventing malnutrition among 6-to 23-month-old children and for treating moderate malnutrition among 6-to 59-month-old children. *Food Nutr. Bull.* **2009**, *30*, S434. [CrossRef] [PubMed]
5. Fleige, L.E.; Moore, W.R.; Garlick, P.J.; Murphy, S.P.; Turner, E.H.; Dunn, M.L.; Van Lengerich, B.; Orthoefer, F.T.; Schaefer, S.E. Recommendations for optimization of fortified and blended food aid products from the United States. *Nutr. Rev.* **2010**, *68*, 290–315. [CrossRef] [PubMed]
6. Webb, P.; Rogers, B.; Rosenberg, I.; Schlossman, N.; Wanke, C.; Bagriansky, J.; Sadler, K.; Johnson, Q.; Tilahun, J.; Reese Masterson, A. *Delivering Improved Nutrition: Recommendations for Changes to US Food Aid Products and Programs*; Tufts University: Boston, MA, USA, 2011.
7. Black, C.T.; Pahulu, H.F.; Dunn, M.L. Effect of preparation method on viscosity and energy density of fortified humanitarian food-aid commodities. *Int. J. Food Sci. Nutr.* **2009**, *60*, 219–228. [CrossRef] [PubMed]
8. Ozcan, S.; Jackson, D.S. Functionality behavior of raw and extruded corn starch mixtures. *Cereal Chem.* **2005**, *82*, 223–227. [CrossRef]

9. Hoppe, C.; Andersen, G.S.; Jacobsen, S.; Molgaard, C.; Friis, H.; Sangild, P.T.; Michaelsen, K.F. The use of whey or skimmed milk powder in fortified blended foods for vulnerable groups. *J. Nutr.* **2008**, *138*, 145S–161S. [CrossRef] [PubMed]
10. USAID. Corn Commodity Fact Sheet. Available online: https://www.usaid.gov/what-we-do/agriculture-and-food-security/food-assistance/resources/corn-commodity-fact-sheet (accessed on 13 January 2017).
11. Tenenbaum, D.J. Food vs. fuel: Diversion of crops could cause more hunger. *Environ. Health Perspect.* **2008**, *116*, A254–A257. [CrossRef] [PubMed]
12. Dicko, M.H.; Gruppen, H.; Traoré, A.S.; Voragen, A.G.; Van Berkel, W.J. Review: Sorghum grain as human food in Africa: Relevance of starch content and amylase activities. *Afr. J. Biotechnol.* **2006**, *5*, 384–395.
13. Mahasukhonthachat, K.; Sopade, P.; Gidley, M. Kinetics of starch digestion and functional properties of twin-screw extruded sorghum. *J. Cereal Sci.* **2010**, *51*, 392–401. [CrossRef]
14. Uzogara, S.; Ofuya, Z. Processing and utilization of cowpeas in developing countries: A review. *J. Food Process. Preserv.* **1992**, *16*, 105–147. [CrossRef]
15. Anglani, C. Sorghum for human food—A review. *Plant Foods Hum. Nutr.* **1998**, *52*, 85–95. [CrossRef] [PubMed]
16. Chanadang, S.; Chambers, E.I.; Kayanda, R.; Alavi, S.; Msuya, W. Novel fortified blended foods: Preference testing with infants and young children in Tanzania and descriptive sensory analysis. *J. Food Sci.* **2018**, *83*, 2343–2350. [CrossRef] [PubMed]
17. Chanadang, S.; Chambers, E.I. Sensory shelf life estimation of novel fortified blended foods under accelerated and real-time storage conditions. *J. Food Sci.* **2019**, *84*. in press.
18. Delimont, N.M.; Chanadang, S.; Joseph, M.V.; Rockler, B.E.; Guo, Q.; Regier, G.K.; Mulford, M.R.; Kayanda, R.; Range, M.; Mziray, Z.; et al. The MFFAPP Tanzania efficacy study protocol: Newly formulated, extruded fortified-blended foods for food aid. *Curr. Dev. Nutr.* **2017**, *1*, e000315. [CrossRef] [PubMed]
19. USDA. USDA Commodity Requirements. Corn-Soy Blend for Use in Export Programs. Available online: https://www.fsa.usda.gov/Internet/FSA_File/csb13_110507.pdf (accessed on 10 January 2017).
20. Chanadang, S.; Chambers, E.I.; Alavi, S. Tolerance Testing for Cooked Porridge made from a Sorghum Based Fortified Blended Food. *J. Food Sci.* **2016**, *81*, S1210–S1221. [CrossRef] [PubMed]
21. Moussa, M.; Qin, X.; Chen, L.F.; Campanella, O.H.; Hamaker, B.R. High- quality instant sorghum porridge flours for the West African market using continuous processor cooking. *Int. J. Food Sci. Technol.* **2011**, *46*, 2344–2350. [CrossRef]
22. Rowe, J.P.; Brodegard, W.C.; Pike, O.A.; Steele, F.M.; Dunn, M.L. Storage, preparation, and usage of fortified food aid among Guatemalan, Ugandan, and Malawian beneficiaries: A field study report. *Food Nutr. Bull.* **2008**, *29*, 213–220. [CrossRef] [PubMed]
23. Mouquet, C.; Greffeuille, V.; Treche, S. Characterization of the consistency of gruels consumed by infants in developing countries: Assessment of the Bostwick consistometer and comparison with viscosity measurements and sensory perception. *Int. J. Food Sci. Nutr.* **2006**, *57*, 459–469. [CrossRef] [PubMed]
24. Bredie, W.L.; Mottram, D.S.; Guy, R.C. Aroma volatiles generated during extrusion cooking of maize flour. *J. Agric. Food Chem.* **1998**, *46*, 1479–1487. [CrossRef]
25. Bredie, W.L.; Mottram, D.S.; Guy, R.C. Effect of temperature and pH on the generation of flavor volatiles in extrusion cooking of wheat flour. *J. Agric. Food Chem.* **2002**, *50*, 1118–1125. [CrossRef]
26. Parker, J.K.; Hassell, G.M.; Mottram, D.S.; Guy, R.C. Sensory and instrumental analyses of volatiles generated during the extrusion cooking of oat flours. *J. Agric. Food Chem.* **2000**, *48*, 3497–3506. [CrossRef] [PubMed]
27. Kobayashi, A.; Tsuda, Y.; Hirata, N.; Kubota, K.; Kitamura, K. Aroma constituents of soybean [Glycine max (L.) Merril] milk lacking lipoxygenase isoenzymes. *J. Agric. Food Chem.* **1995**, *43*, 2449–2452. [CrossRef]
28. Sessa, D.J. Biochemical aspects of lipid-derived flavors in legumes. *J. Agric. Food Chem.* **1979**, *27*, 234–239. [CrossRef]
29. Vara-Ubol, S.; Chambers, E.; Chambers, D.H. Sensory characteristics of chemical compounds potentially associated with beany aroma in foods. *J. Sens. Stud.* **2004**, *19*, 15–26. [CrossRef]
30. Brannan, G.; Setser, C.; Kemp, K.; Seib, P.; Roozeboom, K. Sensory characteristics of grain sorghum hybrids with potential for use in human food. *Cereal Chem.* **2001**, *78*, 693–700. [CrossRef]
31. Anyango, J.O.; de Kock, H.L.; Taylor, J.R. Evaluation of the functional quality of cowpea-fortified traditional African sorghum foods using instrumental and descriptive sensory analysis. *Food Sci. Technol.* **2011**, *44*, 2126–2133. [CrossRef]

32. Iuel-Brockdorf, A.; Draebel, T.A.; Ritz, C.; Fabiansen, C.; Cichon, B.; Christensen, V.B.; Yameogo, C.; Oummani, R.; Briend, A.; Michaelsen, K.F.; et al. Evaluation of the acceptability of improved supplementary foods for the treatment of moderate acute malnutrition in Burkina Faso using a mixed method approach. *Appetite* **2016**, *99*, 34–45. [CrossRef] [PubMed]
33. Gilbertson, T.A.; Baquero, A.F.; Spray-Watson, K.J. Water taste: the importance of osmotic sensing in the oral cavity. *J. Water Health* **2006**, *4*, 35–40. [CrossRef] [PubMed]
34. Teillet, E.; Schlich, P.; Urbano, C.; Cordelle, S.; Guichard, E. Sensory methodologies and the taste of water. *Food Qual. Prefer.* **2010**, *21*, 967–976. [CrossRef]
35. Drewnowski, A.; Gomez-Carneros, C. Bitter taste, phytonutrients, and the consumer: A review. *Am. J. Clin. Nutr.* **2000**, *72*, 1424–1435. [CrossRef]
36. Rackis, J.; Sessa, D.; Honig, D. Flavor problems of vegetable food proteins. *J. Am. Oil Chem. Soc.* **1979**, *56*, 262–271. [CrossRef]
37. Kobue-Lekalake, R.I.; Taylor, J.; De Kock, H.L. Effects of phenolics in sorghum grain on its bitterness, astringency and other sensory properties. *J. Sci. Food Agric.* **2007**, *87*, 1940–1948. [CrossRef]
38. Chambers, E.I.; Maughan, C.; Padmanabhan, N.; Alavi, S.; Adedeji, A. Sensory analysis of 20% solids fortified blended porridge. *Br. Food J.* **2019**, *121*, 633–641. [CrossRef]
39. Joseph, M.V. Extrusion, Physico-Chemical Characterization and Nutritional Evaluation of Sorghum-Based High Protein, Micronutrient Fortified Blended Foods (Chapter 3—Characterization of Physico-Chemical Properties of High Protein Extrudates from Binary Blends of Cereal and Legume Flours). Ph.D. Thesis, Kansas State University, Manhattan, KS, USA, 2016.
40. Feng, Y.; Lee, Y. Effect of Specific Mechanical Energy on In-Vitro Digestion and Physical Properties of Extruded Rice-Based Snacks. *Food Nutr. Sci.* **2014**, *5*, 1818–1827.

Article

Use of Almond Skins to Improve Nutritional and Functional Properties of Biscuits: An Example of Upcycling

Antonella Pasqualone [1,*], Barbara Laddomada [2], Fatma Boukid [3], Davide De Angelis [1] and Carmine Summo [1]

1 Department of Soil, Plant and Food Science (DISSPA), University of Bari Aldo Moro, Via Amendola, 165/a, I-70126 Bari, Italy; davide.deangelis@uniba.it (D.D.A.); carmine.summo@uniba.it (C.S.)

2 Institute of Sciences of Food Production (ISPA), CNR, via Monteroni, 73100 Lecce, Italy; barbara.laddomada@ispa.cnr.it

3 Institute of Agriculture and Food Research and Technology (IRTA), Food Safety Programme, Food Industry Area, Finca Camps i Armet s/n, 17121 Monells, Catalonia, Spain; fatma.boukid@irta.cat

* Correspondence: antonella.pasqualone@uniba.it

Received: 23 October 2020; Accepted: 16 November 2020; Published: 20 November 2020

Abstract: Upcycling food industry by-products has become a topic of interest within the framework of the circular economy, to minimize environmental impact and the waste of resources. This research aimed at verifying the effectiveness of using almond skins, a by-product of the confectionery industry, in the preparation of functional biscuits with improved nutritional properties. Almond skins were added at 10 g/100 g (AS10) and 20 g/100 g (AS20) to a wheat flour basis. The protein content was not influenced, whereas lipids and dietary fiber significantly increased ($p < 0.05$), the latter meeting the requirements for applying "source of fiber" and "high in fiber" claims to AS10 and AS20 biscuits, respectively. The addition of almond skins altered biscuit color, lowering L^* and b^* and increasing a^*, but improved friability. The biscuits showed sensory differences in color, odor and textural descriptors. The total sum of single phenolic compounds, determined by HPLC, was higher ($p < 0.05$) in AS10 (97.84 μg/g) and AS20 (132.18 μg/g) than in control (73.97 μg/g). The antioxidant activity showed the same trend as the phenolic. The *p*-hydroxy benzoic and protocatechuic acids showed the largest increase. The suggested strategy is a practical example of upcycling when preparing a health-oriented food product.

Keywords: almond skins; by-product; upcycling; biscuits; health claims; fiber; nutritional composition; sensory properties; phenolic compounds

---

1. Introduction

Recently, the reuse of food industry by-products has become a particularly important research topic, in order to develop systems capable of minimizing environmental impact and the waste of resources. The confectionery industry, in the production of blanched almonds, generates large quantities of almond skins as a by-product, which are mostly destined to cattle feeding [1] and composting [2]. However, almond skins can be considered functional food ingredients because they contain several bioactive phenolic compounds, namely flavonoids, phenolic acids, and tannins, the latter both hydrolysable and condensed [3–7]. The phenolic content of fresh almond skins comprises between 11.1 and 17.7 mg/g, depending on the extraction protocol [7], whereas 0.25–0.85 mg/g d.m. (dry matter) were quantified in dried almond skins, with the lowest amount in sun-dried skins and the highest in skins oven-dried at a temperature of 45–60 °C [7].

The polyphenols of almond skins are bioavailable and possess in vitro and in vivo antioxidant activity, able to reduce plasmatic oxidative stress [8] and to protect LDL (low-density lipoprotein) from oxidation [4,9]. The bioactive compounds of almond skins display also antibacterial and antiviral effects [10,11]. Recently, an extract of almond skins has been proposed for use in intestinal inflammatory diseases [12]. Furthermore, almond skins are also a rich source of fiber and therefore have a prebiotic effect, favorably influencing the gut microbiome [13,14]. The recommended daily intake of fiber ranges from 18 g to 38 g for adults and it varies among different countries, but many people do not reach this threshold [15]. Almond skins could hence be used to functionalize foods and to improve their nutritional profile in terms of fiber content. The reuse of almond skins in food products would represent an example of upcycling [16], responding to the need to increase sustainability in the food industries within the framework of the principles of a circular economy [17].

Functional ingredients, such as almond skins, could be easily added to cereal-based products, but any modification of the physico-chemical and sensory characteristics of the end-products should be carefully evaluated so as to fulfill consumer expectations for healthy but pleasant foods. The potential use of almond skins in composite dough with wheat flour was evaluated in a previous study, highlighting significant alterations of alveograph and farinograph indices due to the presence of fibers, which interfere with the gluten network [7]. Therefore, almond skins could be used in those cereal-based products which better tolerate a weak gluten network, such as biscuits.

Biscuits are popular baked goods, eaten daily and characterized by a long shelf-life. These features make biscuits a good recipient for the addition of functional ingredients. To date, however, almond skins are still an underexploited resource and no study has considered their introduction in biscuit formulation, despite many researchers having reformulated biscuits by incorporating an array of new ingredients, mostly of vegetable origin, such as apple peel powder [18], acorn flour [19], grape marc extract [20,21], purple wheat flour [22], inulin [23], soy protein isolate [24], blue berry by-product [25], and green tea extract [26].

Within this framework, the aim of this research has been to verify the effectiveness of almond skin addition in the preparation of functional biscuits with improved nutritional properties.

2. Materials and Methods

2.1. Raw Materials

The ingredients used for preparing the experimental biscuits were: refined wheat flour (0.52 g/100 g ashes) (Molini Spigadoro, Bastia Umbra, Italy), sucrose (Eridania, Bologna, Italy), extra virgin olive oil (Olearia De Santis, Bitonto, Italy), baking powder (sodium bicarbonate and potassium bitartrate, 'Belbake', Lidl Stiftung & Co. KG, Neckarsulm, Germany), all purchased at local retailers, and almond skins. The latter were collected from an almond processing industry (Calafiore S.r.l., Floridia, Italy), then dried at 60 °C for 30 min by a rotary air drier (mod. Scirocco, Società Italiana Essiccatoi, Milano, Italy), milled (Cutting Mill SM 100, Retsch, Haan, Germany) and sieved on a sieve with 0.6 mm holes. Moisture, a_w, phenolic compounds, antioxidant activity, color, and odor notes of almond skins are reported in a previous paper [7].

2.2. Preparation of Biscuits

The formulation of biscuits is reported in Table 1. Two levels of addition of almond skins were considered: 10 g/100 g (AS10) and 20 g/100 g (AS20) on a wheat flour basis, which were compared with control biscuits prepared without adding almond skins. The amount of water was defined in preliminary trials in order to achieve the same dough workability in the three types of biscuits. The process consisted in: kneading for 3 min sucrose, extra virgin olive oil and baking powder by an electric mixer with flat beater (Kitchen Aid, Antwerp, Belgium), then adding flour (pure wheat flour or blended with almond skin powder as in Table 1) and kneading for 3 min, finally adding water and kneading for about 10 min to form a homogeneous dough. The dough was then rolled out with a rolling pin to a thickness of 4 mm and cut into 6 cm diameter disks with the aid of a circular biscuit cutter with scalloped edges. The disks of dough were placed on a baking tray, mixing them in a randomized block pattern to minimize any effect of tray location during baking, then were baked in an electric oven (mod. Ignis ACF961IX, Whirlpool Italia S.r.l., Pero, Italy) at 175 °C for 15 min. Two independent production trials were carried out. Biscuits were finely crushed for analysis, except for the textural, colorimetric and sensory analyses.

Table 1. Formulation of the experimental biscuits (per 100 g of flour). Control = Biscuits without Almond Skins; AS10 and AS20 = Biscuits prepared by adding 10 g and 20 g Almond Skin Powder per 100 g of Wheat Flour, respectively.

	Control	AS10	AS20
Wheat flour (g)	100	90	80
Almond skin powder (g)	-	10	20
Sucrose (g)	28	28	28
Extra virgin olive oil (g)	18	18	18
Water (g)	26	28	30
Baking powder (g)	1	1	1

2.3. Determination of Nutritional Composition

Protein (N × 5.7) and moisture content were determined according to the American Association of Cereal Chemists (AACC) Methods 46–11.02 and 08–01, respectively [27]. The lipid fraction was extracted according to ICC Standard Method no. 136 [28]. Total dietary fiber was determined by the enzymatic-gravimetric procedure according to the AOAC Official Method 991.43 [29]. Carbohydrates were calculated by difference: 100 – (moisture + proteins + lipids + fiber + ash). Energy value (kJ), calculated by using the Atwater general conversion factors, also considered the contribution of 8 kJ/g from total dietary fiber, according to Annex XIV of Regulation (EC) No 1169/2011 [30]. All analyses were carried out in triplicate.

2.4. Determination of Physical Properties

The a^* (red/green balance), b^* (yellow/blue balance), and L^* (lightness) coordinates of the CIELAB color space were determined by a colorimeter (CM-600d Chromameter, Konica Minolta, Tokyo, Japan) under illuminant D65. Five replicated analyses were carried out. Total color difference (ΔE) was calculated as follows [31]:

$$\Delta E = [(\Delta L^*)^2 + (\Delta a^*)^2 + (\Delta b^*)^2]^{1/2}$$

The following scale was considered: ΔE = 0–0.5, very low difference; 0.5–1.5; slight difference; 1.5–3.0, noticeable difference; 3.0–6.0, appreciable difference; 6.0–12.0, large difference; and >12.0, very obvious difference [32].

Water activity (a_w) was analyzed in triplicate by a water activity meter (mod. Aqualab 4TE, Meter group, Pullman, WA, USA).

Textural properties, in terms of breaking strength (N mm^{-2}), were determined by a three-point bending test ("snap test") using a ZI.0 TN texture analyzer (ZwickRoell GmbH & Co. KG, Ulm, Germany), equipped with 1 kN load-cell. The biscuits were placed on the analyzer supports with their top surface down. The distance between the support bars was 4 cm. The downward movement of the probe, set at a speed of 5 mm min^{-1}, was continued until the biscuit was broken. Eight replicated analyses were carried out.

2.5. Baking Induced Variations of Dimensional Parameters and Weight

The weight (W) of biscuits before and after baking was assessed by a balance (Gibertini, Novate Milanese, Italy). The diameter (D) and thickness (T) of biscuits before and after baking were determined by a caliper. The spread factor was calculated as the ratio between D and T of baked biscuits, according to the AACC Method 10-50.05 [27]. The percentage variations in W, D, and T were calculated as follows:

% variation of W (or D, T) = (W (or D, T) after baking—W (or D, T) before baking)/W (or D, T) before baking × 100. Six replicated analyses were carried out.

2.6. HPLC analysis of Phenolic Compounds

The phenolic compounds were extracted from 1 g biscuits according to the procedure reported in Laddomada et al. [33], which involved defatting, alkaline hydrolysis, acidification and double ethyl acetate extraction. The extracts were lyophilized and dissolved in 400 µL of a solution of methanol diluted with 200 mL/L distilled water, then 50 µL were filtered on 0.45 µm polytetrafluoroethylene (PTFE) filters (Teknokroma, Barcelona, Spain) and analyzed by HPLC-DAD (Agilent 1100 Series, Agilent Technologies, Santa Clara, CA, USA) with a reversed phase C18(2) Luna column (Phenomenex, Torrance, CA, USA) (5 µm, 250 × 4.6 mm), as in Pasqualone et al. [7]. Identification of peaks was made by comparison of their UV-Vis spectra, and their retention times to those of authentic phenolic standards. Phenolic acids were quantified via a ratio of 3,5-dichloro-4-hydroxybenzoic acid, used as internal standard, and calibration curves of phenolic acid standards. Other phenolics (flavan-3-ols, flavonol and flavonone glycosides and aglycones) were quantified using calibration curves according to the external standard method [6]. The linear range, correlation coefficient, limit of detection (LOD) and limit of quantification (LOQ) for the phenolic compounds quantified are reported in Table S1.

2.7. Determination of Antioxidant Activity

An amount of 1 g sample, mixed with 10 mL of methanol and shaken at 250 rpm for 2 h in the dark, was centrifuged for 5 min at 5000 × g. The supernatant was submitted to the assessment of the antioxidant activity by the 2,2-diphenyl-1-picrylhydrazyl (DPPH) radical scavenging capacity assay, as in Pasqualone et al. [22]. A calibration curve was prepared with 0.1–100 µM solutions of 6-hydroxy-2,5,7,8-tetramethylchroman-2-carboxylic acid (Trolox) (Sigma–Aldrich Chemical Co., St. Louis, MO, USA) ($y = -0.008x + 0.6087$; $R^2 = 0.9971$).

2.8. Determination of Sensory Properties

Quantitative Descriptive Analysis (QDA) of biscuits was performed by a trained sensory panel of eight people, following the ethical guidelines of the laboratory of Food Science and Technology of the Department of Soil, Plant and Food Science (DISSPA), Dept. of Bari University (Italy). Panelists, regular consumers of biscuits and almonds and free of food intolerances or allergies, were informed about the study aims, and signed an individual written informed consent. Pre-test sessions were carried out, as in Pasqualone et al. [34]. Eight sensory descriptors, defined in Table 2, were rated on a 0–9 score range (0 = minimum; 9 = maximum intensity). The analyses were carried out in triplicate.

2.9. Statistical Analysis

One-way analysis of variance (ANOVA) followed by Tukey's HSD test, was made using the XLSTAT software (Addinsoft SARL, New York, NY, USA).

Table 2. Descriptive terms used for the sensory profiling of biscuits.

Descriptor	Definition	Scale Anchors	
		Min (0)	Max (9)
		Odor	
Caramel odor	Typical odor associated with caramel	Absent	Very intense
Leafy odor	Smell reminiscent of green leaves	Absent	Very intense
		Visual-tactile characteristics	
Color	Color of biscuit surface	Beige	Dark brown
Friability	The way the biscuit fractures, when broken by fingers	Very tough, it breaks with difficulty	Very friable and crumbly, it breaks easily
		Taste	
Sweetness	Basic taste produced by sucrose	Absent	Very intense
Bitterness	Basic taste produced by caffeine	Absent	Very intense
		Texture attributes perceived during chewing	
Dryness	Dryness perceived at the surface of biscuit	Moist	Very dry
Graininess	Graininess perceived at the end of chewing	Not grainy, giving finely sized crumbs	Very grainy, giving differently sized crumbs, medium and large

3. Results and Discussion

3.1. Nutritional and Technological Characteristics

Almond skin powder is particularly rich in fiber (52.6 g/100 g), as shown by the analysis of its nutritional characteristics (Table 3).

Table 3. Nutritional composition of dried almond skin powder and wheat flour used in the preparation of experimental biscuits. Values per 100 g, expressed on fresh weight basis.

	Almond Skin Powder	Refined Wheat Flour
Moisture (g)	10.1 ± 0.2	14.2 ± 0.4
Carbohydrates (g)	5.4 ± 0.5	73.3 ± 0.9
Fats (g)	21.3 ± 0.6	0.9 ± 0.1
Proteins (g)	10.6 ± 0.2	9.8 ± 0.2
Fiber (g)	52.6 ± 0.5	1.8 ± 0.2

This by-product of almond processing also showed a relevant presence of lipids (21.3 g/100 g). The lipid fraction of almond skins, however, is particularly healthy, being composed mainly of mono and polyunsaturated fatty acids (mostly oleic and linoleic acids) [6] associated with high amounts of vitamin E [6]. The composition of the lipid fraction of skins parallels the lipid composition of the whole seed [35]. The protein content of almond skins accounted for about 11 g/100 g, and low amounts of carbohydrates were observed. The overall composition of almond skin powder agreed with the current literature [6,36]. The composition of wheat flour was quite different than that of almond skins, being rich in carbohydrates and poor in fiber, with negligible levels of lipids.

The analysis of the nutritional features of biscuits (Table 4) shows that the protein content was not significantly influenced by the addition of almond skins, the latter having a protein content similar to wheat flour. However, AS20 biscuits had a significantly ($p < 0.05$) higher lipid content than control, due to the relevant contribution of almond skins. The lipid content of all biscuits was in the range of those commonly marketed [37].

Table 4. Nutritional features (values per 100 g, expressed on fresh weight basis) of biscuits enriched by increasing levels of almond skins. Control = biscuits without almond skins; AS10 and AS20 = biscuits prepared by adding 10 g and 20 g of almond skin powder per 100 g of wheat flour, respectively.

	Control	AS10	AS20
Moisture (g)	5.2 ± 0.3 a	5.5 ± 0.3 a	5.6 ± 0.4 a
Carbohydrates (g)	77.8 ± 1.1 a	74.3 ± 1.2 b	70.2 ± 0.7 c
Fats (g)	10.3 ± 0.9 b	11.5 ± 0.4 a,b	12.4 ± 0.4 a
Proteins (g)	5.6 ± 0.1 a	5.6 ± 0.2 a	5.6 ± 0.3 a
Fiber (g)	1.1 ± 0.2 c	3.1 ± 0.1 b	6.2 ± 0.2 a
Energy value (kJ)	1794 ± 9 a	1797 ± 8 a	1789 ± 10 a

Different letters in row indicate significant differences ($p < 0.05$).

As for the content of dietary fiber, it progressively increased with the increase of almond skin addition. EC Regulation n. 1924/2006 [38], relating to nutrition and health claims made on food products, defines that a food is a "source of fiber" only if contains at least 3 g/100 g fiber, or at least 1.5 g/100 kcal fiber, while "high in fiber" applies only if a food contains at least 6 g/ 100 g fiber, or at least 3 g/100 kcal fiber. The level of fiber ascertained in AS10 and AS20 biscuits met the requirements for applying the "source of fiber" and the "high in fiber" claims, respectively.

Moisture content increased, but not significantly, after the addition of almond skins due to their contribution of fiber. The higher the protein and fiber content, the higher the water absorption by the dough and moisture retention are found of the final product [39].

As a consequence of the increase in fats and fiber, the level of carbohydrates significantly decreased in almond skin-added biscuits compared to control. The energy value did not vary significantly by adding almond skins, because the increase of lipids was compensated for by an increase of fiber and a decrease in carbohydrates.

As for the main physical characteristics (Table 5), the a_w of AS10 and AS20 was slightly higher than control, but without a significant difference. The a_w values observed in all biscuits agreed with moisture content and showed that they were conveniently dry and stable from the microbiological point of view ($a_w < 0.6$).

Table 5. Physical characteristics of biscuits enriched by increasing levels of almond skins. Control = biscuits without almond skins; AS10 and AS20 = biscuits prepared by adding 10 g and 20 g of almond skin powder per 100 g of wheat flour, respectively.

	Control	AS10	AS20
a_w	0.24 ± 0.02 a	0.27 ± 0.01 a	0.28 ± 0.02 a
Colorimetric data			
a^*	9.85 ± 1.09 b	12.35 ± 0.43 a	12.61 ± 0.36 a
b^*	35.12 ± 1.25 a	24.94 ± 0.66 b	22.66 ± 0.64 c
L^*	68.7 ± 2.41 c	49.35 ± 0.85 b	46.15 ± 1.01 a
$\Delta E_{vs\ Control}$	-	21.08 ± 0.51	25.76 ± 0.69
Texture			
Fracture strength (N/mm^2)	8.82 ± 0.66a	7.63 ± 0.49ab	6.87 ± 0.31b

Different letters in row indicate significant differences ($p < 0.05$).

The addition of almond skins, which were brown colored, resulted in an expected substantial alteration of biscuit color (Figure 1), with a significant decrease of L^* and b^*, and an increase of a^* in AS10 and AS20 compared to the control (Table 5). The total color difference (ΔE) of AS10 and AS20 biscuits compared to the control was greater for AS20 than for AS10, but in both cases with very high values, confirming that the control had a distinct color [40]. ΔE values >12.0, in fact, indicate a very obvious color difference [32].

Figure 1. Biscuits enriched by increasing levels of almond skins. From left to right: Control = biscuits prepared without adding almond skins; biscuits prepared by adding 10 g of almond skin powder per 100 g of wheat flour (AS10); biscuits prepared by adding 20 g of almond skin powder per 100 g of wheat flour (AS20).

The textural analysis showed that the addition of almond skins caused a decrease in the strength necessary to break the biscuits, i.e., an increase of friability, which is a particularly important characteristic. Friability is a salient textural characteristic for biscuits [41,42]. Tough, non-crumbly biscuits have low acceptance values in consumer tests [43]. This variation of breaking strength was significant when comparing control with AS20 and was due to the high presence of fiber in the almond skins. Fibers are highly hygroscopic and interfere with the formation of a strong and complete gluten network [44]. Preliminary work, in fact, showed that the rheological properties of the dough [7] significantly worsened after the addition of almond skin powder. However, among baked goods, biscuits are the most suitable for being reformulated with the addition of fibrous raw materials, since for their production a weak gluten network is not only sufficient but even necessary. In addition, although the difference in lipid content with control was significant only for AS20, the lipid fraction deriving from almond skins could have positively influenced the friability [45]. Therefore, the addition of almond skins did not harden biscuits at all; on the contrary, it gave a crumblier texture.

As for the dimensional variations induced by baking (Table 6), due to the thermal expansion of gases (carbon dioxide developed by the baking powder, dough moisture, and air entrapped during kneading), all biscuits increased more in thickness than in diameter. This result, commonly observed in biscuit baking [19], is due to the retaining effect of gluten, which tends to limit enlargement, whereas the upward thrust of the oven heat (oven rise) is less opposed [45]. AS10 and AS20 showed a greater diameter increase than control, which was significantly different for AS20, but had a lower increase in thickness. The easier enlargement observed in almond-skin added biscuits was due to the coupled effect of the dilution of gluten by a non-gluten raw material and the interference with gluten formation by the fiber and lipids of the same material. These findings agreed with studies where other fibrous and gluten-free ingredients were added to biscuit dough [18,19]. In addition, better expanded biscuits are usually less compact and more friable than those which expand less, in agreement with the observed textural data.

Table 6. Baking induced variations of dimensional parameters of biscuits enriched by increasing levels of almond skins. Control = biscuits without almond skins; AS10 and AS20 = biscuits prepared by adding 10 g and 20 g of almond skin powder per 100 g of wheat flour, respectively.

	Control	AS10	AS20
Diameter variation (%)	3.45 ± 0.91 b	4.51 ± 0.72 a,b	6.21 ± 0.89 a
Thickness variation (%)	40.23 ± 3.41 a	38.12 ± 2.98 a	33.65 ± 1.04 b
Spread factor	10.83 ± 0.81 b	11.26 ± 0.38 a,b	12.45 ± 0.22 a
Weight loss (%)	15.04 ± 0.37 a	14.78 ± 0.43 a,b	14.09 ± 0.31 b

Different letters in row indicate significant differences ($p < 0.05$).

The spread factor increased progressively as the amount of almond skins increased, with a significant difference between control and AS20. A higher spread factor indicates a better quality and is linked to an increase in consumer acceptability [46]. The observed values were higher than those reported for biscuits enriched with pure fiber of various cereals [47].

Weight loss, primarily due to the moisture loss from dough during baking, decreased by increasing the amount of almond skins as a consequence of the greater hygroscopicity of fibers, which limited water migration. The values ascertained were in the range of other researches [48–50].

The sensory profiles of the biscuits showed significant differences in odor, color and textural descriptors (Table 7). As for taste, the bitter note was negligible in the biscuits investigated, while sweetness was moderately intense, both without significant difference among formulations.

Table 7. Sensory properties of biscuits enriched by increasing levels of almond skins. Control = biscuits without almond skins; AS10 and AS20 = biscuits prepared by adding 10 g and 20 g of almond skin powder per 100 g of wheat flour, respectively.

	Control	**AS10**	**AS20**
Caramel odor	2.3 ± 0.2 a	1.9 ± 0.1 a	2.1 ± 0.2 a
Leafy odor	0.0 ± 0.0 c	0.9 ± 0.2 b	1.6 ± 0.2 a
Color	4.3 ± 0.5 c	7.8 ± 0.4 b	8.9 ± 0.4 a
Friability	3.5 ± 0.2 b	3.7 ± 0.2 a,b	4.2 ± 0.3 a
Sweetness	4.4 ± 0.3 a	4.7 ± 0.2 a	4.6 ± 0.2 a
Bitterness	0.2 ± 0.1 a	0.1 ± 0.1 a	0.2 ± 0.1 a
Dryness	4.7 ± 0.3 a	4.8 ± 0.3 a	5.0 ± 0.4 a
Graininess	1.7 ± 0.1 b	2.5 ± 0.2 a	2.9 ± 0.3 a

Different letters in row indicate significant differences ($p < 0.05$).

A slight odor note of caramel, derived from sugar caramelization and Maillard reaction, was perceived by the panelists in all biscuit types, without statistically significant differences between them. Instead, differences between the samples were found in the intensity of leafy odor. This odor note, absent in the control, was perceived with low intensity in biscuits formulated with almond skins, with the highest perception in AS20 and with an intermediate value in AS10. In previous research [7] this characteristic smell note was observed in the dried almond skins used in biscuit-making, albeit much more pronounced than in the finished product.

The color of biscuits became progressively and significantly darker as the level of addition of almond skin powder increased, as already indicated by colorimeter determinations.

As for friability, evaluated as the way biscuit fractured when broken by finger, the sensorial results were similar to those obtained instrumentally by the texture analyzer (snap test). AS20 was significantly more friable than control.

Dryness and graininess, on the other hand, were evaluated during chewing. Dryness did not show significant differences, whereas graininess was scored higher in almond-skin added biscuits, due to their granular and fibrous crumbles.

3.2. Functional Characteristics

Almond skins are rich in phenolic compounds [7], therefore the content of these bio-actives was evaluated in biscuits, as well as antioxidant activity (Table 8). The total sum of phenolic compounds, determined by HPLC, was significantly higher in AS10 and AS20 than in control.

Table 8. Phenolic compounds and antioxidant activity of biscuits enriched by increasing levels of almond skins. Control = biscuits without almond skins; AS10 and AS20 = biscuits prepared by adding 10 g and 20 g almond skin powder per 100 g of wheat flour, respectively.

	Control	AS10	AS20
AA (DPPH) (μmol TE/g)	1.89 ± 0.16 c	6.11 ± 0.61 b	9.76 ± 0.74 a
Single phenolic compounds (μg/g)			
Vanillic acid	1.43 ± 0.02 c	2.77 ± 0.18 b	4.53 ± 0.10 a
Syringic acid	2.69 ± 0.02 c	6.04 ± 0.45 b	9.71 ± 0.30 a
p-Coumaric acid	0.36 ± 0.01 c	0.79 ± 0.12 b	1.14 ± 0.08 a
Ferulic acid	63.72 ± 0.52 a	55.21 ± 1.47 b	55.96 ± 1.50 b
Sinapic acid	5.33 ± 0.04 a	4.30 ± 0.15 b	4.25 ± 0.20 b
p-Hydroxybenzoic acid	0.44 ± 0.02 c	5.49 ± 0.19 b	12.96 ± 0.44 a
Protocatechuic acid	0.00 ± 0.00 c	3.55 ± 0.06 b	12.67 ± 0.31 a
(+)-Catechin	0.00 ± 0.00 c	11.17 ± 0.06 b	19.52 ± 1.06 a
(-)-Epicatechin	0.00 ± 0.00 c	8.54 ± 0.02 b	11.45 ± 0.21 a
Total sum	73.97 ± 0.54 c	97.84 ± 2.55 b	132.18 ± 1.63 a

AA = antioxidant activity; TE = Trolox equivalents; DPPH = 2,2-diphenyl-1-picrylhydrazyl radical. Different letters in row indicate significant differences ($p < 0.05$).

In more detail, the variation of the single phenolics did not show the same trend for all the compounds, which showed different behavior according to the phenolic composition of the raw materials. In particular, among the phenolic acids, the *p*-hydroxy benzoic and protocatechuic acids showed a relevant increase after the addition of almond skins. The flavan-3-ols catechin and epicatechin also followed the same trend, being not detectable in control and showing a concentration-effect increment between the AS10 and AS20. In fact, these phenolic compounds are the most abundant in almond skins [7]. A smaller increase, but always statistically significant, was observed for syringic acid, vanillic and *p*-coumaric acids.

Instead, the most abundant phenolic acid, namely the ferulic acid, followed by the sinapic acid, decreased when comparing control biscuits with the almond-skin added, because these phenolic acids are typically present in wheat [33,51], but not in almond.

The flavonol glycosides and their aglycones, as well as the flavanone glycosides and their aglycones, despite their presence in almond skins [7], were not detected in biscuits. Probably, since their starting quantity was not remarkably high, they became undetectable in the biscuits, due to the dilution effect of wheat flour. In addition, oxidation and other degradation phenomena could not be excluded during processing (kneading and baking) since a decrease in phenolic compounds had already been observed when raw almond skins were thermally dried [7]. In any case, the total phenolic compounds of AS20 were approximately double that of the control, indicating that the addition of almond skins in the formulation can concretely contribute to enhance the nutritional value and the potential health benefits of the end products.

The antioxidant activity followed the same trend as the phenolic and showed higher values in the almond skin supplemented biscuits, compared to the control, also highlighting a concentration effect. Indeed, in the AS20, the antioxidant activity was about five times higher than the control. The observed values of antioxidant activity were consistent with those of the almond skins added [7].

4. Conclusions

The increasing sensibility of modern consumers towards the potential benefits of food on human health has led to a strong demand for functional products.

To date, almond skins, in spite of having high fiber content and antioxidant substances, are a by-product of almond processing usually addressed to animal feed and/or composting. This study, instead, demonstrates that almond skins can be effectively used for the production of functional biscuits, addressing the needs of both producers, who require the reduction of waste production, and consumers, who increasingly demand healthier food. For this latter purpose, the nutritional claims "source of fiber" and "high in fiber", defined in EC Regulation n. 1924/2006, were applicable to the AS10 and AS20 biscuits, respectively.

Therefore, using almond skins in biscuit-making is a feasible way to convert a low-value by-product into a valuable resource, providing to the almond processing industry an efficient and environment-friendly solution for waste disposal. This is a practical example of upcycling while preparing a health-oriented food product.

Supplementary Materials: The following is available online at http://www.mdpi.com/2304-8158/9/11/1705/s1, Table S1: Regression equation, linear range, correlation coefficient (R^2), limit of detection (LOD), and limit of quantification (LOQ) of the HPLC analysis of phenolic compounds quantified in the experimental biscuits.

Author Contributions: Conceptualization, A.P.; Formal analysis, A.P., B.L., C.S.; Funding acquisition, C.S.; Methodology, A.P., B.L., D.D.A.; Writing—original draft, A.P.; Writing—review & editing, A.P., B.L., F.B., D.D.A. and C.S. All authors have read and agreed to the published version of the manuscript.

Funding: This research was funded by the Regione Puglia, Project "Almond Management Innovations – AMI", within the call "PSR 2014-2020, MIS 12, Cooperazione. Sottomisura 16.2: Sostegno a progetti pilota e allo sviluppo di nuovi prodotti, pratiche, processi e tecnologie".

Acknowledgments: The authors acknowledge Virgilio Giannone for helpful discussions and for kindly furnishing the samples of almond skins. The authors acknowledge also Leone D'Amico for skillful assistance in HPLC analyses.

Conflicts of Interest: The authors declare no conflict of interest.

References

1. Grasser, L.; Fadel, J.G.; Garnett, I.; Depeters, E. Quantity and economic importance of nine selected by-products used in California dairy rations. *J. Dairy Sci.* **1995**, *78*, 962–971. [CrossRef]
2. González, J.F.; Gañan, J.; Ramiro, A.; González-García, C.; Encinar, J.; Sabio, E.; Román, S. Almond residues gasification plant for generation of electric power. Preliminary study. *Fuel Process. Technol.* **2006**, *87*, 149–155. [CrossRef]
3. Bolling, B.W. Almond polyphenols: Methods of analysis, contribution to food quality, and health promotion. *Compr. Rev. Food Sci. Food Saf.* **2017**, *16*, 346–368. [CrossRef]
4. Chen, C.-Y.; Milbury, P.E.; Lapsley, K.; Blumberg, J.B. Flavonoids from almond skins are bioavailable and act synergistically with vitamins C and E to enhance hamster and human LDL resistance to oxidation. *J. Nutr.* **2005**, *135*, 1366–1373. [CrossRef]
5. Garrido, I.; Monagas, M.; Gómez-Cordovés, C.; Bartolomé, B. Polyphenols and antioxidant properties of almond skins: Influence of industrial processing. *J. Food Sci.* **2008**, *73*, C106–C115. [CrossRef]
6. Mandalari, G.; Tomaino, A.; Arcoraci, T.; Martorana, M.; Lo Turco, V.; Cacciola, F.; Rich, G.T.; Bisignano, C.; Saija, A.; Dugo, P.; et al. Characterization of polyphenols, lipids and dietary fiber from skins of almonds (*Amygdalus communis L.*). *J. Food Comp. Anal.* **2010**, *23*, 166–174. [CrossRef]
7. Pasqualone, A.; Laddomada, B.; Spina, A.; Todaro, A.; Guzmàn, C.; Summo, C.; Mita, G.; Giannone, V. Almond by-products: Extraction and characterization of phenolic compounds and evaluation of their potential use in composite dough with wheat flour. *LWT* **2018**, *89*, 299–306. [CrossRef]
8. Chen, C.-Y.O.; Milbury, P.E.; Blumberg, J.B. Polyphenols in almond skins after blanching modulate plasma biomarkers of oxidative stress in healthy humans. *Antioxidants* **2019**, *8*, 95. [CrossRef]
9. Amarowicz, R. Natural phenolic compounds protect LDL against oxidation. *Eur. J. Lipid Sci. Technol.* **2016**, *118*, 677–679. [CrossRef]
10. Arena, A.; Bisignano, C.; Stassi, G.; Mandalari, G.; Wickham, M.S.; Bisignano, G. Immunomodulatory and antiviral activity of almond skins. *Immunol. Lett.* **2010**, *132*, 18–23. [CrossRef]
11. Musarra-Pizzo, M.; Ginestra, G.; Smeriglio, A.; Pennisi, R.; Sciortino, M.T.; Mandalari, G. The antimicrobial and antiviral activity of polyphenols from almond (*Prunus dulcis L.*) Skin. *Nutrients* **2019**, *11*, 2355. [CrossRef] [PubMed]

12. Lauro, M.R.; Marzocco, S.; Rapa, S.F.; Musumeci, T.; Giannone, V.; Picerno, P.; Aquino, R.; Puglisi, G. Recycling of almond by-products for intestinal inflammation: Improvement of physical-chemical, technological and biological characteristics of a dried almond skins extract. *Pharmaceutics* **2020**, *12*, 884. [CrossRef] [PubMed]
13. Liu, Z.; Lin, X.; Huang, G.; Zhang, W.; Rao, P.; Ni, L. Prebiotic effects of almonds and almond skins on intestinal microbiota in healthy adult humans. *Anaerobe* **2014**, *26*, 1–6. [CrossRef] [PubMed]
14. Dhillon, J.; Li, Z.; Ortiz, R.M. Almond snacking for 8 wk increases alpha-diversity of the gastrointestinal microbiome and decreases *Bacteroides fragilis* abundance compared with an isocaloric snack in college freshmen. *Curr. Dev. Nutr.* **2019**, *3*, nzz079. [CrossRef]
15. Jones, J.M. CODEX-aligned dietary fiber definitions help to bridge the 'fiber gap'. *Nutr. J.* **2014**, *13*, 34. [CrossRef]
16. Rondeau, S.; Stricker, S.M.; Kozachenko, C.; Parizeau, K. Understanding motivations for volunteering in food insecurity and food upcycling projects. *Soc. Sci.* **2020**, *9*, 27. [CrossRef]
17. Jurgilevich, A.; Birge, T.; Kentala-Lehtonen, J.; Korhonen-Kurki, K.; Pietikäinen, J.; Saikku, L.; Schösler, H. Transition towards circular economy in the food system. *Sustainability* **2016**, *8*, 69. [CrossRef]
18. Nakov, G.; Brandolini, A.; Hidalgo, A.; Ivanova, N.; Jukić, M.; Komlenić, D.K.; Lukinac, J. Influence of apple peel powder addition on the physico-chemical characteristics and nutritional quality of bread wheat cookies. *Food Sci. Technol. Int.* **2020**, *26*, 574–582. [CrossRef]
19. Pasqualone, A.; Makhlouf, F.Z.; Barkat, M.; Difonzo, G.; Summo, C.; Squeo, G.; Caponio, F. Effect of acorn flour on the physico-chemical and sensory properties of biscuits. *Heliyon* **2019**, *5*, e02242. [CrossRef]
20. Pasqualone, A.; Bianco, A.M.; Paradiso, V.M.; Summo, C.; Gambacorta, G.; Caponio, F. Physico-chemical, sensory and volatile profiles of biscuits enriched with grape marc extract. *Food Res. Int.* **2014**, *65*, 385–393. [CrossRef]
21. Pasqualone, A.; Bianco, A.M.; Paradiso, R. Production trials to improve the nutritional quality of biscuits and to enrich them with natural anthocyanins. *CyTA-J. Food* **2013**, *11*, 301–308. [CrossRef]
22. Pasqualone, A.; Bianco, A.M.; Paradiso, V.M.; Summo, C.; Gambacorta, G.; Caponio, F.; Blanco, A. Production and characterization of functional biscuits obtained from purple wheat. *Food Chem.* **2015**, *180*, 64–70. [CrossRef] [PubMed]
23. Giarnetti, M.; Paradiso, R.; Caponio, F.; Summo, C.; Pasqualone, A. Fat replacement in shortbread cookies using an emulsion filled gel based on inulin and extra virgin olive oil. *LWT* **2015**, *63*, 339–345. [CrossRef]
24. Mohsen, S.M.; Fadel, H.H.M.; Bekhit, M.A.; Edris, A.E.; Ahmed, M.Y.S. Effect of substitution of soy protein isolate on aroma volatiles, chemical composition and sensory quality of wheat cookies. *Int. J. Food Sci. Technol.* **2009**, *44*, 1705–1712. [CrossRef]
25. Perez, C.; Tagliani, C.; Arcia, P.; Cozzano, S.; Curutchet, A. Blueberry by-product used as an ingredient in the development of functional cookies. *Food Sci. Technol. Int.* **2017**, *24*, 301–308. [CrossRef]
26. Gómez-Mascaraque, L.G.; Hernández-Rojas, M.; Tarancón, P.; Tenon, M.; Feuillère, N.; Ruiz, J.F.V.; Fiszman, S.; López-Rubio, A. Impact of microencapsulation within electrosprayed proteins on the formulation of green tea extract-enriched biscuits. *LWT* **2017**, *81*, 77–86. [CrossRef]
27. AACC International. *Approved Methods of Analysis*, 10th ed.; American Association of Cereal Chemists: St. Paul, MN, USA, 2009.
28. International Association for Cereal Chemistry. *ICC Standard Method no. 136: Cereals and Cereal Products-Determination of Total Fat Content*; ICC: Vienna, Austria, 1984.
29. AOAC. *Official Methods of Analysis*, 16th ed.; Association of Official Analytical Chemists: Arlington, VA, USA, 1995.
30. European Parliament and Council. Regulation (EU) No 1169/2011 of the European Parliament and of the Council of 25 October 2011 on the provision of food information to consumers, amending Regulations (EC) No 1924/2006 and (EC) No 1925/2006 of the European Parliament and of the Council, and repealing Commission Directive 87/250/EEC, Council Directive 90/496/EEC, Commission Directive 1999/10/EC, Directive 2000/13/EC of the European Parliament and of the Council, Commission Directives 2002/67/EC and 2008/5/EC and Commission Regulation (EC) No 608/2004. *Off. J. Eur. Union* **2001**, *L304*, 18–63.
31. Summo, C.; De Angelis, D.; Difonzo, G.; Caponio, F.; Pasqualone, A. Effectiveness of oat-hull-based ingredient as fat replacer to produce low fat burger with high beta-glucans content. *Foods* **2020**, *9*, 1057. [CrossRef]
32. Chen, X.D.; Mujumdar, A. *Drying Technologies in Food Processing*; Blackwell Publishing Ltd: Oxford, UK, 2008; pp. 26–29.

33. Laddomada, B.; Durante, M.; Mangini, G.; D'Amico, L.; Lenucci, M.S.; Simeone, R.; Piarulli, L.; Mita, G.; Blanco, A. Genetic variation for phenolic acids concentration and composition in a tetraploid wheat (*Triticum turgidum L.*) collection. *Genet. Resour. Crop. Evol.* **2016**, *64*, 587–597. [CrossRef]
34. Pasqualone, A.; De Angelis, D.; Squeo, G.; Difonzo, G.; Caponio, F.; Summo, C. The effect of the addition of Apulian black chickpea flour on the nutritional and qualitative properties of durum wheat-based bakery products. *Foods* **2019**, *8*, 504. [CrossRef]
35. Summo, C.; Palasciano, M.; De Angelis, D.; Paradiso, V.M.; Caponio, F.; Pasqualone, A. Evaluation of the chemical and nutritional characteristics of almonds (Prunus dulcis (Mill). D.A. Webb) as influenced by harvest time and cultivar. *J. Sci. Food Agric.* **2018**, *98*, 5647–5655. [CrossRef]
36. Song, Y.; Wang, W.; Cui, W.; Zhang, X.; Zhang, W.; Xiang, Q.; Liu, Z.; Li, N.; Jia, X. A subchronic oral toxicity study of almond skins in rats. *Food Chem. Toxicol.* **2010**, *48*, 373–376. [CrossRef] [PubMed]
37. Caponio, F.; Summo, C.; Delcuratolo, D.; Pasqualone, A. Quality of the lipid fraction of Italian biscuits. *J. Sci. Food Agric.* **2005**, *86*, 356–361. [CrossRef]
38. European Parliament and Council. Regulation (EC) No 1924/2006 of the European Parliament and of the Council of 20 December 2006 on nutrition and health claims made on foods. *Off. J. Eur. Union* **2006**, *L404*, 9–25.
39. Rodrigues Batista, J.E.; de Morais, M.P.; Caliari, M.; Júnior, M.S.S. Physical, microbiological and sensory quality of gluten-free biscuits prepared from rice flour and potato pulp. *J. Food Nutrition Res.* **2016**, *55*, 101–107.
40. Mokrzycki, W.S.; Tatol, M. Colour difference ΔE. A survey. *Mach. Graph. Vis.* **2011**, *20*, 383–411.
41. Piazza, L.; Bringiotti, R.; Masi, P. The influence of moisture content on the failure mode of biscuits. In *Developments in Food Engineering*; Springer Science and Business Media LLC: Berlin/Heidelberg, Germany, 1994; pp. 134–136.
42. Piazza, L.; Schiraldi, A. Correlation between fracture of semi-sweet hard biscuits and dough viscoelastic properties. *J. Texture Stud.* **1997**, *28*, 523–541. [CrossRef]
43. Rojo-Poveda, O.; Barbosa-Pereira, L.; Orden, D.; Stévigny, C.; Zeppa, G.; Bertolino, M. Physical properties and consumer evaluation of cocoa bean shell-functionalized biscuits adapted for diabetic consumers by the replacement of sucrose with tagatose. *Foods* **2020**, *9*, 814. [CrossRef]
44. Pasqualone, A.; Laddomada, B.; Centomani, I.; Paradiso, V.M.; Minervini, D.; Caponio, F.; Summo, C. Bread making aptitude of mixtures of re-milled semolina and selected durum wheat milling by-products. *LWT* **2017**, *78*, 151–159. [CrossRef]
45. Maache-Rezzoug, Z.; Bouvier, J.-M.; Allaf, K.; Patras, C. Effect of principal ingredients on rheological behaviour of biscuit dough and on quality of biscuits. *J. Food Eng.* **1998**, *35*, 23–42. [CrossRef]
46. Klunklin, W.; Savage, G. Effect of substituting purple rice flour for wheat flour on physicochemical characteristics, in vitro digestibility, and sensory evaluation of biscuits. *J. Food Qual.* **2018**, *2018*, 1–8. [CrossRef]
47. Sudha, M.L.; Vetrimani, R.; Leelavathi, K. Influence of fiber from different cereals on the rheological characteristics of wheat flour dough and on biscuit quality. *Food Chem.* **2007**, *100*, 1365–1370. [CrossRef]
48. Hossain, A.K.M.M.; Brennan, M.; Mason, S.L.; Guo, X.; Zeng, X.-A.; Brennan, C.S. The effect of astaxanthin-rich microalgae "*Haematococcus pluvialis*" and wholemeal flours incorporation in improving the physical and functional properties of cookies. *Foods* **2017**, *6*, 57. [CrossRef] [PubMed]
49. Cronin, K.; Preis, C. A statistical analysis of biscuit physical properties as affected by baking. *J. Food Eng.* **2000**, *46*, 217–225. [CrossRef]
50. Lara, E.; Cortés, P.; Briones, V.; Perez, M. Structural and physical modifications of corn biscuits during baking process. *LWT* **2011**, *44*, 622–630. [CrossRef]
51. Pasqualone, A.; Summo, C.; Laddomada, B.; Mudura, E.; Coldea, T.E. Effect of processing variables on the physico-chemical characteristics and aroma of borş, a traditional beverage derived from wheat bran. *Food Chem.* **2018**, *265*, 242–252. [CrossRef] [PubMed]

Publisher's Note: MDPI stays neutral with regard to jurisdictional claims in published maps and institutional affiliations.

Article

Evaluation of the Use of a Coffee Industry By-Product in a Cereal-Based Extruded Food Product

Elisa A. Beltrán-Medina [1], Guadalupe M. Guatemala-Morales [1], Eduardo Padilla-Camberos [1,*], Rosa I. Corona-González [2], Pedro M. Mondragón-Cortez [1] and Enrique Arriola-Guevara [2,*]

1 Tecnología Alimentaria, Biotecnología Médica y Farmacéutica, Centro de Investigación y Asistencia en Tecnología y Diseño del Estado de Jalisco, A.C. (CIATEJ), Normalistas 800, C.P. 44270 Guadalajara, Jalisco, Mexico; es_ebeltran@ciatej.mx (E.A.B.-M.); gguatemala@ciatej.mx (G.M.G.-M.); pmondragon@ciatej.mx (P.M.M.-C.)

2 Departamento de Ingeniería Química, Centro Universitario de Ciencias Exactas e Ingenierías, Universidad de Guadalajara. Blvd. Marcelino García Barragán #1421, esq. Calzada Olímpica. C.P. 44430 Guadalajara, Jalisco, Mexico; rosa.corona@academicos.udg.mx

* Correspondence: epadilla@ciatej.mx (E.P.-C.); enrique.arriola@academicos.udg.mx (E.A.-G.); Tel.: +33-33455200 (ext. 1640) (E.P.-C.)

Received: 23 June 2020; Accepted: 23 July 2020; Published: 27 July 2020

Abstract: The evaluation of by-products to be added to food products is complex, as the residues must be analyzed to demonstrate their potential use as safe foods, as well as to propose the appropriate process and product for recycling. Since coffee is a very popular beverage worldwide, the coffee industry is responsible for generating large amounts of by-products, which include the coffee silverskin (CS), the only by-product of the roasting process. In this work, its characterization and food safety were evaluated by chemical composition assays, microbiological determinations, aflatoxin measurements and acute toxicity tests. The results showed that CS is safe for use in food, in addition to providing dietary fiber, protein and bioactive compounds. An extruded cereal-based ready-to-eat food product was developed through an extreme vertices mixture design, producing an extruded food product being a source of protein and with a high fiber content. Up to 15% of CS was incorporated in the extruded product. This work contributes to the establishment of routes for the valorization of CS; nevertheless, further research is necessary to demonstrate the sustainability of this food industry by-product.

Keywords: coffee silverskin; chemical characterization; toxicological analysis; extrusion; extreme vertices mixture design; product development

---

1. Introduction

Today, there is a considerable emphasis on the recovery, recycling and upgrading of wastes, particularly in the food processing industry, in which wastes, effluents, residues and by-products can be recovered. They can often be upgraded into useful products and value-added food supplements that can provide dietary fiber and bioactive compounds [1,2]. The possibility of the utilization of these food processing by-products for manufacturing various human foods has created enormous scope for waste reduction, indirect income generation, the reduction of raw material costs [1,3] and even the potential for them to be considered as novel foods with beneficial properties [4]. However, for the development of future sustainable industrial processes centered on the valorization of food waste, aspects such as technical feasibility, an analysis of their-economic potential and a life-cycle-based environmental assessment need to be considered [5].

Coffee is one of the most consumed beverages in the world and is the second largest traded commodity after petroleum [6]. The coffee production chain begins with the harvest of the ripe coffee berries that are to be treated in order to separate the pulp from the coffee bean by one of two processes—(a) a wet process or (b) a dry process—where the green coffee bean is obtained. Finally, the bean is heat treated by a process called roasting, thus producing the coffee that will be used for the preparation of the drink [4,7]. In the world, during the 2018/2019 season, 10.3 million tons of green coffee was produced [8]. Since coffee is a very popular and appreciated beverage around the world, the coffee industry is responsible for generating large amounts of wastes, which include the coffee silverskin (CS), the only waste obtained during the roasting process [7,9]. The CS represents about 4.2% (w/w) of the coffee beans [9]. Despite the produced quantity being low compared to that of other coffee by-products, it has been reported that for 120 tons of roasted coffee, about 1 ton of CS is produced [10]. It can be considered that if all the green coffee produced worldwide during the 2018/19 season had been roasted, it would be equivalent to having produced around 71,822 tons of CS (conversion factor: 1.19 tons of green coffee = 1 ton of roasted coffee [8]). CS is a yellowish transparent endosperm that covers each green coffee bean (Figure 1) [4,7,9] and is currently used as a fuel and fertilizer [11]. However, coffee wastes have been reported to possess bioactive compounds, mainly secondary metabolites such as phenolic acids, for example, hydroxycinnamic acids and flavonoids, desired for their beneficial antioxidant properties [12,13]. 5-caffeoylquinic acid (5CQA) belongs to the family of the chlorogenic acids (hydroxycinnamic acids). It is one of the most abundant polyphenolic compounds in the human diet and is produced by certain plant species; it is an important component in coffee and in the CS [14,15]. 5CQA is of special interest due to the wide spectrum of its potentially beneficial effects on health, including antidiabetic, anti-obesity, antioxidant, anti-hypertension, anti-inflammatory and antibacterial effects [16,17].

Figure 1. Coffee silverskin from coffee roasting process.

CS has been reported as a source of chlorogenic acids; however, to date, there are few reports concerning the content of 5CQA in CS, and those that exist show controversial results, since the reported concentrations are in the range of 1000 to 11,678 mg of 5CQA/kg of CS [11,18–20]. Different studies have shown the functional properties of CS such as a high dietary fiber content (54.11 to 74.15 g/100 g of CS) [9,21,22] and a total phenolic content in the range of 4.6 to 46.65 mg/g, depending on the extraction method employed [11,21,23,24]. The principal constituents of its fibrous tissues are cellulose (24%) and hemicellulose (17%). It is a source of minerals such as potassium (21,100 mg/kg dry basis (db)), iron (843 mg/kg db), sodium (57 mg/kg db), manganese (50 mg/kg db) and zinc (22 mg/kg db), among others [9]. The enzyme inhibitory properties of CS extracts and peptide composition of CS protein hydrolysates have been investigated, from the perspective of their application in the pharmaceutical and nutraceutical industry [24,25].

The holistic concept of food production tries to connect differing goals, such as the highest product quality and safety, highest production efficiency and the integration of environmental aspects into product development and food production. Vegetable residues mostly contain considerable amounts of potentially interesting compounds [1]. However, the benefits of recycling should not be undermined by the environmental impacts caused by new production processes [26]. Food extrusion is a versatile process in food engineering as it combines various unit operations such as transport, thermomechanical and degradation changes, mixing and molding. It is a technology widely used in the food industry due to its versatility, high productivity and energy efficiency [27,28]. Extrusion cooking is increasingly used in the food industries for the development of new cereal-based snacks, baby foods and breakfast cereals [29]. The incorporation of by-products from different fruit and vegetable processing industries into extruded products has led to hope for their utilization as well as the development of nutritionally healthy extruded products [30]. The extrusion process has been used to develop new products in which 2% to 20% of various by-products of the agri-food industry have been incorporated, such as barley-fruit and cauliflower by-products and red lentil-carrot pomace, among others [30–33]. The by-product incorporation in extruded food has been reported at the lab scale, and no industrial-level studies have been shown. Nevertheless, extrusion is a mature and scalable technology; even scale-up considerations and mathematical models for extrusion cooking are available [34,35].

Cereal-based food products have been the basis of the human diet since ancient times. Cereals contain all the macronutrients (protein, fats and carbohydrates) we need for support and maintenance [36]. They contain only low levels of micronutrients, most of which are lost during processing for food [37], bringing the possibility to incorporate new raw materials that provide these micronutrients. To the best of our knowledge, there are few reports on the use of CS in a cereal-based product. A treatment of CS with alkaline hydrogen peroxide before being added to Barbari bread to improve its properties has been described [38]. In another study, CS was added to cake as a fat replacer [39]. Some authors investigated the use of CS in biscuits, where it was incorporated as a sugar replacer or to enhance the phenolic content and antioxidant capacity of the product, employing a standardized formulation [40,41]. However, no studies regarding the formulation design under official standards to create a product with specific requirements have been reported. Hence, the development of a cereal-based food product adding CS using extrusion technology is proposed.

In all those articles in which CS was added in cereal-based products, wheat flour was used as the cereal basis [38–41]; nonetheless, in present work, corn and popped amaranth were chosen as the cereal product base. Corn is undoubtedly part of the identity of Mexico; it is present in the daily life of its inhabitants [42], which will allow the obtaining of a familiar flavor and better acceptance of the developed product. Amaranth is a popular snack in Mexico of pre-Hispanic origin [42] and is characterized by an excellent nutritional composition (*A. hypochondriacus*, protein, 15.9%; lysine, 4.9 g/100 g of protein; fat, 6.1%; tocopherols, 5.5 mg/100 g; starch, 62.4%; sucrose, 1.4%; ash, 3.3% [43]); nevertheless, it is difficult to produce expanded products directly by the extrusion cooking of amaranth grain alone because of its high fat content. Therefore, the extrusion cooking of amaranth flour in combination with other cereals produces well-accepted forms of expanded extrudates [43]. The combination of these cereals with the CS could allow the obtaining of a product with good nutritional quality.

Therefore, as CS appears to be a potential new low-cost ingredient, the aim of this work was to evaluate whether its consumption was harmless to humans, demonstrating the food safety of CS by microbiological tests and the determination of aflatoxins and the Lethal Dose (LD50) by the acute oral toxicity test. Its potential use as a food ingredient was evaluated by determining its nutritional contribution and by developing a cereal-based extruded food product with this new ingredient added.

2. Materials and Methods

2.1. Materials

CS produced by roasting coffee beans (*C. Arabica* 100%) was obtained from two states of Mexico (Chiapas and Jalisco). Popped amaranth was purchased from Nutriactivate Company (Puebla, Puebla, Mexico), and white corn was obtained from the food market of Guadalajara city (Jalisco, Mexico). CS, popped amaranth and white corn were milled prior to the extraction and extrusion process (Average Particle Size = 0.28 ± 0.01 mm).

5-caffeoylquinic acid powder reference standard (USP 12601), 2,2-diphenyl-1-picrylhydrazil (DPPH), 2,2′-azino-bis(3-ethylbenzthiazoline-6-sulphonic acid) (ABTS) and 6-hydroxy-2,5,7,8-tetramethyl chroman-2-carboxylic acid (Trolox) were purchased from Sigma-Aldrich (St. Louis, MO, USA). Methanol (MeOH) was obtained at HPLC grade (Sigma-Aldrich, St. Louis, MO, USA). Phosphoric acid was obtained at reagent grade (Karal, Leon, Guanajuato, Mexico).

The standard of 5CQA was diluted in MeOH to obtain a stock solution at 1000 μg/mL, from which the calibration curve was prepared. All the solutions remained refrigerated at 4 °C in amber vials.

2.2. CS Characterization and Microbiological Quality

2.2.1. Bromatological and Microbiological Analysis

The bromatological analyses were performed according to the following Mexican Regulations: moisture, NMX-F-083-1986 [44]; protein, NMX-F-608-NORMEX-2011 [45]; ashes, NMX-F-607-NORMEX-2013 [46]; fats (ethereal extract), NOM-086-SSA1-1994 (Regulatory Appendix C, Number 1) [47]; and, carbohydrates, method 986.25 A.O.A.C. Volume 1 [48].

The microbiological analysis was performed according the Official Mexican Regulations: aerobic mesophilic bacteria, NOM-092-SSA1-1994 [49]; total coliforms, NOM-113-SSA1-1994 [50]; molds and yeasts, NOM-111-SSA1-1994 [51]; *Salmonella*, NOM-114-SSA1-1994 [52]; *Escherichia coli*, CCAYAC-M-004 [53]; and *Staphylococcus aureus*, NOM-115-SSA1-1994 [54].

2.2.2. Total Dietary Fiber (TDF)

The TDF was estimated by the enzymatic gravimetric method according to the Mexican Regulation NMX-F-622-NORMEX-2008 [55]. Briefly, one gram of sample suspended in phosphate buffer solution was sequentially digested by heat stable α-amylase for 30 min in a boiling water bath, after which a 0.275 M NaOH solution and protease were added and incubated for 30 min at 60 °C. Then, a 0.325 M HCl solution and amyloglucosidase were added and incubated for 30 min at 60 °C. After filtration, the insoluble dietary fiber was recovered from enzyme digestate, dried at room temperature and then weighed. Soluble dietary fiber in the filtrate was precipitated with ethanol and filtered. The precipitate was dried and weighed. Insoluble and soluble dietary fiber contents were corrected for residual protein and ash content. The TDF content was the sum of both fibers.

2.2.3. Extraction Method

This extraction method was used for DPPH, ABTS, total polyphenol and HPLC analysis. The extraction method was adapted from Del Río et al. (2014) [56]; 0.5 g of sample (CS or extruded product) was weighed, and 5 mL of MeOH/water 3:1 (v/v) was added. The mixture was sonicated (Branson 5800, Dansbury, CT, USA) at a 40 kHz frequency for 15 min, removed and stirred in a Vortex-Genie (Scientific Industries, Bohemia, NY, USA) for another 15 min; afterwards, it was centrifuged at 3400 rpm for 10 min. The supernatant was transferred to another container, and the residue was re-extracted. The second extract was added to the first, and it was filtered (0.45 μm). All the extracts were kept in amber vials, under refrigeration, until analysis.

2.2.4. Total Polyphenol Determination

The quantification of total polyphenols was carried out by the Folin–Ciocalteu method proposed by Singleton and Rossi (1965) [57]. The extracts (30 μL) were mixed with 150 μL of Folin–Ciocalteu reagent (1:10), followed by the addition of 120 μL of 20% (w/v) sodium carbonate. After 1 h, the absorbance at 760 nm was read in the spectrophotometer. The results are expressed as g gallic acid equivalents (GAE)/100 g sample.

2.2.5. Antioxidant Activity

The antioxidant activity of the extracts was determined by two methods: the ABTS and DPPH (free radical scavenging) assays. The ABTS assay was based on a method developed by Nenadis et al. (2004) [58]. A solution of 7 mM ABTS, 2.5 mM potassium persulfate and 10 mL of distilled water were mixed and incubated in the dark at room temperature for 16 h before use. This solution was diluted with MeOH to an absorbance of 0.7 ± 0.02 at 734 nm. After the addition of 20 µL of extract or Trolox standard to 200 µL of diluted ABTS solution, the absorbances were recorded at 6 min after mixing. Methanolic solutions of known Trolox concentrations were used for calibration. The results are expressed as mg Trolox equivalents (eq)/g sample.

The DPPH antioxidant activity assay was performed by the Brand-Williams et al. (1995) [59] method with slight modifications. A MeOH solution containing 500 µmol of DPPH was prepared. After adjusting the blank with MeOH, an aliquot of 20 µL of extract was added to 200 µL of this solution. After 30 min in the dark, the absorbance at 515 nm was read with the spectrophotometer. The results are expressed as mg Trolox eq/g sample.

2.2.6. HPLC Analysis

Sample analysis was performed on a liquid chromatograph Alliance 2695, equipped with a 2998 Diode Array Detector (Waters, Milford, MA, USA) and Software Empower 3. The separation was carried out on a 5 micron (100 Å, 250 × 4.6 mm) C-18 reverse phase Kromasil column (Ale, Bohus, Sweden) at room temperature. The mobile phase was phosphoric acid at 5 mM (solvent A) and MeOH (solvent B), at a flow rate of 1 mL/min. The elution gradient was as follows: a linear gradient of 85–80% solvent B (0–5 min), 60% B (6–10 min), 70% B (11–20 min), 80% B (21–25 min) and, finally, 85% B (26–30 min). The injection volume was 20 µL, and the 5CQA was detected at a wavelength of 325 nm. This method was adapted from Fujioka and Shibamoto (2008) [60]. Sample chromatograms were compared with those of the 5CQA standard for identification. The measurements were carried out in triplicate. Instrumental calibration: Eight different levels of concentration were employed for 5CQA. The Pearson correlation coefficient (r) was calculated to estimate the type of adjustment of the experimental points in the calibration curve, and subsequently, statistical analyses with Student's t-test [61] and variance analysis were performed, to verify its significance.

2.3. *CS Toxicological Analysis*

Aflatoxins B1, B2, G1 and G2 were quantified by the method of QuEChERS extraction and ultra-high liquid chromatography tandem mass spectrometry (UPLC-MS/MS) detection [62].

An acute oral toxicity test was performed following the procedure described in the OECD 425 guidelines [63]. Briefly, five female mice, Balb-c strain, 9 weeks old, were used. They were administered 2000 mg/kg of body weight of the aqueous extract of CS, in a single dose, with a cannula, with a 4 h food fast but not water fast. Under the conditions of a temperature of 24 ± 1 °C and photoperiod of 12 h light/12 h darkness, mortality and toxicity signs were registered daily, and weight was measured weekly. Animal experimentation was carried out in accordance with the Official Mexican Method NOM-062-ZOO-1999 [64]; in addition, the protocol was authorized and reviewed by the Internal Committee for the Care and Use of Laboratory Animals of CIATEJ (code 2019-002A).

2.4. *Product Development*

2.4.1. Extrusion Cooking

Ingredient mixes of cornmeal (CM), amaranth flour (AF) and CS were weighed and then mixed in a Kitchen Aid mixer (St. Joseph, Michigan, MI, USA). The mixtures were conditioned to adjust them to 21.0 ± 1.0% of moisture content, placed into plastic bags and maintained under refrigeration for 48 h before processing. Sixteen samples in total were prepared. In each treatment, 300 g of sample was used.

Extrusion trials were performed using a Brabender single screw extruder (Plasti-Corder 815808, Duisburg, Germany). The barrel diameter and D/L ratio were 475 mm and 19/25, respectively. A screw configuration with a 3:1 compression ratio was used. The exit diameter of the circular die was 2 mm. A vertical dosing screw feeder (628456, Duisburg, Germany) was used for feeding the conditioned mixtures. The process conditions were set as follows: a feed rate of 40 g/min, a screw speed of 80 rpm, and three barrel temperatures—120 °C at the feed entry, 130 °C at the middle and 140 °C at the die exit. The pressure, material temperature and torque were monitored during the extrusion runs. The extrusion conditions were obtained by preliminary tests (data not shown).

Extrudates were left to cool at room temperature for about 30 min. Moisture content was determined [44]. The extruded products were subsequently baked at 60 °C for 2 h, until a moisture content of 5.4 ± 0.3% was achieved, packaged in plastic bags and stored at room temperature until analysis.

2.4.2. Water Solubility Index (WSI)

The method of Anderson et al. (1970) [65] was used. In brief, 2.5 g of sample was added to 30 mL of distilled water at 30 °C, in centrifuge tubes, and shaken on a rotary shaker (Roto-Shake Genie, Bohemia, New York, NY, USA) for 30 min. They were then placed in a centrifuge (Universal 320 R Hettich, Tuttlingen, Germany) run at 4000 rpm for 10 min. The supernatant liquor from each tube was transferred into aluminum trays to be oven dried at 80 °C for 24 h. As the WSI, the amount of dried solids recovered by evaporating the supernatant from the water absorption test just described is expressed as the percentage of dry solids. Analyses were carried out in triplicate.

2.4.3. Experimental Design

An extreme vertices mixture design [66] was used, varying the amounts of AF (50–98%), CM (0–45%) and CS (2–15%). The proposed range of CS was determined according to previous studies in which agri-food industry by-products were incorporated into extruded foods [30–33]. Figure 2 shows the experimental region of the extreme vertices mixture design, where the 16 points of the experimental runs are indicated.

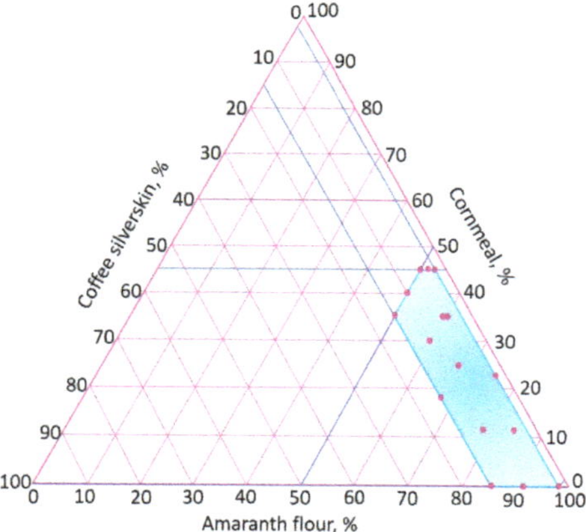

Figure 2. Experimental region of the extreme vertices mixture design.

2.5. Statistical Analysis

The STATGRAPHICS Centurion XV package (Statpoint Technologies; Warrengton, VA, USA) was used for the data analysis of the analytical method, as well as for design and analysis of the extreme vertices mixture design. Statistically significant differences between values were determined at the $p < 0.05$ level [66]. The results are expressed as the mean values ± standard errors of the three separate determinations.

3. Results and Discussion

3.1. CS Characterization and Microbiological Quality

3.1.1. Bromatological Composition

The results of the bromatological analyses for CS are shown in Table 1. The protein content in CS (15.09% *w/w*) was minor compared to previous values reported, 18.6% *w/w* [22], and 18.69% *w/w* [9]. The low fat content in CS (1.99% *w/w*) was similar to a value cited before, 2.2% *w/w* [22], and lower than those presented by other authors, 3.78% *w/w* [9]. The ash, carbohydrate and moisture values obtained were similar to those published before [6,9,22]. The dietary fiber content was 67.6% *w/w*, which is superior to that published, 54.11% *w/w* [9] and 62.4% *w/w* [22]. The differences in the contents of the nutrients could be due to the origins of the coffee beans.

Table 1. Chemical composition of coffee silverskin.

Parameters	CS %
Moisture	7.2
Protein	15.09
Ashes	5.55
Fats	1.99
Carbohydrates	70.17
Dietary Fiber	67.6

3.1.2. Microbiological Determinations

There is no regulation for the microbiological parameters for CS, as it is a coffee industry by-product; however, in the Mexican Regulation for roasted coffee [67], less than 3 CFU/g of *E. coli* is specified; thus, by comparison, the CS result shows an acceptable level, and when comparing the results obtained for the remaining microbiological determinations with the Official Mexican Regulation for cereals and their products [68]—because the chemical composition of CS is similar to that of cereals—the results obtained are within the parameters, as shown in Table 2.

Table 2. Microbiological results (CFU/g).

Parameters	CS	Roasted Coffee [1]	Wholemeal Flour [2]
Aerobic mesophilic bacteria	1400	ns	500,000
Total coliforms	40	ns	500
Molds	45	ns	500
Yeasts	<10	ns	ns
Salmonella spp	Absence	ns	ns
Escherichia coli	<3	<3	ns
Staphylococcus aureus	<100	ns	ns

[1] Maximum level suggested for roasted coffee for Mexico by the Secretaría de Economía. [2] Maximum level for cereal additives for Mexico by the Secretaría de Salud. ns—not specified.

3.1.3. Antioxidant Capacity and Total Polyphenol Content

The DPPH assay is based on the change of the blue-violet color towards pale yellow due to the reaction with antioxidant substances. The antioxidant capacity of the sample according to the DPPH method was 33.23 ± 0.02 µM Trolox/g of CS (dry basis, db). In another study, 21.35 ± 0.39 µM eq Trolox/g (db) was reported [9]. The antioxidant capacity according to the ABTS method was 3.45 ± 0.02 mM

Trolox/100 g of CS (db). Some authors have published values for CS of 1.92 mM Trolox/100 g [22], and 2.12 ± 0.4 mM Trolox/100 g dry matter [15], which are consistent with results found in this work.

The total polyphenol assay provides an approximation of the total amount of polyphenols in the sample. In present work was obtained 16.48 ± 6.6 mg GAE/g of CS (db). This was similar to what has been already reported, 16.1 ± 1.2 mg/g of CS [11]. These results suggest the possibility of recycling CS in a new food product as a contribution of bioactive compounds.

3.1.4. Quantification of 5CQA

Method Performance

According to Regulation (EC) No. 333/2007 [69], if an analytical method includes an extraction step, the result of the analysis must be corrected based on the recovery, so the level of recovery was calculated. The efficiency of the extraction of 5CQA was 87.01% of recovery. The determination coefficient (r^2) was 0.99, which demonstrates the linearity of the calibration curve for the 5CQA at eight concentration levels in the range of 10–500 μg/mL. The instrumental detection (LOD) and quantification (LOQ) limits for 5CQA were determined based on the signal-to-noise ratios of 3 and 10, respectively, using the weighted parameters [61], thus obtaining an LOD of 3.311 μg/mL and LOQ of 11.037 μg/mL.

5CQA Content in CS

The concentration of 5CQA extracted from CS was 499.03 ± 7.45 mg of 5CQA/kg of CS (db). An amount of 198.9 ± 6.6 mg of chlorogenic acid/100 g of CS was reported [11], which is four times higher than the concentration obtained in this work. Meanwhile, others studies have shown contents of 1.0 ± 0.0 to 1.7 ± 0.1 mg of chlorogenic acid/g of CS [18], 9.4 ± 2.6 mg of 5CQA/g extract of CS [19] and 89.83 ± 0.64 mg of 5CQA/g of dry extract of CS [20]. The difference in the contents of 5CQA in the CS could be due to the nature of the coffee beans, their origins, the extraction methods, and the processes of coffee roasting. During this process, when the temperature is higher than 160 °C, a series of exothermic and endothermic reactions take place; the bean become light brown, its volume increases considerably and the detachment of CS occurs. The chemical reactions responsible for the aroma and flavor of roasted coffee are triggered at approximately 190 °C. These reactions are interrupted at the desired point based on the bean color or programmed time [70–72]. At temperatures between 150 °C and 170 °C the decrease in 5CQA content starts to speed up [72]. Therefore, as the beans (and the CS) stay longer in the roaster, where high temperatures are present, the content of 5CQA considerably diminishes. This could explain the concentration of 5CQA obtained in the CS.

3.2. *Toxicological Aspects*

The negative impact on human health of aflatoxins, especially because of their carcinogenicity, shows the importance of carrying out their quantification [73]. The aflatoxin quantification yielded the following results: aflatoxin B1 < 0.20 ppb, B2 < 0.06 ppb, G1 < 0.20 ppb and G2 < 0.06 ppb. The maximum admissible levels in food, in the European Union, for the sum of the four aflatoxins (B1, B2, G1 and G2) have been set from 4 μg/mL to 15 μg/mL, depending on the type of food (peanuts, nuts, dried fruits and their by-products, and cereals and their by-products) [73]; thus, the sum of the four aflatoxins for CS was below these limits.

For the acute oral toxicity test [63], a single dose administration of aqueous extract (CS) at 2000 mg/kg, was provided by esopharingeal cannulation. Normal behavior was recorded daily in the mice, with normal postural reflex and hygiene habits as well as food and water consumption as appropriate for the species. There were no clinical abnormalities. During the test period (14 days), no signs of evident toxicity or mortality of the experimental mice were observed. The results obtained allow us to affirm that the LD50 is above 2000 mg of CS/kg body weight.

The characterization of the CS and its toxicological evaluation allowed the evaluation of its potential as an ingredient for the food industry, confirming that it is a source of bioactive compounds (including 5CQA), dietary fiber and protein, and low in fat, and that its consumption is safe, so it can be considered for food development.

3.3. Product Development

The CS was totally incorporated into a food product to recycle this by-product without generating a new by-product derived from the subsequent process; therefore, an extruded ready-to-eat cereal-based food was developed.

3.3.1. Product Formulation

The product formulation was developed using the parameters obtained from the bromatological analyses of the three raw materials. The composition obtained for corn was 7.57% protein, 1.24% ashes, 2.22% fats, 77.46% carbohydrates and 7.49% corrected dietary fiber [74], and that for popped amaranth was 15.60% protein, 2.88% ashes, 7.97% fats, 73.55% carbohydrates and 9.41% dietary fiber; the CS composition is described in Section 3.1.1.

An extreme vertices mixture design was used to determine the best combination of the three raw materials—CS, CM and AF—that minimized the WSI and maximized the 5CQA content. The proposed formulations were designed to be classified as foods with high fiber contents and sources of protein, in accordance with the established Regulation (EC) No. 1924/2006 [75], where a food with a high fiber content is one that has a minimum of 6 g of fiber/100 g of product, and a food source of protein is one in which protein contributes at least 12% of the total energy value, which was verified by performing a theoretical calculation using the values obtained from the bromatological analysis of the raw materials, according to the formulations obtained through the mixture design. Table 3 shows the formulations proposed by the mixture design and the results for the dietary fiber and corresponding percentage of energy contributed by proteins, satisfying both requirements.

Table 3. Extreme vertices mixture design. Theoretical values for dietary fiber and protein energy contribution. WSI and 5CQA concentration determined in extruded product.

CS:CM:AF [1] %	Dietary Fiber [2] g/100 g Product	Protein Contribution to the TEV [2,3] %	WSI [4,6]	5CQA [5,6] mg/kg
2:0:98	10.3	14.6	39.85 ± 0.86	13.64 ± 0.24
15:35:50	20.7	12.9	27.01 ± 0.11	63.60 ± 1.48
15:0:85	16.1	14.9	46.34 ± 0.88	62.94 ± 1.12
5:45:50	17.5	12.1	21.73 ± 0.79	28.77 ± 0.73
2:45:53	16.2	12.1	26.59 ± 0.87	27.01 ± 0.75
4.9:12.5:82.6	13.2	14.0	37.76 ± 0.16	25.50 ± 0.68
11.4:30:58.6	18.4	13.1	29.68 ± 0.22	47.35 ± 1.08
11.4:12.5:76.1	16.1	14.1	41.45 ± 0.09	51.58 ± 1.49
6.4:35:58.6	16.8	12.8	29.91 ± 1.42	33.09 ± 0.61
4.9:35:60.1	16.2	12.7	23.59 ± 0.62	29.58 ± 0.84
8.5:0:91.5	13.2	14.7	44.79 ± 0.52	34.11 ± 0.90
2:22.5:75.5	13.2	13.4	33.23 ± 0.71	13.57 ± 0.12
15:17.5:67.5	18.4	13.9	31.86 ± 0.30	59.82 ± 0.51
10:40:50	19.1	12.5	22.54 ± 0.61	49.39 ± 0.78
3.5:45:51.5	16.8	12.1	23.52 ± 0.84	23.18 ± 0.39
7.8:25:67.2	16.2	13.4	31.55 ± 0.37	34.70 ± 0.18

[1] CS:CM:AF, coffee silverskin/cornmeal/amaranth flour; [2] calculated values for each mixture; [3] TEV, Total Energy Value; [4] WSI, Water Solubility Index; [5] 5CQA, 5-caffeoylquinic acid; mg/kg, mg of 5CQA/kg extruded product; [6] Measured values.

3.3.2. WSI and 5CQA Content in Extruded Products

Extrusion cooking was accomplished. The WSI and 5CQA content were determined for the extrudates; the results are exhibited in Table 3.

The WSI is related to the quantity of soluble molecules, which is related to dextrinization. Thus, the WSI can be used as an indicator for the degradation of molecular compounds and measures the degree of starch conversion during extrusion [29]. The WSI of the extrudates was influenced by the quadratic effect of the raw materials. The adjusted R-square value was 0.90. The CM effect was more important for the decrease in this property. In Figure 3, a decrease in the WSI with an increase in the CM content can be observed. The reduction in starch degradation lowers the WSI, which increases the bowl life of breakfast cereals and reduces the undesirable powdery mouthfeel of extruded snacks [28].

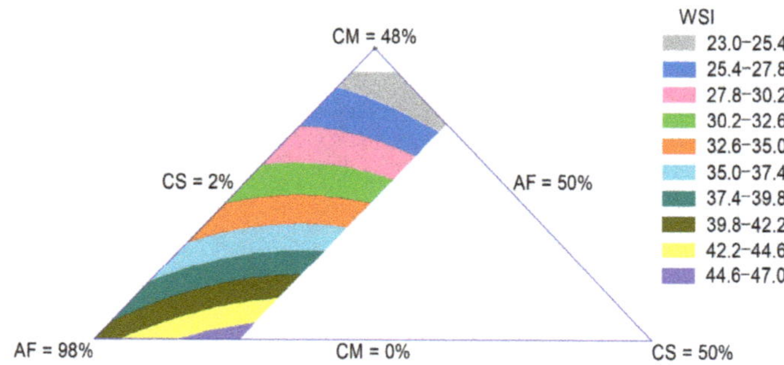

Figure 3. Contour plot for WSI of extruded product formulated with corn meal (CM), amaranth flour (AF) and coffee silverskin (CS).

The 5CQA content of extrudates was influenced by the quadratic effect of the raw materials. The adjusted R-square value was 0.97. The CS effect was more important for the increase in this property. In Figure 4, an increase in the 5CQA content with an increase in the CS content can be observed, as expected since this ingredient is the source of this bioactive compound, as shown in Section 3.1.4.

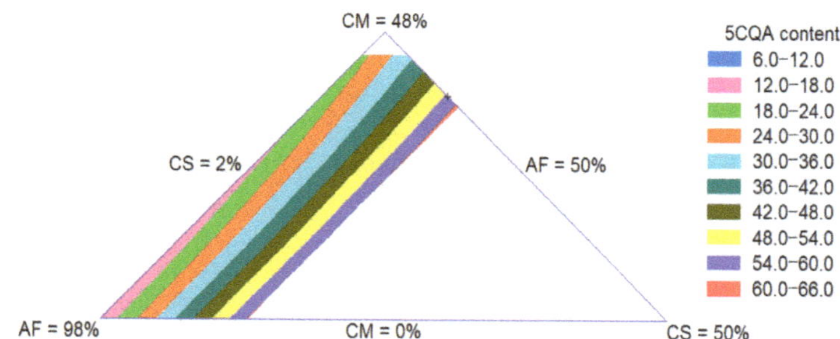

Figure 4. Contour plot for 5CQA content of extruded product formulated with corn meal (CM), amaranth flour (AF) and coffee silverskin (CS).

The optimization of the formulation with the response variables WSI and 5CQA content, using the desirability function [66] in the indicated region, maximizing the 5CQA content and minimizing the WSI response, showed that the optimal values for the studied components were 35% CM, 50% AF and 15% CS. The overall desirability was 0.937. The desirability function predicts the response values of the WSI at 24.67 and 5CQA content at 63.41 mg of 5CQA/kg of extruded product, which are similar to the values determined in Table 3 for this formulation. Figure 5 shows the optimized extruded product.

Figure 5. Optimized extruded product.

Further research is suggested in order to study the texture properties and sensory evaluation of the extrudates; also, consumer acceptability needs to be explored.

4. Conclusions

The CS can be considered as a new food ingredient, which can increase the protein and dietary fiber content of food, in addition to providing bioactive compounds such as 5CQA, which has been shown to exert benefits in the human organism. Up to 15% of CS was incorporated into the extruded product.

To ensure the safety of CS as a food ingredient, the application of good manufacturing and storage practices are recommended, from its collection in the coffee industry. A grinding process is suggested for easy handling.

Although this study contributes to establishing routes for the valorization of CS, it is important to note that studies are still needed to demonstrate the sustainability of this food industry by-product. To incorporate CS in the food production chain, it is suggested to carry out shelf-life tests, studies on logistics issues and/or business opportunity studies for coffee roasters. An assessment of consumer acceptability is also necessary. Furthermore, economic feasibility studies are required.

Author Contributions: Conceptualization, G.M.G.-M., E.A.-G., R.I.C.-G. and E.A.B.-M.; methodology, E.A.B.-M. and R.I.C.-G.; validation, E.A.B.-M., R.I.C.-G. and P.M.M.-C.; formal analysis, G.M.G.-M., E.A.-G., E.A.B.-M., E.P.-C. and R.I.C.-G.; investigation, G.M.G.-M., E.A.B.-M., E.P.-C., P.M.M.-C. and R.I.C.-G.; resources, G.M.G.-M., E.P.-C. and E.A.-G.; writing—original draft preparation, E.A.B.-M. and P.M.M.-C.; writing—review and editing, G.M.G.-M. and E.A.-G.; visualization, G.M.G.-M.; supervision, E.A.-G.; project administration, G.M.G.-M.; funding acquisition, G.M.G.-M. and E.A.-G. All authors have read and agreed to the published version of the manuscript.

Funding: This research was funded by CONACYT, FORDECYT 292474, the Centro de Investigación y Asistencia en Tecnología y Diseño del Estado de Jalisco, A.C. (CIATEJ), and the Universidad de Guadalajara (UDG).

Acknowledgments: The authors would like to thank CONACYT for scholarship No. 262766.

Conflicts of Interest: The authors declare no conflict of interest.

References

1. Laufenberg, G.; Kunz, B.; Nystroem, M. Transformation of vegetable waste into value added products: (A) the upgrading concept; (B) practical implementations. *Bioresour. Technol.* **2003**, *87*, 167–198. [CrossRef]
2. Elleuch, M.; Bedigian, D.; Roiseux, O.; Besbes, S.; Blecker, C.; Attia, H. Dietary fibre and fibre-rich by-products of food processing: Characterisation, technological functionality and commercial applications: A review. *Food Chem.* **2011**, *124*, 411–421. [CrossRef]
3. Sharma, S.K.; Bansal, S.; Mangal, M.; Dixit, A.K.; Gupta, R.K.; Mangal, A.K. Utilization of Food Processing By-products as Dietary, Functional, and Novel Fiber: A Review. *Crit. Rev. Food Sci. Nutr.* **2016**, *56*, 1647–1661. [CrossRef] [PubMed]

4. Klingel, T.; Kremer, J.I.; Gottstein, V.; Rajcic de Rezende, T.; Schwarz, S.; Lachenmeier, D.W. A Review of Coffee By-Products Including Leaf, Flower, Cherry, Husk, Silver Skin, and Spent Grounds as Novel Foods within the European Union. *Foods* **2020**, *9*, 665. [CrossRef]
5. Caldeira, C.; Vlysidis, A.; Fiore, G.; De Laurentiis, V.; Vignali, G.; Sala, S. Sustainability of food waste biorefinery: A review on valorisation pathways, techno-economic constraints, and environmental assessment. *Bioresour. Technol.* **2020**, *312*, 123575. [CrossRef]
6. Mussatto, S.I.; Machado, E.M.S.; Martins, S.; Teixeira, J.A. Production, Composition, and Application of Coffee and Its Industrial Residues. *Food Bioprocess Technol.* **2011**, *4*, 661–672. [CrossRef]
7. Belitz, H.D.; Grosch, W.; Schieberle, P. Coffee, Tea, Cocoa. In *Food Chemistry*, 4th ed.; Springer: Berlin/Heidelberg, Germany, 2009; pp. 938–942.
8. International Coffee Organization. Available online: www.ico.org (accessed on 1 July 2020).
9. Ballesteros, L.; Teixeira, J.; Mussatto, S. Chemical, functional, and structural properties of spent coffee grounds and coffee silverskin. *Food Bioprocess Technol.* **2014**, *7*, 3493–3503. [CrossRef]
10. Alves, R.C.; Rodrigues, F.; Nunes, M.A.; Vinha, A.F.; Oliveira, M.B.P.P. State of the art in coffee processing by-products. In *Handbook of Coffee Processing By-Products*; Galanakis, C.M., Ed.; Academic Press Elsevier: London, UK, 2017; pp. 1–21.
11. Bresciani, L.; Calani, L.; Bruni, R.; Brighenti, F.; Del Rio, D. Phenolic composition, caffeine content and antioxidant capacity of coffee silverskin. *Food Res. Int.* **2014**, *61*, 196–201. [CrossRef]
12. Bondesson, E. A Nutritional Analysis on the By-Product Coffee Husk and its Potential Utilization in Food Production. Bachelor's Thesis, Swedish University of Agricultural Sciences, Uppsala, Sweden, 2015.
13. Abdeltaif, S.A.; SirElkhatim, K.A.; Hassan, A.B. Estimation of Phenolic and Flavonoid Compounds and Antioxidant Activity of Spent Coffee and Black Tea (Processing) Waste for Potential Recovery and Reuse in Sudan. *Recycling* **2018**, *3*, 27. [CrossRef]
14. Meng, S.; Cao, J.; Feng, Q.; Peng, J.; Hu, Y. Roles of chologenic acid on regulating glucose and lipids metabolism: A review. *Evid. Based Compl. Alt.* **2013**, *1*, 1–11. [CrossRef]
15. Murthy, P.S.; Naidu, M.M. Sustainable management of coffee industry by-products and value addition-A review. *Resour. Conserv. Recy.* **2012**, *66*, 45–58. [CrossRef]
16. Tajik, N.; Tajik, M.; Mack, I.; Enck, P. The potential effects of chlorogenic acid, the main phenolic components in coffee, on health: A comprehensive review of the literature. *Eur. J. Nutr.* **2017**, *56*, 2215–2244. [CrossRef] [PubMed]
17. Naveed, M.; Hejazi, V.; Abbas, M.; Kamboh, A.; Khan, G.J.; Shumzaid, M.; Ahmad, F.; Babazadeh, D.; FangFang, X.; Modarresi-Ghazani, F.; et al. Review Chlorogenic acid (CGA): A pharmacological review and call for further research. *Biomed. Pharm.* **2018**, *97*, 67–74. [CrossRef] [PubMed]
18. Narita, Y.; Inouye, K. High antioxidant activity of coffee silverskin extracts obtained by the treatment of coffee silverskin with subcritical water. *Food Chem.* **2012**, *135*, 943–949. [CrossRef] [PubMed]
19. Iriondo-DeHond, A.; Aparicio García, N.; Fernandez-Gomez, B.; Guisantes-Batan, E.; Velázquez Escobar, F.; Blanch, G.P.; San Andres, M.I.; Sanchez-Fortun, S.; del Castillo, M.D. Validation of coffee by-products as novel food ingredients. *Innov. Food Sci. Emerg.* **2019**, *51*, 194–204. [CrossRef]
20. Regazzoni, L.; Saligari, F.; Marinello, C.; Rossoni, G.; Aldini, G.; Carini, M.; Orioli, M. Coffee silver skin as a source of polyphenols: High resolution mass spectrometric profiling of components and antioxidant activity. *J. Funct. Foods* **2016**, *20*, 472–485. [CrossRef]
21. Toschi, T.G.; Cardenia, V.; Bonaga, G.; Mandrioli, M.; Rodriguez-Estrada, M.T. Coffee Silverskin: Characterization, Possible Uses, and Safety Aspects. *J. Agric. Food Chem.* **2014**, *62*, 10836–10844. [CrossRef]
22. Borrelli, R.C.; Esposito, F.; Napolitano, A.; Ritieni, A.; Fogliano, V. Characterization of a New Potential Functional Ingredient: Coffee Silverskin. *J. Agric. Food Chem.* **2004**, *52*, 1338–1343. [CrossRef]
23. Fernandez-Gomez, B.; Lezama, A.; Amigo-Benavent, M.; Ullate, M.; Herrero, M.; Martin, M.A.; Mesa, M.D.; del Castillo, M.D. Insights on the health benefits of the bioactive compounds of coffee silverskin extract. *J. Funct. Foods* **2016**, *25*, 197–207. [CrossRef]
24. Zengin, G.; Sinan, K.I.; Mahomoodally, M.F.; Angeloni, S.; Mustafa, A.; Vittori, S.; Maggi, F.; Caprioli, G. Chemical Composition, Antioxidant and Enzyme Inhibitory Properties of Different Extracts Obtained from Spent Coffee Ground and Coffee Silverskin. *Foods* **2020**, *9*, 713. [CrossRef]

25. Perez-Miguez, R.; Marina, M.L.; Castro-Puyana, M. High resolution liquid chromatography tandem mass spectrometry for the separation and identification of peptides in coffee silverskin protein hydrolysates. *Microchem. J.* **2019**, *149*, 103951. [CrossRef]
26. Mirabella, N.; Castellani, V.; Sala, S. Current options for the valorization of food manufacturing waste: A review. *J. Clean. Prod.* **2014**, *65*, 28–41. [CrossRef]
27. Tacer-Caba, Z.; Nilufer-Erdil, D.; Ai, Y. Chemical Composition of Cereals and Their Products. In *Handbook of Food Chemistry*; Cheung, P., Mehta, B., Eds.; Springer: Berlin/Heidelberg, Germany, 2015; pp. 323–325.
28. Altan, A.; Maskan, M. Development of Extruded Food by Utilizing Food Industry By-Products. In *Advances in Food Extrusion Technology*; Maskan, M., Altan, A., Eds.; CRC Press: Boca Raton, FL, USA, 2012; pp. 121–159.
29. Tiwari, A.; Jha, S.K. Extrusion cooking technology: Principal mechanism and effect on direct expanded snack–an overview. *Int. J. Food Stud.* **2017**, *6*, 113–128. [CrossRef]
30. Alam, M.S.; Kumar, S. Optimization of Extrusion Process Parameters for Red Lentil-carrot Pomace Incorporated Ready-to-eat Expanded Product Using Response Surface. *Food Sci. Technol.* **2014**, *2*, 106–119. [CrossRef]
31. Altan, A.; McCarthy, K.L.; Maskan, M. Effect of extrusion process on antioxidant activity, total phenolics and β-glucan content of extrudates developed from barley-fruit and vegetable by-products. *Int. J. Food Sci. Tech.* **2009**, *44*, 1263–1271. [CrossRef]
32. Stojceska, V.; Ainsworth, P.; Plunkett, A.; Ibanoğlu, E.; Ibanoğlu, Ş. Cauliflower by-products as a new source of dietary fibre, antioxidants and protein in cereal based ready-to-eat expanded snacks. *J. Food Eng.* **2008**, *87*, 554–563. [CrossRef]
33. Karkle, E.L.; Keller, L.; Dogan, H.; Alavi, S. Matrix transformation in fiber-added extruded products: Impact of different hydration regimens on texture, microstructure and digestibility. *J. Food Eng.* **2012**, *108*, 171–182. [CrossRef]
34. Gimenez-Escalante, P.; Rahimifard, S. *Innovative Food Technologies for Redistributed Manufacturing*; Report; Centre for Sustainable Manufacturing and Recycling Technologies, Loughborough University: Loughborough, UK, 2016.
35. Janssen, L.P.B.M.; Mościcki, L. Scale-Up of Extrusion-Cooking in Single-Screw Extruders. In *Extrusion-Cooking Techniques*; Mościcki, L., Ed.; Wiley-VCH: Weinheim, Germany, 2011; pp. 191–199.
36. Borneo, R.; León, A.E. Whole grain cereals: Functional components and health benefits. *Food Funct.* **2012**, *3*, 110–119. [CrossRef]
37. Poletti, S.; Gruissem, W.; Sautter, C. The nutritional fortification of cereals. *Curr. Opin. Biotech.* **2004**, *15*, 162–165. [CrossRef]
38. Pourfarzad, A.; Mahdavian-Mehr, H.; Sedaghat, N. Coffee silverskin as a source of dietary fiber in bread-making: Optimization of chemical treatment using response surface methodology. *LWT Food Sci. Technol.* **2013**, *50*, 599–606. [CrossRef]
39. Ateş, G.; Elmacı, Y. Coffee silverskin as fat replacer in cake formulations and its effect on physical, chemical and sensory attributes of cakes. *LWT Food Sci. Technol.* **2018**, *90*, 519–525. [CrossRef]
40. Garcia-Serna, E.; Martinez-Saez, N.; Mesias, M.; Morales, F.J.; del Castillo, M.D. Use of coffee silverskin and stevia to improve the formulation of biscuits. *Pol. J. Food Nutr. Sci.* **2014**, *64*, 243–251. [CrossRef]
41. Gocmen, D.; Sahan, Y.; Yildiz, E.; Coskun, M.; Aroufai, İ.A. Use of coffee silverskin to improve the functional properties of cookies. *J. Food Sci. Technol.* **2019**, *56*, 2979–2988. [CrossRef] [PubMed]
42. Servicio de Información Agroalimentaria y Pesquera. *Panorama Agroalimentario*, 1st ed.; SIAP: Mexico City, Mexico, 2019; pp. 30–31, 98–100.
43. Berghofer, E.; Schoenlechner, R. Grain Amaranth. In *Pseudocereals and Less Common Cereals*, 1st ed.; Belton, P.S., Taylor, J.R.N., Eds.; Springer: Berlin/Heidelberg, Germany, 2002; pp. 219–252.
44. Secretaría de Economía. NMX-F-083-1986 Foods–Moisture in food products determination. In *Diario Oficial de la Federación*; Direccion General de Normas: Mexico City, México, 1986.
45. Secretaría de Economía. NMX-F-608-NORMEX-2011 Alimentos–Determinacion de proteínas en alimentos –Método de ensayo (prueba). In *Diario Oficial de la Federación*; Direccion General de Normas: Mexico City, México, 2011.
46. Secretaría de Economía. NMX-F-607-NORMEX-2013 Alimentos–Determinacion de cenizas en alimentos–Método de prueba. In *Diario Oficial de la Federación*; Direccion General de Normas: Mexico City, México, 2013.

47. Secretaría de Salud. NOM-086-SSA1-1994 Alimentos y bebidas no alcohólicas con modificaciones en su composición. In *Diario Oficial de la Federación*; Direccion General de Normas: Mexico City, México, 1994.
48. AOAC International. *A.O.A.C. Method 986.25 Carbohydrates, Determination by difference, a Proximate Analysis. Official Methods of Analysis*, 15th ed.; Association of official analytical chemist: Arlington, VA, USA, 1990; Volume I.
49. Secretaría de Salud. NOM-092-SSA1-1994 Método para la cuenta de bacterias aerobias en placa. In *Diario Oficial de la Federación*; Direccion General de Normas: Mexico City, México, 1994.
50. Secretaría de Salud. NOM-113-SSA1-1994 Cuenta de organismos coliformes en placa en alimentos. In *Diario Oficial de la Federación*; Direccion General de Normas: Mexico City, México, 1994.
51. Secretaría de Salud. NOM-111-SSA1-1994 Método de conteo de hongos y levaduras en alimentos. In *Diario Oficial de la Federación*; Direccion General de Normas: Mexico City, México, 1994.
52. Secretaría de Salud. NOM-114-SSA1-1994 Método para la determinación de Salmonella en alimentos. In *Diario Oficial de la Federación*; Direccion General de Normas: Mexico City, México, 1994.
53. Secretaría de Salud. CCAYAC-M-004 Método de prueba para la estimación de la densidad microbiana por la técnica del número más probable (NMP), detección de coliformes totales, coliformes fecales y Escherichia coli. In *Diario Oficial de la Federación*; Direccion General de Normas: Mexico City, México, 1994.
54. Secretaría de Salud. NOM-115-SSA1-1994 Método para la determinación de Staphylococcus aureus en alimentos. In *Diario Oficial de la Federación*; Direccion General de Normas: Mexico City, México, 1994.
55. Secretaría de Economía. NMX-F-622-NORMEX-2008 Alimentos–Determinación de fibra dietética, fracción insoluble y fracción soluble (Método gravimétrico enzimático)–Método de prueba. In *Diario Oficial de la Federación*; Direccion General de Normas: Mexico City, México, 2008.
56. Del Rio, D.; Calani, L.; Dall'Asta, M.; Brighenti, F. Polyphenolic composition of hazelnut skin. *J. Agric. Food Chem.* **2011**, *59*, 9935–9941. [CrossRef] [PubMed]
57. Singleton, V.L.; Rossi, J.J.A. Colorimetric of total phenolics with phosphomolybdic-phosphotungstic acid reagents. *Am. J. Enol. Vitic.* **1965**, *16*, 144–158.
58. Nenadis, N.; Wang, L.; Tsimidou, M.; Zhang, H. Estimation of scavenging activity of phenolic compounds using the ABTS+ assay. *J. Agric. Food Chem.* **2004**, *52*, 4669–4674. [CrossRef]
59. Brand-Williams, W.; Cuvelier, M.E.; Berset, C. Use of a free radical method to evaluate antioxidant activity. *Lebenson Wiss Technol.* **1995**, *28*, 25–30. [CrossRef]
60. Fujioka, K.; Shibamoto, T. Chlorogenic acid and caffeine contents in various commercial brewed coffees. *Food Chem.* **2008**, *106*, 217–221. [CrossRef]
61. Miller, J.N.; Miller, J.C. *Estadística y Quimiometría para Química Analítica*, 4th ed.; Prentice Hall: Madrid, Spain, 2002; pp. 111–152.
62. Bravo, S.D.; Zavala, K.G.; Puch, J.L.; Rodríguez, J.; Lugo, O.Y. Desarrollo y validación de un método UPLC-MS/MS para la determinación y cuantificación de micotoxinas. *Av. De Investig. En Inocuidad De Aliment.* **2019**, *2*, 1–4.
63. Organisation for Economic Co-operation and Development. *OECD 425 Guidelines for the Testing of Chemicals*; Organisation for Economic Co-operation and Development: Paris, France, 2008.
64. Secretaría de Agricultura, Ganadería, Desarrollo rural, Pesca y Alimentación. NOM-062-ZOO-1999 Especificaciones técnicas para la producción, cuidado y uso de los animales de laboratorio. In *Diario Oficial de la Federación*; Direccion General de Normas: Mexico City, México, 2001.
65. Anderson, R.A.; Conway, H.F.; Peplinski, A.J. Gelatinization of corn grits by roll cooking, extrusion cooking and steaming. *Die Stärke* **1970**, *4*, 130–135. [CrossRef]
66. Guitérrez, H.; de la Vara, R. Optimización simultánea de varias respuestas. Diseño de experimentos con mezclas. In *Análisis y Diseño de Experimentos*, 3rd ed.; Toledo, M.A., Roig, P.E., Rocha, M.I., Delgado, A.L., García, Z., Eds.; McGraw Hill: Mexico City, México, 2012; pp. 385–396, 439–455.
67. Secretaría de Economía. NMX-F-013-SCFI-2010 Pure roasted coffee, whole beans or ground, decaffeinated or not decaffeinated-Specifications and test methods. In *Diario Oficial de la Federación*; Direccion General de Normas: Mexico City, México, 2010.
68. Secretaría de Salud. NOM-247-SSA1-2008 Cereales y sus productos. Cereales, harinas de cereal, sémolas o semolinas. Alimentos a base de: Cereales, semillas comestibles, de harinas, sémolas o semolinas o sus mezclas. Productos de panificación. Disposiciones y especificaciones sanitarias y nutrimentales. Métodos de prueba. In *Diario Oficial de la Federación*; Direccion General de Normas: Mexico City, México, 2009.

69. Unión Europea. Reglamento (CE) No. 333/2007. In *Comisión de las Comunidades Europeas*; Unión Europea: Brussels, Belgium, 2007.
70. Farah, A. Coffee Constituents. In *Coffee: Emerging Health Effects and Disease Prevention*, 1st ed.; John Wiley & Sons: Hoboken, NJ, USA, 2012; pp. 21–58.
71. Narita, Y.; Inouye, K. Chlorogenic Acids from Coffee. In *Coffee in Health and Disease Prevention*; Elsevier: Amsterdam, The Netherlands, 2015; pp. 189–199.
72. Ruiz-Palomino, P.; Guatemala-Morales, G.; Mondragón-Cortéz, P.M.; Zúñiga-González, E.A.; Corona-González, R.I.; Arriola-Guevara, E. Empirical model of the chlorogenic acid degradation kinetics during coffee roasting in a spouted bed. *Rev. Mex. Ing. Quim.* **2019**, *18*, 387–396. [CrossRef]
73. Martínez, M.M.; Vargas del Río, L.M.; Gómez, V.M. Aflatoxinas: Incidencia, impactos en la salud, control y prevención. *Biosalud* **2013**, *12*, 89–109.
74. Sotelo, A.; Argote, R.M.; Cornejo, L.; Escalona, S.; Ramos, M.; Nava, A.; Palomino, D.; Carreón, O. Medición de fibra dietética y almidón resistente: Reto para alumnos del Laboratorio de Desarrollo Experimental de Alimentos (LabDEA). *Educ. Quím.* **2008**, *19*, 42–49. [CrossRef]
75. Unión Europea. Reglamento (CE) No. 1924/2006. In *Comisión de las Comunidades Europeas*; Unión Europea: Brussels, Belgium, 2006.

Article

Agave Syrup as an Alternative to Sucrose in Muffins: Impacts on Rheological, Microstructural, Physical, and Sensorial Properties

César Ozuna [1,2,*], Eugenia Trueba-Vázquez [1], Gemma Moraga [3], Empar Llorca [3] and Isabel Hernando [3,*]

1 Departamento de Alimentos, División de Ciencias de la Vida, Campus Irapuato-Salamanca, Universidad de Guanajuato, Carretera Irapuato-Silao km 9, 36500 Irapuato, Guanajuato, Mexico; etruvaz@hotmail.com

2 Posgrado en Biociencias, División de Ciencias de la Vida, Campus Irapuato-Salamanca, Universidad de Guanajuato, Carretera Irapuato-Silao km 9, 36500 Irapuato, Guanajuato, Mexico

3 Grupo de Investigación Microestructura y Química de Alimentos, Departamento de Tecnología de Alimentos, Universitat Politècnica de València, Camino de Vera s/n, 46022 Valencia, Spain; gemmoba1@tal.upv.es (G.M.); emllomar@tal.upv.es (E.L.)

* Correspondence: cesar.ozuna@ugto.mx (C.O.); mihernan@tal.upv.es (I.H.); Tel.: +52-462-624-18-89 (ext. 5234) (C.O.); +34-96-387-73-63 (ext. 77363) (I.H.)

Received: 19 May 2020; Accepted: 19 June 2020; Published: 8 July 2020

Abstract: Natural sweeteners, such as agave syrup, might be a healthy alternative to sucrose used in sweet bakery products linked to obesity. We evaluated the effect of sucrose replacement by agave syrup on rheological and microstructural properties of muffin batter and on physical and sensorial properties of the baked product. Muffins were formulated by replacing 25%, 50%, 75%, and 100% of sucrose by agave syrup (AS) and partially hydrolyzed agave syrup (PHAS), and by adding xanthan gum and doubled quantities of leavening agents. Rheological and microstructural properties of batter during baking were analyzed over the range of 25–100 °C. In the muffins, the structure, texture, color, and sensory acceptance were studied. The combination of agave syrup with xanthan gum and doubled quantities of leavening agents affected ($p < 0.05$) rheological and microstructural properties of the batters and textural properties of the low-sucrose muffins compared to the controls. The increase in agave syrup levels resulted in a darker crumb and crust. Sensory evaluation showed that AS-75 and PHAS-75 were the best alternatives to the control samples. Our results suggest a plausible substitution of up to 75% of sucrose by agave syrup in preparation of muffins, with physical and sensorial characteristics similar to those of their sucrose-containing counterparts.

Keywords: inulin; bakery products; xanthan gum; leavening agent

1. Introduction

Nowadays, the eating habits of the world's population include high ingestion of foods rich in sugars and fats, such as sweet bakery products [1,2]. Due to their practicality and pleasant taste, muffins are widely consumed by people of all ages, although mainly children [3,4]. The excessive consumption of these products has contributed to an increase in a series of non-communicable diseases and comorbidities, such as overweight, obesity [5], and dental caries [6]. According to the World Health Organization, almost 40% of the adult world population was overweight in 2016 and 13% were obese, whereas 40 million children under 5 years old were overweight or obese in 2018 [7].

In addition to providing sweetness to bakery products, sucrose contributes to the development of their structure, texture, and color [8]. Therefore, the replacement of sucrose content by artificial or natural sweeteners in the production of bakery products represents a challenge for the food industry [1,9,10]. Intense or non-caloric sweeteners (sucralose, saccharin, cyclamate, stevia, etc.) have great sweetening power; however, they do not contribute to the formation of the body of the bakery product [11]. On the other hand, energy sweeteners (monosaccharides, disaccharides, polyalcohols, etc.) can give rise to bakery products with stable structure, but they tend to lack in flavor [11]. Nevertheless, some natural agents may combine the best qualities of both groups of sweeteners: good sweetening power and a stable structure of the final bakery product; this group of sweeteners includes agave syrup (AS) [12–14].

Agave syrup is the sweet substance obtained by the hydrolysis of fructans present in the *Agave* spp. heads. In Mexico, where *Agave* spp. is endemic, there are about 205 species; however, agave syrup is obtained mostly from *Agave tequilana* Weber var. blue [15]. This natural sweetener, composed of fructose and fructooligosaccharides, has proven to have beneficial properties on human health, such as high prebiotic capacity and a low glycemic index score, and could prevent obesity and type II diabetes mellitus [16,17].

In bakery products, the use of agave syrup as a partial or total sucrose replacer has been employed in the elaboration of cookies [12], gluten-free cakes [13], and cereal bars [14]. However, the effects of agave syrup on the rheological, microstructural, and sensorial properties of bakery products remain unknown, both in the batter and in the final product.

The aim of this research was to evaluate the effects of sucrose replacement by agave syrup on rheological and microstructural properties of muffin batter, as well as on physical and sensorial properties of the baked product. Additionally, our study aimed to find added value in a natural sweetener that is currently underused in bakery products due to the lack of knowledge on its behavior during the baking process.

2. Materials and Methods

2.1. Muffin Batter Ingredients

The muffin batter ingredients included wheat flour (Cerealien Bischheim GmbH, Bischheim, Germany; composition provided by the supplier: 15% moisture, 12% protein); sugar (Pfeifer & Langen GmbH and Co., Cologne, Germany); whole liquid egg (Huevos Guillen S. L., Valencia, Spain); skimmed milk powder (Capsa Food, Asturias, Spain); refined sunflower oil (Sovena España S.A., Sevilla, Spain); agave syrup (Mieles Campos Azules S. A. de C. V., Amatitlan, Mexico; specifications of moisture and total sugars provided by the supplier: 23.20% moisture, 92.86% fructose, 0.15% glucose, 0.12% sucrose, 6.71% inulin, and 0.16% other carbohydrates); partially hydrolyzed agave syrup (PHAS) (Mieles Campos Azules S. A. de C. V., Amatitlan, Mexico; specifications of moisture and total sugars provided by the supplier: 23.30% moisture, 85.52% fructose, 0.40% glucose, 0.25% sucrose, 13.58% inulin, and 0.25% other carbohydrates); xanthan gum (Cargill France SAS, Puteaux, France); leavening agents, including sodium bicarbonate, malic acid, and tartaric acid (Hacendado, Valencia, Spain); and salt (Salinas Del Odiel, S. L., Huelva, Spain). The sucrose equivalent (SE) of agave syrup (AS) and partially hydrolyzed agave syrup (PHAS) was calculated using the equation proposed by Koehler and Kays [18] and Belščak-Cvitanović et al. [19]; the SE values for AS and PHAS were 1.61 and 1.49, respectively.

2.2. Batter and Muffin Preparation

Nine muffin batters were prepared according to batter formulations in Table 1. In the case of batters elaborated with sucrose replacement by AS and PHAS (25% replacement: AS-25 and PHAS-25; 50% replacement: AS-50 and PHAS-50; 75% replacement: AS-75 and PHAS-75; and 100% replacement: AS-100 and PHAS-100), the concentration of leavening agents was doubled and xanthan gum was used as a loading agent according to Martínez-Cervera et al. [3].

Table 1. Formulations of the different muffin batters (percentage on wheat flour basis).

Ingredients	CONTROL	AS-25	AS-50	AS-75	AS-100	PHAS-25	PHAS-50	PHAS-75	PHAS-100
Wheat flour	100	100	100	100	100	100	100	100	100
Sucrose	100	75	50	25	-	75	50	25	-
AS	-	15.48	30.95	46.42	61.90	-	-	-	-
PHAS	-	-	-	-	-	16.68	33.36	50.05	66.73
Xanthan gum	-	0.5	0.5	0.5	0.5	0.5	0.5	0.5	0.5
Pure water	-	30	30	30	30	30	30	30	30
Whole liquid egg	81	81	81	81	81	81	81	81	81
Skimmed milk	100	100	100	100	100	100	100	100	100
Sunflower oil	46	46	46	46	46	46	46	46	46
Sodium bicarbonate	4	8	8	8	8	8	8	8	8
Malic and tartaric acid	3	6	6	6	6	6	6	6	6
Salt	1.5	1.5	1.5	1.5	1.5	1.5	1.5	1.5	1.5

AS: Agave syrup; PHAS: Partially hydrolyzed agave syrup.

The all-in mixing procedure reported by Rodríguez-García et al. [20] was employed in batter preparation, with some modifications. First, the liquid ingredients (including 0.5 g of xanthan gum dissolved in 30 g of water), except for the sunflower oil, were introduced into the commercial kneader (Thermomix, TM31, Wuppertal, Germany) and mixed for 1 min at a speed of 200 rpm. Then, the solid ingredients were added into the same container and mixed for an additional 2 min at the same speed. Finally, sunflower oil was added and mixed in for 3 min at 500 rpm. Once the smooth batter was obtained, 45 g were placed in paper molds and subsequently deposited in silicone trays. Finally, they were introduced into an electric oven (Electrolux, EOC3430DOX, Stockholm, Sweden) that had been preheated to 180 °C. Samples were baked for 30 min at 180 °C. Each baking batch consisted of 12 muffins and each formulation was carried out in triplicate, representing 36 samples per formulation. All analyses were performed 24 h after baking in order to ensure stability in the samples.

2.3. *Rheology and Microstructure of Batters*

Rheological and microstructural analyses were performed in batters with 0% (control), 50%, and 100% sucrose replacement by both AS and PHAS. Rheological measurements were carried out using a rotational rheometer (Kinexus Pro+, Malvern Panalytical, Malvern, UK) equipped with a Peltier plate cartridge. A series of tests were performed at 20 °C with parallel plate geometry (Φ = 40 mm), with the geometry gap set at 1500 µm. Before the rheological test, the batters were all kept at 25 °C for 60 min post-preparation. Flow measurements were conducted (shear rate 1 s^{-1} to 100 s^{-1}, temperature = 25 °C), along with frequency sweeps (stress = 10 Pa, frequency = 0.1–10 Hz, temperature = 25 °C) and temperature sweeps in the linear viscoelastic region (frequency = 1 Hz, stress = 100 Pa, temperature = 25–100 °C). Vaseline oil was applied to the exposed surfaces of the samples to prevent sample drying during testing.

Microscopical examination of muffin batters during simulated micro-baking was carried out following the methodology proposed by Rodríguez-García et al. [20]. One drop of the sample was placed in the concavity of a glass slide and placed into a temperature-controlled vault. The temperature in the vault rose steadily from 25 °C to 105 °C at the rate of 1.5 °C/min. Batter samples were observed at 4× magnification (objective lens 4 × /0.13∞/ – WD 17.1, Nikon, Tokyo, Japan). A camera (ExWaveHAD, model no. DXC-190, Sony, Tokyo, Japan) was attached to the microscope and connected to the video entry port of a computer. Images were captured and stored in the format of 640 × 540 pixels using the microscope software (Linksys 32, Linkam, Surrey, UK).

2.4. *Muffin Height and Crumb Structure*

Muffins were cut vertically with a stainless steel knife and scanned by means of a conventional scanner (Epson Perfection 1250; Epson America, Inc., Long Beach, CA, USA) at a resolution of 300 dpi. The maximum muffin height and the crumb structure were measured using ImageJ software (National

Institute of Health, Bethesda, MA, USA). Each image was cropped to a 30 × 30 mm section on which the crumb analysis was performed. The image was split into color channels, the contrast was enhanced, and the image was binarized at the grey scale, resulting in air cells being colored in black and the rest of the crumb in white. Four macroscopic characteristics of the crumb cell structure were calculated, namely the cell area (mm^2), cell circularity, cell density, and total cell area within the crumb (%). The data for the muffin height and crumb structure were obtained from 18 different images for each formulation.

2.5. Muffin Texture

Muffin textural properties were evaluated using a texture analyzer (Stable Micro System, TA-XT plus, Godalming, UK) and the Texture Exponent Lite 32 program (version 6.1.4.0, Stable Micro Systems, Godalming, UK). For texture profile analysis (TPA), cubes were cut from the central area of the muffin (1.5 cm per side). Double compressions of 40% deformation at a speed of 1 mm/s were performed, with a resting time of 5 s between the two compressions. Compression was performed with a 35-mm diameter aluminum plate. After the two compression cycles, the following parameters were recorded: hardness, elasticity, cohesiveness, and chewiness [21].

2.6. Muffin Crust and Crumb Color

Color measurements of both the muffin crust and crumb were carried out with a CR-400 chroma meter (Konica Minolta Sensing Americas, Inc., Ramsey, MN, USA). The results were expressed in accordance with the International Commission on Illumination (CIELAB) system with reference to illuminant C and a visual angle of 2°. The determined parameters were L^* (lightness, L^* = 0 (black) and L^* = 100 (white)), a^* ((a^* negative values = greenness and a^* positive values = redness)), and b^* ((b^* negative values = blueness and b^* positive values = yellowness)). The chroma (C^*_{ab}), hue angle (h^*_{ab}), and total color difference (ΔE^*) were calculated using Equations (1)–(3):

$$C^*_{ab} = \sqrt{a^{*2} + b^{*2}} \quad (1)$$

$$h^*_{ab} = tan^{-1}\frac{(b^*)}{(a^*)} \quad (2)$$

$$\Delta E^* = \sqrt{\left(L^* - L^*_0\right)^2 + \left(a^* - a^*_0\right)^2 + \left(b^* - b^*_0\right)^2} \quad (3)$$

where L^*_0, a^*_0, and b^*_0 represent the values of the chromatic coordinates of the control muffin.

2.7. Sensory Evaluation of Final Product

Seventy untrained judges (18–80 years old, recruited from the Life Sciences Division of the University of Guanajuato, Irapuato, Mexico) simultaneously evaluated the sensory characteristics (appearance, flavor, texture, color, and general acceptability) of the control and the eight muffin formulations made with AS and PHAS. The product acceptability was determined using a nine-point hedonic scale (9 = like; 1 = dislike) [22].

2.8. Statistical Analysis

Two-way analyses of variance (ANOVAs) with sucrose replacement (0%, 25%, 50%, 75%, and 100%) and agave syrup type (AS and PHAS) as between-subject factors were carried out on all dependent variables in SPSS 18.0 software (SPSS Inc., Chicago, IL, USA). Tukey post-hoc tests were performed for all significant main effects and interactions.

3. Results and Discussion

3.1. Rheological Properties of Muffin Batters

Figure 1 shows the effect of sucrose replacement (at 0%, 50%, and 100% reduction levels) by AS and PHAS and a combination of xanthan gum and doubled quantities of leavening agents on flow curves of the muffin batters.

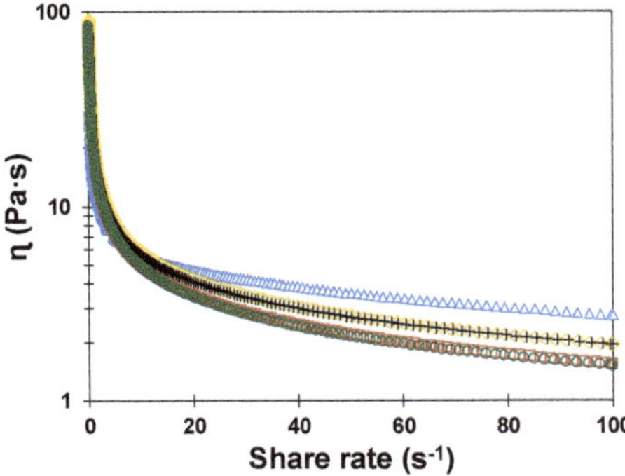

Figure 1. The impact of sucrose replacement by agave syrup (AS) and partially hydrolyzed agave syrup (PHAS) on flow curves of the muffin batters: control (blue triangles), PHAS-50 (black crosses), AS-50 (yellow diamonds), PHAS-100 (green circles), and AS-100 (red squares).

A typical shear thinning behavior was observed in all batters (Figure 1). Despite having used a combination of xanthan gum and doubled quantities of leavening agents, the batters were less viscous when AS and PHAS were employed as sucrose replacers. This behavior is quite possibly related to the moisture content of agave syrups (23.20% and 23.30% for AS and PHAS, respectively) which would have induced the decrease in AS and PHAS batter viscosities. At a shear rate of 60 s^{-1}, the apparent viscosity significantly ($p < 0.05$) decreased as more sucrose was replaced by AS and PHAS, from 3.35 (0.01; the values in parentheses refer to the standard deviation of the mean) Pa·s in the control sample to 2.48 (0.04) Pa·s and 1.98 (0.04) Pa·s in 50% and 100% sucrose-replaced batters, respectively. There was no significant ($p > 0.05$) effect of agave syrup type, nor an interaction between the two factors. Other authors also stated the decrease in the batter viscosity due to the sucrose replacement by syrups resulting in a low product volume and poor cell structure [2].

Figure 2A,B show the frequency dependence of the different batters on the elastic modulus (G′), the viscous modulus (G″), and the phase angle (δ) at 25 °C. At low frequencies (0.1–1 Hz), the control batter behaved as a solid (G′ > G″; δ < 45), while at high frequencies (>1 Hz) this batter exhibited a liquid-like behavior (G′ < G″; δ > 45). On the other hand, batters formulated with AS (AS-50 and AS-100) and PHAS (PHAS-50 and PHAS-100), using a combination of xanthan gum and doubled quantities of leavening agents, showed a solid-like behavior (G′ > G″; δ < 45), regardless of the frequencies tested, caused by the gelling effect of xanthan gum in the batter. At 1 Hz of frequency, sucrose replacement significantly ($p < 0.05$) affected the G′ modulus of batters, with higher sucrose substitution levels producing higher G′ values. However, there was no significant effect of agave syrup type ($p > 0.05$), nor an interaction between the two factors.

The competition for the available water between sucrose and xanthan gum led to a less elastic semi-solid network, with the G′ values obtained for batters with 0, 50, and 100% of sucrose replacement being 42.54 (5.43), 84.24 (4.90), and 98.42 (10.55) Pa, respectively. On the other hand, the G″ modulus was not significantly affected by the factors studied or their interaction (all $ps > 0.05$). At 1 Hz, the G″ values ranged from 45.26 (2.09) to 46.63 (1.95) Pa for the control and the 100% sucrose replacement batter, respectively. As for the δ values, these were significantly affected by sucrose substitution ($p < 0.05$), with higher sucrose replacement levels producing lower δ values, indicating a more solid-like behavior. The actual δ values obtained at 1 Hz for 0, 50, and 100% sucrose replacement were 46.98° (2.35), 28.31° (0.45), and 25.50° (1.68), respectively. There was no significant main effect of agave syrup type ($p > 0.05$), nor an interaction between the two factors.

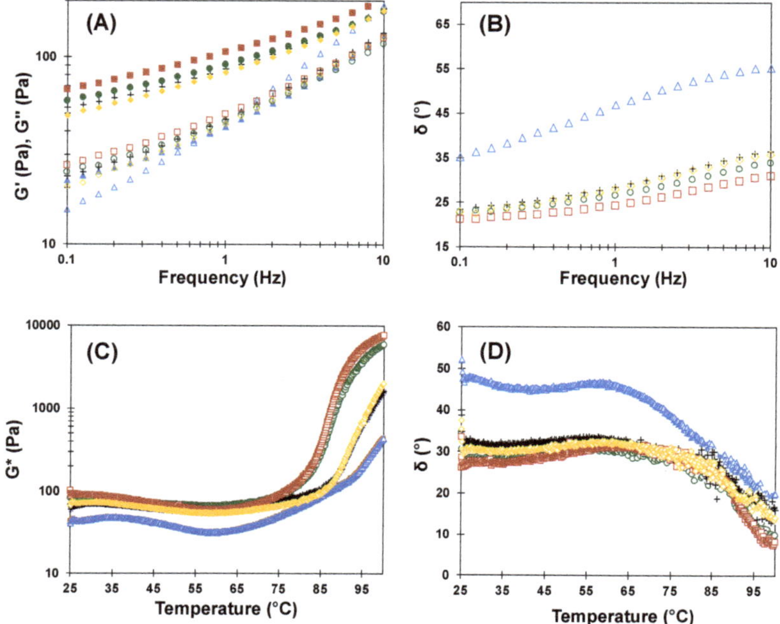

Figure 2. The impact of sucrose replacement by AS and PHAS on frequency (**A**,**B**) and temperature (**C**,**D**) sweeps of the muffin batters: control (blue triangles), PHAS-50 (black crosses), AS-50 (yellow diamonds), PHAS-100 (green circles), and AS-100 (red squares). In Figure 2A, two variables are shown: G′ (full data points) and G″ (empty data points).

To analyze the structural changes provoked by heat in muffin batters, linear viscoelastic properties were studied in the range of 25–100 °C in order to simulate batter behavior during baking. The effects of temperature on the complex modulus G* and δ during batter heating are shown in Figure 2C,D, respectively. The G* modulus, a measure of batter stiffness, was lower in the control sample compared to the samples with sucrose replacement along all the sweep temperatures tested. As the temperature increased, batter stiffness increased as well; this was probably caused primarily by starch gelatinization and protein denaturation [23]. The temperature at which the G* value started to increase was 60 °C for the control batter, however it shifted to 75 °C and 85 °C as sucrose was replaced at 100% and 50%, respectively. In this sense, a synergic effect between sucrose and xanthan gum was observed, delaying the starch gelatinization, and the thermosetting temperature increased in batters with 50% replacement. Most probably, the moisture content of the agave syrups contributed to higher values of the thermosetting temperature in replaced muffins than in the control, since a higher water content needed to be evaporated for the structure of the replaced muffins to be built. A decrease in δ values above those temperatures was also detected, reflecting the predominance of the elastic behavior versus the viscous behavior during starch gelatinization and protein denaturation [24]. In summary, the sucrose replacement by AS and PHAS and a combination of xanthan gum and doubled quantities of leavening agents increased the thermosetting temperature in batters. This fact is important for the correct formation of water vapor and CO_2 in the batter, as well as their diffusion and expansion into occluded air cells during the baking process [3,25].

3.2. Microstructural Properties of Muffin Batters

Figure 3A shows the effect of sucrose replacement (at 0%, 50%, and 100% reduction levels) by AS and PHAS in combination with the addition of xanthan gum and doubled quantities of leavening agents on batter microstructures during simulated micro-baking. These microstructural images were analyzed to quantify the bubble area (Figure 3B).

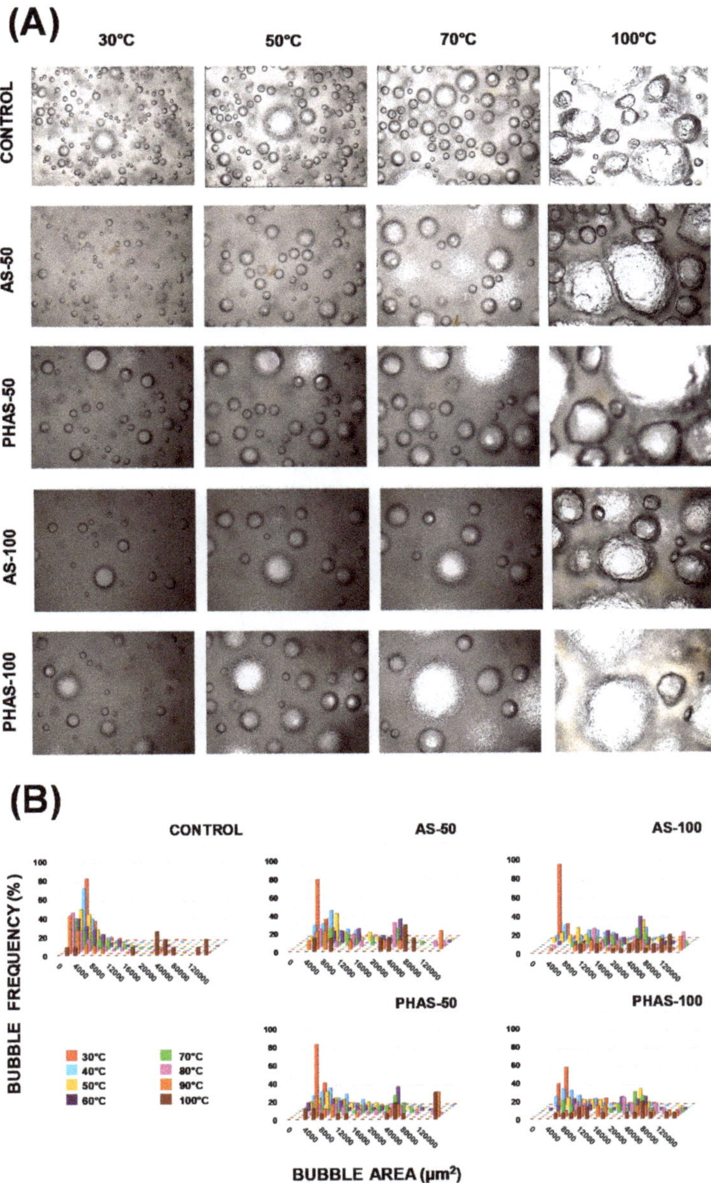

Figure 3. Changes in batters during micro-baking by baking temperature: (**A**) light microscopy images (4×) of bubble expansion; (**B**) bubble size distribution histograms.

When the temperature increased, there was an expansion of the bubbles in all samples and the bubble size distribution tended to widen (Figure 3A). In the control batter, the number of CO_2 bubbles was higher if compared to replaced batters and the bubble size increased at a controlled and uniform rate. This behavior may have been due to the high viscosity of this batter (as observed previously in Figure 1), which may have reduced bubble movement in the batter and slowed down the coalescence phenomena. Thus, regardless of temperature, the control batters showed a higher frequency of small bubble sizes (0–10,000 μm^2) in comparison with replaced batters (Figure 3B).

On the other hand, a lower viscosity of the replaced batters could have aided in bubble coalescence, giving place to bigger bubbles. When sucrose was totally replaced, a high frequency of big bubble sizes (over 60,000 μm^2) could be observed for AS-100 and PHAS-100 when compared to the other batters. Regarding the influence of agave syrup type, PHAS samples showed a higher frequency of big bubbles (over 140,000 μm^2) than AS samples at both 50% and 100% sucrose replacement (Figure 3B).

3.3. *Muffin Crumb Structure*

Figure 4 shows the effect of sucrose replacement by AS and PHAS on the macroscopic characteristics of the crumb cell structure, measured by image analysis.

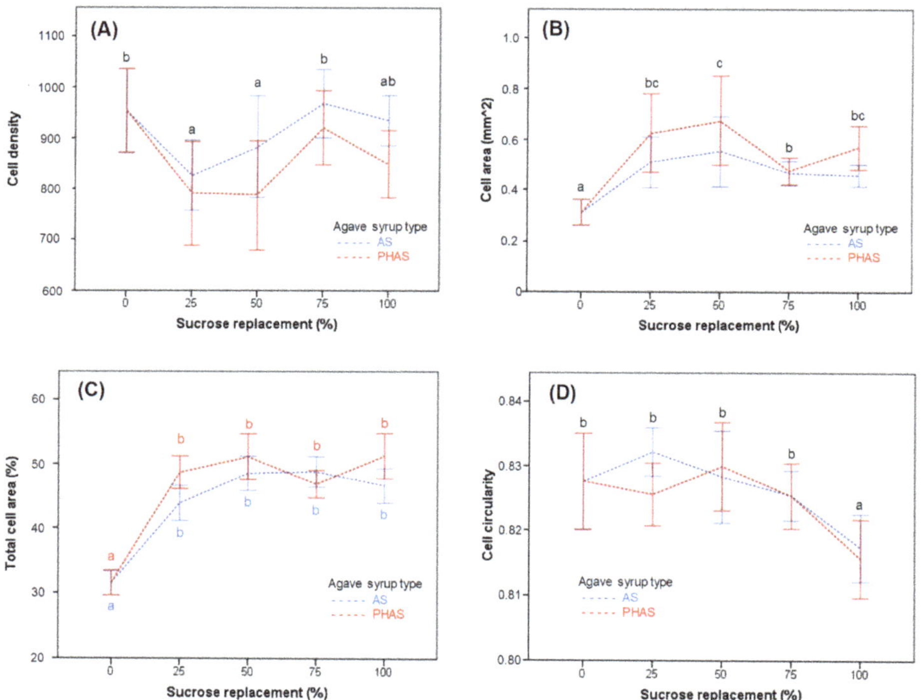

Figure 4. The impact of sucrose replacement by AS and PHAS on the muffin crumb structure, namely cell density (**A**), cell area (**B**), total cell area (**C**), and circularity (**D**). Error bars represent 95% confidence intervals. The different subscript letters of the same color represent Tukey's homogeneous groups, with significant ($p < 0.05$) differences between different sucrose replacement levels for AS (blue letters), PHAS (red letters), or for both types of agave syrup when there was no interaction between the sucrose replacement and agave syrup type (black letters).

Regarding cell density, the main effect of the sucrose replacement was significant ($p < 0.05$), with lower cell density for samples with 25% and 50% sucrose substitution in comparison to control and samples with 75% and 100% sucrose substitution (Figure 4A). There was a significant main effect ($p < 0.05$) of agave syrup type, which did not interact with sucrose replacement ($p > 0.05$). The main effect of sucrose replacement on cell area was significant ($p < 0.05$). As sucrose substitution increased, the average air cell size was higher than the control muffin (Figure 4B). There was also a significant main effect ($p < 0.05$) of agave syrup type, which did not interact with sucrose replacement ($p > 0.05$). A higher average cell area was obtained when substituting sucrose by PHAS than by AS.

As for total cell area (%), there were significant main effects ($p < 0.05$) of sucrose replacement and agave syrup type, with higher levels of sucrose replacement and PHAS reaching higher percentages of total cell area than control muffin (Figure 4C). There was a significant interaction ($p < 0.05$) between the effects of the two variables, caused by an inversion in the general trend at 75% of sucrose substitution by PHAS. Sucrose replacement affected cell circularity, but a significant ($p < 0.05$) reduction was only observed for the formulations with 100% substitution (Figure 4D). However, there was no main effect of agave syrup type on cell circularity, nor an interaction between the two factors ($p > 0.05$).

The obtained results are in accordance with observations of the bubble expansion during micro-baking (Figure 3), where a high frequency of small bubble sizes in control batters could be observed. The variations that occurred in the muffin crumb as sucrose substitution increased are due mainly to the fact that with the addition of xanthan gum and doubled quantity of leavening agents, the batters with sucrose substitution comprised a greater amount of gas. In addition, the low viscosity of the batters allowed for the gas bubbles to have more mobility, provoking their coalescence. The gases did not reach the surface since their exit was hindered by an early formation of the crust, resulting in the setting of the bubbles in the form of diffusion pathways. Thus, the samples with over 25% sucrose replacement by both AS and PHAS reached a height of over 57.33 (2.10) mm. When compared to the average height of 50.93 (2.27) mm obtained from the control samples, significant ($p < 0.05$) differences in height can be observed for all samples with sucrose replacement by AS and PHAS.

3.4. Muffin Texture

Figure 5 shows the effect of sucrose replacement by AS and PHAS on the texture profile analysis of the final products. Since the textural properties of muffins depend greatly on their crumb structure, all samples with over 25% sucrose substitution by either AS or PHAS presented significant differences ($p < 0.05$) with the control for all textural properties evaluated. As a result, all the replaced muffins considerably differed from the control regardless of syrup type or replacement level.

Regarding hardness, the main effect of sucrose replacement was significant ($p < 0.05$), with lower hardness values recorded for samples with sucrose replacement than control, decreasing from 1.27 (0.30) N in the control sample to a range of 0.65 (0.27) to 0.85 (0.32) N for the rest of the samples. Even though agave syrup type did not have a significant main effect ($p > 0.05$) on this attribute, it did interact with sucrose replacement ($p < 0.05$) (Figure 5A).

The main effect of sucrose replacement on springiness was significant ($p < 0.05$). As sucrose substitution increased, the average values tended to increase as well. There was also a significant main effect of agave syrup type on springiness ($p < 0.05$), which did not interact with sucrose replacement ($p > 0.05$). Moreover, a significant difference ($p < 0.05$) was observed between the control sample and the formulations with over 25% sucrose substitution, where the values increased from 0.32 (0.03) in the control formulation to a range of values between 0.38 (0.02) and 0.41 (0.02) for the other formulations. On the other hand, slightly higher springiness values were obtained when substituting sucrose with AS than PHAS (Figure 5B).

Cohesiveness was significantly affected ($p < 0.05$) by sucrose replacement, as well as by agave syrup type. However, there was no significant interaction between these variables ($p > 0.05$). As sucrose substitution levels escalated, cohesiveness significantly increased ($p > 0.05$) as well, from 0.71 (0.02) in the control formulation to at least a value of 0.74 (0.02) for the other samples. Similarly to springiness, slightly higher cohesiveness values were obtained when substituting sucrose with AS than PHAS (Figure 5C). As for chewiness, the results are very similar to hardness. The main effect of sucrose replacement was significant ($p < 0.05$), while the main effect of agave syrup type was not ($p > 0.05$). There was a significant interaction between the effect of sucrose replacement and agave syrup type ($p < 0.05$) caused by an inversion in the general trend at 50% sucrose substitution by AS (Figure 5D).

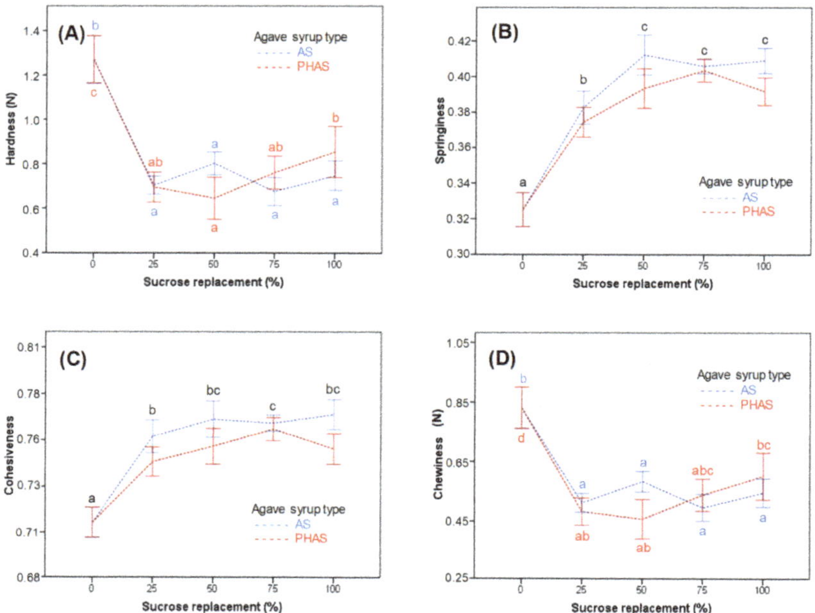

Figure 5. The impact of sucrose replacement by AS and PHAS on the textural profile analysis of muffins: (**A**) hardness, (**B**) springiness, (**C**) cohesiveness, and (**D**) chewiness. The different subscript letters of the same color represent Tukey's homogeneous groups, with significant ($p < 0.05$) differences between different sucrose replacement levels for AS (blue letters), PHAS (red letters), or for both types of agave syrup when there was no interaction between sucrose replacement and agave syrup type (black letters).

In replaced batters, the gases were not able to fully exit the product due to the increase in the amount of gas production during baking, which was caused by the doubling of leavening agents and the low permeability of the crust. Along with this, the low viscosity of the batter allowed for the air bubbles to move and coalesce, causing the muffin crumb to contain large air cells resembling diffusion pathways, as well as compact areas [26]. This particular structure of the crumb led to increases in cohesiveness and springiness, due to the fact that the presence of air cells allowed the muffins to easily recover their original shape and size after being compressed.

Regarding the decrease in hardness shown by the sucrose-substituted muffins, this was probably caused by the increase in total cell area (%). The products became softer because the air cells did not give any resistance during their first compression. Since chewiness refers to the difficulty to chew and create food bolus, its value depends on the product hardness, which is why the products were easier to chew as they became less hard [3]. The decrease in muffin chewiness could be also attributed to the increase in cohesiveness, since this characteristic allows for the formation of a bolus instead of fracturing or crumbling upon mastication [27]. According to Gao et al. [28], partial and total replacement of sucrose in sweetened baked products usually results in higher hardness and chewiness values, as well as in lower springiness. This is credited to the ability of sucrose to delay starch gelatinization. Nonetheless, due to the incorporation of double quantities of leavening agents, our muffins showed opposite results [29].

3.5. Muffin Crust and Crumb Color

In general terms, the control muffin showed a golden-brown crust and a yellow crumb, which are both characteristic of bakery products. However, as the percentage of sucrose substitution increased, the muffin color in general tended to be darker than in control samples, with the muffin crust turning towards red hues and the crumb losing some of its yellowness.

The main effect of sucrose replacement on L^* was significant ($p < 0.05$) in both the muffin crust and crumb. As sucrose substitution increased, the L^* values (ranging from 46.27 to 42.50 for the crust and from 61.34 to 53.23 for the crumb) significantly decreased ($p < 0.05$) in comparison with the control samples (58.81 and 69.16 for the crust and crumb, respectively).

Regarding the C^*_{ab} of the crust, significant effects ($p < 0.05$) were observed for sucrose replacement, as well as for agave syrup type, with a significant interaction ($p < 0.05$) between these factors. As for the crumb C^*_{ab}, only sucrose substitution had a significant effect ($p < 0.05$). As sucrose replacement increased, the color intensity of both the crust and crumb of the muffins showed different tendencies; while the crusts values decreased (40.59 for control and 35.99–32.22 for replaced muffins), the crumb values increased (21.87 for control and 30.27–27.50 for replaced muffins). Both presented significant differences ($p < 0.05$) between the control and the rest of the samples.

The main effect of sucrose substitution on h^*_{ab} was significant ($p < 0.05$) on both the muffin crust and crumb. Replaced muffins exhibited lower h^*_{ab} values of both the crumb and crust (ranging from 65.02 to 62.47 for the crust and from 84.07 to 76.10 for the crumb) than control samples (77.55 and 94.79 for the crust and crumb, respectively). Moreover, as percentages of sucrose replacement increased, the crumb h^*_{ab} values decreased. This shows that sucrose replacement by agave syrup provokes hue changes from yellow to orange-red, both in the muffin crumb and crust.

As for ΔE^*, the main effect of the two factors was significant ($p < 0.05$) for the crust and the crumb. In general, the samples in which sucrose was replaced with PHAS syrup showed higher ΔE^* values in the crust (20.45), whereas crumb ΔE^* values were similar for both AS and PHAS (ranging from 14.66 to 18.50). According to Bodart et al. [30], when the values obtained for total color difference are over three, the color variations between the analyzed samples are visible to the human eye. In this study, all samples with sucrose replacement showed ΔE^* values of over three when compared to the control formulation, meaning the variations in color were easily noticeable to the naked eye.

The differences in color exhibited by the different formulations can mainly be attributed to the intensification of non-enzymatic browning (Maillard reaction) that happened as the amount of incorporated agave syrup increased [31,32]. Similar results were reported by Kocer et al. [33]—as the amount of sucrose replaced by polydextrose increased, the baked products became darker. When substituting sucrose by other sweeteners such as erythritol, sorbitol, maltitol, xylitol, and mannitol, the opposite tends to occur, as these sweeteners do not contribute to the Maillard reaction [3,34]. On a smaller scale, the muffin color was also affected by the dark color of the syrups in comparison to sucrose. Temperature fulfills an important role regarding the color of the crust and crumb of bakery products. Since the Maillard reaction is temperature-dependent, its effect will vary according to the maximum temperature reached. The crust tends to present a greater change due to the fact that it is exposed to higher temperatures, while the crumb's exposure to the heat is more limited [32].

3.6. Sensory Evaluation of Final Product

Figure 6 shows the effects of sucrose replacement by AS and PHAS on sensory acceptance of muffins. The main effect of sucrose replacement on the appearance and flavor of the final products (Figure 6A,B, respectively) was significant ($p < 0.05$). There was also a significant main effect of agave syrup type ($p < 0.05$), which interacted with sucrose replacement ($p < 0.05$). Significant differences ($p < 0.05$) between AS and PHAS were observed at 100% sucrose substitution. Throughout all samples, PHAS maintained higher average scores than AS. The interaction in the case of appearance was caused by the decrease of scores that occurred at 50% sucrose substitution by PHAS, the point at which the tendency for the samples with AS kept rising.

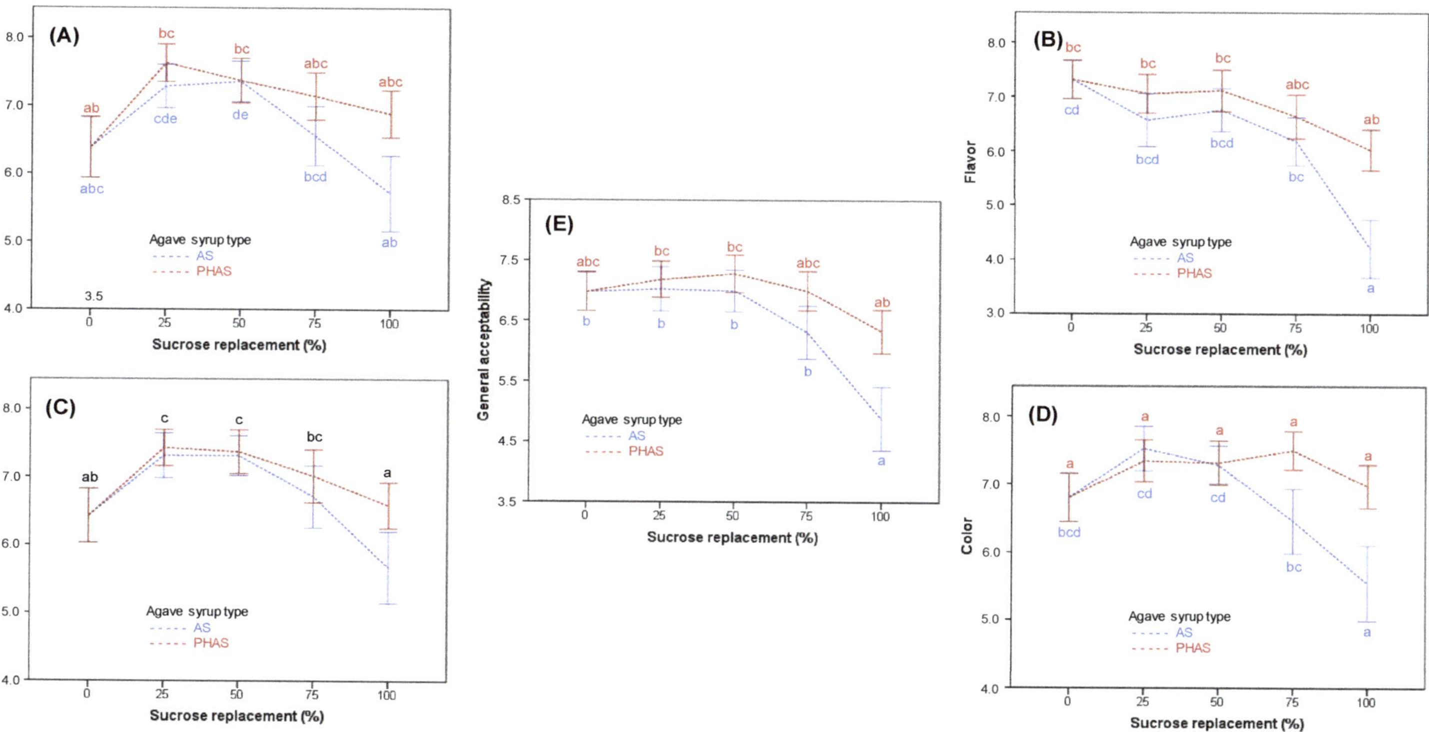

Figure 6. The impact of sucrose replacement by AS and PHAS on sensory properties in muffins, namely appearance (**A**), flavor (**B**), texture (**C**), color (**D**), and general acceptability (**E**). The letters above each data point represent Tukey's homogenous groups ($p < 0.05$) calculated for the sucrose replacement factor. Error bars represent 95% confidence interval. The different subscript letters of the same color represent Tukey's homogeneous groups, with significant ($p < 0.05$) differences between different sucrose replacement levels for AS (blue letters), PHAS (red letters), or for both types of agave syrup when there was no interaction between sucrose replacement and agave syrup type (black letters).

Muffin texture was significantly affected by sucrose substitution ($p < 0.05$), as well as by agave syrup type ($p < 0.05$). However, there was no significant interaction between these factors ($p > 0.05$). The texture of samples with AS and PHAS had a similar tendency for all substitution levels, with muffins made with 25% and 50% sucrose substitution yielding significantly ($p < 0.05$) better texture acceptance than the control and 75% and 100% substitution levels (Figure 6C). In general, PHAS had better texture acceptance than AS.

Regarding color and general acceptance, the main effects of both sucrose replacement and agave syrup type were significant ($p < 0.05$). There was a significant interaction between these factors ($p < 0.05$). Although the samples where sucrose was substituted by PHAS did not show significant differences ($p < 0.05$) between replacement levels, the muffins with AS showed less consistency in their scores, presenting significant differences ($p < 0.05$) for the samples with 100% sucrose substitution.

As sucrose substitution increased, the muffin color became darker. This factor favored the color score up to 75% of sucrose substitution by AS. Some judges singled out the fact that the darker color gave the muffins a more "rustic" and "home-made" appearance, which they liked. However, once the color became too dark, the change became unfavorable. On the other hand, when PHAS was used as sucrose replacement, there were no significant differences in the color score of the different samples (Figure 6D). Taking all aspects into consideration, it is possible to say that the samples made by substituting 75% of the sucrose with either AS or PHAS syrups were the best alternative to the control samples (Figure 6E).

4. Conclusions

Based on the results of our study, both AS and PHAS can be used as alternatives for sucrose in muffin formulations in combination with the addition of xanthan gum and doubled quantities of leavening agents. The sucrose replacement by AS and PHAS reduced the batter viscosity and increased the thermosetting temperature in batters from 60 °C for the control to 75–85 °C for replaced batters. The textural properties of muffins depend greatly on their crumb structure; all the replaced muffins considerably differed from the control regardless of syrup type or replacement level. As for color properties, as the percentage of sucrose substitution increased, muffin color in general tended to be darker than in control samples, with the muffin crust turning towards red hues and the crumb losing some of its yellowness. Even though the percentage of sucrose replacement affected the rheological and microstructural properties of the batters and physical parameters analyzed in the baked products, the sensory evaluation of the muffins suggested that both types of agave syrup can be good alternatives up to 75% sucrose substitution. When 100% of the sucrose was replaced, muffins formulated with PHAS showed better texture, flavor, color, and general acceptability than those formulated with AS.

Author Contributions: Conceptualization, I.H. and C.O.; methodology, I.H., G.M., and E.L.; investigation, C.O. and E.T.-V.; writing—original draft preparation, C.O. and E.T.-V.; writing—review and editing, I.H., G.M., and E.L. All authors have read and agreed to the published version of the manuscript.

Funding: This research was funded by Dirección de Apoyo a la Investigación y Posgrado (Universidad de Guanajuato, Mexico), grant number 1366/2019.

Acknowledgments: The authors acknowledge to the Mexican company Mieles Campos Azules S.A. de C.V. for supplying the agave syrups used in this work. The authors would also like to thank Stanislav Mulík, MA (Applied Linguistics), for his valuable contribution in writing the English version of this paper.

Conflicts of Interest: The authors declare no conflict of interest.

References

1. Luo, X.; Arcot, J.; Gill, T.; Louie, J.C.Y.; Rangan, A. A review of food reformulation of baked products to reduce added sugar intake. *Trends Food Sci. Technol.* **2019**, *86*, 412–425. [CrossRef]
2. Peris, M.; Rubio-Arraez, S.; Castelló, M.L.; Ortolá, M.D. From the laboratory to the kitchen: New alternatives to healthier bakery products. *Foods* **2019**, *8*, 660. [CrossRef]

3. Martínez-Cervera, S.; de la Hera, E.; Sanz, T.; Gómez, M.; Salvador, A. Effect of using erythritol as a sucrose replacer in making spanish muffins incorporating xanthan gum. *Food Bioprocess. Technol.* **2012**, *5*, 3203–3216. [CrossRef]
4. Othman, N.A.; Manaf, M.A.; Harith, S.; Ishak, W.R.W. Influence of avocado purée as a fat replacer on nutritional, fatty acid, and organoleptic properties of low-fat muffins. *J. Am. Coll. Nutr.* **2018**, *37*, 583–588. [CrossRef] [PubMed]
5. Bucher Della Torre, S.; Keller, A.; Laure Depeyre, J.; Kruseman, M. Sugar-sweetened beverages and obesity risk in children and adolescents: A systematic analysis on how methodological quality may influence conclusions. *J. Acad. Nutr. Diet.* **2016**, *116*, 638–659. [CrossRef] [PubMed]
6. Moynihan, P.J.; Kelly, S.A.M. Effect on caries of restricting sugars intake: Systematic review to inform WHO guidelines. *J. Dent. Res.* **2014**, *93*, 8–18. [CrossRef] [PubMed]
7. OMS. Obesity and Overweight. Available online: https://www.who.int/news-room/fact-sheets/detail/obesity-and-overweight (accessed on 30 March 2020).
8. Ghosh, S.; Sudha, M.L. A review on polyols: New frontiers for health-based bakery products. *Int. J. Food Sci. Nutr.* **2012**, *63*, 372–379. [CrossRef] [PubMed]
9. Di Monaco, R.; Miele, N.A.; Cabisidan, E.K.; Cavella, S. Strategies to reduce sugars in food. *Curr. Opin. Food Sci.* **2018**, *19*, 92–97. [CrossRef]
10. Sahin, A.W.; Zannini, E.; Coffey, A.; Arendt, E.K. Sugar reduction in bakery products: Current strategies and sourdough technology as a potential novel approach. *Food Res. Int.* **2019**, *126*, 108583. [CrossRef]
11. Struck, S.; Jaros, D.; Brennan, C.S.; Rohm, H. Sugar replacement in sweetened bakery goods. *Int. J. Food Sci. Technol.* **2014**, *49*, 1963–1976. [CrossRef]
12. Liang, S.; Were, L.M. Chlorogenic acid oxidation-induced greening of sunflower butter cookies as a function of different sweeteners and storage conditions. *Food Chem.* **2018**, *241*, 135–142. [CrossRef] [PubMed]
13. Rothschild, J.; Rosentrater, K.A.; Onwulata, C.; Singh, M.; Menutti, L.; Jambazian, P.; Omary, M.B. Influence of quinoa roasting on sensory and physicochemical properties of allergen-free, gluten-free cakes. *Int. J. Food Sci. Technol.* **2015**, *50*, 1873–1881. [CrossRef]
14. Zamora-Gasga, V.M.; Bello-Pérez, L.A.; Ortíz-Basurto, R.I.; Tovar, J.; Sáyago-Ayerdi, S.G. Granola bars prepared with agave tequilana ingredients: Chemical composition and in vitro starch hydrolysis. *LWT Food Sci. Technol.* **2014**, *56*, 309–314. [CrossRef]
15. Mellado-Mojica, E.; López-Pérez, M.G. Análisis comparativo entre jarabe de agave azul (*Agave tequilana weber* Var. *Azul*) y otros jarabes naturales. *Agrociencia* **2013**, *47*, 233–244.
16. Santiago-García, P.A.; Mellado-Mojica, E.; León-Martínez, F.M.; López, M.G. Evaluation of *Agave angustifolia* fructans as fat replacer in the cookies manufacture. *LWT* **2017**, *77*, 100–109. [CrossRef]
17. Hooshmand, S.; Holloway, B.; Nemoseck, T.; Cole, S.; Petrisko, Y.; Hong, M.Y.; Kern, M. Effects of Agave nectar versus sucrose on weight gain, adiposity, blood glucose, insulin, and lipid responses in mice. *J. Med. Food* **2014**, *17*, 1017–1021. [CrossRef]
18. Koehler, P.E.; Kays, S.J. Sweet Potato Flavor: Quantitative and qualitative assessment of optimum sweetness. *J. Food Qual.* **1991**, *14*, 241–249. [CrossRef]
19. Belščak-Cvitanović, A.; Komes, D.; Dujmović, M.; Karlović, S.; Biškić, M.; Brnčić, M.; Ježek, D. Physical, bioactive and sensory quality parameters of reduced sugar chocolates formulated with natural sweeteners as sucrose alternatives. *Food Chem.* **2015**, *167*, 61–70. [CrossRef]
20. Rodríguez-García, J.; Salvador, A.; Hernando, I. Replacing fat and sugar with inulin in cakes: Bubble size distribution, physical and sensory properties. *Food Bioprocess. Technol.* **2014**, *7*, 964–974. [CrossRef]
21. Diez-Sánchez, E.; Quiles, A.; Llorca, E.; Reiβner, A.-M.; Struck, S.; Rohm, H.; Hernando, I. Extruded flour as techno-functional ingredient in muffins with berry pomace. *LWT* **2019**, *113*, 108300. [CrossRef]
22. Nieto-Mazzocco, E.; Saldaña-Robles, A.; Franco-Robles, E.; Rangel-Contreras, A.K.; Cerón-García, A.; Ozuna, C. Optimization of sorghum, rice, and amaranth flour levels in the development of gluten-free bakery products using response surface methodology. *J. Food Process. Preserv.* **2020**, *44*, e14302. [CrossRef]
23. Wilderjans, E.; Luyts, A.; Brijs, K.; Delcour, J.A. Ingredient functionality in batter type cake making. *Trends Food Sci. Technol.* **2013**, *30*, 6–15. [CrossRef]
24. Carrascal, A.; Rasines, L.; Ríos, Y.; Rioja, P.; Rodríguez, R.; Alvarez-Sabatel, S. Development of reduced-fat muffins by the application of jet-impingement microwave (JIM) technology. *J. Food Eng.* **2019**, *262*, 131–141. [CrossRef]

25. Alvarez, M.D.; Herranz, B.; Fuentes, R.; Cuesta, F.J.; Canet, W. Replacement of wheat flour by chickpea flour in muffin batter: Effect on rheological properties. *J. Food Process. Eng.* **2017**, *40*, e12372. [CrossRef]
26. Fye, A.M.; Setser, C.S. Optimizing texture of reduced-calorie yellow layer cakes. *Cereal Chem.* **1991**, *69*, 338–343.
27. Encina-Zelada, C.R.; Cadavez, V.; Monteiro, F.; Teixeira, J.A.; Gonzales-Barron, U. Combined Effect of xanthan gum and water content on physicochemical and textural properties of gluten-free batter and bread. *Food Res. Int.* **2018**, *111*, 544–555. [CrossRef] [PubMed]
28. Gao, J.; Brennan, M.A.; Mason, S.L.; Brennan, C.S. Effects of Sugar Substitution with "Stevianna" on the Sensory Characteristics of Muffins. Available online: https://www.hindawi.com/journals/jfq/2017/8636043/ (accessed on 9 February 2020).
29. Gómez, M.; Ronda, F.; Caballero, P.A.; Blanco, C.A.; Rosell, C.M. Functionality of different hydrocolloids on the quality and shelf-life of yellow layer cakes. *Food Hydrocoll.* **2007**, *21*, 167–173. [CrossRef]
30. Bodart, M.; de Peñaranda, R.; Deneyer, A.; Flamant, G. Photometry and colorimetry characterisation of materials in daylighting evaluation tools. *Build. Environ.* **2008**, *43*, 2046–2058. [CrossRef]
31. Peressini, D.; Sensidoni, A. Effect of soluble dietary fibre addition on rheological and breadmaking properties of wheat doughs. *J. Cereal Sci.* **2009**, *49*, 190–201. [CrossRef]
32. Conforti, F.D.; Strait, M.J. The effects of liquid honey as a partial substitute for sugar on the physical and sensory qualities of a fat-reduced muffin. *J. Consum. Stud. Home Econ.* **1999**, *23*, 231–237. [CrossRef]
33. Kocer, D.; Hicsasmaz, Z.; Bayindirli, A.; Katnas, S. Bubble and pore formation of the high-ratio cake formulation with polydextrose as a sugar- and fat-replacer. *J. Food Eng.* **2007**, *78*, 953–964. [CrossRef]
34. Kim, J.-N.; Park, S.; Shin, W.-S. Textural and Sensory characteristics of rice chiffon cake formulated with sugar alcohols instead of sucrose. *J. Food Qual.* **2014**, *37*, 281–290. [CrossRef]

Article

Development of Durum Wheat Breads Low in Sodium Using a Natural Low-Sodium Sea Salt

Elena Arena [1], Serena Muccilli [2], Agata Mazzaglia [1], Virgilio Giannone [3], Selina Brighina [1], Paolo Rapisarda [2], Biagio Fallico [1], Maria Allegra [2] and Alfio Spina [4,*]

1 Di3A—Dipartimento di Agricoltura, Alimentazione e Ambiente, University of Catania, via S. Sofia 100, 95123 Catania, Italy; earena@unict.it (E.A.); agata.mazzaglia@unict.it (A.M.); selina.brighina@unict.it (S.B.); bfallico@unict.it (B.F.)

2 CREA—Consiglio per la ricerca in agricoltura e l'analisi dell'economia agraria, Centro di Ricerca Olivicoltura, Frutticoltura e Agrumicoltura, Corso Savoia 190, 95024 Acireale (Catania), Italy; serenamuccilli@hotmail.com (S.M.); paolo.rapisarda@crea.gov.it (P.R.); maria.allegra@crea.gov.it (M.A.)

3 DSAAF—Dipartimento di Scienze Agrarie, Alimentari e Forestali, University of Palermo, Viale delle Scienze, Ed. 4, 90128 Palermo, Italy; virgiliogiannone@hotmail.com

4 CREA—Consiglio per la ricerca in agricoltura e l'analisi dell'economia agraria, Centro di Ricerca Cerealicoltura e Colture Industriali, Corso Savoia 190, 95024 Acireale (Catania), Italy

* Correspondence: alfio.spina@crea.gov.it

Received: 11 May 2020; Accepted: 4 June 2020; Published: 5 June 2020

Abstract: Durum wheat is widespread in the Mediterranean area, mainly in southern Italy, where traditional durum wheat breadmaking is consolidated. Bread is often prepared by adding a lot of salt to the dough. However, evidence suggests that excessive salt in a diet is a disease risk factor. The aim of this work is to study the effect of a natural low-sodium sea salt (Saltwell®) on bread-quality parameters and shelf-life. Bread samples were prepared using different levels of traditional sea salt and Saltwell®. The loaves were packaged in modified atmosphere conditions (MAPs) and monitored over 90 days of storage. No significant differences ($p \leq 0.05$) were found in specific volumes and bread yield between the breads and over storage times, regardless of the type and quantity of salt used. Textural data, however, showed some significant differences ($p \leq 0.01$) between the breads and storage times. 5-hydroxymethylfurfural (HMF) is considered, nowadays, as an emerging ubiquitous processing contaminant; bread with the lowest level of Saltwell® had the lowest HMF content, and during storage, a decrease content was highlighted. Sensory data showed that the loaves had a similar rating ($p \leq 0.05$) and differed only in salt content before storage. This study has found that durum wheat bread can make a nutritional claim of being "low in sodium" and "very low in sodium".

Keywords: *Triticum turgidum* L. subsp. *durum* Desf.; bread; NaCl; low-sodium sea salt; Na^+ reduction; physico-chemical and textural attributes; sensory evaluation

---

1. Introduction

There is much evidence suggesting that excessive salt intake endangers our health [1–3], and reducing its consumption is one of the first steps to preventing noncommunicable diseases [4]. Dietary habits are often developed during childhood [5–7], so nutritional education towards a low-sodium diet with adequate potassium intake should be encouraged [8,9]. In Italy, salt consumption by children and adolescents suggests that the average daily sodium consumption exceeds the official recommendations [10].

Natural foods contain modest amounts of sodium [11], and approximately two-thirds of salt intake come from its addition during food preparation [12]. Eighty food categories were identified as significant contributors to salt intake, and targets were set for the food industry to meet in each category within a certain period [13].

The WHO member states have agreed to reduce the global population's intake of salt by a relative 30 % by 2025, and several strategies have been undertaken to improve the consumer's understanding of healthy eating recommendations [14–19].

Nutrition claims of "low sodium/salt", "very low sodium/salt", and "sodium/salt-free" for foods containing 1.2, 0.4, and 0.05 g kg^{-1} of sodium, respectively, (or the equivalent value for salt) on food labels, informs consumers about salt content [20–22].

Salt is an essential ingredient in breadmaking: it retards gas production, enhances bread flavor, affects the rheological properties of dough, controls fermentation (decreasing yeast activity in the dough), and it can affect the quality parameters of bread [23,24]. Furthermore, NaCl has a strengthening effect on gluten, increasing its resistance or elasticity, and decreasing the extensibility of the dough [25,26].

The strategies to reduce sodium in bread include the use of reduced-sodium sea salt [27], the partial replacement of sodium chloride with potassium chloride and yeast extract [28], the use of a salt substitute with 57% of sodium chloride [29], and heterogeneous NaCl distribution, leading to enhanced saltiness by taste contrast [30].

In bread wheat (*Triticum aestivum* L.), salt is generally used at levels of about 1–2% based on flour weight [31]. A survey of salt content in artisan and industrial bread produced in all Italian regions was conducted in 2009/2010, its data having been recently published [32]. Artisan breads contained between 0.7% and 2.3% g/100 g of salt, while industrial bread, on average, contained 1.6% salt, most samples (56%) having a very high content. In the Mediterranean area, the cultivation of durum wheat (*Triticum turgidum* L. subsp. *durum* Desf.) is widespread compared to that of bread wheat [33] as it has a greater tolerance to drought, high temperatures, and fungal diseases, but less resistance to winter and spring cold. According to traditional uses, mainly in Southern Italy, bread is prepared from remilled durum wheat semolina [34]. Durum wheat milling products are characterized by peculiar chemical, rheological, colorimetric, and baking properties [35–39]. The bread has a compact texture, being in some cases excessively dense, with lower specific volume and harder crumbs than white bread [39], and the characteristic taste and flavor are generally enhanced by adding a high percentage of salt, from 20 to 25 g kg^{-1} [28].

In the last few years, the use of low-sodium salts in foods has been recommended.

However, almost all the commercially available low-sodium salts are produced by blending purified potassium chloride with ordinary table salts to achieve a reduced sodium content.

Natural low-sodium sea salt not only provides less sodium and does not affect the taste profile, but contains lots of essential trace minerals such as magnesium, potassium, calcium, and other nutrients the body requires.

Therefore, the aim of this paper is to evaluate the effect of substituting salt with low-sodium sea salt to measure the quality parameters of durum wheat bread over long storage.

2. Materials and Methods

2.1. Materials

Durum wheat (*Triticum turgidum* L. subsp. *durum* Desf.) remilled semolina for breadmaking [39] was provided by "Valle del Dittaino Società Cooperativa Agricola" (Assoro, Italy), an industrial bakery with a durum wheat mill.

The bread ingredients were food grade. Compressed yeast (AB Mauri, Casteggio, Italy) and traditional sea salt (99.5% NaCl; Mulino S. Giuseppe, Catenanuova, Italy) employed in the breadmaking process were purchased in a local retailer. Saltwell® (Salinity Group, Saltwell AB Göteborg Sweden) is a natural low-sodium sea salt (less than 35%) extracted from an underground sea below the Atacama desert (Chile). This natural sea salt contains 65 ± 1% NaCl, 30 ± 1.5% KCl, 1.0 ± 0.1% of

$MgSO_4$, 0.5 ± 0.1% of $CaSO_4$, and traces of other salts and minerals. Saltwell® was kindly donated by Medsalt—Mediterranean Salt Company S.r.l. (Rome, Italy).

Various levels (1.70%, 0.35%, 0.15% on semolina basis) of traditional sea salt and Saltwell® were used in dough formulations, as listed in Table 1.

Table 1. Bread type code and percentage of two salts on remilled semolina basis.

Bread Types Code	Salt Added (% *w/w* Remilled Semolina)
Control A	1.70% Traditional sea salt
Control B	1.70% Saltwell®
1A	0.35% Traditional sea salt
1B	0.35% Saltwell®
2A	0.15% Traditional sea salt
2B	0.15% Saltwell®

2.2. Methods

2.2.1. Physico-Chemical and Rheological Analyses of Remilled Semolina

The physico-chemical analyses of remilled semolina were carried out following the methods indicated by Giannone et al. (2018) and Palumbo et al. (2002) [39–41]: moisture content was determined according to the AACC 08-01 method (AACC, 2000) [42]. Protein content was determined by means of the Infratec 1241 Grain Analyzer (Foss Tecator, Höganäs, Sweden), based on near infrared transmittance. Ash content was determined according to the AACC 44-19 method (AACC, 2000) [42]. The particle size distribution was determined by a LabSifter (KBF7SN, Buhler, Switzerland). Remilled semolina for breadmaking was sieved for exactly 5 min on sieves with openings of 300, 200, 180, and 160 μm. Wet and dry gluten and gluten index were obtained by using a Glutomatic System (Glutomatic 2200, Centrifuge 2015, Glutork 2020; Perten Instruments AB, Huddinge, Sweden), according to the UNI 10690 method (UNI, 1979) [43]. The α-amylase activity was obtained by using the Falling Number 1500 apparatus (Perten Instruments AB, Huddinge, Sweden), according to the ISO 3093:2009 method (ISO, 2009) [44].

The CIELAB color parameters (L^*, a^*, b^*) were determined by Chromameter CR-300 (Minolta, Osaka, Japan), using the illuminant D_{65}. Alveograph indices were determined according to the AACC method 54-30A (AACC, 2000) [42] using an alveograph model MA 87 equipped with the software Alveolink NG (Tripette et Renaud, Villeneuve-la-Garenne, France). Farinograph parameters were obtained according to the AACC 54-21 method (AACC, 2000) [42] using a Farinograph, equipped with the software Farinograph® (Brabender instrument, Duisburg, Germany).

2.2.2. Bread Sample Production and Packaging

Bread samples were produced in a local breadmaking company ("Valle del Dittaino Società Cooperativa Agricola", Assoro, Italy), according to a proven industrial formulation: remilled durum wheat semolina (65 kg), compressed yeast (0.9% on semolina basis), water (66.0% on semolina basis) and the corresponding amount of salt. Six bread formulations containing different levels of traditional sea salt and low-sodium sea salt were produced (Table 1). The dough was mixed for 17 min in a high-speed mixer (San Cassiano, Italy). The final dough temperature was 26 ± 1 °C. The dough was left to rest in bulk for 15 min, divided into 980 ± 20 g portions (100 loaves for each production), proofed for 150 min at 32 ± 1 °C and 66 ± 2% relative humidity (RH) and baked at 220 °C for 60 min, in industrial tunnel ovens measuring 33 × 3 m (Pavailler Engineering, Galliate, Italy). The baked loaves, with an approximate weight of 1 kg each, were automatically transported to a cooling chamber (Tecnopool, Italy) set at 20 ± 2 °C for 120 min. After cooling, the loaves were sliced by means of an automatic slicing machine (Brevetti Gasparin, Marano Vicentino, Italy) to 11 ± 1 mm thickness. About 450 g of sliced bread per loaf was packaged under modified atmosphere conditions (MAPs) using inert gas

(70:30 N_2:CO_2). The bread packaging materials consisted of two plastic films provided by Cryovac Sealed Air (Elmwood Park, NJ, USA).

The samples were stored for up to 90 days at 20 ± 2 °C and 60 ± 2% RH. The quality parameters were determined at regular intervals in triplicate for each batch.

The following parameters and properties were tested for each bread sample during each sampling: volume, height, weight, diameter basis, crumb porosity, internal structure, top and base crust thickness, texture profile analysis, water activity, moisture, pH, 5-hydroxymethylfurfural (HMF) content, crust and crumb color, and sensory evaluation.

2.2.3. Bread Quality Evaluation

Determination of the Physico-Chemical Properties of the Breads

The volume was determined in a loaf volume meter by measuring the volume of rapeseed displaced by the bread, according to the AACC method 10.05.01 (AACC, 2000) [42]. The specific volume (mL/g) was calculated as a ratio of the loaf volume and the bread weight. The specific weight was calculated as the ratio of the loaf weight and bread volume. The h/d ratio was obtained as the ratio of the bread height and bread diameter of the loaf base. The crumb porosity was estimated using the Mohs scale. The CIELAB space L^* a^* b^* color parameters were measured for the crumb, in the transversely cut bread, and on the crust surface, averaging ten distinct points in each case, using a chromameter (CR-200, Konica Minolta, Osaka, Japan) with illuminant D_{65}.

Bread samples were analyzed for Na^+ (mg Kg^{-1}) content by inductively coupled plasma optical emission spectrometry (ICP-OES Optima 2000DV, Perkin Elmer, Italy). The samples were first ground to a powder, and oven-dried at 105 °C for 4 h until constant weight, then an aliquot equal to 0.5 g was weighed and placed in a muffle furnace at 600 °C for 12 h. After mineralization, the ashes were dissolved in 4 mL of distilled water and 0.5 mL of nitric acid at 69.5% (Superpure; Merck, Darmastadt, Germany). The solutions were poured into 50 mL flasks and brought to volume with distilled water before the analyses.

Water activity (a_w) was determined by a Hygropalm 40 AW (Rotronic Instruments Ltd., Crawley, UK) according to the manufacturer's instructions. Three bread slices (11 ± 1 mm thickness) were used, after removal of the crust. For each set of determinations, separate loaves were used.

The moisture content of bread crumb was determined by oven drying at 105 °C until constant weight, according to AOAC method no. 945.15 [45]. The pH was measured according to [46] using a pH meter (Mettler Toledo, MP 220).

2.3. *Texture Profile Analysis of Breads*

The texture profile analysis (TPA) of bread was determined using a Universal testing machine (model 3344, Instron, Norwood, MA, USA.) equipped with a cylindrical probe of 50 mm of diameter and a 2000 N load cell. Data were acquired through Bluehill® 2 software (Instron, Norwood, MA, USA). Cyclic compression tests (a 30-s gap between first and second compression) were set up: the crosshead speed was 3.3 mm/s, the force required to compress the samples by 40% was recorded on 5-cm side square portions of 24-mm thick slices, and the average value of five replicates was taken. The TPA profile recorded four primary parameters: hardness (N), springiness (mm), resilience, gumminess, and one derived parameters (chewiness, N mm).

2.4. *HMF Extraction and HPLC Analysis*

HMF was extracted and determined following the methodology proposed by [28]. Ground bread samples (5 g; La Moulinette, Moulinex, 2002) and 25 mL of water (J.T. Baker, Deventer, Holland) were put into a volumetric flask (50 mL) and stirred for 10 min. Then the sample was diluted up to 50 mL with water (JT. Baker, Deventer, Holland) and centrifuged for 45 min at 5000 rpm. An aliquot of the supernatant was filtered through a 0.45-µm filter (Albet) and injected into an HPLC system (Shimadzu

Class VP LC-10ADvp) equipped with a DAD (Shimadzu SPD-M10Avp). The column was a Gemini NX C18 (150 × 4.6 mm, 5 μm; Phenomenex) fitted with a guard cartridge packed with the same stationary phase. The HPLC conditions were the following: isocratic mobile phase, 90% water (J.T. Baker) at 1% acetic acid (Merck), and 10% methanol (Merck); flow rate, 0.7 mL/min; injection volume, 20 μL. The wavelength range was 220–660 nm, and the chromatograms were monitored at 283 nm. HMF was identified by splitting the peak of the HMF from the bread-solution sample with a standard of HMF ($p > 98\%$ Sigma-Aldrich, St. Louis, MO, USA) and by comparing the UV spectra of the HMF standard with that of the bread samples. All analyses were performed in duplicate, including the extraction procedure, and the reported HMF concentration was, therefore, the average of four values. The results were expressed as mg of HMF per kilogram of dry matter.

2.5. Sensory Evaluation

The sensory profile [28,47] was defined by a trained [48] panel of 12 judges (six females and six males, 28–40 years old). The judges, recruited for their individual abilities, had more than five years of experience in the sensory analysis of bread and bakery products, and they were submitted to further training over 4 weeks to generate attributes using handmade and industrial breads and to familiarize themselves with the scales and procedures. The judges, using a discontinuous scale between 1 (absence of the sensation) and 9 (extremely intense), have evaluated the intensity of the 11 sensory attributes selected on the basis of frequency (≥60%), following the definitions given by [49–51] (Table 2).

Table 2. Descriptive terms used for sensory profiling of bread.

	Attributes	Definition	Scale Anchors	
Crumb appearance	Crumb color	Color intensity of crumb	Whitish	Light yellow
	Alveolar structure	Porosity of crumb	Fine and uniform	Coarse and poorly homogeneous
Visual-tactile	Elasticity	Ability of the crumb to recover from compression exerted by fingers	Slow and partial recovery	Fast and complete recovery
	Humidity	Humidity perceived at the surface of bread crumb	Dry	Humid
Aroma/Flavor	Bread	The typical aroma/flavor of bread just taken out of the oven	Weak	Strong
	Yeasty	The aroma/flavor of a fermented yeast-like	None	Strong
	Wheat	The typical aroma/flavor of wheat	None	Strong
	Off-odour/ Off-flavour	Aroma/Flavor unpleasant, not characteristic of bread perceived through taste and smell when swallowing	None	Strong
Taste	Sweet	A basic taste factor produced by sugars	None	Strong
	Salty	A basic taste factor produced by sodium chloride	None	Strong
	Sour	A basic taste factor produced by acids	None	Strong
	Bitter	A basic taste factor produced by caffeine	None	Strong
Mouthfeel	Astringent	Sensory perception in the oral cavity that may include drying sensation and roughing of the oral tissue	None	Strong
Texture	Softness	Force required to compress the product with the molars	Hard	Soft
	Overall evaluation	An overall assessment expressed by considering all of the attributes	Low	High

The evaluation sessions, performed at 0, 15, 30, 60, and 90 days of storage, were conducted in the sensory laboratory [52] of Di3A (University of Catania, Italy) from 11:00 a.m. to 12:00 a.m. in individual booths illuminated with white light. The sliced bread samples were served on plates, coded with three-digit numbers, and water was provided to judges for rinsing between samples. The order presentation was randomized among judges and sessions using a randomized complete block. All data were acquired by a direct computerized registration system (FIZZ Biosystems. ver. 2.00 M, Couternon, France).

2.6. Statistical Analysis

The statistical analysis was performed using the Statgraphics® *Centurion XVI* software package (Statpoint Technologies, INC.). A two-way analysis of variance (ANOVA), followed by Tukey's HSD test ($p \leq 0.001$; $p \leq 0.01$; $p \leq 0.05$), was carried out on physico-chemical and textural attributes. The data were expressed as means ± standard deviations. The sensory data for each attribute were submitted to one-way ANOVA. The significance was tested by means of the F-test. A principal component analysis (PCA) was performed using PAST, Paleontological Statistics software package, 2011 [53].

3. Results and Discussion

3.1. Physico-Chemical and Rheological Characterization of the Durum Wheat Remilled Semolina

Physico-chemical characteristics of remilled semolina were moisture 13.8 ± 0.07 g/100 g, protein 12.2 ± 0.10 g/100 g, and ash 0.87 ± 0.01 g/100 g. These quality parameters met the Italian legal requirements [54]. Particle size distribution was >300 μm: 11.0 ± 1.73 g/100 g; between 200–300 μm: 26.3 ± 1.15 g/100 g; between 180–200 μm: 22.0 ± 2.00 g/100 g; between 160–180 μm: 20.0 ± 1.00 g/100 g; <160 μm: 20.7 ± 4.04 g/100 g. These findings agreed with those reported by other authors for remilled semolina [39]. Dry gluten content was 10.0 ± 0.1 g/100 g. The gluten index value was 80.7 ± 4.0, and the value of amylase activity at the falling number was low (577 ± 3.0 s). Regarding dry gluten content and relative qualitative index, the sample exhibited regular gluten quantities and high gluten tenacity. Similar values were reported by [28,39,55].

As regards color parameters, the values were lightness (L^*) 71.0 ± 0.3, red index (a^*) 2.12 ± 0.02, yellow index (b^*) 18.52 ± 0.05.

Rheological behavior was ev ergy (W) was 209 ± 4 $10^{-4} \times$ J, while the tenacity/extensibility (P/L) value showed a tenacious dough (value = 2.5). Strong gluten is expected in remilled durum wheat semolina [24].

Mixing behavior was evaluated by a farinograph apparatus. The semolina sample indicated the quantity of water absorbed at 500 BU (Brabender Unit), and the dough consistency was 60.6 ± 0.04% due to high protein content. The values of dough development time (1 min, 48 s ± 3.0 s), dough stability (4 min ± 12 s), and softening index (58 ± 1 BU) agreed with those reported by other authors on remilled semolina [28,38,39,55].

3.2. Sodium Content in Bread

The levels of the two salts used in the loaves, the sodium content, and the minimum limits established by EU regulations [20,21] applying to nutritional claims are shown in Table 3.

Table 3. Percentage of two salts in bread, sodium content and limits established by EU regulations [21,22] (data are means ± standard deviations).

Type	Salt in Experimental Bread (%)	Na^+ Content (g/100 g)	Regulations (EU) No. 1924/2006 and No. 1047/2012—Nutritional Claims
Control A	1.22% Traditional sea salt	0.430 ± 0.014A	-
Control B	1.22% Saltwell®	0.240 ± 0.014B	-
1A	0.25% Traditional sea salt	0.087 ± 0.001C	0.12 g of Na^+—low in sodium
1B	0.25% Saltwell®	0.064 ± 0.001C	0.12 g of Na^+—low in sodium
2A	0.11% Traditional sea salt	0.048 ± 0.001C	-
2B	0.11% Saltwell®	0.035 ± 0.000C	0.04 g of Na^+—very low in sodium

Different capital letters in the same column indicate significant difference ($p \leq 0.001$).

3.3. The Quality Parameters of Breads and Their Evolution during Storage

The p-values for all the physical and textural parameters of the bread types with respect to storage time are reported in Table 4.

The specific volumes and weights of the loaves were significant for each of the two factors of variability (type (A), storage time (B), and their interaction (A × B), even with different p levels ($p \leq 0.001$ for storage time, $p \leq 0.01$ A × B interaction, and $p \leq 0.05$ per type; see Table 4).

The results of the physical and textural properties of the industrial breads in the MAP conditions during 90 days of storage are shown in Tables 5 and 6.

No significant differences in specific volumes were shown among the bread samples, regardless of the type and level of sea salt (Table 4).

These findings agree with those reported by [23], but they disagree with those reported by [24]. Additionally, no significant differences in specific weight were observed among the controls and other bread samples or during storage time. The addition of different types and quantities of sea salt did not decrease bread yield. After 60 days of storage, the specific weight decreased.

The ratio between the height and diameter of the loaves used in the baking industry to parametrize possible dough failure was significant ($p \leq 0.001$) for all the factors and their interaction (Table 4). At time 0, control A was found to have the greatest h/d ratio (approximately 4.5) due to the addition of ordinary sea salt (Table 5). The other bread samples, as expected, showed a lower ratio during storage, especially the bread samples containing less traditional sea salt and sea salt with reduced Na^{+}. These findings agree with those reported by [28].

Significant differences were found for loaf porosity among the types ($p \leq 0.001$) and the A × B interaction ($p \leq 0.05$), but not for storage time (B) (Table 4). After baking (t0), almost all the types, except for 2B, showed proper development of crumb porosity. Starting from 15 days of storage, the performance of 2A also slightly decreased (Table 5).

Significant differences were found between the types ($p \leq 0.001$) and storage times ($p \leq 0.01$ and $p \leq 0.001$, respectively) but not for A × B interaction as regards internal structure and top crust thickness (Table 4). As for internal structure, only control A had an irregular structure over the whole storage time. Similar results were reported by [28].

As for top crust thickness, for up to 30 days of storage, no remarkable differences were recorded among the types (mean value of 3.8 mm); after 60 days, the values decreased up to 2.67 mm for control B.

No significant difference was highlighted for basis crust thickness between the types, the different storage times, and their interactions (Table 4). Almost all the bread samples exhibited a mean value of basis crust thickness of 4 mm. These findings agree with those reported by [28].

Three of the five parameters of texture profile analysis (hardness, gumminess, and chewiness) were always significant ($p \leq 0.001$), while resilience and springiness were significant per type and storage time ($p \leq 0.001$), but not for A × B interaction (Table 4). The two control breads (1A and 1B), as expected, showed lower values for the first three parameters. Starch retrogradation (i.e., the recrystallization of polysaccharide in gelatinized starch) is believed to be the main cause of crumb firmness change during storage [56].

Textural data highlighted high values of hardness, with significant differences among the samples, as reported by [39], and storage time (Table 6).

The hardness values, as expected, increased as the storage period progressed. As regards the bread samples, control A reported the lowest values during the entire storage period. Up to t30, the two controls, albeit with statistically different values, recorded the lowest hardness values. From t60, the control A values remained low, while the control B values increased until reaching about 55 N at the end of storage.

Table 4. Analysis of variance of the physical and textural parameters studied on the loaves (p-values).

Factors of Variability	Degrees of Freedom	Specific Volume	Specific Weight	h/d Ratio	Porosity	Internal Structure	Top Crust Thickness	Basis Crust Thickness	Hardness	Springiness	Resilience	Gumminess	Chewiness
Type (A)	5	0.002	0.014	0.000	0.000	0.000	0.000	0.444	0.000	0.000	0.000	0.000	0.000
Storage time (B)	4	0.000	0.000	0.000	0.156	0.007	0.000	0.571	0.000	0.000	0.000	0.000	0.000
A × B	20	0.011	0.008	0.000	0.021	0.824	0.064	0.568	0.000	0.529	0.088	0.000	0.000

Table 5. Evaluation of physical properties during storage of the bread samples produced using different types and levels of sea salt (data are means ± standard deviations).

Days of Storage	Type	Specific Volume (mL/g)	Specific Weight (g/mL)	h/d Ratio	Porosity (1-8) [a]	Internal Structure (1-2) [b]	Top Crust Thickness (mm)	Basis Crust Thickness (mm)
0	Control A	3.03 ± 0.09 gh	0.33 ± 0.01 ab	4.46 ± 0.25 a	6.00 ± 0.00 abc	2.00 ± 0.00	3.50 ± 0.00	4.50 ± 0.00
	Control B	3.14 ± 0.20 cdefgh	0.32 ± 0.02 abcd	3.75 ± 0.11 bcd	6.00 ± 0.00 abc	1.00 ± 0.00	3.00 ± 0.00	4.17 ± 0.29
	1A	3.06 ± 0.04 fgh	0.33 ± 0.00 abc	3.60 ± 0.16 bcdefg	6.00 ± 0.00 abc	1.00 ± 0.00	3.17 ± 0.29	4.00 ± 0.00
	1 B	3.09 ± 0.04 defgh	0.32 ± 0.00 abcd	3.58 ± 0.06 bcdefg	6.00 ± 0.00 abc	1.00 ± 0.00	3.83 ± 0.29	4.50 ± 0.50
	2 A	3.09 ± 0.04 efgh	0.32 ± 0.00 abcd	3.23 ± 0.11 defgh	6.00 ± 0.00 abc	1.33 ± 0.58	3.83 ± 0.29	4.67 ± 0.29
	2 B	2.96 ± 0.07 h	0.34 ± 0.01 a	3.11 ± 0.04 gh	7.00 ± 0.00 a	1.00 ± 0.00	4.00 ± 0.00	4.83 ± 0.29
15	Control A	3.20 ± 0.09 cdefgh	0.31 ± 0.01 abcdef	3.77 ± 0.09 bc	5.67 ± 0.00 bc	2.00 ± 0.00	3.50 ± 0.50	4.33 ± 0.29
	Control B	3.13 ± 0.18 cdefgh	0.32 ± 0.02 abcd	3.70 ± 0.12 bcde	6.00 ± 0.00 abc	1.33 ± 0.58	3.17 ± 0.29	5.00 ± 0.00
	1A	3.33 ± 0.03 abcdefgh	0.30 ± 0.00 abcdefg	3.42 ± 0.16 bcdefgh	6.00 ± 0.00 abc	1.00 ± 0.00	3.83 ± 0.29	4.83 ± 0.29
	1 B	3.23 ± 0.23 bcdefgh	0.31 ± 0.02 abcdefg	3.65 ± 0.20 bcdef	6.00 ± 0.00 abc	1.00 ± 0.00	4.33 ± 0.29	4.67 ± 0.29
	2 A	3.35 ± 0.11 abcdefgh	0.30 ± 0.01 abcdefg	3.29 ± 0.27 bcdefgh	6.50 ± 0.00 ab	1.00 ± 0.00	4.33 ± 0.29	5.50 ± 0.50
	2 B	3.43 ± 0.05 abcdefgh	0.29 ± 0.00 bcdefg	3.23 ± 0.04 defgh	7.00 ± 0.00 a	1.00 ± 0.00	4.67 ± 0.29	5.17 ± 0.29
30	Control A	3.33 ± 0.31 abcdefgh	0.30 ± 0.03 abcdefg	3.77 ± 0.22 bc	5.67 ± 0.00 bc	2.00 ± 0.00	3.67 ± 0.00	4.50 ± 0.00
	Control B	3.59 ± 0.04 abc	0.28 ± 0.00 defg	3.83 ± 0.03 b	5.67 ± 0.00 bc	1.67 ± 0.58	3.67 ± 0.29	4.50 ± 0.00
	1A	3.52 ± 0.03 abcdef	0.28 ± 0.00 cdefg	3.39 ± 0.06 bcdefgh	6.00 ± 0.00 abc	1.67 ± 0.58	3.67 ± 0.29	5.00 ± 0.00
	1 B	3.30 ± 0.09 abcdefgh	0.30 ± 0.01 abcdefg	3.27 ± 0.07 cdefgh	6.00 ± 0.00 abc	1.67 ± 0.58	4.33 ± 0.29	4.67 ± 0.29
	2 A	3.45 ± 0.01 abcdefg	0.29 ± 0.00 bcdefg	3.32 ± 0.13 bcdefgh	7.00 ± 0.00 a	1.33 ± 0.58	4.17 ± 0.29	5.50 ± 0.50
	2 B	3.23 ± 0.06 bcdefgh	0.31 ± 0.01 abcdefg	3.34 ± 0.09 bcdefgh	7.00 ± 0.00 a	1.00 ± 0.00	3.83 ± 0.29	4.83 ± 0.29
60	Control A	3.17 ± 0.17 cdefgh	0.32 ± 0.02 abcde	3.62 ± 0.08 bcdefg	5.33 ± 0.00 c	2.00 ± 0.00	3.33 ± 0.58	4.83 ± 0.76
	Control B	3.51 ± 0.19 abcdef	0.29 ± 0.02 bcdefg	3.71 ± 0.09 bcde	5.67 ± 0.00 bc	1.33 ± 0.58	3.33 ± 0.58	5.17 ± 0.29
	1A	3.40 ± 0.10 abcdefgh	0.29 ± 0.01 abcdefg	3.52 ± 0.14 bcdefgh	5.67 ± 0.00 bc	2.00 ± 0.00	3.50 ± 0.50	5.17 ± 0.29
	1 B	3.42 ± 0.05 abcdefgh	0.29 ± 0.00 abcdefg	3.55 ± 0.09 bcdefgh	5.33 ± 0.00 c	1.00 ± 0.00	3.33 ± 0.29	5.17 ± 0.29
	2 A	3.54 ± 0.09 abcde	0.28 ± 0.01 cdefg	3.16 ± 0.16 fgh	7.00 ± 0.00 a	1.67 ± 0.58	3.50 ± 0.50	5.00 ± 0.00
	2 B	3.42 ± 0.08 abcdefgh	0.29 ± 0.01 abcdefg	3.18 ± 0.04 efgh	7.00 ± 0.00 a	1.33 ± 0.58	3.50 ± 0.00	4.50 ± 0.50
90	Control A	3.57 ± 0.19 abcd	0.27 ± 0.03 g	3.49 ± 0.16 bcdefgh	5.33 ± 0.00 c	2.00 ± 0.00	3.00 ± 0.50	4.50 ± 0.00
	Control B	3.54 ± 0.12 abcde	0.28 ± 0.01 cdefg	3.74 ± 0.30 bcd	5.67 ± 0.00 bc	1.33 ± 0.58	2.67 ± 0.58	4.00 ± 0.00
	1A	3.71 ± 0.15 a	0.27 ± 0.01 fg	3.33 ± 0.07 bcdefgh	5.67 ± 0.00 bc	2.00 ± 0.00	3.33 ± 0.58	4.33 ± 0.58
	1 B	3.54 ± 0.15 abcde	0.28 ± 0.01 cdefg	3.40 ± 0.07 bcdefgh	5.33 ± 0.00 c	1.33 ± 0.58	3.17 ± 0.29	4.00 ± 0.00
	2 A	3.68 ± 0.17 ab	0.27 ± 0.01 efg	3.25 ± 0.09 cdefgh	7.00 ± 0.00 a	1.33 ± 0.58	3.17 ± 0.29	4.00 ± 0.00
	2 B	3.28 ± 0.11 abcdefgh	0.30 ± 0.01 abcdefg	3.03 ± 0.25 h	7.00 ± 0.00 a	1.00 ± 0.00	3.17 ± 0.29	3.83 ± 0.29

[a] 1, most porous; 8, least porous. [b] 1, regular; 2, irregular. Different letters in the same column indicate significant difference ($p \leq 0.01$).

Table 6. Evaluation of textural parameters of bread samples produced using different types and levels of sea salt during storage (data are means ± standard deviations).

Days of Storage	Type	Hardness (N)	Springiness mm)	Resilience	Gumminess	Chewiness (N × mm)
0	Control A	10.57 ± 0.43 o	5.10 ± 0.60	0.91 ± 0.02	9.58 ± 0.31 q	50.36 ± 2.94 o
	Control B	16.97 ± 0.68 mn	4.76 ± 0.46	0.91 ± 0.01	10.26 ± 0.58 pq	52.44 ± 4.99 o
	1A	28.11 ± 0.63 jkl	5.32 ± 0.10	0.85 ± 0.01	20.14 ± 0.32 mno	108.70 ± 0.90 m
	1 B	34.10 ± 2.40 hi	5.12 ± 0.04	0.81 ± 0.01	27.58 ± 2.03 ijk	145.49 ± 1.70 l
	2 A	23.70 ± 3.06 l	6.12 ± 0.22	0.80 ± 0.05	25.72 ± 1.60 ijkl	163.56 ± 2.74 k
	2 B	37.70 ± 0.46 h	5.92 ± 0.58	0.79 ± 0.04	35.42 ± 0.93 h	202.35 ± 4.35 h
15	Control A	25.85 ± 0.14 kl	5.07 ± 0.68	0.87 ± 0.03	14.87 ± 0.60 opq	79.40 ± 1.59 n
	Control B	28.84 ± 0.28 ijkl	5.58 ± 0.39	0.89 ± 0.01	15.53 ± 0.41 nopq	76.20 ± 2.44 n
	1A	47.76 ± 2.36 g	6.26 ± 1.14	0.82 ± 0.03	27.92 ± 1.19 ij	168.22 ± 3.37 ijk
	1 B	34.19 ± 0.78 hi	5.85 ± 0.56	0.81 ± 0.01	29.90 ± 0.97 i	177.55 ± 1.94 i
	2 A	62.25 ± 0.57 cd	5.81 ± 0.40	0.76 ± 0.03	29.65 ± 2.21 i	176.42 ± 2.52 ij
	2 B	53.51 ± 3.96 ef	6.47 ± 0.22	0.77 ± 0.02	37.91 ± 2.97 fgh	243.61 ± 3.80 g
30	Control A	17.45 ± 1.39 m	5.05 ± 0.86	0.89 ± 0.02	22.39 ± 0.81 klm	114.59 ± 1.88 m
	Control B	26.34 ± 1.14 kl	5.29 ± 0.40	0.88 ± 0.02	23.00 ± 1.47 jklm	117.35 ± 0.24 m
	1A	46.85 ± 0.46 g	5.91 ± 0.15	0.81 ± 0.01	36.73 ± 2.35 gh	206.66 ± 1.86 h
	1 B	32.11 ± 1.30 hij	5.78 ± 0.34	0.80 ± 0.03	38.22 ± 1.36 fgh	210.03 ± 3.49 h
	2 A	62.16 ± 1.93 cd	6.26 ± 0.32	0.70 ± 0.02	38.02 ± 2.69 fgh	237.42 ± 2.09 g
	2 B	36.85 ± 0.63 h	6.61 ± 0.24	0.71 ± 0.02	40.27 ± 0.76 efgh	256.17 ± 1.89 f
60	Control A	11.32 ± 1.65 no	6.85 ± 0.15	0.87 ± 0.07	20.90 ± 3.12 lmn	157.87 ± 2.66 k
	Control B	35.16 ± 1.04 h	6.90 ± 0.18	0.84 ± 0.03	24.70 ± 0.71 ijklm	165.73 ± 1.66 jk
	1A	66.44 ± 1.10 cd	7.14 ± 0.44	0.77 ± 0.01	43.36 ± 1.40 cdef	285.83 ± 4.77 e
	1 B	60.56 ± 0.34 de	7.49 ± 0.06	0.70 ± 0.01	42.46 ± 1.91 def	283.67 ± 2.31 e
	2 A	51.12 ± 1.76 fg	7.20 ± 0.15	0.70 ± 0.01	41.50 ± 0.85 defg	314.36 ± 3.39 d
	2 B	74.27 ± 1.81 b	6.86 ± 0.23	0.76 ± 0.03	49.66 ± 1.01 b	372.13 ± 2.97 b
90	Control A	29.43 ± 1.00 ijkl	6.36 ± 0.79	0.82 ± 0.02	23.13 ± 0.31 jklm	159.13 ± 5.34 k
	Control B	54.60 ± 0.99 ef	6.90 ± 0.11	0.83 ± 0.04	29.17 ± 1.11 i	200.12 ± 4.55 h
	1A	45.72 ± 2.10 g	7.17 ± 0.32	0.74 ± 0.04	45.23 ± 0.60 bcde	332.35 ± 3.61 c
	1 B	58.08 ± 1.59 de	7.00 ± 0.19	0.79 ± 0.05	48.01 ± 1.73 bc	342.25 ± 4.40 c
	2 A	51.04 ± 1.27 fg	7.57 ± 0.32	0.70 ± 0.01	46.21 ± 1.20 bcd	362.72 ± 4.90 b
	2 B	80.69 ± 0.02 a	7.41 ± 0.11	0.71 ± 0.01	57.97 ± 0.04 a	425.86 ± 1.38 a

Different letters in the same column indicate significant differences ($p \leq 0.01$).

No significant differences in springiness or resilience were shown among the bread samples and during the storage times, whatever the type and level of salt (Table 6). Up to 30 days of storage, no remarkable differences were recorded among the breads (mean value of 5.7 mm); after 60 days, the values of springiness increased by up to 7.0 mm. These findings do not agree with those reported by [39].

As for resilience, the average value was around 0.80. During the entire storage period, the two controls showed higher resilience values. From the end of the baking to the end of storage, resilience values decreased slightly. These findings agree with those reported by [39].

With regard to gumminess and chewiness, they increased progressively with increasing storage times and with decreasing salt content, regardless of type, until they reach the maximum at t90 for 2B (58.0 and 426.0). During the entire storage time, the two controls always showed the lowest values, and were similar to each other, except for t90.

Water activity (a_w) and moisture content were significant compared to all the factors of variability (Table 7). As for pH and HMF, they were significant compared to all the factors of variability ($p \leq 0.001$; Table 7).

Table 7. Analysis of variance of the chemical and color parameters studied on the loaves (p-values).

Factors of Variability	Degrees of Freedom	a_w	Moisture	pH	HMF	Crumb			Crust		
						L^*	a^*	b^*	L^*	a^*	b^*
Type (A)	5	0.000	0.000	0.000	0.000	0.005	0.000	0.000	0.000	0.000	0.000
Storage time (B)	4	0.000	0.032	0.000	0.000	0.020	0.000	0.000	0.009	0.000	0.116
A × B	20	0.000	0.000	0.000	0.000	0.011	0.033	0.069	0.002	0.007	0.133

Crumb lightness and redness were significant compared to all the factors of variability. Crumb yellowness was significant for bread (A) and storage time (B) ($p \leq 0.001$) but not for their interaction (A × B) (Table 7). The effect of the addition of sea salt with reduced Na^+ on the L^* parameter of crumb during the entire time storage was not significant (Table 7).

Chemical properties of the breads during the storage time are reported in Table 8.

Crumb a_w is an important parameter of food processing and conservation technologies that comes into play for food stability and safety. It indicates the amount of free water not linked by bonds with the soluble constituents of the food, i.e., the water that can participate in chemical, physical, biological, and enzymatic reactions.

In general, water activity is a relatively easy parameter to measure, which can be an advantage, especially in the food industry [57].

The a_w value ranged from about 0.88 for Control A at t90, to 0.93 for 2A at t0 (Table 7). Similar values have been reported by [55].

After baking, and up to t15, there is no difference among the breads. From t30, water activity decreases for both controls. From t60 to the end of storage, a_w decreases slightly for all the types. At t90, only the a_w value of Control A is lower than the other types. Moisture content ranged from about 35.5–38.4% at the beginning (Table 8). Bread samples containing natural low Na^+ sea salt show the highest moisture content, and significant differences were found between all the breads. During storage, the breads with NaCl generally show the highest levels of moisture, and at 90 days of storage, the moisture content decreased, ranging from 35.3–32.4%. No significant differences were found between control B and samples 1B (1.22% and 0.25% Saltwell®) and the bread samples with the lowest levels of salt (2A and 2B).

The pH ranges from 5.36 to 5.93 at the beginning; at 90 days of storage, it ranges from 5.73 to 5.82 (Table 8). The variability seems to be more related to the storage time rather than to the different levels and salts used in the recipe. Similar trends were reported both for durum wheat bread with yeast extract and fortified with fiber [28,50].

Table 8. Evaluation of the chemical characteristics of the bread samples produced using different types and levels of sea salt during storage (data are means ± standard deviations).

Days of Storage	Type	a_w	Moisture (%)	pH	HMF (mg/kg Dry Matter)
0	Control A	0.92 ± 0.00 abc	35.5 ± 0.07 n	5.65 ± 0.01 i	28.9 ± 1.73 fghi
	Control B	0.92 ± 0.00 a	37.1 ± 0.02 g	5.81 ± 0.00 efg	32.6 ± 2.18 de
	1A	0.92 ± 0.00 ab	36.1 ± 0.05 l	5.88 ± 0.01 cde	34.3 ± 1.40 cd
	1B	0.92 ± 0.00 ab	38.4 ± 0.04 b	5.94 ± 0.01 abc	38.2 ± 1.19 b
	2A	0.93 ± 0.00 a	35.8 ± 0.07 m	5.36 ± 0.00 l	39.2 ± 2.81 b
	2B	0.92 ± 0.00 a	36.7 ± 0.07 h	5.93 ± 0.02 abc	23.0 ± 1.07 mnopqr
15	Control A	0.91 ± 0.00 bcdef	36.6 ± 0.02 hi	5.80 ± 0.00 g	37.6 ± 0.58 bc
	Control B	0.92 ± 0.00 abcde	35.7 ± 0.08 mn	5.88 ± 0.01 cdef	29.2 ± 0.51 efgh
	1A	0.92 ± 0.00 abcd	39.5 ± 0.04 a	5.92 ± 0.01 abc	30.3 ± 0.98 efg
	1B	0.92 ± 0.00 abcd	37.9 ± 0.06 cd	5.93 ± 0.02 abc	30.8 ± 1.61 def
	2A	0.92 ± 0.00 abcd	37.12 ± 0.04 g	5.92 ± 0.01 abc	29.2 ± 0.64 efgh
	2B	0.92 ± 0.01 abcd	36.7 ± 0.08 h	5.96 ± 0.01 ab	27.9 ± 1.37 fghij
30	Control A	0.91 ± 0.00 defgh	35.9 ± 0.04 lm	5.73 ± 0.01 h	24.9 ± 0.27 jklmnop
	Control B	0.91 ± 0.01 cdefg	35.8 ± 0.02 m	5.78 ± 0.00 gh	21.0 ± 0.14 qr
	1A	0.92 ± 0.00 abcd	37.7 ± 0.05 de	5.99 ± 0.04 a	21.5 ± 0.03 pqr
	1B	0.92 ± 0.00 abcde	36.4 ± 0.07 i	5.88 ± 0.02 cdef	24.1 ± 1.14 lmnopqr
	2A	0.92 ± 0.00 abcd	37.3 ± 0.04 fg	5.84 ± 0.02 defg	37.8 ± 0.09 bc
	2B	0.92 ± 0.00 abcd	35.9 ± 0.03 lm	5.93 ± 0.02 abc	16.3 ± 0.03 s
60	Control A	0.90 ± 0.00 j	37.3 ± 0.07 fg	5.82 ± 0.03 efg	45.8 ± 0.07 a
	Control B	0.90 ± 0.00 ghij	35.0 ± 0.02 p	5.81 ± 0.01 fg	22.4 ± 0.20 nopqr
	1A	0.91 ± 0.00 efghij	37.5 ± 0.16 ef	5.83 ± 0.01 efg	27.7 ± 0.05 fghijk
	1B	0.91 ± 0.00 efghij	38.1 ± 0.06 c	5.78 ± 0.01 gh	26.0 ± 0.15 hijklm
	2A	0.91 ± 0.00 defghi	36.1 ± 0.02 l	5.83 ± 0.00 efg	25.0 ± 0.03 jklmno
	2B	0.91 ± 0.00 defgh	38.1 ± 0.08 c	5.91 ± 0.01 bcd	26.8 ± 0.08 ghijkl
90	Control A	0.88 ± 0.01 k	34.8 ± 0.06 p	5.73 ± 0.04 h	25.5 ± 0.13 ijklmn
	Control B	0.90 ± 0.00 ij	34.2 ± 0.02 q	5.81 ± 0.01 fg	24.2 ± 0.30 klmnopq
	1A	0.90 ± 0.00 hij	35.3 ± 0.07 o	5.78 ± 0.00 gh	20.6 ± 0.13 r
	1B	0.90 ± 0.00 fghij	34.2 ± 0.01 q	5.82 ± 0.08 efg	22.6 ± 0.03 mnopqr
	2A	0.90 ± 0.00 fghij	32.4 ± 0.04 r	5.80 ± 0.01 gh	21.5 ± 0.27 opqr
	2B	0.90 ± 0.00 fghij	32.4 ± 0.01 r	5.82 ± 0.00 efg	21.8 ± 0.00 opqr

Different letters in the same column indicate significant difference ($p \leq 0.01$).

HMF is a widely used compound as heat induces the chemical index generally used for monitoring thermal abuse [58–61]. In bread and in other baking products, HMF is used to monitor the heating process, and several factors influence its formation, such as manufacturing conditions and recipe [57–59]. Even if the toxicity risk of HMF is still debated, nowadays, HMF is under evaluation as an emerging ubiquitous processing contaminant since there is evidence to suggest that HMF and its metabolites may have harmful effects on human health [60–63].

Among foods, coffee and bread contribute the most HMF exposure, about 85% of total intake [64].

The HMF parameter was significant compared to all the factors of variability ($p \leq 0.001$; Table 7). HMF levels at the beginning ranged from about 23 to 39 mg/kg of dry matter (Table 8), and significant differences were found between all samples. These levels were lower than those reported for durum wheat bread with KCl and taste enhancer [28], and it is known that differences in water content in the leavening and/or baking time and the ratio between crumb and crust of the loaf could influence HMF content [58]. Bread samples with the lowest levels of natural low Na^+ sea salt (2 B) had the lowest HMF content. During storage, a decrease in HMF amount was highlighted, though the trend in decrease was not regular. Generally, the bread samples with the lowest levels of salt had the lowest HMF content due to the effects of a high level of NaCl on starch degradation and yeast growth, resulting, in both cases, in higher levels of Maillard indicators [65]. At 90 days of storage, this parameter ranged from about 20.6 to 25.5 mg/kg of dry matter. The HMF trend during storage was similar to those reported by [28,50], suggesting that HMF decrease is more related to storage time rather than recipe.

During storage, crumb redness in the traditional sea salt (control A) test slowly decreased. After t15, the a^* value begins to decrease for all breads (Table S1).

3.4. Sensory Evaluation

The addition of different types and quantities of sea salt had little effect on the sensory characteristics of the bread sample. Table 9 reports the ANOVA results of sensory data and the bread attributes, which significantly differentiated at different p-levels ($p \leq 0.05$; $p \leq 0.01$; $p \leq 0.001$), at each sampling. Mean values were reported only for significantly different attributes.

Table 9. Influence of type of bread (6) on the attributes and mean scores of the significant sensory attributes (comparison of formulations). Data expressed as means.

Days of Storage	Attributes	F Values	Type					
			Control A	Control B	1A	1B	2A	2B
0	Salty	12.08 ***	4.2 b	4.3 b	1.6 a	1.6 a	1.6 a	1.5 a
15	Sweet	5.23 ***	2.9 a	3.0 a	4.3 ab	5.2 bc	5.8 c	5.0 bc
	Salty	7.49 ***	4.4 b	4.1 b	1.9 a	1.9 a	1.4 a	1.7 a
	Bread flavor	2.98 *	6.1 b	6.2 b	4.9 ab	4.1 a	4.6 ab	3.6 a
	Overall evaluation	5.01 ***	6.5 b	6.3 b	4.1a	3.9 a	4.2 a	3.4 a
30	Sweet	5.30 ***	3.1 ab	2.8 a	4.3 abc	4.6 bc	6.4 d	5.3 cd
	Salty	7.40 ***	3.9 b	4.2 b	2.4 a	2.2 a	1.6 a	1.7a
	Bread flavor	2.45 **	5.4 b	5.5 b	4.6 ab	3.8 ab	3.3 a	3.5 a
	Overall evaluation	3.48 **	5.8 b	5.7 b	4.6 ab	3.8 a	3.3 a	3.6 a
60	Sweet	3.25 *	2.9 a	3.4 a	4.1 ab	4.4 ab	5.6 b	5.0 b
	Salty	9.45 ***	5.2 b	4.8 b	2.8 b	3.2 b	1.9 ab	1.3 a
	Overall	3.17 *	5.4 bc	5.6 c	4.0 ab	4.3 abc	3.7 a	3.2 a
90	Sweet	6.45 ***	5.4 bc	5.6 c	3.8 ab	4.1 abc	3.5 a	3.3 a
	Salty	12.45 ***	5.0 b	4.5 b	2.3 a	2.4 a	1.8 a	2.4 a
	Overall evaluation	2.87 *	5.4 bc	5.6 c	3.8 ab	4.1 abc	3.5 a	3.3 a

Different letters in the same row indicate significant differences at $p \leq 0.05$ *, $p \leq 0.01$ **, $p \leq 0.001$ ***.

At t0, the bread samples were evaluated similarly by panellists, with the exception of the "salty" attribute. Obviously, the control breads (Control A and Control B) had the highest value of saltiness.

At 15 and 30 days of storage, the samples were significantly different for the attributes sweet, salty, bread flavor, and overall evaluation. The 0.15 NaCl sample showed the highest intensity of sweet taste, while the control samples, as expected, had the highest score of salt, bread flavor, and overall evaluation.

At 60 and 90 days of storage, the attributes of sweet, salty, and overall significantly differentiated the bread samples. The 0.15 NaCl and 0.15 Saltwell® bread samples had the highest intensity of sweet and the lowest of the attributes salt and overall. The control samples showed the highest intensity of the attribute overall.

The different levels of sea salt did not influence the attributes of texture (i.e., softness), as reported by [28].

Table 10 reports the sensory attributes which significantly differentiated ($p \leq 0.05$) during the 90 days of storage.

Table 10. Mean values of the significantly different sensory attributes (comparison during storage). Three bread loaves were collected at each sampling.

Attribute	Days of Storage	Control A	Control B	1A	1B	2A	2B
Elasticity	0		6.1 ab				
	15		7.5 b				
	30		6.5 ab				
	60		5.0 a				
	90		5.5 a				
Humidity	0	7.2 b	6.5 b			7.2 b	
	15	6.7 b	6.1 b			5.7 ab	
	30	4.3 a	5.6 b			5.1 a	
	60	4.4 a	3.8 a			4.0 a	
	90	4.8 a	5.2 ab			4.4 a	
Softness	0	6.5 bc	6.3 b				
	15	6.8 c	6.8 b				
	30	5.2 abc	5.4 ab				
	60	4.4 a	4.2 a				
	90	4.9 ab	5.4 ab				

Different letters in the same column indicate significant difference at $p \leq 0.05$.

Control A showed a significant decrease during storage but only for the attributes of humidity and softness. At 0 and 15 days of storage, Control A had the highest intensity of these two sensory attributes.

Control B showed a significant decrease during storage for the attributes of elasticity, humidity, and softness. These attributes began to decrease after 30 days of storage.

Sample 2A showed a significant decrease only for the attribute humidity, while bread samples 1A, 1B, and 2B did not show any significant differences during the 90 days of storage.

During storage, the bread samples did not develop off-odors or off-flavors in agreement with those reported by [28].

3.5. Multivariate Statistical Analysis

Principal component analysis (PCA) is a multivariate analysis that allows the reduction and interpretation of large multivariate datasets with some underlying linear structure. In this trial, it was carried out to determine if and which salt (type and concentration) had an influence on the qualitative and sensory traits of the breads. The PCA included the following 24 dependent variables: specific volume, specific weight, h/d ratio, crumb porosity, hardness, gumminess, chewiness, springiness, resilience, water activity, moisture, pH, HMF, acidity, and crust and crumb color parameters (as L^*, a^*, b^*, h, C).

The two main factors accounting for 56.92% of the total variance were PC1 and PC2 at 37.08% and 19.84% (Figure 1).

There are two types of trends on the first axis: (1) based on salt content, the groups shift from the negative to the positive section, from the breads with minimum salt concentrations (2A and 2B), to those with more (Control A and Control B) (Figure 1); (2) based on days of storage, from the longest (t90) to the shortest (t0) (Figure 1). Convex hulls were used to highlight these trends. They can be defined as the intersection of all convex sets containing a given subset of a Euclidean space. The convex hull of a set of data is the smallest convex set that contains it.

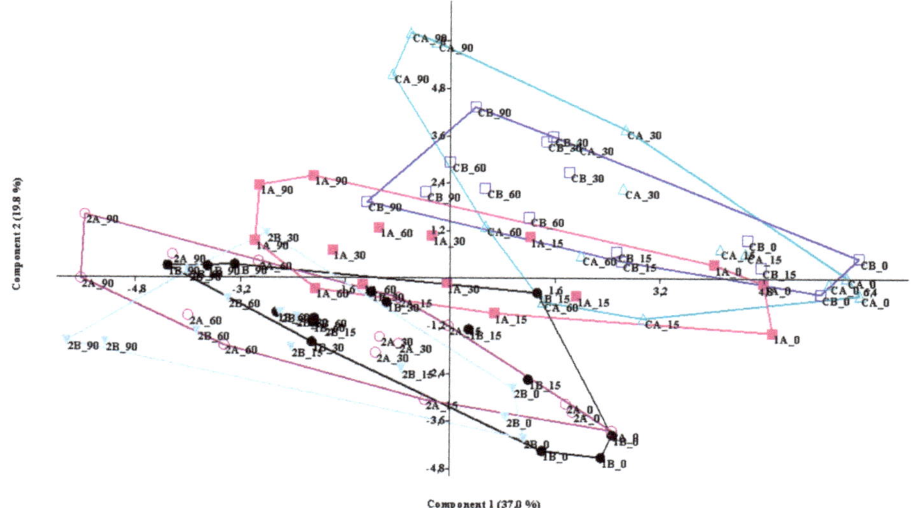

Figure 1. Principal component analysis (PCA) scatter diagram defined by the first two principal components (i.e., PC1, PC2) and convex hulls for the measured physico-chemical and textural traits of the breads, grouped by storage time.

The variables that determined these trends were resilience, crust color (as a^*, C), h/d ratio, crumb color (h), and moisture, which showed the highest positive loading values (0.272, 0.230, 0.227, 0.228, 0.218, and 0.209 respectively), chewiness, hardness, gumminess, and springiness, with the highest negative loadings (−0.314, −0.311, −0.309, −0.263, respectively).

The groups also showed a gradient with respect to the days of storage, if PC2 is observed: from the positive scores of the longer storage time to the gradually lower scores of the shorter ones (Figure 1).

The variables that positively correlated with PC2 were crust color parameters (L^*, h, b^*) and moisture (loading values, 0.367, 0.334, 0.271, 0.292, respectively). Moreover, PC2 negatively correlated with a_w, specific volume, and crust hardness (−0.294, −0.266, −0.248, respectively).

In summary, sorting the data according to the first two axes distributes the groups in relation to the lowest salt concentration with the maximum storage time, and so on, up to the breads with the highest salt concentrations with the shortest storage times.

PCA loadings did not have the necessary strength to affect the net separation of groups, but this seems to support the hypothesis that the different breads and salt concentrations do not lead to substantial differences in the overall qualitative characteristics and acceptability of the product.

4. Conclusions

The results of this study showed that replacing traditional sea salt with Saltwell® in durum wheat bread is a possible strategy for reducing sodium intake while maintaining the quality and sensorial characteristics of the bread.

There were no significant differences in the specific volume and bread yield among bread samples and during storage times, regardless of the type and level of sea salt used. The textural data showed high hardness and chewiness values, with significant differences between samples and storage times.

Sensory data showed that the different levels of sea salt did not influence the attributes of softness.

Principal component analysis (PCA) seems to support these findings since, overall, the parameters analyzed were unable to differentiate groups effectively.

Natural low sodium sea salt has made it possible to obtain durum wheat bread with the nutritional claim "low in sodium" (<0.12 g/100 g) and/or "very low in sodium" (<0.04 g/100 g) on food labels, in accordance with EU regulations [20–22]. However, the breads showed good taste and flavor.

These results should encourage the opportunity to produce low-sodium or very low-sodium bread because of consumers' increasing interest in durum wheat bread in accordance with the guidelines for a healthy diet.

Supplementary Materials: The following are available online at http://www.mdpi.com/2304-8158/9/6/752/s1, Table S1: Colour parameters of the bread samples produced using different types and levels of sea salt during storage (data are means ± standard deviations). Three bread loaves were collected at each sampling..

Author Contributions: Conceptualization, E.A., S.M., V.G., and A.S.; data curation, S.M., A.M., and M.A.; formal analysis, S.M., A.M., V.G., and S.B.; funding acquisition, P.R. and B.F.; investigation, E.A. and A.S.; methodology, E.A., S.M., A.M., V.G., S.B., and A.S.; project administration, P.R. and B.F.; resources, E.A., P.R., B.F., and A.S.; software, S.M. and M.A.; supervision, E.A., P.R., B.F., and A.S.; validation, E.A., S.M., A.M., V.G., S.B., M.A., and A.S.; visualization, A.S.; writing—original draft, E.A., S.M., A.M., V.G., and A.S.; writing—review and editing, E.A., M.A. and A.S. All authors have read and agreed to the published version of the manuscript.

Funding: Part of this research was funded by the Regional Operational Program of "Regione Siciliana" PO FESR 2007–2013—Asse IV, Obiettivo 4.1.1—Linea di Intervento 4.1.1.2 with the title of the research program "Impiego e valutazione di fibre e sostanze nutraceutiche per l'ottenimento di prodotti da forno salutistici", grant number 5787/3 of the 14/12/2011.

Acknowledgments: The authors wish to thank Nick Field for scientific English language editorial assistance. The authors thank Medsalt—Mediterranean Salt Company S.r.l. (Rome, Italy) for kindly donating a sample of Saltwell®.

Conflicts of Interest: The authors declare no conflict of interest. The funders had no role in the design of the study; in the collection, analyses, or interpretation of data; in the writing of the manuscript, or in the decision to publish the results.

References

1. Whelton, P.K.; He, J.; Gail, T.L. *Lifestyle Modifications for the Prevention and Treatment of Hypertension*; Whelton, P.K., He, J., Gail, T.L., Eds.; Marcel Dekker, Inc. Publisher's Location: New York, NY, USA, 2003; p. 403.
2. Appel, L.J.; Frohlich, E.D.; Hall, J.E.; Pearson, T.A.; Sacco, R.L.; Seals, D.R.; Sacks, F.M.; Smith, S.C.; Vafiadis, D.K.; Van Horn, L.V. The importance of population-wide sodium reduction as a means to prevent cardiovascular disease and stroke: A call to action from the American Heart Association. *Circulation* **2011**, *123*, 1138–1143. [CrossRef]
3. Strazzullo, P.; D'Elia, L.; Kandala, N.B.; Cappuccio, F.P. Salt intake, stroke and cardiovascular disease: Meta-analysis of prospective studies. *Br. Med. J.* **2009**, *339*, b4567. [CrossRef] [PubMed]
4. Beaglehole, R.; Bonita, R.; Horton, R.; Adams, C.; Alleyne, G.; Asaria, P.; Baugh, V.; Bekedam, H.; Billo, N.; Casswell, S.; et al. Priority actions for the non-communicable disease crisis. *Lancet* **2011**, *377*, 1438–1447. [CrossRef]
5. Beauchamp, G.K.; Mennella, J.A. Early flavour learning and its impact on later feeding behaviour. *J. Pediatr. Gastroenterol. Nutr.* **2009**, *48*, S25–S30. [CrossRef] [PubMed]
6. Stein, L.J.; Cowart, B.J.; Beauchamp, G.K. The development of salty taste acceptance is related to dietary experience in human infants: A prospective study. *Am. J. Clin. Nutr.* **2012**, *94*, 123–129. [CrossRef] [PubMed]
7. Strazzullo, P.; Campanozzi, A.; Avallone, S. Does salt intake in the first two years of life affect the development of cardiovascular disorders in adult hood? *Nutr. Metab. Cardiovasc. Dis.* **2012**, *22*, 787–792. [CrossRef] [PubMed]
8. Strazzullo, P.; Cairella, G.; Campanozzi, A.; Carcea, M.; Galeone, D.; Galletti, F.; Giampaoli, S.; Iacoviello, L.; Scalfi, L. Population based strategy for dietary salt intake reduction: Italian initiatives in the European framework. *Nutr. Metab. Cardiovasc. Dis.* **2012**, *22*, 161–166. [CrossRef] [PubMed]
9. Donfrancesco, C.; Ippolito, R.; Lo Noce, C.; Palmieri, L.; Iacone, R.; Russo, O.; Vanuzzo, D.; Galletti, F.; Galeone, D.; Giampaoli, S.; et al. Excess dietary sodium and inadequate potassium intake in Italy: Results of the MINISAL study. *Nutr. Metab. Cardiovasc. Dis.* **2013**, *23*, 850–856. [CrossRef]

10. Campanozzi, A.; Avallone, S.; Barbato, A.; Iacone, R.; Russo, O.; De Filippo, G.; D'Angelo, G.; Pensabene, L.; Malamisura, B.; Cecere, G.; et al. High Sodium and Low Potassium Intake among Italian Children: Relationship with Age, Body Mass and Blood Pressure. *PLoS ONE* **2015**, *10*, e0121183. [CrossRef]
11. Eaton, S.B.; Konner, M. Paleolithic nutrition. A consideration of its nature and current implications. *N. Engl. J. Med.* **1985**, *312*, 283–289. [CrossRef]
12. Leclercq, C.; Ferro-Luzzi, A. Total and domestic consumption of salt and their determinants in three regions of Italy. *Eur. J. Clin. Nutr.* **1991**, *45*, 151–159. [PubMed]
13. European Food Safety Authority (EFSA). Opinion of the scientific panel on dietetic products, nutrition and allergies on a request from the commission related to the tolerable upper intake level of sodium. *EFSA J.* **2005**, *209*, 1–26.
14. WASH World Action on Salt & Health. Available online: http://www.worldactiononsalt.com/ (accessed on 3 May 2020).
15. CASH Consensus Action on Salt and Health. Available online: http://www.actiononsalt.org.uk/about (accessed on 3 May 2020).
16. Food Standard Agency. *Food. Using Traffic Lights to Make Healthier Choices*; Ancient House Printing Group: Ipswich, UK. Available online: http://www.resourcesorg.co.uk/assets/pdfs/foodtrafficlight1107.pdf (accessed on 3 May 2020).
17. Wyness, L.A.; Butriss, J.L.; Stanner, S.A. Reducing the population's sodium intake: The UK Food Standards Agency's salt reduction programme. *Public Health Nutr.* **2012**, *15*, 254–261. [CrossRef] [PubMed]
18. WHO & FAO—World Health Organisation & Food and Agriculture Organisation. *Diet, Nutrition and the Prevention of Chronic Diseases*; WHO Technical Report: Geneva, Switzerland, 2003; Volume 916, p. 149.
19. WHO World Health Organisation. Reducing Salt Intake in Populations. Report of a WHO Forum and Technical Meeting 5–7 October 2006, Paris, France. Geneva, Switzerland. 2007. Available online: http://www.who.int/dietphysicalactivity/Salt_Report_VC_april07.pdf (accessed on 3 May 2020).
20. European Parliament & Council of the European Union. Regulation (EC) No. 1924/2006 of the European Parliament and of the Council of 20 December 2006 on nutrition and health claims made on foods. *Off. J. Eur. Union (OJEU)* **2006**, *L404*, 9–25.
21. Commission Regulation (EU) No. 1047/2012 of 8 November 2012 amending Regulation (EC) No 1924/2006 with regard to the list of nutrition claims (Text with EEA relevance). *Off. J. Eur. Union (OJEU)* **2012**, *L310*, 36–37.
22. European Parliament & Council of the European Union. Regulation (EU) No. 1169/2011 of the European Parliament and of the Council of 25 October 2011 on the provision of food information to consumers. *Off. J. Eur. Union (OJEU)* **2011**, *L304*, 18–63.
23. Lynch, E.J.; Dal Bello, F.; Sheehan, E.M.; Cashman, K.D.; Arendt, E.K. Fundamental studies on the reduction of salt on dough and bread characteristics. *Food Res. Int.* **2009**, *42*, 885–891. [CrossRef]
24. Pasqualone, A.; Caponio, F.; Pagani, M.A.; Summo, C.; Paradiso, V.M. Effect of salt reduction on quality and acceptability of durum wheat bread. *Food Chem.* **2019**, *289*, 575–581. [CrossRef]
25. Noort, J.H.F.; Bult, M.; Stieger, M. Saltiness enhancement by taste contrast in bread prepared with encapsulated salt. *J. Cereal Sci.* **2012**, *55*, 218–225. [CrossRef]
26. Simsek, S.; Martinez, M.O. Quality of dough and bread prepared with sea salt or sodium chloride. *J. Food Process Eng.* **2016**, *39*, 44–52. [CrossRef]
27. Miller, R.A.; Jeong, J. Sodium reduction in bread using low-sodium sea salt. *Cereal Chem.* **2014**, *91*, 41–44. [CrossRef]
28. Spina, A.; Brighina, S.; Muccilli, S.; Mazzaglia, A.; Rapisarda, P.; Fallico, B.; Arena, E. Partial Replacement of NaCl in Bread from Durum Wheat (*Triticum turgidum* L. subsp. *durum* Desf.) with KCl and Yeast Extract: Evaluation of Quality Parameters During Long Storage. *Food Bioprocess Technol.* **2015**, *8*, 1089–1101. [CrossRef]
29. Raffo, A.; Carcea, M.; Moneta, E.; Narducci, V.; Nicoli, S.; Peparaio, M.; Sinesio, F.; Turfani, V. Influence of different levels of sodium chloride and of a reduced-sodium salt substitute on volatiles formation and sensory quality of wheat bread. *J. Cereal Sci.* **2018**, *79*, 518–526. [CrossRef]
30. Sinesio, F.; Raffo, A.; Peparaio, M.; Moneta, E.; Civitelli, E.S.; Narducci, V.; Turfani, V.; Ferrari Nicoli, S.; Carcea, M. Impact of sodium reduction strategies on volatile compounds; sensory properties and consumer perception in commercial wheat bread. *Food Chem.* **2019**, *301*, 125252. [CrossRef] [PubMed]

31. Delcour, J.; Hoseney, R.C. *Principles of Cereal Science and Technology*; AACC International: St. Paul, MN, USA, 2010.
32. Carcea, M.; Narducci, V.; Turfani, V.; Aguzzi, A. A survey of sodium chloride content in Italian artisanal and industrial bread. *Foods* **2018**, *7*, 181. [CrossRef] [PubMed]
33. Fiore, M.C.; Mercati, F.; Spina, A.; Blangiforti, S.; Venora, G.; Dell'Acqua, M.; Lupini, A.; Preiti, G.; Monti, M.; Pè, M.E.; et al. High-throughput genotyping, morphological and quality traits to assess genetic diversity of wheat landraces from Sicily. *Plants* **2019**, *8*, 116. [CrossRef] [PubMed]
34. Pasqualone, A. Italian Durum Wheat Breads. In *Bread Consumption and Health*; Pedrosa Silva Clerici, M.T., Ed.; Nova Science Publisher Inc.: Hauppauge, NY, USA, 2012; pp. 57–79.
35. Boyacioglu, M.H.; D'Appolonia, B.L. Characterization and utilization of durum wheat for breadmaking. I. Comparison of chemical, rheological, and baking properties between bread wheat flour and durum wheat flours. *Cereal Chem.* **1994**, *71*, 21–28.
36. Boyacioglu, M.H.; D'Appolonia, B.L. Characterization and utilization of durum wheat for breadmaking. II. Study of flour blends and various additives. *Cereal Chem.* **1994**, *71*, 28–34.
37. Liu, C.Y.; Shepherd, K.W.; Rathjen, A.J. Improvement of durum wheat pastamaking and breadmaking qualities. *Cereal Chem.* **1996**, *73*, 155–166.
38. Pasqualone, A.; De Angelis, D.; Squeo, G.; Difonzo, G.; Caponio, F.; Summo, C. The Effect of the Addition of *Apulian black* Chickpea Flour on the Nutritional and Qualitative Properties of Durum Wheat-Based Bakery Products. *Foods* **2019**, *8*, 504. [CrossRef]
39. Giannone, V.; Giarnetti, M.; Spina, A.; Todaro, A.; Pecorino, B.; Summo, C.; Caponio, F.; Paradiso, V.M.; Pasqualone, A. Physico-chemical properties and sensory profile of durum wheat Dittaino PDO (Protected Designation of Origin) bread and quality of re-milled semolina used for its production. *Food Chem.* **2018**, *241*, 242–249. [CrossRef] [PubMed]
40. Pasqualone, A.; Piergiovanni, A.R.; Caponio, F.; Paradiso, V.M.; Summo, C.; Simeone, R. Evaluation of the technological characteristics and bread-making quality of alternative wheat cereals in comparison with common and durum wheat. *Food Sci. Technol. Int.* **2011**, *17*, 135–142. [CrossRef] [PubMed]
41. Palumbo, M.; Spina, A.; Boggini, G. Bread-making quality of Italian durum wheat (*Triticum durum* Desf.) cultivars. *Ital. J. Food Sci.* **2002**, *14*, 123–134.
42. AACC International. *Approved Methods of Analysis*, 10th ed.; American Association of Cereal Chemists: St Paul, MN, USA, 2000.
43. UNI. *UNI Method No. 10690. Durum Wheat and Semolina. Determination of Gluten Quality. Gluten Index Method*; UNI: Milan, Italy, 1979.
44. ISO. *ISO 3093:2009. Wheat, Rye and Their Flours, Durum Wheat and Durum Wheat Semolina. Determination of the Falling Number according to Hagberg-Perten*; ISO: Geneva, Switzerland, 2009.
45. AOAC. *Official Methods of Analysis*, 17th ed.; The Association of Official Analytical Chemists: Gaithersburg, MD, USA, 2000.
46. Lefebvre, D.; Gabriel, V.; Vayssier, Y.; Fontagne'-Faucher, C. Simultaneous HPLC determination of sugars, organic acids and ethanol in sourdough process. *Lebensm.-Wiss. u.-Technol.* **2002**, *35*, 407–414. [CrossRef]
47. Lanza, C.M.; Mazzaglia, A.; Scacco, A.; Pecorino, B. Changes in sensory and instrumental features of industrial Sicilian bread during storage. *Ital. J. Food Sci.* **2011**, *23*, 6–12.
48. UNI EN ISO 8586:2014. *Sensory Analysis. General Guidelines for the Selection, Training and Monitoring of Selected Assessors and Expert Sensory Assessors*; UNI, Ente Nazionale Italiano di Unificazione: Milano, Italy, 2012.
49. Raffo, A.; Pasqualone, A.; Sinesio, F.; Paoletti, F.; Quaglia, G.; Simeone, R. Influence of durum wheat cultivar on the sensory profile and staling rate of Altamura bread. *Eur. Food Res. Technol.* **2003**, *218*, 49–55. [CrossRef]
50. Spina, A.; Brighina, S.; Muccilli, S.; Mazzaglia, A.; Fabroni, S.; Fallico, B.; Rapisarda, P.; Arena, E. Wholegrain durum wheat bread fortified with citrus fibers: Evaluation of quality parameters during long storage. *Front. Nutr.* **2019**, *6*, 13. [CrossRef]
51. Ficco, D.B.M.; Muccilli, S.; Padalino, L.; Giannone, V.; Lecce, L.; Giovanniello, V.; Del Nobile, M.A.; De Vita, P.; Spina, A. Durum wheat breads 'high in fibre'and with reduced in vitro glycaemic response obtained by partial semolina replacement with minor cereals and pulses. *J. Food Sci. Technol.* **2018**, *55*, 4458–4467. [CrossRef]
52. UNI EN ISO 8589:2014. *Sensory Analysis—General Guidance for the Design of Test Rooms*; UNI, Ente Nazionale Italiano di Unificazione: Milano, Italy, 2012.

53. Hammer, Ø.; Harper, D.A.T.; Ryan, P.D. PAST: Paleontological statistics software package for education and data analysis. *Palaeontol. Electron.* **2001**, *4*, 9.
54. *D.P.R. 187. Decreto del Presidente Della Repubblica 9 febbraio 2001, n. 187, "Regolamento per la Revisione Della Normativa Sulla Produzione e Commercializzazione di Sfarinati e Paste Alimentari, a Norma dell'articolo 50 Della Legge 22 Febbraio 1994, n. 146"*; Gazzetta Ufficiale Della Repubblica Italiana: Roma, Italy, 2001; p. 13.
55. Giannone, V.; Lauro, M.R.; Spina, A.; Pasqualone, A.; Auditore, L.; Puglisi, I.; Puglisi, G. A novel α-amylase-lipase formulation as anti-staling agent in durum wheat bread. *LWT-Food Sci. Technol.* **2016**, *65*, 381–389. [CrossRef]
56. Siswoyo, T.A.; Tanaka, N.; Morita, N. Effect of lipase combined with alpha-amylase on retrogradation of bread. *Food Sci. Technol. Res.* **1999**, *5*, 356–361. [CrossRef]
57. Schmidt, S.J. Water and solids mobility in foods. In *Advances in Food and Nutrition Research*; Steve, L.T., Ed.; Elsevier Academic Press 525 B Street, Suite 1900: San Diego, CA, USA, 2004; Volume 48, pp. 1–89.
58. Arena, E.; Fallico, B.; Maccarone, E. Thermal damage in blood orange juice: Kinetics of 5-hydroxymethyl-2-furancarboxaldehyde formation. *Int. J. Food. Sci. Technol.* **2001**, *36*, 145–151. [CrossRef]
59. Fallico, B.; Zappalà, M.; Arena, E.; Verzera, A. Effect of conditioning on HMF content in unifloral honeys. *Food Chem.* **2004**, *85*, 305–313. [CrossRef]
60. Capuano, E.; Fogliano, V. Acrylamide and 5-hydroxymethylfurfural (HMF): A review on metabolism, toxicity, occurrence in food and mitigation strategies. *LWT Food Sci. Technol.* **2011**, *44*, 793–810. [CrossRef]
61. Mesias, M.; Delgado-Andrade, C.; Morales, F.J. Process contaminants in battered and breaded foods prepared at public food service establishments. *Food Control* **2020**, *114*, 107217. [CrossRef]
62. Abraham, K.; Gürtler, R.; Berg, K.; Heinemeyer, G.; Lampen, A.; Appel, K.E. Toxicology and risk assessment of 5-Hydroxymethylfurfural in food. *Mol. Nutr. Food. Res.* **2011**, *55*, 667–678. [CrossRef]
63. Choudhary, A.; Kumar, V.; Kumar, S.; Majid, I.; Aggarwal, P.; Suri, S. 5-Hydroxymethylfurfural (HMF) formation, occurrence and potential health concerns: Recent developments. *Toxin Rev.* **2020**, 1–17. [CrossRef]
64. Rufián-Henares, J.A.; de la Cueva, S.P. Assessment of hydroxymethylfurfural intake in the Spanish diet. *Food Addit. Contam. A.* **2008**, *25*, 1306–1312. [CrossRef]
65. Moreau, L.; Bindzus, W.; Hill, S. Influence of salts on starch degradation: Part II—salt classification and caramelisation. *Starch-Starke* **2011**, *63*, 676–682. [CrossRef]

MDPI
St. Alban-Anlage 66
4052 Basel
Switzerland
Tel. +41 61 683 77 34
Fax +41 61 302 89 18
www.mdpi.com

Foods Editorial Office
E-mail: foods@mdpi.com
www.mdpi.com/journal/foods

www.ingramcontent.com/pod-product-compliance
Lightning Source LLC
LaVergne TN
LVHW071527170726
843515LV00008B/2076

* 9 7 8 3 0 3 6 5 0 7 0 6 4 *